GROUP THEORETICAL METHODS IN PHYSICS

Volume II

GROUP THEORETICAL METHODS IN PHYSICS

Proceedings of the Third Yurmala Seminar

Yurmala, USSR, 22–24 May 1985

Edited by
M. A. Markov, V. I. Man'ko and V. V. Dodonov

CRC Press is an imprint of the
Taylor & Francis Group, an **informa** business

CRC Press
Taylor & Francis Group
6000 Broken Sound Parkway NW, Suite 300
Boca Raton, FL 33487-2742

Visit the Taylor & Francis Web site at
http://www.taylorandfrancis.com

and the CRC Press Web site at
http://www.crcpress.com

PREFACE

In recent years, group-theoretical methods have been widely developed in various branches of physics; particularly in quantum field theory, elementary particle physics, nuclear physics, quantum optics and electronics, solid-state physics, and mathematical physics. Therefore, the subdiscipline "Group theoretical methods in physics" emerged. In 1972, the colloquium "Group theoretical methods in physics" was organized. This became traditional and meets annually. To date 14 meetings have been held, of which the proceedings have been published. The colloquium has an International Standing Committee which meets each year.

In additional to Colloquia, seminars on group theoretical methods in physics are held in various countries in coordination with the Standing Committee.

In November 1979, the Nuclear Physics Department and the P. N. Lebedev Physical Institute of the USSR Academy of Sciences organized the first international seminar on group theoretical methods in physics which was held in Zvenigorod. Proceedings of the seminar were published in two volumes (Nauka publishers, Moscow, 1980). The second Zvenigorod seminar was held in November 1982. The proceedings of that seminar were published in two volumes by Nauka publishers (1983) and in one volume by Gordon & Breach (1985). The third such seminar was held in Yurmala in 1985. The organizers of the seminar were the Nuclear Physics Department and the P. N. Lebedev Physical Institute of the USSR Academy of Sciences and the Institute of Physics of the Latvian Academy of Sciences. This seminar was attended by more than 200 participants. The seminars in Zvenigorod and Yurmala may be considered to be the Soviet counterpart of the International Colloquia project.

The material of the Yurmala seminar is published in two volumes. The first volume contains the chapters: Gravitation, Cosmology, Supersymmetry and Supergravity; Quantum Field Theory; Superalgebras and their Applications; Nonlinear Integrable Equations; Group Representations and Special Functions; Quantum and Classical Mechanics. The second volume contains the chapters: Dynamical Symmetries and Symmetries of Equations; Representation Theory; Gauge Theories; Space Groups and Solid State Physics; Symmetries in Optics, Atomic, Molecular, and Nuclear Physics. As the reader will see, the range of subjects covered is diverse.

The editors thank the USSR Translation Centre and Dr D. A. Leites for their help in preparing English texts of the papers for the publication in the Proceedings of the Yurmala seminar.

THE EDITORS

CONTENTS

Part I

DYNAMICAL SYMMETRIES AND SYMMETRIES OF EQUATIONS

TWENTY YEARS OF DYNAMICAL GROUPS AND SPECTRUM GENERATING ALGEBRAS

A.Bohm, Y.Ne'eman

Center for Particle Theory, The University of Texas at Austin, Austin, Texas 78712

1. Introduction: An Anniversary. The method of Dynamical Groups (DG) or of Spectrum-Generating Algebras (SGA) was introduced in Physics - and in particular in the Physics of Particles and Fields - twenty years ago, in 1965 [1]. The mood was one of unfettered exploration. Relativistic Quantum Field Theory appeared to have failed us, a real disappointment after the spectacular success of Quantum Electrodynamics. Strong interactions did not fit a perturbation treatment, as long as the fundamental couplings were believed to be the large $g_{\bar{N}N\pi} \sim 14$ or somewhat smaller $g_{\bar{N}N\rho} \sim 1$. S-matrix methods appeared too ambiguous to yield definite results, especially since they were presented as requiring almost no input. Experience in real life appeared to contradict these expectations of a "free lunch". Weak interactions, as given by the Sudarshan-Marshak-Feynman-Gell Mann effective Hamiltonian, were plagued with a dimensional coupling and were thus non-renormalizable.

Group theory had just reaped spectacular successes, with ("flavour") SU(3) and a first glimpse at quarks, an intriguing possibility since the more recent success of SU(6) [3]. This in itself was a rather unorthodox symmetry, perhaps a degeneracy algebra for rest-states, which did not commute with the Lorentz group. Attempts at finding a relativistic version either involved a merger with the Lorentz group in SU(6,6), a non-compact group, or pointed to current algebra. The latter was a revival of the Heisenberg matrix mechanics, using the saturation of commutators by complete sets of intermediate states. The algebra was either thar of the global generators or of their local densities. Field theory having failed, this was a return to the previous stage, a new beginning. It was felt that this could usher in an alternative to field theory, with the current densities as an example of local "quasi-fields" that could be determined through the measurement of their matrix elements in the non-strong interactions.

Dynamical Degeneracy Groups (DDG) - symmetries of the Hamiltonian - had been used in Quantum Mechanics from the very beginning [3)]. Pauli had solved the Kepler problem in its quantum version algebraically and an SO(4) turned out to be behind it. In Nuclear Physics, the shell model was in the nature of a DDG [4)] and so was Elliot's SU(3) or Wigner's "Supermultiplet" theory. The Gursey-Radicati-Sakita SU(6) and its $\frac{1}{2}(1\pm\gamma_5^0)$ extension as U(6) x U(6) were of a similar narure [5)]. It was thus natural that two groups - at Trieste and at Caltech - should be looking at the old methods of Quantum Mechanics and trying to abstract new lessons from old results. This is how both groups discovered [1)] the Spectrum Generating Algebras [2)]: beyond the DDG there are Lie groups or Lie algebras, generally non-compact whose generators change the energy or the mass and whose irreducible representation gives the entire spectrum of solutions for a quantum-mechanical system. Whenever the set of solutions is infinite - the SGA is non-compact, with infinite-dimensional irreducible unitary representations. Some times the problems can be approximated by finite srts of solutions and the SGA should be compact in those special cases.

The sanctified hydrogen atom itself provided an interesting example. The DDG SO(4) is the compact subgroup of a SGA, either SO(4,1) or SO(4,2)~SU(2,2), depending upon the optional inclusion of another compact subgroup, the SO(2)~U(1) whose eigenvalues correspond to the Principal Quantum Number. This was soon understood by both groups at Trieste and Caltech [6], but the full explicit structure of the dynamical generators of this SO(4,2) were only given some time later [7].

The evolution of the subject since 1965 has included the following stages:

a) A great boost in the study of the unitary infinite-dimensional irreducible representations (unirreps) of non-compact groups and the development of simplified constructions for cetrain types.

In recent years, these methods have been applied to the unirreps of the exceptional non-compact groups (such as $E_{7,7}$) which appear in the super-unification schemes of 11-dimensional supergravity and the various superstrings (including the most recent $E_8 \otimes E_8$ heterotic superstring).

b) Finding the SGA for many of the standard problems in classical and quantum mechanics and improving our understanding of what they stand for.

c) Studying hadrons as spatially-extended structures:

c1) rotators, oscillators, spinning tops, pulsators, etc.

c2) strings (with some interest in extending the ideas to membranes and lumps)

c3) identifying regularities in the hadron spectrum (Regge trajectories etc.) and using generalizations abstracted from c1 or c2 to classify and predict the patterns.

c4) calculating decay rates, magnetic moments and other structure parameters of hadrons when the flavour group is considered as a mass changing SGG [8]

d) In nuclear physics, the "Interacting Boson Model" (IBM) [9] has uncovered dynamical group reduction chains, explaining the interweaving pattern of vibrational and rotational excitations and predicting the existence of a new shape of nuclei without axial symmetry.

e) With the discovery of supersymmetry [10] and progress in superalgebras and supergroups [11] these were applied in dynamical versions or as SGA in particle, nuclear and atomic physics [12].

Topic (a), the unitary irreducible representations (unirreps) of the classical and other non-compact groups has been in the forefront both in mathematics and in physics throughout these years. I.M.Gelfand, M.A.Naimark and the Moscow group., V.Bargmann, Iwasawa and Harish-Chandra at Princeton, B.Kostant and the M.I.T. group, Langlands and his school, and others have all brought about great advances in listing and constructing the unirreps of the main non-compact groups. Nevertheless much still has to be done. One example of a task almost untouched by mathematicians is the problem of the double-covering (i.e. spinorial representations) of the $SL(n, \mathbb{R})$ [13]. This appeared less compelling to mathematicians, although it was obviously important to the physics applications.

It needed the intervention of physicists such as D.Joseph and the original 1965 Caltech group, L.Biedenharn, Dj.Sijacki [14] and others of the Duke University group, V.I.Ogievetsky and the Dubna group - to complete the task just for $\overline{SL}(3, \mathbb{R})$. The most recent mathematical survey [15] by B.Speh left the issue unresolved, but Rownsley and Sternberg of Harvard have recently observed some highly special features displayed by the multiplicity - free spinorial representation - from the mathematician's viewpoint, this time! As for $\overline{SL}(4, \mathbb{R})$, it is still far from having achieved the status reached by $\overline{SL}(3, \mathbb{R})$, although much work by physicists has been extended in this program [16], We shall not discuss these mathematical issues any further in this article.

the theoretical foundations were only touched upon. Dothan [17)] showed how any SGA could be regarded as a symmetry with an explicit dependence of the generators on the time coordinate. Similar results using time dependent integrals of the motion as generators of state generating algebras have been obtained by Malkin and Man' ko [17b] and are reviuved in their monograph [17c] . The model is the Lorentz group: it does not commute with the Hamiltonian, and yet it is a symmetry of the S-matrix. The total time-derivative of the generator does vanish,

$$\frac{d}{dt}\vec{K} = \frac{\partial}{\partial t}\vec{K} = i\left[H,\vec{K}\right] = 0 \quad .$$

The other advance on the theoretical side has consisted in developments relating to infinite-component wave equations. Majorana's pioneering work in the thirties was revived and continued by Dirac and others [18)] .

2. <u>The idea of a SGG</u>. "By a dynamical group we mean a group (in general a non-compact one) which gives the actual energy or mass spectrum of a quantum mechanical component system" [1)] . The simplest example of an SGA is the non-relativistic rotator (dumbbell). It could be worked out in detail in [1)] , because the non-compact group representations needed for it were completely known and well presented in [19)] . It describes a diatomic rotating molecule.

The spectrum of the di-atom dumbbell is given in Fig. 2.1. To the energy level with angular momentum j the space R^j of the irreducible representation of SO(3) corresponds (the space of physical states for a di-atom dumbbell with angular momentum j is R^j). "All" the states of the di-atomic dumbbell are given by the direct sum

$$\mathcal{H} = \sum_{j=0,1,2...} \oplus \; R^j \tag{2.1}$$

where stands for whatever is needed for each particular molecule - sometimes of the order 10^2 other times 10^1 - unit this simple model becomes inadequate or breaks down for the diatomic molecule. Ideally, is infinite. The energy operator and the energy spectrum are given by

$$H = \frac{1}{2I}\vec{S}^2 , \quad E_j = \frac{1}{2I} j(j+1) . \qquad (2.2)$$

Groups which have (1) as a unitary representation space are $SG = SO(4)$ (if ω is finite), $SG = SO(3,1)$ and $SG = E_3$. Thus, the rotator has the following complete chain of subgroups

$$SG \supset SO(3)_{S_i} \supset SO(2)_{S_3} \qquad (2.3)$$

and H contains only invariant (Casimir) operators of the complete chain. For non-relativistic systems, this simple idea is generalized [9,20] to more complicated SGA's having larger subgroup chains:

$$SG \supset G' \supset G'' \supset G''' \supset \ldots \qquad (2.3')$$

and more complicated energy operators

$$H = a_o + \sum_i a_i' C_i' + \sum_i a_i'' C_i'' + \ldots \qquad (2.2')$$

Here a_o, a_i' , a_i'' ... are numbers (free parameters fitted from experiment and characteristic of the particular physical system, e.g., $a_i' = \frac{1}{2} I$, I = moment of inertia of the diatomic molecule), and $C_i^{(n)}$ are the Casimir operators of $G^{(n)}$ in (3') .

The larger groups in the chain, in particular the SGG's, introduce new operators, which together with the operators of the symmetry subgroup (the S_i in (3)) form the algebra of the SGG. In the case of (3) these additional generators are the vector operators Γ_i

$$[S_i, \Gamma_j] = i\,\varepsilon_{ijk}\,\Gamma_k \tag{2.4}$$

which in addition satisfy the commutation relation (c.r.)

$$[S_i, \Gamma_j] = -i\,\varepsilon_{ijk} S_k \times \begin{cases} (-1) & \text{for SGG} = SO(4) \\ 0 & \text{for SGG} = E(3) \\ (+1) & \text{for SGG} = SO(3,1) \end{cases}$$

They (or rather their spherical components $_{(k)}$:

$$\Gamma_{(0)} = \Gamma_3 \,, \quad \Gamma_{(\pm 1)} = \mp \frac{1}{\sqrt{2}} \,(\Gamma_1 \pm i\Gamma_2)\,) \tag{2.5}$$

transform between different energy levels as shown in Fig. 2.1. What should their physical interpretation be?

The simplest case would be to set SSG = E(3) and

$$\Gamma_i = mc\,\xi_i \quad \underline{\text{or}} \quad \Gamma_i = mc\,\alpha'\,\pi_i \tag{2.6}$$

where ξ_i is the dipole operator or the operator of an intrinsic position and π_i is the operator of an intrinsic momentum; (mc) and $mc\,\alpha'$ are some constants which make the dimensions correct (mc has dimension cm^{-1} and α' has dimension $(mc)^{-1}$ for $\hbar = 1$);

$$[\xi_i, \xi_j] = 0 \qquad [\pi_i, \pi_j] = 0 \quad [\xi_i, \xi_j] = i\delta_{ij} \tag{2.7}$$

One can now approach the problem from another direction: one starts with three components of "intrinsic position" π_i and "intrinsic momentum" π_i which satisfy (7) and defines $S^i = \varepsilon^{ijk} S^{jk}$ by:

$$S^{ij} = \xi^i \wedge \pi^j = -i(\alpha_-^i \alpha_+^j - \alpha_-^j \alpha_+^i)\,, \quad i = 1,2,3 \tag{2.8}$$

where the $\alpha_{\pm}^i$ are the creation and annihilation operators connected with ξ_i, π_j by

$$\xi^i(\tau) = \sqrt{\frac{\alpha'}{2}} \; (\alpha_+^i e^{-i} \; + \alpha_-^i e^{+i\tau})$$
$$\xi^i(\tau) = \frac{-i}{\sqrt{2\alpha'}} \; (\alpha_+^i e^{-i} \; - \alpha_-^i e^{+i\tau}) \qquad (2.9)$$

which satisfy - as a consequence of (7) - the commutation relations (c.r.):

$$[\alpha_+^i , \alpha_-^j] = +\delta^{ij} I = -g^{ij} I \, . \qquad (2.7a)$$

We then have not one but two E(3)-groups, the $E(3)_{\xi_i S_{ij}}$ and the $E(3)_{\pi_i S_{ij}}$. Either one coild be interpreted as the spectrum generating group of angular momentum and dipole operator (note that ξ_i and π_i in (7) differ just by a phase convention and are equivalent, Born reciprocity).

The rigid rotator - responsible for the rotational excitations - is of course not the whole story of the diatom; it can also perform vibrations, (Fig. 2.2)

Therefore, following Iachello et al. [20] we construct the vibron model by introducing in addition to the 3 components of the vector boson operator α_-^i (creation, called π^{i+} in [20] and α_+^i (annihilation, called π^i in [20,21]) scalar boson operators σ^+, σ^- : with

$$[\sigma, \sigma^+] = \pm 1 = g^{00} 1 \quad \text{where} \quad g^{00} = +1 \quad \text{or} \quad g^{00} = -1 \qquad (2.10)$$

In [20] $g^{00} = +1$; we will later also set $g^{00} = -1$. The difference between -1 in (10) and +1 is that the former leads to $SO(3,1)_{D^i S^{ij}}$ as the SGG of angular momentum and the dipole operators, whereas the latter leads to SO(4) [20].

Setting $\alpha_-^0 = \sigma^+ \quad \alpha_+^0 = \sigma$ we can write (7a) and

(10) together as

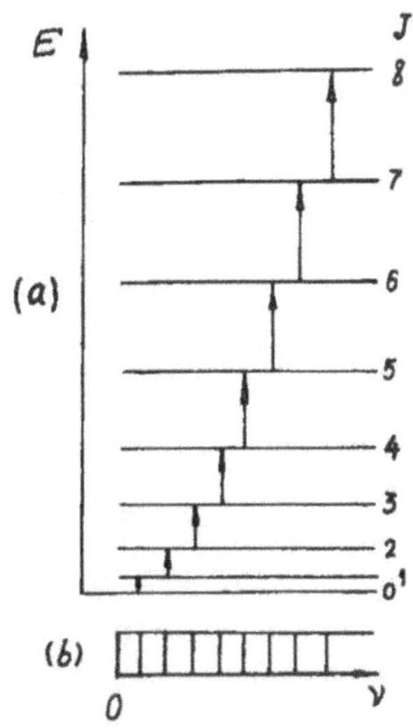

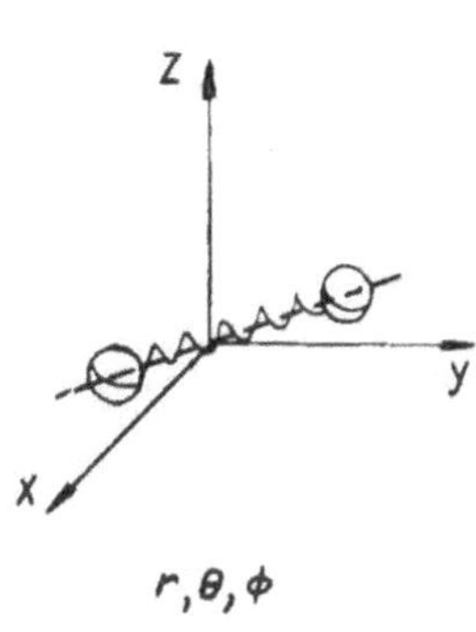

Fig. 2.1. Energy levels and infraded transitions of a rigid rotator: (a) The energy-level diagram, (b) the resulting spectrum (schematic).

Fig. 2.2. Schematic representation of the geometrix structure of diatomic molecules (b)

$$[\alpha_+^{\mu}, \alpha_-^{\nu}] = -g^{\mu\nu}\, 1 \qquad \mu,\nu = 0,1,2,3 \tag{2.11}$$

where $g^{\mu\mu} = (-1,-1,-1,-1)$ according to [20] and $g^{\mu\mu} = (+1,-1,-1,-1)$ according to our later choice.

The dimensionless dipole operator of the vibron model is defined in terms of the $\alpha^{\mu}_{\mp}$ by

$$D^i = (\alpha^i_- \sigma - \alpha^i_+ \sigma^+) = (\alpha^i_- \alpha^0_+ - \alpha^0_- \alpha^i_+) \tag{2.12'}$$

It is not yet Hermitian, its Hermitian equivalent is

$$S^{i0} = -iD^i = -i(\alpha^i_- \alpha^0_i - \alpha^0_i \alpha^i_+) \tag{2.12}$$

The new dipole operator of the vibron model differs from the old dipole operator of the rigid rotator in the following way:

Table 1

	Rigid Rotator	Vibron	Timelike Vibron
Dipole op:	i	S_{4i}	$\frac{1}{mc} S_{0i}$
SGG	E(3)	SO(4)	SO(3,1)
	$-imc\sqrt{\frac{\alpha^7}{2}}\, e^{-i}$	σ^+, σ	α^0_-, α^0_+
	no oscillations	(4 space-like oscill.)	(3 space, 1 time-like oscill.)
Spectrum	Fig.2.1	Fig.2.1 finite j^{max}	Fig.2.1 infinite j^{max}

One may think that it would be easy to decide from physics whether the spectrum of j is finite or infinite. But it is not, because one can have either of two points of view: 1) (Original idea of SGG and dynamical symmetry [1]): The spectrum is infinite - the group, therefore, non-compact. But the SGG, like any theoretical description, has a limited applicability so that for large j one first needs correction terms (centrifugal correction) and then for still larger j the model breaks down. At which value of j it breaks down depends on the molecule and its surroundings.

2) (Prevalent in nuclear physics and recent molecular physics applications [9,20]. The spectrum is finite - the group, therefore, compact. The highest value j^{max} determines the representation (of SO(4)), which in turn determines the value of the Casimir operators and thus through (2') the energy spectrum. Whether there is a well-defined j^{max} and therewith a well-defined highest SO(4) label $(k_0, k_1^{max} = j^{max})$ for each particular molecule or correspondingly for each particular nucleus is

physics, for instance, the angular momenta extend to very high values (order of 100) but somewhere around $j=10$ a phase transition occurs, the moment of inertia increases (from J to J_{rigid}) which is interpreted as a breakup of the boson. At this breakup energy the model (IBM) is no longer valid and one takes the corresponding critical value j as the j^{max}.

Thus at present one should keep an open mind about this question and not force a decision in favor of 1) or 2), at least for the non-relativistic case. The conventional non-relativistic dipole operator is commuting and leads to an infinite spectrum described by the SGG $E(3)_{S_{ij}\xi_i}$. The relativistic dipole operator has most likely non-commuting components, and most likely still leads to an infinite spectrum. But there is the area of non-relativistic physics where scalar and vector (or tensor)-oscillators combine to lead to non-commuting dipole operators with finite spectrum.

The timelike vibron is related to the lowest mode of the relativistic string. (8) and (12) combine to

$$S^{\mu\nu} = -i(\alpha_-^{\mu}\alpha_+^{\nu} - \alpha_-^{\nu}\alpha_+^{\mu}), \quad \mu,\nu = 0,1,2,3; \qquad (2.13)$$
$$g^{\mu\nu} = \eta^{\mu\nu} = (+1,-1,-1,-1)$$

which form an $SO(3,1)_{S^{\mu\nu}}$ and S^{ij} is the intrinsic angular momentum (the angular momentum in the center of charge or position rest frame) [23], not the spin (which is the angular momentum in the center of mass rest frame). $SO(3,1)_{S^{\mu\nu}}$ is the intrinsic Lorentz group not the group of physical Lorentz transformation which we denote by $SO(3,1)_{J_{\mu\nu}} \subset P_{P_\mu J_{\mu\nu}}$ and which is the subgroup of the Poincaré group P. The justification to call S^{0i} the dipole operator (non-commuting) in this relativistic case is not the analogy to the S^{4i} of the vibron model but the Inönü-Wigner Contraction limit of

$$p \to G \quad \text{(Galilei group)} \tag{2.14}$$

in which the Lorentz boost J^{0_i} goes into the non-relativistic position operator $\frac{1}{cM} J^{0_i} \to Q^i$. In this non-relativistic limit [24] the intrinsic Lorentz boost $\frac{1}{cM} S^{0_i}$ goes into the commuting intrinsic position operator

$$\frac{1}{cM} S^{0_i} \to \xi^i \ . \tag{2.15}$$

Before we discuss the relativistic case we review some of the applications to non-relativistic systems.

3. Some Applications in Non-relativistic Physics. The spectrum of a vibrating and rotating diatomic molecule is shown in Fig. 3.1 [22]. To see more of their group structure in it we rearrange it slightly in Fig. 3.2. v is the usual vibrational quantum number (q.n.) and j is the rotational q.n.

Starting with the four spacelike oscillators $\sigma^+ = \alpha_-^4 \alpha_-^i$ one is naturally lead to an SU(4):

$$Q^{\mu\nu} = -i\alpha_-^{\mu} \alpha_+^{\nu} \tag{3.1}$$

obey as a consequence of (1.11) with $g_{\mu\nu} = -\delta_{\mu\nu}$ the c.r.

$$[Q^{\mu\nu}_{\mu}, Q^{\rho\sigma}] = ig^{\nu\rho} Q^{\mu\sigma} - ig^{\mu\sigma} Q^{\nu\rho} \tag{3.2}$$

$$Q^{\mu\nu +} = -Q^{\nu\mu} \tag{3.3}$$

which defines (a unitary representation of) SU(4).

Splitting the $Q^{\mu\nu}$ up into

$$S^{\mu\nu} = Q^{\mu\nu} - Q^{\nu\mu} \qquad S^{\mu\nu +} = S^{\mu\nu} \tag{3.4}$$

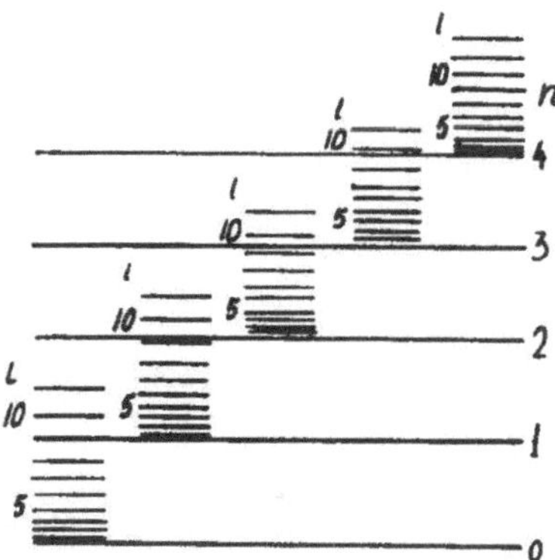

Fig. 3.1. Energy levels of the vibrating rotator. For each of the first five vibrational levels, a number of rotational levels are drawn (short horisontal lines)

$$T^{\mu\nu} = \tfrac{1}{2}(Q^{\mu\nu}+Q^{\nu\mu}) - \delta^{\mu\nu}\,\tfrac{1}{4}Q^{\rho}_{\ \rho} \qquad T^{\mu\nu +} = -T^{\mu\nu} \tag{3.5}$$

leads to the reduction chain

$$U(4)_{Q^{\mu\nu}} \supset SO(4)_{S^{\mu\nu}} \supset SO(3)_{S^{ij}} \tag{MII}$$

Fig. 3.2 displays the reduction with respect to this reduction chain if j^{max} is finite. If e.g. the total vibron number characterizing the U(4) irrep is N=29, then it contains irreps of SO(4) characterized by $(k_0=0,k_1)$ $k_1 = 29,27,25... 3, 1$ as shown in the lower part of Fig. 3.3. The upper part of Fig. 3.3 gives the experimental spectrum of H_2 (in the $^1\Sigma_g^+$ electronic state). Most diatomic molecules have spectra like those shown in Figs. 3.1, 3.2, 3.3 .

In Fig. 3.2 we introduce a new quantum number $\nu = v+j$. The lines $\nu = 0,1,2...$ are shown on the right of Fig. 3.2b. We now distort the Figure by bending up the lines ν = const. This leads to Fig. 3.4b, where ν is plotted vertically and j horizontally. We have divided these levels into two classes, those drawn by a solid li-

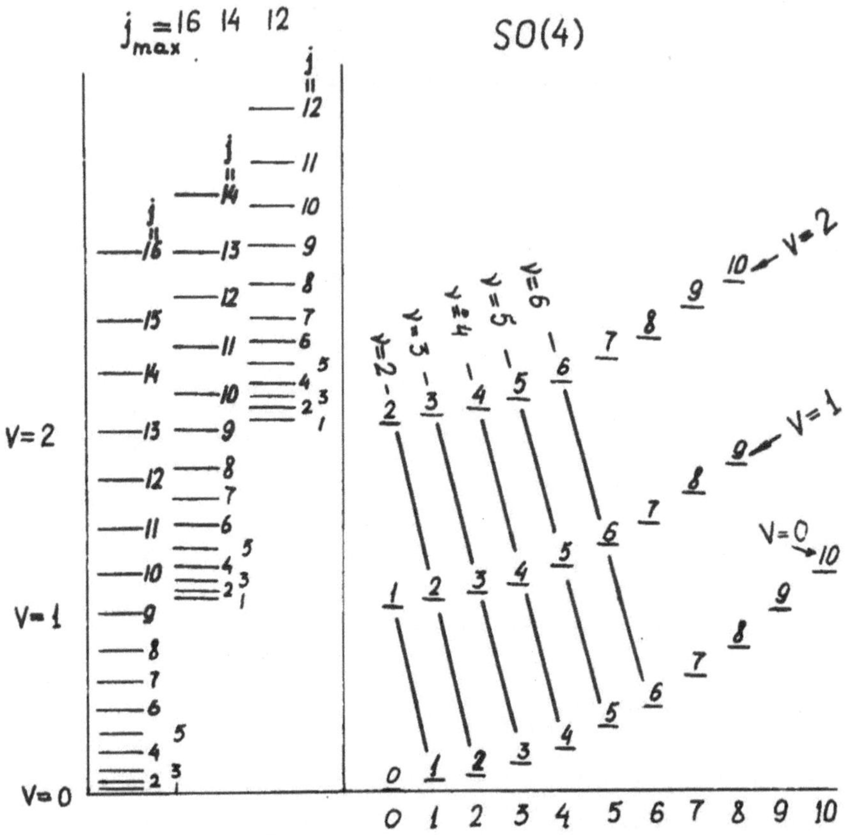

Fig. 3.2. (a) shows the same spectrum as Fig. 3.1 for the case $j_{max}=16$ (total vibron number N=16). v is the usual vibrational quantum number. (b) gives a rearangement of the levels in (a) such that the new quantum number $\nu = v+j$ is displayed

ne and the others drawn by a broken line. The solid levels, which are redrawn in Fig. 3.4a, represent the energy levels of a three dimensional oscillator. Each value of ν belongs to an energy

$$E_\nu = \hbar\omega\,(\nu + 3/2) \qquad \nu = 0,1,2,\ldots\ , \tag{3.6}$$

and

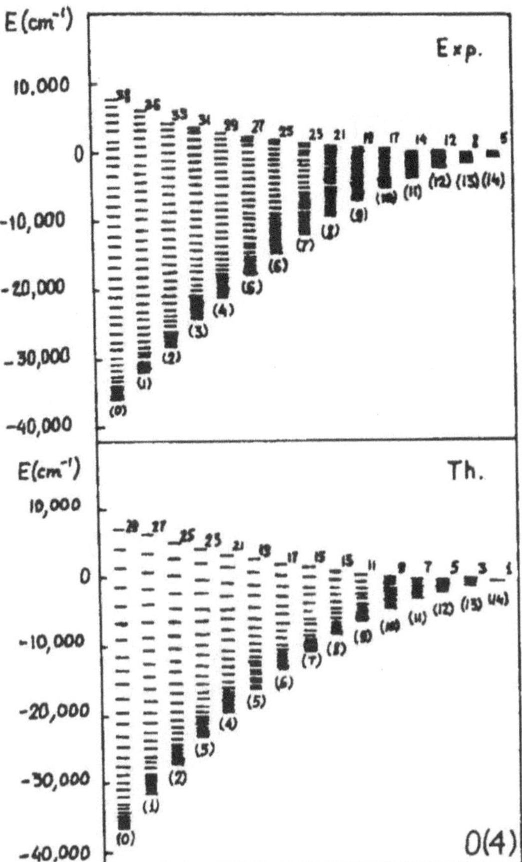

Fig. 3.3. The upper part gives the experimental spectrum of the hydrogen molecule in the $^1\Sigma_g^+$ electronic state. The lower part gives the theoretical energies calculated from $E=Ak_1(k_1+2)+ +Bj(j+1)$ with N=29, where the empirical parameters A and B have been obtained from a fit to the experimental data

$j = 0,2,4,\ldots \nu$ for even ν

$j = 1,3,5,\ldots \nu$ for odd ν

are the angular momenta with the same energy. Each ν-level belongs to an irrep of U(3). Fig. 3.4a gives the

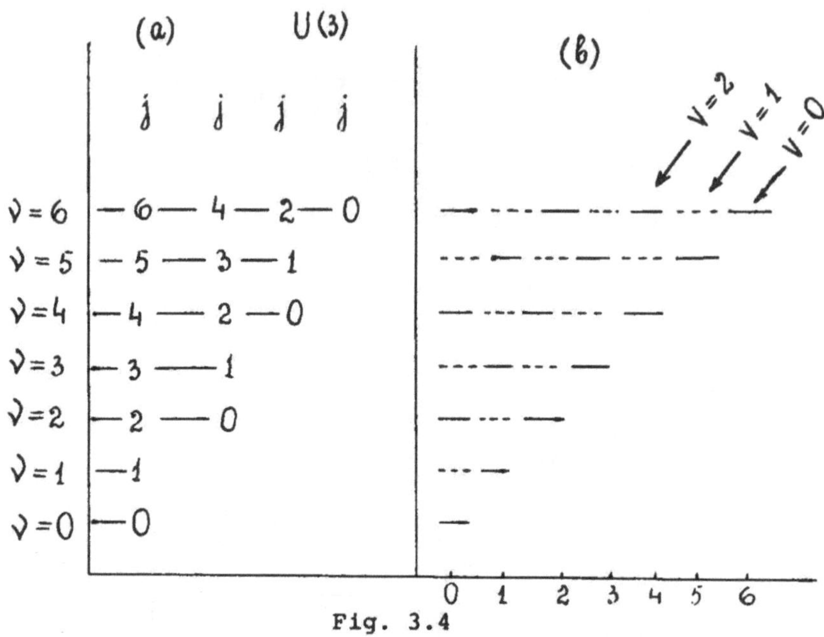

Fig. 3.4

reduction of an U(4) representation with respect to the reduction chain

$$U(4)_{Q^{\mu\nu}} \supset U(3)_{Q^{ij}} \supset SO(3)_{S^{ij}} \qquad i,j = 1,2,3 \qquad \text{(M1)}$$

According to our above deduction the dynamical group chain II corresponds to a rotator which in addition performs scalar (σ) oscillations like the classical mechanical object of Fig. 2.2. The dynamical group chain I corresponds to a three-dimensional oscillator which can also have vibrational angular momentum.

In molecular physics (in particular for diatomic molecules) case II is realized.

For time-like vibrons, which we expect in the relativistic case, the reduction chains I and II have to be replaced by the corresponding noncompact group chains:

$$SL(4,R)_{Q_{\mu\nu}} \supset SL(3,R)_{Q^{ij}} \supset SO(3)_{S^{ij}} \qquad (I^n)$$

$$SL(4,R)_{T_{\mu\nu}, S_{\mu\nu}} \supset SO(3,1)_{S^{\mu\nu}} \supset SO(3)_{S^{ij}} \qquad (II^n)$$

$Q^{\mu\nu}$, $T^{\mu\nu}$, $S^{\mu\nu}$ are again as in (2), (4), and (5) only that now $g^{\mu\nu} = \eta^{\mu\nu}$ and the Hermiticity condition (for unitary irrep) is $Q^{\mu\nu +} = Q^{\mu\nu}$, $S^{\mu\nu +} = S^{\mu\nu}$, $T^{\mu\nu +} = T^{\mu\nu}$.

Whereas the simple rotator has as basic observables the angular momentum S_{ij} and one additional vector operator ξ^i specifying the direction of the dumbbell axis (which is also the direction of the electric dipole moment) the symmetric top is specified by two additional vector operators ξ^i specifying the direction of the symmetry axis and η^i specifying the direction orthogonal to it (any of the equivalent directions). If

these directions are rigid (rigid symmetric top) then - as in the case of the rigid rotator - these observables commute: $[\xi^i, \xi^j] = 0$, $[\eta^i, \eta^j] = 0$, $[\xi^i, \eta^j] = 0$. If these directions have vibrations they - like in the case of the vibrating rotator - satisfy $[\xi^i, \xi^j] = \pm i\ell^2 S^{ij}$, $[\eta^i, \eta^j] = \pm i\ell^2 S^{ij}$, where ℓ^2 is a parameter of dimension cm^2. In the former case the SGG is the semidirect product of $SO(3)_{S^{ij}}$ with two translation groups, in the latter case the SGG is SO(5) or its non-compact form SO(3,2) (for timelike vibrators). For the asymmetric top one has one more vector operator ζ^i and (in the non-commuting case) SO(6) as the SGG. For the symmetric top molecule the subgroup reduction chain is

$$SO(5) \supset SO(4)_{\xi^i, S^{ij}} \supset SO(3)_{S^{ij}} \quad . \qquad (8)$$

For the infinite dimensional case the reduction chain is

$$SO(3,2) \supset SO(3,1)_{\xi^i, S^{ij}} \supset SO(3)_{S^{ij}} \quad . \qquad (8^n)$$

The Hamiltonian is

$$H = \frac{1}{2I_B} (\vec{S}^2 - \frac{I_A - I_B}{I_A} (\vec{\xi}^2)^{-1} (\vec{\xi} \cdot \vec{S})^2 \qquad (9)$$

where I_A, I_B are the moments of inertia and $\vec{\xi} \cdot \vec{S} = = \frac{1}{8} \epsilon_{\mu\nu\rho\sigma} S^{\mu\nu} S^{\rho\sigma}$ is one of the Casimir operators of SO(4) or SO(3,1) (or $E(3)_{\xi^i S^{ij}}$) with the eigenvalue $\sim k|\vec{\xi}|$. The spectrum of S_k is $k = = 0,1,2,3,\ldots$, with maximum value k^{max} for SO(4) and inbounded dor SO(3,1) and E(3). The energy spectrum is thus the well-known

$$E_{jk} = \frac{1}{2I_B} (j(j+1) - \frac{I_A - I_B}{I_A} k^2) \quad . \qquad (3.9')$$

The spectrum corresponding to the reduction chain (8^n) is shown in Fig. 3.5. Each vertical tower corresponds to an irrep of SO(3,1) extended by parity (or of the parity extended SO(4), in which case the towers are finite). And each box represents an SO(3) irrep with parity. Fig. 3.6 shows the spectrum of the prolate (Fig. 3.6a) and oblate (Fig. 3,6 b) symmetric top molecule [25]. The agreement between group theory (Fig. 3.5) and mole cular physics (Fig. 3.6) is apparent.

In nuclear physics the dynamical symmetries are based on the interacting boson model [9,27]. There are six bosons, one s-boson with angular momentum L=0 and five d-bosons with angular momentum L=2, hence the dynamical group U(6) (or a non-compact form there of if some of them would be timelike). In its original version the s and d bosons were identified with N correlated nucleon pairs outside a closed shell or with N hole pairs.

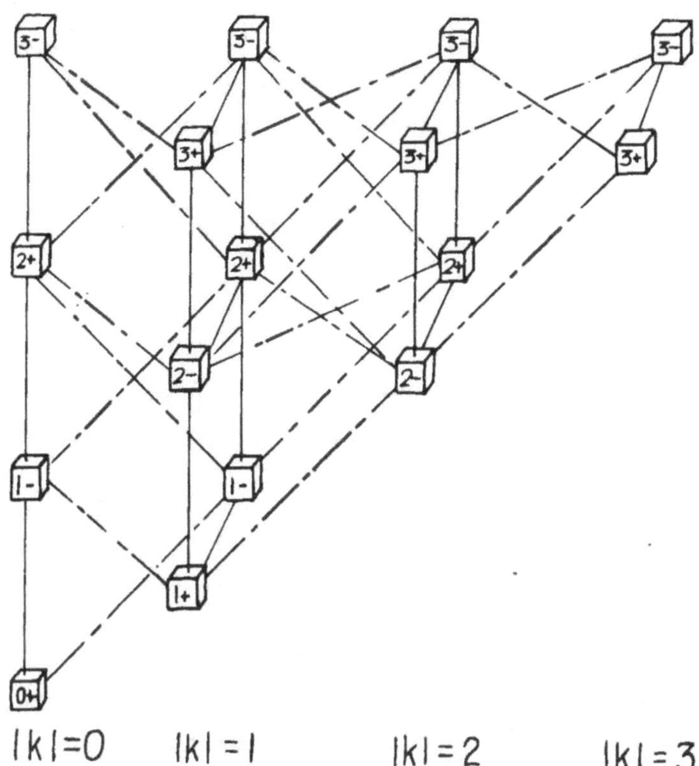

Fig.3.5. The operator ^{i},describing the symmetry axis of a symmetric top transforms between the boxes connected by solid lines,the operator ^{i} transforms along the dashed lines

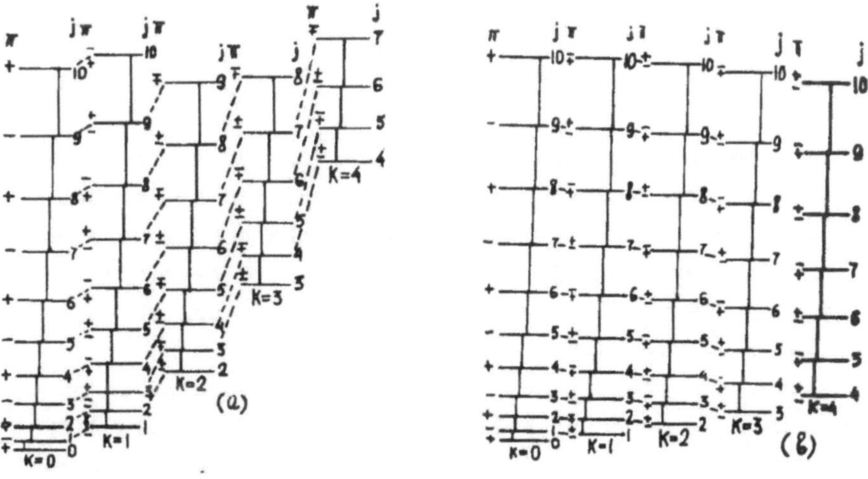

It thus described an even-even nucleus as a system of N interacting bosons. The Hamiltonian describing the system of interacting bosons, with creation operators b^+ and annihilation operators b , = 1,2,3,4,5,6, is in general

$$H = \sum_{\mu,\nu} \alpha_{\mu\nu} Q^{\mu\nu} + \sum_{\varrho,6} \beta_{\mu\nu\varrho 6} Q^{\mu\nu} Q^{\varrho 5},$$
$$Q^{\mu\nu} = -ib^+_{\mu} b_{\nu}$$

are generators of U(6), where $\alpha_{\mu\nu}$, $\beta_{\mu\nu\varrho 6}$ are empirical parameters. But it could happen that the coefficients α and β take some appropriate values and that H can be written only in terms of Casimir operators of subgroup chains as in eq. (1,2'). This indeed happens to a very good approximation for the low energy spectra of nuclei between major shell, which previously appeared very complicated.

There are three subgroup reduction chains of U(6) which contain the physical angular momentum group SO(3) as a subgroup

$$U(6) \supset U(5) \supset SO(5) \supset SO(3) \qquad \text{(NI)}$$

$$U(6) \supset SU(3) \supset SO(3) \qquad \text{(NII)}$$

$$U(6) \supset O(6) \supset O(5) \supset SO(3) \qquad \text{(NIII)}$$

In contract to molecular physics, in nuclear physics all the three subgroup chains are realized by various nuclei, though nuclei of the latter chain NIII were observed experimentally only after they had been predicted by this SGG model. Examples of each chain are shown in Figs. 3.7, 3.8, 3.9.

After the theoretical model has been found in terms of the algebra of observables connected with the SGG U(6), their origin in terms of nucleon pairs is really of se-

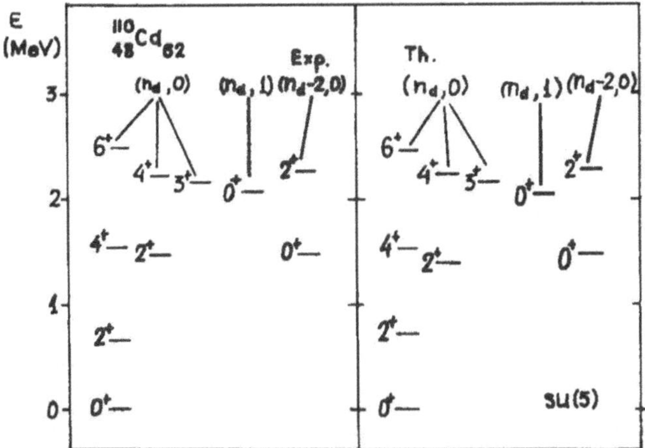

Fig. 3.7. An example of a spectrum with U(5) symmetry $^{110}_{48}Cd_{62}$, N =7

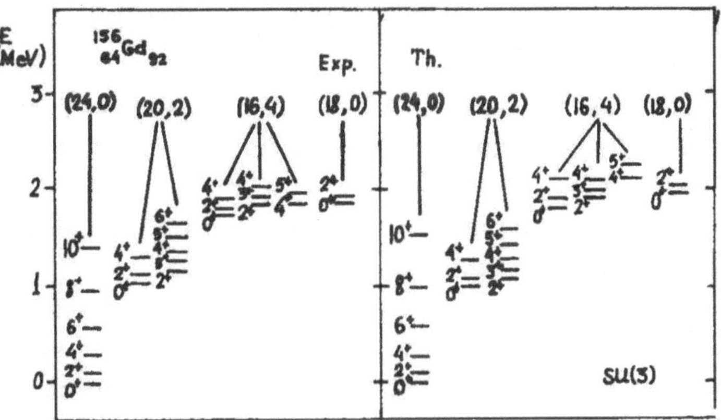

Fig. 3.8. An example of a spectrum with SU(3) symmetry $^{156}_{64}Gd_{92}$, N = 12

condary importance. Usually the spectroscopy of transitional nuclei uses the collective model [28] in which the nucleus is assigned a geometrical shape which undergoes rotations and vibrations. The dynamical group structure is then analyzed in terms of shape variables [29] and each of the subgroup chains NI-NIII is associated with

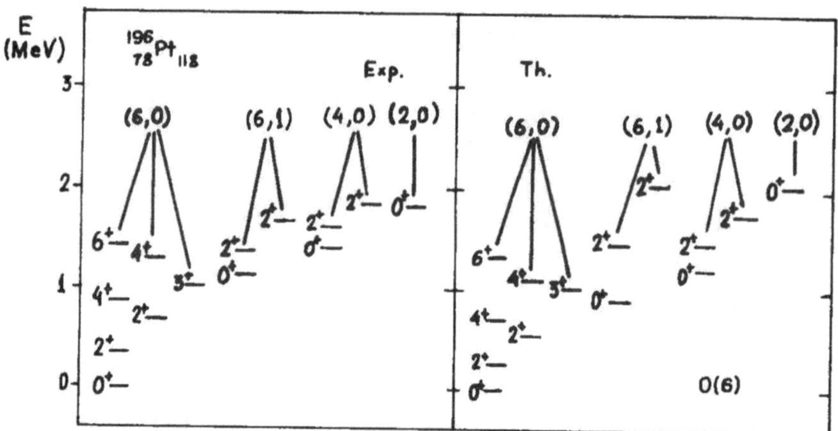

Fig. 3.9. An example of a spectrum with O(6) symmetry: $^{196}_{78}Pt_{118}$, N = 6

a particular shape and its motion.

Chain NI is the analogue of MI and describes vibrations about a spherical shape.

Chain NII is the analogue of MII and describes rotations of a symmetric top about an axis perpendicular to the symmetry axis.

Chain NIII corresponds to an asymmetric top performing rotations and vibrations about this shape .

The dynamical group approach of the IBM can thus be viewed as a quantized version of the Bohr-Mottelson collective model. The idea of proton and neutron constituents recedes behind the idea of vibrational and rotational motion.

Dynamical groups have also been used in the study of atomic spectra; as mentioned above, the reduction chain SO(4,2) SO(4,1) SO(4) SO(3) for the hydrogen atom was historically the second example of a SGA and its DDG, SO(4) played an important role in the discovery of SGG. This can be extended to the doubly excited states of Helium and other inert gases [26,22].

4. Relativistic SGG. What the orbits are for the under-

standing of classical physics the energy levels are for the understanding of non-relativistic quantum physics. In relativistic quantum physics energy levels should be replaced by mass levels. Still, the question of the mass spectrum of hadrons need not necessarily be a relevant question in relativistic quantum physics, but as it has a clear experimental answer - as long as one neglects the widths - , it most likely is. The relativistic version of the dynamical group method for the description of the hadron spectrum was from the very beginning the principal motivation for this approach [1,30]. The problem is thus to understand hadrons as extended relativistic oscillations and relativistic rotations. As in molecular [25] and nuclear [28] physics this approach is complementary (in the Bohr sense) to the understanding in terms of constituents given by the quark model.

A good candidate for the relativistic spectrum generating group is SL(4,R) because on the one hand it could be represented in a phenomenological field theory [31] and on the other hand it could describe shape pulsations of a three-dimensional "lump" (i.e. it follows from a Lagrangian given by $\mathcal{L} = \sqrt{-g}$) [32]. The commutation relation of $SL(4,R)_{T_{\mu\nu},S_{\mu\nu}} \supset SO(3,1)_{S_{\mu\nu}}$ follows from the definition (3.4) and (3.5) and (3.1) with $g_{\mu\nu} = \eta_{\mu\nu}$. But the representations of SL(4,R) given by (3.1) with (2.11) are very special. If one wants a larger class of representations one could define $Q^{\mu\nu}$ in terms of a large number N of 4-dimensional (1 time like, 3 space like) oscillators corresponding to a large number of modes in the relativistic string [33] $Q^{\mu\nu} = -i \sum_{n=1}^{N} \times$ $\times\, \alpha^{\mu}_{-n} \alpha^{\nu}_{+n}$, perhaps even taking $N \to \infty$. But one would still not obtain half-integer spin representations if one does not also use anticommuting quantities. It is much more useful to liberate oneself from the definition in terms of creation and annihilation operators and just define the $T^{\mu\nu}$ and $S^{\mu\nu}$ by the various representations of $\overline{SL}(4,R)$ (the covering group of $SL(4,R)$) obtaining yet another quantization of the relativistic string.

In Figure 4.1 we display the level diagrams (weight diagrams, K-types) for some of the simple representations of $SL(4,R)$. Such levels correspond to a definite value of j and each value of j that occurs is infinitely degenerate with the degeneracy label given by (k_0, k_1), the two numbers that specify the maximal compact subgroup $K = SO(4)$. The levels ($SO(3)$ irreps) that belong to one irrep of $SO(4)$ are on the same horizontal line. The levels along the diagonal $k_1 = j$ and along the lines parallel to it belong to an irrep of $SL(3,R)$ in the non-compact reduction chain $SL(4,R) \supset SL(3,R)_{T^{ij}S^{ij}} \supset SO(3)_{S^{ij}}$.

If $SL(4,R)$ is considered as a relativistic spectrum generating group each level should correspond to one hadron.

Instead of taking the many mode system as the starting point for a SGG model one can confive oneself to the one mode configurations [34]. For this simpler case, which would correspond to a vibrating di-quark (oscillator potential in the non-relativistic limit) rather than a pulsating "lump", there is already a simpler relativistic SGG than $SL(4,R)$. This is the group [30]

$$SO(3,2)_{S_{\mu\nu}\Gamma_\nu} \supset SO(3,1)_{S_{\mu\nu}} \supset SO(3)_{S_{ij}}$$

which is an infinite dimensional generalization of the $SO(3,2)_{\frac{1}{2}\sigma_{\mu\nu}, \frac{1}{2}\gamma_\mu}$ for the Dirac electron. For this case the theory has been worked out in all detail [35].

The main progress in the application of the SGG idea has been in the non-relativistic domain - particularly in nuclear physics, because in the relativistic domain the SGG approach was lacking a dynamical framework. This has now been provided by the quantum version of relativistic constraint Hamiltonian mechanics [36]. It leads to a new intrinsic dynamics which goes into the well-known non-relativistic quantum mechanics only in the non-relativistic limit.

The relativistic model should be conjectured such that in the non-relativistic contraction limit, (2.14), of the Poincaré Galilei group the non-relativistic oscillator model is reobtained.

The usual way to make (2.7), (2.8) with (2.9) relativistic is (the lowest mode of the relativistic string):

$$\begin{aligned} & S_{ij} \to S_{\mu\nu} && \mu\nu = 1,2,3,0 \quad \text{with} \quad SO(3,1)_{S^{\mu\nu}} \\ & \pi^{i} \to \pi^{\mu} \quad \text{with} && [\pi_\mu, \xi_\nu] = i g_{\mu\nu} \\ & \xi^{i} \to \xi^{\mu} && [\xi_\mu, \xi_\nu] = 0 \quad [\pi_\mu, \pi_\nu] = 0. \end{aligned} \tag{4.1}$$

This leads to plenty of difficulties [37]. It also contradicts our experience with a non-commuting intrinsic position of the vibron model in Table 1, according to which S_{0i} should becomes something like the new dipole operator (intrinsic position). We will therefore look for a new relativistic intrinsic position operator ξ^{rel}_{μ} which is a covariant formulation of $\frac{1}{mc} S^{0i}$.

In addition to the intrinsic observables $S_{\mu\nu}$, ξ^{μ}, π^{ν} one has the center of mass (c.m.) observables

$$P_\mu, \; J_{\mu\nu} = P_\mu \wedge Q_\nu + S_{\mu\nu} = P_\mu \wedge Y_\nu + \Sigma_{\mu\nu}, \; P_\mu P^\mu = M^2 \tag{4.2}$$

which generate the Poincaré group of the relativistic c.m. motion and go in the non-relativistic limit into the generators of the Galilei group for the c.m. motion. Like the ξ^{i}, π^{i}, S^{ij} in the non-relativistic case the ξ^{μ}, π^{ν},

$S^{\mu\nu}$ describe the intrinsic motion.

For ξ^{rel}_{μ} one then takes [38] the quantum mechanical Mathisson vector:

$$\xi^{rel}_{\mu} = -S_{\mu\nu} \frac{P^{\nu}}{c^2 M^2} \tag{4.3}$$

$$\left[\xi^{rel}_{\mu}, \xi^{rel}_{\nu}\right] = -i \frac{1}{c^2 M^2} \Sigma_{\mu\nu} ; \tag{4.4}$$

where

$$\Sigma_{\mu\nu} = S_{\mu\nu} + \xi_{\mu} \wedge P_{\nu} \tag{4.5}$$

is the spin tensor, i.e.:

$$-\hat{W}_{\mu} \hat{W}^{\mu} = \frac{1}{2} \Sigma_{\mu\nu} \Sigma^{\mu\nu}, \quad (W_{\mu} = \frac{1}{2} \varepsilon_{\mu\nu\varrho\sigma} \hat{P}^{\nu} J^{\varrho\sigma};$$

$$\hat{P}_{\mu} = P_{\mu} M^{-1}). \tag{4.6}$$

A simple relativistic Hamiltonian from which this motion of the observables can be derived is

$$\mathcal{H} = \phi \, (P_{\mu} P^{\mu} - \frac{1}{\alpha'} \hat{P}_{\mu} \Gamma^{\mu}), \quad \hat{P}_{\mu} = P_{\mu} M^{-1} \tag{4.7}$$

ϕ is the Lagrange parameter (generalized velocity) which will be fixed by a gauge constraint to $\phi = -\frac{1}{2M}$. α' is a constant which makes the units correct (elementary length squared, it will turn out to be phenomenologically given by approximately the slope of the Regge trajectories).

(7) is a conjecture and there are many other possible choices for $\mathcal{H}$ which lead to similar motions. It has a form familiar from the Dirac theory: $P_{\mu}\Gamma^{\mu} \leftrightarrow \bar{\psi}\gamma^{\mu}\partial_{\mu}\chi$, $\Gamma^{\mu} \leftrightarrow \bar{\psi}\gamma^{\mu}\psi$. It leads to the relativistic mass-point Hamiltonian in the limit $1/\alpha' \to 0$. But its main point of distinction in the present connection is that

in the non-relativistic limit $\frac{1}{c} \to 0$ it yields the non-relativistic energy:

$$H^{OSC} = \frac{\vec{P}^2}{2m} + \frac{1}{2\mu}\vec{\pi}^2 + \frac{2\mu}{(4m\alpha')^2}\vec{\xi}^2 , \tag{4.8}$$

which is the energy operator of the non-relativistic oscillator for the intrinsic motion plus the kinetic energy for the center of mass motion.

Starting out with the operator of the intrinsic position (3) (whose choice has been motivated by 1) the non-relativistic limit of the Inönü-Wigner group contraction $p \to G$, 2) the non-commuting dipole operators of the vibron model and 3) the Zitterbewegunf of the classical relativistic spinning object) and using the relativistic Hamiltonian (7) one is led to the following definition of the "conjugate momenta" of ξ^{rel}_{μ} :

$$\pi^{rel}_{\mu} = -\frac{1}{\alpha'}\frac{1}{CM}\check{g}^{\sigma}_{\mu}\Gamma_{\sigma} \tag{4.9}$$

where

$$\check{g}^{\sigma}_{\mu} = \eta^{\sigma}_{\mu} - \hat{P}_{\mu}\hat{P}^{\sigma} \; ; \quad \hat{P}_{\mu} = P_{\mu}/M \tag{4.10}$$

is the projector on the plane perpendicular to P_{μ}. π^{rel}_{μ} is thus essentially Γ_{μ} projected into the space like plane perpendicular to P_{μ}.

In the non-relativistic limit (extension of the Inönü-Wigner group contraction to the whole algebra) the new relativistic position and momentum go into

$$\begin{aligned} \xi^{rel}_{\mu} &\xrightarrow{\frac{1}{c}\to 0} \begin{cases} 0 & \text{for } \mu = 0 \\ \xi_i & \text{for } \mu = 1,2,3 \end{cases} \\ \pi^{rel}_{\mu} &\xrightarrow{\frac{1}{c}\to 0} \begin{cases} 0 & \text{for } \mu = 0 \\ \pi_i & \text{for } \mu = i \end{cases} \end{aligned} \tag{4.11}$$

where ξ_i and π_i are the usual non-relativistic intrinsic position and momentum operators with the usual 3-dimensional Heisenberg c.r. (2.7). They are the same operators appearing in (4.8).

The surprising feature of the new "conjugate pair" of intrinsic relativistic position ξ_μ^{rel} and momentum π_μ^{rel} is that they do not satisfy the usual 4-dimensional) relativistic Heisenberg c.r. (1), but the following new relativistic Heisenberg c.r. [39]

$$[\xi_\mu^{rel}, \pi_\nu^{rel}] = -i\check{g}_{\mu\nu} \qquad [\xi_\mu^{rel}, \xi_\nu^{rel}] = \frac{i}{c^2 M^2} \Sigma_{\mu\nu}$$

$$[\pi_\mu^{rel}, \pi_\nu^{rel}] = -i \frac{1}{\alpha'^2 c^2 M^2} \Sigma_{\mu\nu} . \tag{4.12}$$

One can already see (and the detailed calculation confirms) that New relativ. Heisenberg c.r. $\xrightarrow{(1/c)\to 0}$ non-rel. Heisenberg c.r. (2.7). Therefore these new Heisenberg c.r. are as valid relativistic generalizations as the usual ones; in fact they have the added advantage that the non-sensical π_0 and ξ_0 vanish for $\frac{1}{c}\to 0$.

Summing, the algebra of the new Quantum Relativistic Oscillator (QRO) goes in the non-relativistic limit into the three dimensional non-relativistic oscillator algebra.

As the algebras are related by group contraction, the representation spaces must also be connected. There is indeed a representation of SO(3,2) whose reduction with respect to SO(3) is given by the level diagram in Fig. 3.4 ν is now the eigenvalue of Γ_0 .

In the non-relativistic case, to each level there corresponds an irrep. of the Galilei group characterized by [40] j and E_ν . Instead of E_ν one can use the principal quantum number ν related to E_ν by (3.6). To each level for the relativistic case corresponds an irrep of p characterized by j and m. But m is now related to the eigenvalue

$$\mu \equiv \text{eigen}(\Gamma_0) = \text{eigen}(\hat{P}_\nu \Gamma^\nu)^{\text{rest}} = \nu_0 + \nu, \quad \nu = 0,1,2,.. \tag{4.13}$$

by the constraint relation that follows from (7)

$$P_\mu P^\mu - \frac{1}{\alpha'} \hat{P}_\mu \Gamma^\mu = 0 \tag{4.14}$$

and which leads to

$$m^2_{(\nu)} = m^2_0 + \frac{1}{\alpha'} \nu. \tag{4.15}$$

Thus to a (ν, j) level there corresponds an irrep space $\mathcal{H}^\nu(m_\nu, j)$ of p which is obtained from the irrep of $SO(2)_{\Gamma_0} \times SO(3)_{S_{ij}} \subset SO(2,3)_{S_{\mu\nu}\Gamma_\mu}$ by something like an induction [41].

In addition to the simple representation of SO(3,2) with the level diagram (weight diagram or K-type) given by solid levels in Fig. 3.4b there are other representations of SO(3,2) that are suitable for the relativistic SGG. E.g. the broken levels of Fig. 3.4b form another irrep, so that all of Fig. 3.4b depicts a direct sum of two irreps of SO(3,2). One often denotes the irreps of SO(3,2) [42] by $D(\mu_{min}, j_{min} = s)$, where μ_{min} = lowest (or highest) eigenvalue of Γ_0 and j_{min} is the smallest value of j. (This fully characterizes the irreps with semi bounded spectrum of Γ_0). Each series of levels on the same horizontal line in Figs. 3.4b, solid lines and dashed lines belong to an irrep. of SO(4) (like the horizontals in Fig. 4.1), though SO(4) is not a subgroup, because these are also weight diagrams of SO(4,2) = SU(2,2) representations which remain irreducible under SO(3,2).

The SO(3,2) representation with weight diagram 3.4 gives in the non-relativistic limit the spinless oscillator, i.e. the oscillator whose angular momentum (had-

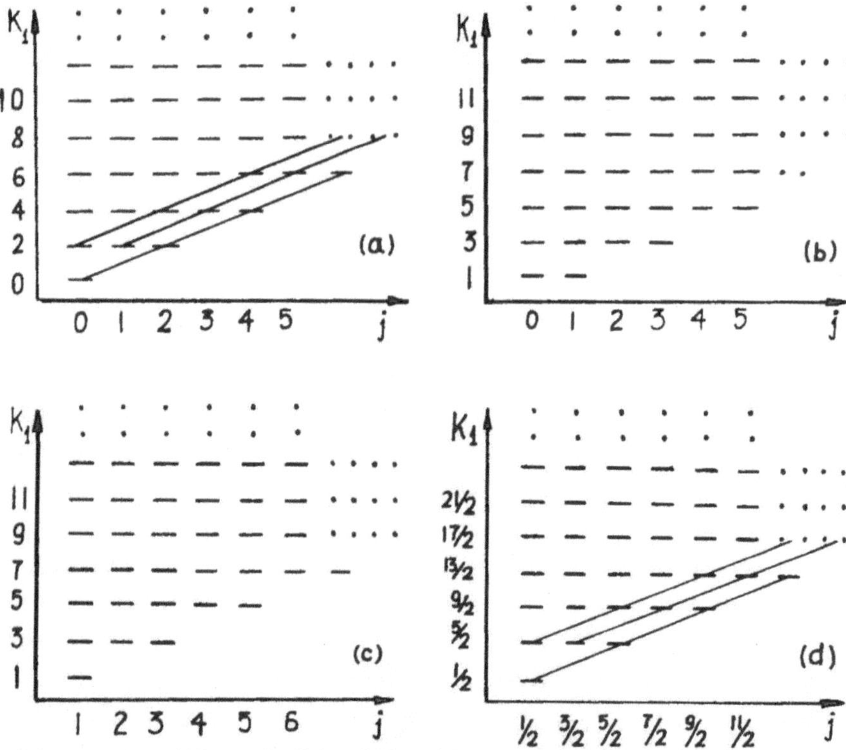

Fig. 4.1. (a) and (b) show the two simplest irreps $D(0,0,p_2)$ and $D(1/2,1/2,p_2)$ respectively [68] ("Ladder representations"). (c) and (d) show the "lowest" states of two discrete series representations (for the discrete series only half of the levels are shown, the others are mirror symmetric to the k axis). ($k_0 = j^{min}, k_1$) characterize the irreps of SO(4) which are contained in the irrep of SL(4,R)

ron spin) comes entirely from the intrinsic orbital angular momentum $\xi_i \wedge \pi_j$. A di-quark vibrator with quarks of spin $\frac{1}{2}$ and total quark spin 1, as required for the vector mesons etc. according to the quark model, should be described by a representation which has in addition to the intrinsic orbital angular momentum an intrinsic spin with value 1. For the baryon representation the total intrinsic spin should be $\frac{1}{2}$. The intrinsic spins are related to the value of the 4-th order Casimir operator of SO(3,2) which in the non-relativistic contraction limit $\frac{1}{c} \rightarrow 0$ leads to the eigenvalue equation [35)]

$$(S_{mn} - \xi_m \wedge \pi_n)^2 = s(s+1)\ 1 \qquad (4.16)$$

where $s = 0,1/2,1,3/2,\ldots$ is one of the numbers that characterize the SO(3,2) representation. For the irreps of Fig. 3.4, $s = 0$.

S_{mn} is the intrinsic total angular momentum which in the rest frame (but not otherwise) is the spin of the extended object; $\xi_m \wedge \pi_n$ is the intrinsic orbital angular momentum coming from the orbital motions of the constituents.

$$\tilde{S}_{mn} \equiv S_{mn} - \xi_m \wedge \pi_n \qquad (4.17)$$

must therefore be the total spin of the constituents. All this is only defined in the non-relativistic limit. The $s = 1$ representation will thus describe the hadron towers as quantum relativistic oscillator levels which have total quark spin 1, like $\rho, \omega, A_2 \ldots$. The $s= \frac{1}{2}$ representation leads to a total quark spin $\frac{1}{2}$ in the non-relativistic limit, as needed for the baryon tower.

As the mass splitting for mesons and baryons is the same and the two representations D(3/2, 1/2) and D(2,1) combine into the irreducible representation of the superalgebra $Osp(1,4) \supset SO(3,2)$:

$$D(3/2,1/2) \oplus D(2,1) \qquad (4.18)$$

this may be some evidence of dynamical supersymmetries in the low mass hadron spectra, quite analogous to the non-relativistic case for nuclear supersymmetry [12)].

We have considered the dynamical group for varuous physical objects which are made of many different things: Molecules are made of electrons and nuclei, nuclei are made of hadrons, hadrons are made of quarks. Yet, from the rotational bands of the molecules with level splittings of 10^{-14} GeV over the rotational and vibrational levels of nuclei with energies of $1¾^{-5}$ - 10^{-3} GeV to the mass differences between hadrons of the order of 1 GeV, similar features appear again and again. The reason for this is that not the constituents but the motions the "parts" complementary to the constituents - determine these features and they are the same. As for the algebra of observables, given by the generators of the dynamical group, the constituents are of little relevance (determining the parameters), the SGG approach is the most appropriate method for this fascinating display of unity in physics. It certainly will go on over many more orders of magnitude.

5. Strings (Second Quantized). As explained in our introduction, the search for an understanding of the hadron spectrum in the sixties combined three approaches: (a) internal quantum numbers such as strangeness - SU(3) etc., (b) S-matrix methods, mainly effective in the first attempts to describe the angular momentum excitations; and (c) Spectrum Generating Groups. In principle, the string is very much in the spirit of the SGG, but it entered physics in 1968 through route (b). S-matrix theory had brought in the notion of Regge trajectories and analytical continuation in angular momenta. Yukawa exchange models were favoured, and the predictions were of short Regge trajectories with just one or two excited levels on them. Experiments gave a different picture, with linear trajectories going on "indefinitely" (5-6 poles at such energies give that impression...) .

The S-matrix methods were a algebraized by the Fubini school ("superconvergence"), with commutators being saturated by intermediate states as in matrix mechanics. The idea of the bootstrap - particles interact through the exchange of the same particles - culminated in the "Finite Energy Sum Rules".

1985 is a highlight year for strings, perhaps the third such peaking in the evolution of the idea. Excitement first peaked in 1968, when the work of the Rehovoth-Tel Aviv group on those Finite Energy Sum Rules (FESR) - simultaneously discovered in the USSR and in Japan - culminated in the Veneziano representation, an extremely aesthetic solution of the FESR. It was treated with some of the most advanced tools of SGG. It gave a highly stylized description of the hadron spectrum, in the context of the on-mass-shell S-matrix techniques, prevalent at the period. Nambu and Susskind soon showed that the Veneziano amplitude could be reinterpreted as the spectrum of excitations of a rotating and oscillating one-dimensional body, a string. We refer the reader to appropriate reviews [33,43] of the various ideas and the growth of the concept especially from the symmetry standpoint. We only note here that unitarity (no negative-norm states) required the string to be embedded in a 26-dimensional Minkowski manifold, or more precisely - a manifold with 24 "transverse" Euclidean dimensions. However, even in 26 dimensions the theory is still plagued by a spinless tachyonic state.

The second peak in our expectations from strings came in 1971, when P. Ramond and independently A. Neveu and J.Schwarz constructed the first "superstrings". The advance consisted in fact in an independent introduction of superalgebras, simultaneous with the Golfand-Likhtman discovery of 4-dimensional supersymmetry. In the string, we deal with infinite superalgebras representing the system of constraints. For the Veneziano amplitude, Virasoro had used an algebra constructed around an SL(2,R) "kernel", with the entire algebra nehaving as an SL(2,R) module. The algebra itself has a center, crucial in fixing the dimensionality of the embedding manifold and in other applications. The Virasoro algebra commutation relations are:

$$\left[L_m, L_n\right] = (n-m)L_{m+n} + \left(\frac{D}{12}\right) m(m^2-1)\, \delta_{m,-n}\,, \quad m,n \in \mathbb{Z} \tag{5.1}$$

Without the center, this coincides with the algebra of analytical diffeomorphisms in one dimension, following Ogievetsky's expansion [44). Indeed,

$$L_m \sim i x^{m+1} \frac{\partial}{\partial x} \tag{5.2}$$

Satisfy the commutation relations. This is also the conformal algebra in 2-dimensions (two copies, L_m and $\bar{L}_m$ are needed for a complex manifold and its conjugate), where the conformal group is infinite. The L_m obey a Hermiticity condition

$$(L_m)^+ = L_{-m} \tag{5.3}$$

and serve (as an algebra of constraints) to remove unphysical states. The spectrum obeys:

$$\left.\begin{aligned} L_m |0\rangle &= 0 \qquad \text{for} \quad m>0 \\ L_0 |0\rangle &= 0 \end{aligned}\right\} \tag{5.4}$$

In the superstring the constraints [45) make a superalgebra; the new operators G_r $(r \in \frac{1}{2}\mathbb{Z})$ are fermionic and the algebra preserves a $\frac{1}{2}\mathbb{Z}$ grading:

$$\left\{G_r, G_s\right\} = 2L_{r+s} + \frac{D}{2}(r^2 - \tfrac{1}{4})\, \delta_{m,-n}\,, \quad \left[L_m, G_r\right] = \left(\frac{m}{2} - r\right) G_{m+r}$$

$$\left[L_m, L_n\right] = (n-m)L_{m+n} + \frac{D}{8}\, m(m^2-1)\, \delta_{m,-n}\,. \tag{5.5}$$

The critical dimension is D=10, or 8 transverse dimensions. The search thus went on for a model that would have D=4 - apparently inexistent - but in the meanwhile great things had been happening in Particle Physics: 't Hooft had proved the renormalizability of the Yang-Mills theory (with or without spontaneous symmetry breakdown) and Relativistic Quantum Field Theory (RQFT) was back. Attention was drawn to RQFT and especially to Gauge Theories; after the highly successful $[SU(2) \otimes U(1)]_{W+SM}$ and the almost-as-successful $SU(3)_{colour}$ (remember that by 1985 there is as yet no proof of confinement in QCD, as plausible as it be, beyond the proofs on a lattice) interest moved in the direction of Unification and Superunification, including Supergravity.

The string was neglected. However, right then in 1974, Scherk and Schwarz [46)] noticed that the superstring spectrum contains in the zero-slope limit a massless spin 2 state. Identifying this state with the graviton, they suggested that superstrings might be regarded as fundamental off-mass-shell entities, something like a "manifield". The lowest level in the spectrum yields a "normal" field-theory: either Supergravity or a Yang-Mills theory, depending on the model, but again in D=10, of course. Note that the higher states in the superstring spectrum are massive. These could make Gravity renormalizable, just as the exchange of W mesons resolved the Weak Hamiltonian non-renormalizability. They could cancel the dimensional features of Newton's constant,

Between 1975 and 1984, Schwarz and his collaborators continued to study these possibilities. After 1982, it turned out that some of the possible models yield finite one-loop diagrams. In 1984, Green and Schwarz [47)] discovered that non-orientable type I superstrings, both open and closed (with N=1 supersymmetry) are finite and free of anomalied if the spin 1 field they contain is made to gauge SO(32), and J.Thierry-Mieg [48)] showed that this is also true of E(8) x E(8) (both groups have a rank 16 algebra). "Type 1" implies a supersymmetry whose spinorial charges are both self-conjugate ("Majorana") and chiral ("Weyl"). It is a chiral theory in 10 dimensions, but it is not yet clear whether in D=4, after a dimensional reduction, the theory will still be chiral. An open type I theory has for its massless sector a Super-Yang-Mills RQFT (in D=10). This is a divergent model, but its 4-dimensional reduction is the N=4 super-Yang-Mills theory that was shown to be finite to all orders of perturbation theory. A closed type I superstring has as massless sector N=1 10-dimensional supergravity. Following the proof of vanishing anomalies, Gross et al. [49] have constructed a "heterotic" superstring which we shall discuss, this being the third "peak" we were greeting in our opening paragraph. In particular, the $E_8 \times E_8$ model has a good chance of providing answers that would fit the phenomenology [50]. This is not the phenomenology of the hadron mass-spectrum, which was the center of interest in sect. 4 . Here the aim is to produce the fields we observe, i.e. at least 3 generations of 15 chiral constituent fields each: 3 isodoublet pairs of left-chiral quarks, two sets of 3 isosinglet right-chiral quarks, an isodoublet lepton (ν_ℓ, ℓ) left-chiral pair and an isosinglet right-chiral charged lepton ("iso" here related quantum numbers to $SU(2) \otimes U(1)$).

Part of this advance has resulted from progress in the mathematical background, especially in the understanding of Kac-Moody algebras [51] and of mathematical lattice theory [52]. We have also witnessed a demonstration of the Unity of Physics: the algebra (5.1) can be applied to the critical behaviour of two-dimensional physical lattice systems [53]: the physics there is local (neighbour interactions) and becomes scale-invariant at the critical temperature. The D in (5.1) is related to the corresponding critical exponents (for $D < 1$ here): $D = \frac{1}{2}$ for the Ising model, $\frac{7}{10}$ for the tricritical Ising

(He absorbed in Kr-plated graphite), $\frac{4}{5}$ for the 3-state Potts model, $\frac{6}{7}$ for the tricritical tri-state Potts model. The existence of a $D < 1$ region for (5.1) with precisely such values constituted in itself progress in the relevant representation theory.

Physicists have been using Kac-Moody algebras since the 1965-69 search for the representations of the algebra of the local current densities [54). A Kac-Moody (or Affine) algebra is constructed over a kernel formed by any "single-laced" semi simple Lie algebra; the term refers to Dynkin diagrams in which any two contiguous primitive roots are connected by a single line (i.e. an angle of $2\pi/3$) and all such primitive roots have the same length. These are also the only simple Lie algebras with positive-definite, symmetric Cartan "cos α_{ij}" -matrices, α_{ij} being the angles between primitive roots. The root diagrams of such algebras therefore also constitute bases for all possible lattices under 24 dimensions, in the theory of mathematical lattices. These algebras are the A_n (generating the SU(n+1) etc....), the D_n (generating the SO(2n) etc ...) and the exceptionals E_6, E_7, E_8. The length of the primitive root vectors is normalized to $(e_i)^2=2$.

The bridge between mathematics and physics carring "traffic" both ways. I.B.Frenkel and V.G.Kac [55) used the construction developed for the Veneziano model ("vertex operators") and adapted it to the construction of the basic representation (the non-trivial highest weight) for all Affine algebras. Affine Lie algebras, together with the Lie algebra vector fields on the circle constitute the entire set of infinite-dimensional Lie algebras which admit a $\mathbb{Z}$-grading with subspaces of bounded dimensions

$$(\dim\{L_{m+1}\} / \dim\{L_m\} \sim m^{g-1}, \quad g \leq 2) \tag{5.6}$$

and without graded ideals. Their structure is very close to that of simple finite Lie algebra.

In inserting internal quantum numbers in a string,the method [56)] had been to attach some representation space (e.g. quarks) to the ends of the strings. The QCD-oriented picture in which the string was constituted by the compressed flux-lines fitted this construction, as the string would then have the quantum numbers of a gluon. However, it was shown that this Paton-Chan symmetry insertion [56)] would only allow SO(n) and USp(2n). The Frenkel-Kac construction changed this situation, the mathematicians now providing the physicists with an entirely new type of inclusion of a semi simple group (singly-laced) in the superstring. This is how $E_8 \times E_8$ could now be built into a physical model [48,49].

A word about mathematical lattices [52)]. This branch of mathematics has been in the background of Supersymmetry, Supergravity, Strings and Superstrings. The connection is clear for all the "supers" in this list, since it relates to the spinors: the question of which Euclidean or Minkowskian spaces allow spinors to be Majorana, Weyl or both is the same as asking which spaces allow self-dual, even unimodular lattice. The Majorana+Weyl type exists only in Euclidean spaces with 8n dimensions (n=1,2,...) or in Minkowskian with 2+8n dimensions. These are spaces allowing even unimodular self-dual lattices. For n=1 , the lattice is that of E_8, for n=2 there are two possibilities, D_{16} (i.e. SO(32)) and $E_8 \otimes E_8$. For n=3 there are 24 such lattices, one of which has all its vectors with $(e_i)^2 \geq 4$. This is the Leech lattice [58)], whose finite Weyl-type symmetry is given [58)] by the "Monster" (or "Friendly Giant"), the largest simple finite group.

It is interesting that the bosonic string lives in n=3. The relationship in this case is not really clear, though some points have been made by Goddard and Olive [52].

We now return to the heterotic [49] string. It uses the fact that in a light-plane gauge the eight right-moving $\chi^i(\tau - \sigma)$ (i=1...8) transverse local frame vectors of the superstring (D=10) and the Majorana-Weyl fermionic right moving/local spinor frames $\chi^a(\tau - \sigma)$ (a=1...8) are in fact propagating independently of the left-moving sector. The latter consists here in 24 transverse (bosonic) local frame vectors $\chi^i(\tau + \sigma)$ and $\chi^I(\tau + \sigma)$ (I=9...24). The variables, and are dimensionless parametrizations of the string time and longitude variables. The string action is that of an Affine Gravity theory in 2 dimensions, with the Lagrangian corresponding to the measure in that (τ, σ) surface, with a large anholonomic group in the tangent, fixed by the dimensionality of the (flat) embedding manifold. Ordinarily in the left-moving sector this would be SO(24) but in this heterotic string, the left-moving string is reduced to the 8-dimensions of $\chi^i(\tau + \sigma)$, the other sixteen χ^I being compactified. Now an important point to remember in the Frenkel-Kac construction [55] is that the Cartan Euclidean submanifold spanned by the commutative. Cartan subalgebra of the kernel group (e.g. a 16-dimensional Euclidean space for $E_8 \times E_8$ or D_{16}) is also the momentum-space generating the infinite structure. In local current algebra [54] the momenta existed in the Fourier transform of space-time, whereas the SU(3) or chiral SU(3) algebra was entirely "internal". In these Kac-Moody algebras, the Fourier transform of space-time is the Cartan subalgebra manifold. This phenomenon was already present in D=11 Supergravity [59]: we could split the 11-dimensions into d=4 of space-time and have as an internal symmetry a non-compact real form of the group E_7; or, alternatively have d=3 with E_8 or d=5 with

E_6 etc. [66].

The heterotic string thus uses the 24-8=16 dimensions left-over in the left-moving sector to construct $E_8 \otimes E_8$ (the kernel of the Kac-Moody algebra) and have it as a symmetry to be gauged by the Yang Mills fields of the super-Yang-Mills basic mode of the superstring section. The $\chi^i(\tau_+ - \sigma)$ and $\chi^i(\tau + \sigma)$ together with the light-plane $\chi^-(\tau, \sigma)$ indeed reconstitute a superstring with effective $N = \frac{1}{2}$ supersymmetry (in terms of the number of fermionic charges). The free heterotic string action in this gauge is given by ($g^{\mu\nu}$ is the 2x2 string metric), $e^i_\mu = \partial_\mu \chi^i$ the "tetrad")

$$S = -\int d\tau \int_c^\pi d\sigma \frac{1}{4\pi\alpha'} \left\{ (e^i_\mu e^j_\nu \eta_{ij} + e^I_\mu e^J_\nu \delta_{IJ}) g^{\mu\nu} + i\bar{\psi}\gamma^- \partial_+ \psi \right\} \tag{5.7}$$

$$\gamma^+ \psi = 0, \quad \tfrac{1}{2}(I + \gamma_{11}) = 0, \quad \partial_- X^I = 0 .$$

The theory is one of closed orientable strings (so as not to mix right and left-movers). It is Lorentz-invariant and supersymmetric; the states are representations of $E_8 \otimes E_8$. There are no anomalies in the low-energy field theory. The heterotic (left-right separate and reinforcing each other) nature of the theory gets rid of the tachyon which was the failing of the bosonic sector. The theory is chiral, hopefully in the reduced d=4 as well (for this it is necessary that the D=10 be chiral). The $E_8 \otimes E_8$ yields an internal E_6 symmetry with 4 generations (27 of E_6, predicting some new types of fields). There is an unbroken N=1 supersymmetry in d=4 dimensions, an essential factor in resolving the "gauge hierarchy" problem. This is the difficulty in understanding how the breakdown of $[SU(2) \times U(1)]_{W+EM} \to U(1)_{EM}$ occurs around 100 GeV and not at 10^{15} GeV, the unification group scale (where $SU(3)_{colour}$ is assumed to separate out). Supergravity appears to provide the only way in which this can occur. Of course, there is also the hope that the superstring would provide the answer to the quantization of gravity (and supergravity).

All of this appears highly promising. It is however too early to know whether some other difficulties won't appear.

Is the string the end of the search? To one of us (Y. N.) it seems this would be frustrating. A theory containing the quanta of gravity, but with a flat 4-dimensional space-time appears unsatisfactory. This calls for moving on from the $SL(2,\mathbb{R})$ invariant string (given by the measure, the determinant representing the surface generated by the evolving string) to a "membrane" with $SL(3,\mathbb{R})$ and a "lump" with $SL(4,\mathbb{R})$. The spinorial representations in both groups have now been constructed [14,16,32] and algebras or superalgebras modeled on the Virasoro algebra can be generated. The actual Ogievetsky algebras may be too large and unwieldy, and some simplified structures with more relaxed growth $g=2$ in (S6) may be sufficient [61]. The construction of spinorial fields corresponding to unitary infinite-dimensional representations of $\overline{SL}(4,\mathbb{R})$ is also a step in this direction. They can be used for the classification of baryons, as discussed in section 4. The motivation - the general success of field theory, with phenomenological fields (with the knowledge that they become composite structures at a different energy level). Also - for QCD confinement, an evolving fixed-volume lump should be a good approximation. Note that the stability subgroup $SL(3,\mathbb{R})$ gave Regge-trajectories [1], and was tested in Nuclear Physics [62].

Acknowledgment. We would like to thank J.Thierry-Mieg, L.C.Biedenharn, T.Tamura, T.Udagawa, D.H.Feng and M.Loewe for discussions, advice and help with various problems of this paper.

References

1) Bohm,A.,Barut,A.O. (1965). Phys. Rev. 139B, 1107; Y.Dothan, M.Gell-Mann, Y.Ne'eman (1965). Phys. Rev. Lett. 17, 145.

2) We shall use the name Dynamical Group (DG) and Spectrum Generating Group (SGG) and Algebra (SGA) interchangeably. So far all enveloping algebras of observables that one has used for an SGA were integrable to unitary representations of a (in general non-compact) group. When we speak of representations of the group G we usually mean the integrable representations of the covering group G . We will say G rather than G only if we want to stress their distinction. In the literature the word dynamical group is also used in a different context: a) as the one-parameter group of time translation; b) as the degeneracy group of the energy levels. The latter we call here DDG,

3) Pauli,W. (1926). Z. Physik 36, 336; Fock,V. (1935). Z.Physik 98, 145; Bargmann,V. (1936); Jauch,J.M., Hill,E.L. (1940). Phys. Rev. 57, 641; Alliluev,A.P. (1958). Sov. Phys. JETP 6, 156; Barut,A.O. (1964). Phys. Rev. 135, B839.

4) G.Racah and other atomic spectroscopists had suggested a shell-model approach very early, from a guess at the pattern. However they were rebuffed with the comment "this is a short-range strong interaction,not a central force" ... (private communication to Y.N.).

5) Dyson,F. (1966). Symmetry Groups, W.A.Benjamin Pub.; Ne'eman,Y. (1967). Algebraic Theory of Particle Physics, W.A.Benjamin Pub.

6) Barut,A.O., Budini,P., Fronsdal,C. (1966). Proc. Roy. Soc. London A291, 106; Bohm,A. (1966). Nuovo Cim. 43, 665; Dothan,Y., Ne'eman,Y. reprinted in ref. 5.

7) Malkin,I.A., Man'ko,V.I. (1965). JETP Letters 2, 230; Barut,A.O., Kleinert,H. (1967). Phys. Rev. 156, 1541; 157, 1180; 160, 1149.

8) Bohm,A., Teese,R.B. (1979). In: Group Theoretical Methods in Physics, Beiglbock,W. (ed.) Springer LN in Phys. 94, 301 ; Phys. Rev. D26, 1103 (1982); Garcia,A., Kielanowski (1985). The Beta Decay of Hyperons, Springer.

9) Arima,A., Iachello,F. (1981). Ann. Rev. Nucl. Part. Sci., 31, 75 and references thereof. Casten,R.F., Feng,D.H. (1984). Physics Today, November, p. 26, and references thereof.

10) Golfand,Y.A., Likhtman,E.P. (1971). JETP Letters 13, 323; Volkov,D.V., Akulov,V.P. (1972). JETP Letters 16,

438; Wess,J., Zumino,B. (1974). Nucl. Phys. B70,39; (1973). Phys. Lett. 46B, 109; Salam,A., Strathdee,J.; (1974). Nucl. Phys. B76, 477.

11) Berezin,F.A., Kac,G.I. (1970). Mat,(USSR Sbornik) 82, 124; Kac,V.G. (1975). Func. Anal. Appl. 9, 91; (1977). Commun. Math. Phys. 53, 31; Corwin,L., Ne'eman,Y., Sternberg,S. (1975). Rev. Mod. Phys. 47, 573; Rittenberg,V. (1978). In: "Group Theoretical Methods in Physics (Proc. Tübingen 1977), Kramer,P., Rieckers,A., eds. Lect. Notes in Phys. 79, Springer Verlag (N.Y.), p. 3.

12) Balantekin,A.B., Bars,I., Iachello,F. (1981). Phys. Rev. Letters 47, 19; Sun,H.-Z., Vallieres,M., Feng,D.H., Gilmore,R., Casten,R.F. (1984). Phys. Rev. C 29, 352. A similar kind of evidence for supersymmetry in atomic physics has recently been reported by Kostelecky,V.A. and Nieto,M.M. (1984). Phys. Rev. Letters 53, 2285.

13) Gelfand,I.M., Graev,M.I. (1953). Izv. Akad. Nauk (USSR) 17, 189; Dothan,Y. et al. (ref.1); Rosen,G. (1966). J. Math. Phys. 7, 1284; Hulthen,I. (1968). Ark. Fys. 38, 175.

14) Joseph,D.W. (1969). "Rep. of the Algebra of SL(3R) with $|\Delta J| = 2$", Univ. of Nebraska reprint, unpub.; Weaver,L., Biedenharn,L.C. (1972). Nucl. Phys. A185, 1; Ogievetsky,V.I., Sokachev,E. (1975). Teor. Math. Phys. 23, 214; Sijacki Dj. (1975). J. Math. Phys. 16, 298; Angelopoulos,E. (1978). J. Math. Phys. 19,2108.

15) Speh,D. (1981). Math. Annalen. 258, 113; Rownsley,J., Sternberg,S. (1982). Am. Jour. Math. 104, 1153.

16) Kihlberg,A. (1966). Ark. Fys. 32,241; Borisov,A.B. (1978). Reports on Math. Phys. 13, 141; Ne'eman,Y., Sijacki,Dj. (1979). Proc. Nat. Acad. Sci. USA 76,561 and (1980). 77, 1761; see also ref. 32; Friedman,J.L., Sorkin,R.D. (1980). J. Math. Phys. 21,1269.

17a) Dothan,Y. (1970). Phys. Rev. D2, 2944.

17b) Malkin,I.A., Man'ko,V.I., Trifonov, D.A. (1969) Phys.Lett.,30A 414

17c) Dynamic symmetries and coherent states of quantum systems,(1979). nauka, Moscow (russian)

18) Nambu,Y. (1967). Proceedings of the 1967 International Conference on Particles and Fields, p. 347, C.R.Hagen et al. (eds.), Interscience Publishers. Stoyanov,D.Tz., Todorov,I.T. (1968). J. Math. Phys. 9, 2146; Bohm,A. (1968). Lect. in Theor. Phys. Gordon and Breach. N.Y., 10B, 483; Dirac,P.A.M. (1972). Proc. Roy. Soc. A328, 1; Staunton,L.P. (1976). Phys. Rev. D13, 3269; Dam,H. van, Biedenharn,L.C. (1979).

Proceedings of the 7th International Group Theory Colloquium, Springer LN in Phys. 94, p. 155.

19) Naimark, M.A. (1964). Linear Representations of the Lorentz group, Pergamon Press; Gelfand,I.M. et al. (1963). Representations of the Rotation Group and the Lorentz Group, Pergamon Press, See also Appendix to Section V.3 of ref. 22 Below.

20) Iachello,F., Levine,R.D., Roosmalen,O.S. van et al. (1982). Chem. Physics 77, 3047 and (1983). J.Chem. Phys. 79, 2515.

21) $\pi^{+}_{\mu=\pm} = \mp \frac{1}{\sqrt{2}}(\pi^{1} \pm i\pi^{2+})$, $\pi_{\mu=0} = \pi^{3}$. The spherical co-ordinates of the dipole operator used in ref. 20: $D^{(1)}_{\mu} = (\vec{\alpha}^{+}\sigma + \sigma^{+}\tilde{\vec{\alpha}})^{(1)}$ are connected with the Cartesian co-ordinates used in equation (2.12) by $D^{(1)}_{\mu=\pm} = \mp \frac{1}{\sqrt{2}}(D^{1}_{\mu} \pm iD^{2})$, $D^{(1)}_{0}=D^{3}$. $k_1 = \omega$ in the notation of reference 20, we use the notation of ref. (19) or Appendix V.3 of ref. 22.

22) Bohm,A. (1979). Quantum Mechanics, Springer N.Y.

23) Corben,H.C. (1968). Classical and Quantum Theory of Spinning Particles, Holden-Day.

24) Aldinger,R.R. et al. (1984). Phys. Rev. D29, 2828; Bohm,A. et al. (1985). Phys. Rev. D to appear.

25) Herzberg,G. (1966). Molecular Spectra and Molecular Structure, D. van Norstrand Publisher.

26) Herrick,D.R. et al.

27) Arima,A., Iachello,F. (1975). Phys. Rev. Lett. 35, 1065; Arima,A., Otsuka,T., Iachello,F., Talmi,I. (1977). Phys. Lett. 66B, 205; (1978). 76B, 139.

28) Bohr,A., Mottelson,B. (1969). Nuclear Structure, 2, Benjamin.

29) Castanos,O., Chacon,E., Frank,E., Moshinsky,M.(1979). JMP 20, 35; Feng,D.M., Gilmore,R., Deans,J.R. (1981). Phys. Rev. C23, 1254.

30) Bohm,A. (1968). Phys. Rev. 175, 1767; (1971). D3,377.

31) Cant,A., Ne'eman,Y. Tel. Aviv Univ. report N156-84, to be published in J. Math. Phys.

32) Ne'eman,Y., Sijacki,Dj. (1979). Annals of Physics (NY) 120, 292; (1985). Tel Aviv University reports N 159, N 160 to be published.

33) For a review and further references see Scherk,J. (1975). Rev.Mod.Phys. 47,123; Schwarz,John M.(1982).

Phys. Rep. 89, 223.

34) Soloviev,L.D. et al. (1984). Proceeding of the XXII International Conference on High Energy Physics, Leipzig; Pron'ko,G.P., Razumov, A.V. (1983). Proc. Soviet Acad. Science 56, No. 2, 192.

35) Bohm,A., Loewe,M., Magnollay,P. (1985). The Quantum Relativistic Oscillator I, II, Univ. of Texas preprint DDE-ER-03992-574, 576, Phys. Rev. D, to appear.

36) Dirac,P.A.M. (1972). Proc. R. Soc. London A328, 1; (1964). Lectures on Quantum Mechanics Yeshiva University Press, N.Y.; (1950). Can J. Math. 2, 129; Hanson, A.J., Regge,T. (1974). Ann. Phys. (N.Y.) 87, 498; Hanson,A.J., Regge,T, Teitelboim,C. (1976). Constrained Hamiltionian Systems. Accademia Nazionale dei Lincei, Roma; Mukunda,N., Dam H.van, Biedenharn,L.C. (1982). Relativistic Models of Extended Hadrons Obeying a Mass-Spin Trajectory Constraint, Springer, N.Y., Chap. V.

37) State vectors with negative norm (ghosts), negative m^2-states (tachyons), incomplete representation theory. See also ref. 33 and 43.

38) In the classical theory of relativistic spinning objects ξ^{rel}_{μ} is a well known variable (M.Mathisson, Acta Physica Polonica 6, 163, 218 (1937) which however has lately usually been constrained to zero following Pryce,M.M.L., (1948). Proc. R.Soc. A195, 62. It describes the extension of the relativistic object and performs a Zitter-bewegung about the direction into which the c.m. Y_{μ} proceeds: $\dot{Y}_{\mu} = P_{\mu}/M$.

39) Equations (4.12) have an appealing form for establishing the correspondence to the non-relativistic case but are otherwise of little practical use. The information contained in them is that provided by $SO(3,2)_{S_{\mu\nu}\Gamma_{\mu}}$ and the Hamiltonian (4.7) (the constraint following from (4.7) has been used in the first of the c.r. (4.12) and is more easily extracted from its origin than form (4.12).

41) The details for the construction of this representation are given in Bohm,A, Loewe,M., Biedenharn,L.C., Dam,H. van (1983). Phys. Rev. D28, 3032 and references thereof. For the case that the constraint relation is trivial - i.e. $P_{\mu}P^{\mu}$ = const instead of (4.14) - this is a special case of a "Relativistic Symmetry"; Budini,P., Fronsdal,C. (1965). Phys. Rev. Lett. 14, 968.

40) In addition to the Galilean mass.

42) Nicolai,H. Lectures at the Spring School on Supergravity and Supersymmetry, Trieste, April 1984, CERN preprint TH 3882-CERN and references thereof. E.Angelopoulos in ref. 12 and references thereof. The irreps of SO(3,2) used here (Figs. 3.4,) have all been derived in Barut,A.O., Bohm,A. (1970). J. Math. Phys. 11, 2938

43) Green,M.B., (1973). Surv. H.E.Phys. 3, 127; Schwarz,J.H. (1984). In: Supersymmetry and Supergravity 84, DeWitt,B. et al. eds, World Scientific Pub. Singapore, p. 426; Ne'eman,Y. (1982). KINAM 4, 403 (Moshinsky Festschrift Ed.).

44) Ogievetsky,V.I. (1973). Lett. Nuovo Cim. 8, 988.

45) Berezin,F.A., Marinov,M.S. (1975). JETP Letters 21, 320.

46) Scherk,J., Schwarz,J.H. (1974). Nucl. Phys. B81, 118.

47) Green,M.B., Schwarz,J.H. Caltech reports CALT-68-1182 and CALT-68-1194.

48) Thierry-Mieg,J. Berkeley report LBL-18464, to be published in Phys. Lett. B.

49) Gross,D.J., Harvey,J.A., Martinez,E., Rohm,R. (1985). Phys. Rev. Lett. 54, 502 and Princeton rep. (Jan. 1985) unpublished,

50) Candelas,P., Horowitz, G.T., Strominger,A., Witten,E. ITP Santa Barbara rep. NSF-ITP-84-170.

51) Kac,V.G. (1968). Math. USSR Izv. 2, 1271; Moody,R.V. (1968). J. of Algebra 10, 211.

52) This is reviewed in Goddard,P., Olive,D. "Algebras, Lattices and Strings", DAMTP report 83/22. In: Proc. MSRI Workshop on Vertex Operators, Lepowsky,J. (ed.), Springer, N.Y. (1985). For the mathematical literature, see for example, Conway,J.H., Sloane,N.J.A. (1982). J. Number Th. 15, 83; Conway,J.H. (1983). J. of Algebra 80, 159.

53) Belavin,A.A., Polyakov,A.M., Zamolodchikov,A.B.,CERN rep. TH-3827 (1984); Friedan,D., Qiu,Z., Shenker,S. in Proc. MSRI Workshop on Vertex Operators, J.Lepowsky (ed.) (Springer); (1984). Phys. Rev. Lett.52, 1575; Goddard,P., Olive,D. DAMTP report 84/16, to be publ.; Goddard,P., Kent,A., Olive,D. (1985). Phys. Lett. B152, 88.

54) Chang,S.J., Dashen,R.F., O'Raifeartaigh,L. (1967). Phys. Rev. 182, 1805; Joseph,A. (1970). Comm. Math. Phys. 19, 106; Sugawara,H. (1968). Phys. Rev. 170, 1659; Sommerfield,C. (1968). Phys. Rev. 176, 2019.

55) Frenkel,I.B., Kac,V.G. (1980). Inv. Math. 62, 23;

Frenkel,I.B, (1981). J. Funct. Analysis. 44, 259.

56) Paton,J., Chan,H.M. (1969). Nucl. Phys. B10,519.

57) Conway,J.H. (1969). Inv. Math. 7, 137; Conway,J.H., Sloane,N.J.A. (1982). Bull. Am. Math. Soc. (New Ser.) 6, 215.

58) Griess,R.L.,Jr. (1982). Inv. Math. 69, 1.

59) Cremmer,E., Julia,B., Scherk,J. (1978). Phys. Lett. 76B, 409,; Cremmer,E., Julia,B. (1978). Phys. Lett. 80B, 48; Anglert,F., (1982). Phys. Lett. 119B,339; see also Duff,M.J., Nilsson,B.E.W., Pope,C.N. (1984). In: Frontiers in Particle Physics 1983 (Dj.Sijacki et al. eds.), World Science Pub., p. 200.

60) Morel,B., Thierry-Mieg,J. (1981). In: Superspace and Supergravity, Hawking,S., Rocek,M. eds.). Cambridge Univ. Press; Julia,B., in the same volume,p. 331.

61) Ne'eman,Y., Sherry,T. (1978). Phys. Lett. 76B, 413; Ne'eman,Y., Sijacki,Dj. (1980). J. Math. Phys. 21, 1312.

62) Weaver,O.L., Biedenharn,L.C. (1972). Nucl. Phys. A185, 1; Cusson,R.Y., (1968). Nucl. Phys. A114, 289; Biedenharn,L.C. (1970). Proceedings of the 15th Solvey Conference, Gordon and Breach.

QUANTUM INVARIANTS AND STATE SYSTEM GENERATING ALGEBRA

V.I. Man'ko

P.N. Lebedev Physical Institute of the USSR Academy of Sciences, Moscow

About 20 years ago almost simultaneously a number of authors had introduced the notion of a dynamic group or dynamic symmetry for quantum systems [1] (also known as spectrum generating algebra [2] and non-invariance group [3]). Introducing this notion required a clearer understanding what is to be called the symmetry of an equation describing a physical system [4] and to investigate concrete examples of quantum systems. In [4-7] it had been shown that the dynamic symmetry of hydrogen atom is O(4,2), as a dynamic group describing Landau levels of an electron in magnetic field can serve either non-homogeneous symplectic group ISp(4,R) or U(2,1) [8,9].

This is a short review of the notion of dynamic symmetry of quantum systems and its generalization to non-stationary quantum system without energy levels (energy is not preserved in such systems). The results of the development of the notion of dynamic symmetries up to the end of '70s are summarized e.g. in [9 , 10].

In [4] two problems are clarified. First the notion of a symmetry of a mathematical equation, introduced differently by different authors until now, have been analyzed. The definition of a symmetry of an equation that we started from [4 , 10] is the following one: a symmetry of an equation is the set of operators which transform the solutions of this equation into its solutions. Symmetry may be not a group. A symmetry of a physical system can be identified with the symmetry of an equation describing this system.

Secondly, as an example of such a definition the symmetry of hydrogen atom had been considered. In the famous Fock's paper [11] it had been shown that an "accidental" degeneracy of levels of the discrete spectrum of hydrogen atom is described by the "hidden" symmetry with respect to the group of 4-dimensional rotations. The "hidden" symmetry of hydrogen atom becomes explicit passing in Schroedinger's equation in momentum presentation to stereographic projection. Then functions corresponding to bounded states of hydrogen atom satisfy, in new variables, the 4-dimensional Laplace equation whose symmetry with respect to rotations is quite manifest. In 1935 Fock had also shown that infinite degeneracy of states of the continuous spectrum of hydrogen atom is described by a "hidden" symmetry of Schroedinger's equation with respect to O(3,1).

In 1965 the next important step generalizing Fock's approach had been performed; the essence of this step is as follows. Before 1965 the symmetry operators were assumed to transform states of an energy level of a system to other states of the same energy level and the set of states (family or multiplet) of one energy level constitute a basis of a representation of the symmetry group of the problem. The idea was to unify in one family (multiplet) states belonging to different energy levels. Idealizing one irreducible representation of the dynamic group should embrace all the stationary states of a physical system. Thus applying dynamic symmetry operators to the unique stationary state would generate the other stationary states of a quantum system. Thus adding to Fock's generators of O(4) generators complementing to O(4,2) one could generate from the ground state of hydrogen atom all the excited states of hydrogen atom's discrete spectrum. In this sense, O(4,2) considered as the symmetry group of hydrogen atom generalizes the Fock symmetry O(4) describing an "accidental" degeneracy of its energy levels and extends its symmetry so that one irreducible degenerate representation of this (non-compact) group contains all the states of the discrete spectrum of hydrogen atom. From what follows it is clear that the dynamic symmetry algebra of a stationary quantum system contains generators both commuting and not commuting with the Hamiltonian, i.e. corresponding to interactions causing transitions between states of different energy levels. Note that for the one-dimensional stationary quantum oscillator the symmetry group O(2,1) containing generators not commuting with Hamiltonian was considered in [12] even in 1959, but then it did not begin a new trend.

Thus, let us sum what do we understand under the dynamic symmetry of a stationary quantum system with Hamiltonian $\hat{H}$ in Schroedinger representation, whose states (wave functions $\Psi_{n,i}$) satisfy the stationary Schroedinger equation

$$\hat{H}\Psi_{n,i} = E_n \Psi_{n,i} \tag{1}$$

where n numbers energy levels, i denotes all the other quantum numbers that distinguish different states with the same energy. Operators $\hat{S}_\alpha$ such that

$$[\hat{S}_\alpha, \hat{H}] = 0 \tag{2}$$

transform the stationary states with the same n into each other:

$$\hat{S}_\alpha \Psi_{n,i} = \sum_{i'} S_\alpha^{ii'} \Psi_{n,i'} \tag{3}$$

and are operators of the symmetry describing **degeneracy** of energy levels of the system. Suppose that $\hat{S}_\alpha$ are complemented by operators $\hat{C}_\ell$ not commuting with $\hat{H}$ such that

$$\hat{C}_\ell \Psi_{n,i} = \sum_{n'} \sum_{i'} C_\ell^{ii',nn'} \Psi_{n',i'} \tag{4}$$

If the set of operators $\hat{C}_\ell$ and $\hat{S}_\alpha$ forms a finite-dimensional Lie algebra such that its complete system of wave functions $\psi_{n,i}$ is a basis of one irreducible representation of this Lie algebra, this algebra is called the spectrum generating algebra [2] or non-invariance algebra [3] or dynamic symmetry (algebra) with Hamiltonian $\hat{H}$ [1,4]. To speak about the dynamic symmetry group one clearly needs the existence of exponents of $\hat{S}_\alpha$ and $\hat{C}_\ell$, but in physical literature they often do not make a distinction between non-invariance algebras and non-invariance groups.

Usually the dynamic symmetry is connected with existence of a finite-dimensional Lie algebra but it is clear from the above that the only what $\hat{S}_\alpha$ and $\hat{C}_\ell$ must do is generate all the states of the system from someone. They could be generators of an infinite dimensional Lie algebra or superalgebra, etc. The tendency is to call all such cases under the label "dynamic symmetry". The analysis of this notion shows that if generators of a "hidden"symmetry $\hat{S}_\alpha$ commuting with the stationary Hamiltonian (the generators themselves do not depend on time) are integrals of motions, then the additional generators $\hat{C}_\ell$ not commuting with $\hat{H}$ are not integrals of motion. Thus the dynamic symmetry of a quantum system determined above is constructed by generators which are not integrals of motion. Is it possible to modify a definition of dynamic symmetry to recover the relation of the symmetry with the integrals of motion, i.e. to make the symmetry operators integrals of motion? This question is the most itching when we try to generalize the notion of dynamic symmetry or a spectrum generating algebra to the case of systems with time dependent Hamiltonian $\hat{H}(t)$. Such Hamiltonians describe e.g. parametric oscillators, charge motion in variable electromagnetic fields and, in general, any system if we take into account that our Universe is not stationary.

For non-stationary quantum system the notion of a dynamic symmetry group connected with the stationary Schroedinger equation (1) and the presence of energy levels cannot be generalized directly since this system has no energy levels themselves. The problem if it is possible to construct dynamic group of a stationary system from generators-quantum integrals had been solved in [13-16]. The generalization of the spectrum generating algebra onto the non-stationary Hamiltonian case, also with generators-integrals of motion had been proposed in [14-16]. This generalization had been called in [10] the state generating algebra of quantum system. Sometimes this generalization is just called dynamic symmetry of a non-stationary quantum system.

To pass to the notion of state generating algebra of quantum system it is necessary to pass from the stationary Schroedinger equation (1) to the Schroedinger equation

$$\left[i\hbar\frac{\partial}{\partial t} - \hat{H}(t)\right]\Psi_{n,i}(t) = 0 \qquad (5)$$

Here $\Psi_{n,i}(t)$ is a time-dependent wave function of the system in Schroedinger's representation. It is important that quantum numbers of a non-stationary system are the same as for a stationary system but their physical meaning is somewhat different. Thus n does not label energy levels (that do not exist now) nor time- dependent eigenvalue $E_n(t)$ of Hamiltonian but is an eigenvalue of an operator -- integral of motion. Now if we have operators $\hat{S}_\alpha(t)$ and $\hat{C}_\ell(t)$ satisfying

$$\left(i\hbar\frac{\partial}{\partial t} - \hat{H}(t)\right)\hat{S}_\alpha(t)\Psi_{n,i}(t) = 0$$
$$\left(i\hbar\frac{\partial}{\partial t} - \hat{H}(t)\right)\hat{C}_\ell(t)\Psi_{n,i}(t) = 0 \qquad (6)$$

i.e. transforming the solutions of (5) into solutions; we call these operators elements of the state generating algebra or operators of dynamic symmetry. It is important that these operators are integrals of motion of a non-stationary quantum system. Thus we have

$$\hat{S}_\alpha(t)\,\Psi_{n,i}(t) \quad \sum_{i'} s_\alpha^{ii'}\,\Psi_{n,i'}(t) \tag{7}$$

$$\hat{C}_l(t)\,\Psi_{n,i}(t) = \sum_{n'}\sum_{i'} C_l^{nn',ii'}\,\Psi_{n',i'}(t)$$

We say that we have found the dynamic symmetry of the system, if the set of operators-integrals of motion enables us to recover all the states of the (non-stationary) system from its unique (non-stationary) state. If $\hat{S}_\alpha(t)$ and $\hat{C}_l(t)$ generate a Lie algebra then the state space of the quantum system is the space of an irreducible representation of this Lie algebra.

Therefore the route from Fock's work to nowadays' is, one could say, almost trivial (though required half a century). First, operators called symmetry operators had transformed into each other wave functions of stationary states of the same energy level. Thirty years after the notion of symmetry had been generalized and the operators transforming into each other wave functions of stationary states of any energy levels were ascribed to symmetry operators.

Five years after that, the scheme had been generalized once more to include non-stationary systems and the dynamic symmetry operators depend in Schroedinger's representation on time and are quantum integrals of motion.

A possible way for further generalizations is to study the totality of several systems with different Hamiltonians and construct operators transforming into each other the wave functions of the states of all these systems. (This is akin to considering matrix Hamiltonians in supersymmetric quantum mechanics.) Knowing such operators and one state of one quantum system one could be able to generate all the other states of all the other systems. For instance, applying the generators of such large dynamic symmetry to the ground state of hydrogen atom we could construct all its states and e.g. the states of a quantum oscillator. This example can be investigated in details with the use of the known relation between hydrogen atom problem with that of 4-dimensional harmonic oscillator. This is an embodiment in quantum mechanics of the definition of a symmetry of a number of mathematical equations given in [17].

Let us give examples of Hamiltonians describing systems for which a more detailed investigation of symmetry (in any of the above definitions) and the explicit search of integrals of motion is of interest. For instance, let

$$\hat{H}(t) = \sum c_i(t)\hat{L}_i \tag{9}$$

be a well-known Hamiltonian in Schroedinger's representation expressed in terms of a Lie algebra generators $\hat{L}_i$

Another example is an operator of the form

$$\hat{H}(t) = A_n P^n(\hat{L}_i, t) \tag{10}$$

where P^n is a n-th degree polynomial in $\hat{L}_i$ with scalar coefficients depending on time. In particular, the stationary Hamiltonian can be a Casimir operator.

We have a more complicated example when we have a set of (finite or infinite-dimensional) Lie algebras G_α $(\alpha = 1, \ldots, N)$ with generators $\hat{L}_i^\alpha$ (of some of their representations):

$$\hat{H}(t) = \sum P_\alpha^n(\hat{L}_i^\alpha, t) P_\beta^m(\hat{L}_{i'}^\beta, t) \tag{11}$$

The cases (8)-(10) can be generalized considering superalgebras and replacing polynomials by other functions (exponents etc.)

Another type of Hamiltonians are those where the energy of interaction is described by

$$\hat{H}(t) = \sum P_\alpha^n(\hat{L}_i^\alpha, t)\, P_\beta^m(\hat{\underline{L}}_i^\beta, t) \qquad (12)$$

$\hat{\underline{L}}_i^\beta$ - superalgebra generators.

Note that almost all the problems of quantum mechanics are described by Hamiltonians of the above type. Thus Hamiltonians of non-stationary quadratic systems [18] belong to type (11) with degrees of polynomials n+m no greater than 2 for Heisenberg-Weyl algebras (quadratic expressions in Bose creation-annihilation operators). These Hamiltonians can be considered as operators of the form (9) with $n+m$ generating a representation of ISp(2N,R) (with $N=\infty$ for the field theory). The problems of quantum field theory and solid state physics are also described by Hamiltonians (11), (12) and the non-trivial results obtained with the help of perturbation theory are obtained considering 4-th degree polynomials in creation-annihilation operators. The realization of $\hat{L}_i^\alpha$ in terms of Bose or Fermi creation-annihilation operators is often used.

There remains, nevertheless, quite a few quantum systems especially infinite-dimensional whose dynamic symmetry and integrals of motion are not yet found. Clearly, in physics it is important to get knowing the dynamic symmetry of a system any predictions comparable with experiment. In this field plenty is to be done. An example showing the usefulness of the dynamic symmetry is still the same hydrogen atom with dynamic symmetry O(4,2). Thus it becomes possible to advance considering hydrogen atom in external electromagnetic field (Stark's effect, etc.) assuming that generating terms in the Hamiltonian are related with generators of the dynamic symmetry [19]. In all the similar cases the considerations in the frames of perturbation theory and finding the exact solutions becomes simpler.

It is also interesting to consider dynamic symmetry in the theory of solitons and other solutions of non-linear equations see [17]. The review of two-decade's development of the notion of dynamic symmetry concerning elementary particle physics see in [20].

Bibliography

1. Barut, A.O., Bohm, A. (1965). Phys. Rev. 139B, 1107.
2. Dothan, Y., Gell-Mann, M., Ne'eman, Y. (1965). Phys. Lett., 17, 148.
3. Mukunda, N., O'Raifeertaigh, L., Sudarshan, E. (1965). Phys. Rev. Lett., 15, 1041.
4. Malkin, I.A., Man'ko, V.I. (1965). ZhETPh Lett., 2, 230. (Russian).
5. Barut, A.O., Kleinert, H. (1967). Phys. Rev., 156, 1541.
6. Nambu, Y. (1966). Progr. Theor. Phys., Suppl., 36-37, 368.
7. Ne'eman, Y. (1967). Algebraic Theory of Elementary Particles.
8. Malkin, I.A., Man'ko, V.I. (1968). Sov. Nucl. Phys., 8, 1264. (Russian).
9. Aronson, E.B., Malkin, I.A., Man'ko, V.I. (1974). Elem. particles and Atomic Spectra, 3, 123. (Russian).
10. Malkin, I.A., Man'ko, V.I. (1979). Dynamic symmetries and coherent states of quantum systems, Nauka, Moscow (Russian).
11. Fock, V.A. (1935). Zs.f.Physik, 98, 145.
12. Goshen, S., Lipkin, H.J. (1959). Ann. Phys., 6, 301.
13. Dothan, Y. (1970). Phys. Rev., V.D2, 2944.
14. Malkin, I.A., Man'ko, V.I., Trifonov, D.A. (1969). Phys. Lett., 30A, 414.
15. Malkin, I.A., Man'ko, V.I., Trifonov, D.A. (1970). Phys. Rev., 2D, 1371.
16. Malkin, I.A., Man'ko, V.I., Trifonov, D.A. (1971). Nuovo Cimento, 4A, 773.
17. Leznov, A.N., Man'ko, V.I., Saveliev, M.V. (1986). Solitons, instantons and operatic quantization.

P. Lebedev Phys. Inst. Trans. 165, 65-206.(Russian).

18. Dodonov, V.V., Man'ko, V.I. (1983). Gravitation, quantization, and group methods in physics. P.Lebedev Phys. Inst. Trans. 152, 145. (Russian).

19. Alliluyev, S.P., Malkin, I.A. (1974). PhETPh, 66, 1283.

20. Bohm, A. Ne'eman, Y. Twenty years of dynamical groups and spectrum generating algebra. (in this proceedings).

DYNAMIC SYMMETRY IN MANY-BODY PROBLEM

Joseph L. Birman, Allan I. Solomon[+]

Physics Department, City College, CUNY, New York, N.Y. 10031, [+]Faculty of Mathematics, Open University, Milton Keynes, U.K.

1. Algebras and Hamiltonians. Theories of the properties of coexisting systems [1] at the mean-field level [2] have had considerable success recently.

We are investigating the Dynamical Symmetry of such mean-field Hamiltonians, and we gave some of the consequences of this symmetry in earlier reports [3].

We introduce a basis for the 2N-electron problem consisting of single particle states labelled $\xi \equiv (k_s, \alpha),\ s = 1 \ldots N,\ \alpha = (\uparrow, \downarrow)$ with creation and destruction operators labelled $a^{+}_{\xi}, \ldots a_{\xi}, \ldots$. The set consisting of: all such single fermion operators plus all pairs generates under commutation the Lie Algebra $B_{2N+1} \sim SO(4N+1)$. The set of all pairs only generates the algebra $D_{4N+1} \sim SO(4N)$, which is a subalgebra of B_{2N+1} . Defining

$$(A_1, A_2 \ldots) \equiv (a_{\xi}, a^{+}_{\bar{\xi}}, a_{\xi}, a^{+}_{\bar{\xi}}, \ldots,) \tag{1.1}$$

with $\xi \equiv (k\uparrow)$, $\bar{\xi} \equiv (-k\downarrow)$, etc. we can then produce the pair operators $X_{ij} = A^{+}_{i} A_{j}$. It is easy to verify that

$$[X_{ij}, X_{\ell m}]_{-} = \delta_{\ell j} X_{im} - \delta_{im} X_{\ell j} \tag{1.2}$$

$$[X_{ij}, A_{s}]_{-} = -\delta_{is} A_{j}\,;\,[A_i, A^{+}_{j}] = \delta_{ij}(1-2X_{is}). \tag{1.3}$$

This shows the closure of the entire set $\{A_j, X_{ss}\} \sim B_{2N+1}$

as well as the subset $\{X_{ss}\} \sim D_{2N+1}$ alone.

The mean field Hamiltonian is a complex form in the operator pairs X_{ij} :

$$\hat{H}_{MF} = \lambda_{ij} X_{ij} \tag{1.4}$$

where λ_{ij} are complex constants. For typical systems the total Hamiltonian is a sum of terms each labelled by k: and

$$\hat{H}_{MF}\,(\text{system}) = \sum_k H_{MF}(k). \tag{1.5}$$

We suppress k. Clearly $\hat{H}_{MF}$ is an element in D_{2N+1} and the system Dynamical Algebra is $\prod_k \otimes D_{2N+1}(k)$. In this work we refer to the single k Dynamical Algebra.

For a homogeneous system the spatial Fourier Transform of the ordinary single particle Thermal Green Function is

$$G(k,\tau) = -\langle T_\tau a_k(\tau) a_k^+ \rangle \tag{1.6}$$

and of the Gor'kov (anomalous) Function

$$F(k,\tau) = \langle T_\tau a_k(\tau) a_{-k} \rangle \tag{1.7}$$

with

$$a_k(\tau) \equiv e^{H\tau} a_k e^{-H\tau} \tag{1.8}$$

$$\langle \hat{O} \rangle = [\mathrm{Tr}(e^{-\beta H} \hat{O})] / \mathrm{Tr}\, e^{-\beta H} \tag{1.9}$$

(spin suppressed); $\beta = (k_B T)^{-1}$. The T=0 functions

$$G(k,t) = (-i) \langle T_t a_k(t) a_k^+ \rangle \tag{1.10}$$

$$F(k,t) = \langle T_t a_k(t) a_{-k} \rangle \tag{1.11}$$

with

$$a_k(t) = e^{iHt} a_k e^{-iHt} \tag{1.12}$$

2. Automorphisms, Structure Constants and Green Functions.

The Hamiltonian is taken as H_{MF} . Then for T=0:

$$a_k(t) \to A_r(t) = [\exp(i\lambda_{ij} X_{ij})] A_r [\exp(-i\lambda_{ij} X_{ij} t)]$$
$$A_r(t) = \sum (e^{-it\lambda})_{rj} A_j . \qquad (2.1)$$

Here λ is the $4N \times 4N$ matrix with elements λ_{rj}. We interpret this as a Heisenberg Automorphism Φ :

$$(A_r) \xrightarrow{\Phi} \{(A_r)\} \qquad (2.2)$$

which maps B_{2N+1} onto itself via "structure constants" $(e^{-it\lambda})_{rj}$.

Now rewrite $\hat{H}_{MF}$ in a more suggestive Cartan-Weyl (C-W) form:

$$H_{MF} = \sum \mu_j h_j + \sum_{\alpha} \mu_{\alpha} e_{\alpha} . \qquad (2.3)$$

Rotate H_{MF} to diagonal form by

$$U H_{MF} U^{-1} = \sum E_j h_j \equiv \hat{H}' \qquad (2.4)$$

where $U = \exp i(\sum_{\alpha} \theta_{\alpha} e_{\alpha})$, θ_{α} are Bogolyubov angles and E_j are eigenvalues depending on $\mu_j, \mu_{\pm\alpha}$. The h_j generate the Cartan subalgebra of D_{2N+1} . Label the eigenstates of h_j as

$$h_j |\{\lambda'_j\}\rangle = \lambda'_j |\{\lambda'_j\}\rangle . \qquad (2.5)$$

Then the ground state of $\hat{H}_{MF}$ is

$$|\Phi'\rangle = U^{-1} |\{\lambda'_j\}\rangle . \qquad (2.6)$$

To compute $G(k,t)$ we need

$$a_k(t) a_k^+ = A_r(t) A_r^+ = \sum_j (e^{-it\lambda})_{rj} A_j A_r^+ =$$
$$= \sum_j (e^{-it\lambda})_{rj} X_{rj}^+ \qquad t > 0 \qquad (2.7)$$

and

$$-A_r A_r^+(t) = \sum_j (e^{-it\lambda})_{rj} X_{rj}^+ \qquad t < 0 . \qquad (2.8)$$

Then since

$$G(k,t)=(i)\langle\Phi'|T_t a_k(t)a_k^+|\Phi'\rangle = \\ =(2)\langle\Phi'|T_t\, U a_k(t)a_k^+ U^{-1}|\Phi'\rangle \tag{2.9}$$

We need

$$U X^+_{z_j} U^{-1} \quad \text{and} \quad U X_{z_j} U^{-1}. \tag{2.10}$$

These are objects in D_{2N+1}. Hence rewriting the bilinear fermion products X_{z_j} as objects in C-W notation, we have an automorphism Φ' :

$$U\{h_\ell, e_\beta\}U^{-1} = \sum_j \gamma_j h_j + \sum_\sigma \gamma_\sigma e_\sigma \tag{2.11}$$

or

$$\{h_\ell, e_\beta\} \xrightarrow{\Phi'} \{h_\ell, e_\beta\} \tag{2.12}$$

with (Φ') being defined by structure constants $(\gamma_j, \gamma_\sigma)$.

Assembling all the above, we take the matrix element and note that only the diagonal value of h_ℓ (i.e. λ_ℓ) survives. Thus

$$G(k,t) \equiv G_2(t) = \begin{cases} -(i)\sum_j (e^{-it\lambda})_{z_j} \sum_\ell \gamma_\ell \lambda_\ell & t>0 \\ +(i)\sum_j (e^{-it\lambda})_{z_j} \sum_\ell \gamma_\ell^* \lambda_\ell & t<0 \end{cases} \tag{2.13}$$

The usual $G(k,\omega)$ is recovered by time Fourier Transform of $G(k,t)$.

Hence $G(k,t)$ is a sum of products of structure constants:

$(e^{-it\lambda})$ of Φ

γ_i of Φ'

and eigenvalues

λ_i of generators h_i.

<u>Mutatis Mutandis</u> we obtain $F(k,t)$ from the same quantities.

For the Matsubara $(T \neq 0)$ Green Functions we proceed in an exactly similar fashion, with result

$$G(k,\tau) = \begin{cases} \tau>0: \sum_j (e^{-\tau\lambda})_{z_j} \left(\sum_{\{\lambda_p\}} \left(\sum_p e^{-\beta E_p \lambda_p} \gamma_p \lambda_p \right) \right) / \sum_{\{\lambda_p\}} e^{-\beta E_p \lambda_p} \\ \tau<0: \sum_j (e^{-\tau\lambda})_{z_j} \left(\sum_{\{\lambda_p\}} \left(\sum_p e^{-\beta E_p \lambda_p} \gamma^*_p \gamma_p \right) \right) / \sum_{\{\lambda_p\}} e^{-\beta E_p \lambda_p} \end{cases} \tag{2.14}$$

Again, a Fourier Transform produces $G(k, i\omega_n)$. We identify $a_k \to A_z$. Likewise $F(k,\tau)$ is obtained by similar steps. In the sum $\{\lambda_p\}$ we take all sets of eigenvalues of the operators $\{\hat{h}_j\}$.

It is to be noted that the Matsubara Green Functions are composed of suitable sums of products of two sets of algebraic structure constants $(e^{-\tau\lambda})$ and $(\gamma_j, \gamma_\varsigma)$ eigenvalues E_p of $\hat{H}'$, and the complete set of eigenvalues $\{\lambda_n\}$ of the $\hat{h}_n$ generators of the Cartan subalgebra. These algebraically determined quantities are the key building blocs of the many-fermion response functions [4].

For a specific system and particular mean-field Hamiltonian with some of the μ_j or μ_α equal to zero, the Dynamical Algebras of the relevant sets of singles and pairs of Fermi operators may be a subalgebra of B_{2N+1} are similarly for the $\hat{H}_{MF}$ we may have to deal with a subalgebra of D_{2N+1}. For example, the reduced BCS singlet superconductor has a Fermi algebra isomorphic to $A_2 \sim SU3$ while the $\hat{H}_{MF}$ belongs to the SU2 algebra, which can be realized by pseudospins [5].

3. <u>State Labelling and Selection Rules</u>. The ground "coherent" state $|\Phi'\rangle$ of $\hat{H}_{MF}$ is obtained from the ground state $|\{\lambda'_j\}\rangle$ which is diagonal in the Cartan generators $\hat{h}_j$ by the inverse Bogolyubov rotation as given in equation (2.6). All excited states $|\Phi\rangle$ can be ob-

tained by the inverse rotation applied to all possible states $|\{\lambda_j\}\rangle$, where the set of eigenvalues $|\{\lambda_j\}\rangle$ are all permitted ones (not only the lowest energy set). If H_{MF} belongs to a rank ℓ Lie Algebra there will be ℓ Casimirs which commute with it, and thus ℓ "good" quantum labels, which are the eigenvalues of the Casimirs. It may be desirable to label states $|\Phi\rangle$ also by the set $\{\lambda_j\}$, and the total energy eigenvalue of H_{MF}. Hence a useful set of state labels includes the eigenvalues of the Casimirs, the eigenvalues of the generators h_j of the Cartan subalgebra, and the energy eigenvalues. Of some importance is the fact that we are able to associate physical attributes - in the sense of type of quasiparticle or excitation - to the state labels.

Some examples can help to clarify these points, more details are given elsewhere [6]. The mean-field (reduced) BCS Hamiltonian which can be realized by the pseudo-spin operators $J_i\ (i=1,2,3)$ is an element in a Dynamical Algebra SU2. This

$$H_k^{BCS} = 2\Delta J_2 - 2\varepsilon_k J_3 + 2\varepsilon_k \tag{3 1}$$

where Δ and ε_k are gap and single particle parameter. Following the above procedure it is easy to show that eigenstates of H_k^{BCS} are given as

$$U^{-1}|\{\lambda_j\}\rangle = \exp(-i\theta\hat{J}_1)|J_3, J^2\rangle \equiv |E_k, J^2\rangle . \tag{3 2}$$

Here $|J_3, J^2\rangle$ is an SU2 eigenket, which is labelled by the eigenvalues of $\hat{J}_3$ (which in C-W notation is the unique h_3), and the eigenvalue of Casimir $\hat{J}^2$. The total ket can be labelled by energy $E_k = J_3(\Delta^2+\varepsilon_k^2)^{1/2}$ and the eigenvalue J^2 of the Cartan $\hat{J}^2$. The operators $\hat{J}_3$ and $\hat{J}^2$ are expressible in terms of number operators:

$$\hat{J}_3 = (-1/2)(\hat{n}_k + \hat{n}_{-k} - 1);\ \hat{J}^2 = (\hat{n}_k \hat{n}_{-k} + \hat{J}_3 + \hat{J}_3^2) \tag{3 3}$$

It immediately follows that $J^2 = 0, 3/4$ only.

States $J^2 = 3/4\ J_3 = \pm 1/2$ belong to irreducible representation $D^{1/2}$ of SU2. These states are excited and ground state <u>pairs</u>. State $J^2 = 0$, is <u>single</u>, belongs to D^0 of SU2. BCS "singles" and "pairs" are thus related to irreducible representation label of the Dynamical Algebra.

The mean-field Hamiltonian for coexisting superconductivity and charge density wave (SC-CDW) in suitable approximation [8] is an element in Dynamical Algebra SO5. A summary of the results for labelling in this case is given below. Note that we used a labelling scheme in which states are labelled by: energy E of the coupled system, eigenvalues of the two Cartan generators $h_3^+ = \hat{J}_3^+$, $h_3^- = \hat{J}_3^-$, and eigenvalues of the two Casimir operators $\hat{J}^{+2}$, $\hat{J}^{-2}$. The latter 4 operators take account of the SU2 x SU2 subgroup labels of SO5. Especially noteworthy is that we again correlate type of physical states (left column) and irreducible representation label $D^{(J^+, J^-)}$. Work is in progress to associate remaining labels [6] and physical properties.

Table: State for $H^{SC-CDW}(k < k_F)$ in $SO(5)$

Type of state		State and Label: $\vert E; J^+_{3k}, J^-_{3k}; J^{+2}_k, J^{-2}_k\rangle$
Two real pairs		$\vert \pm\mu_k, c_2 \pm \lambda_k; \pm 1/2, 3/4, 3/4\rangle$
(One real pair + One single)	or	$\vert \pm 1/4(\mu_k - \lambda_k); 0, \pm 1/2; 0, 3/4\rangle$ $\vert \pm 1/4(\mu_k + \lambda_k); \pm 1/2, 0; 3/4, 0\rangle$
Two singles		$\vert 0; 0, 0; 0, 0\rangle$

For a theory of Raman Scattering for simple BCS system or the coupled SC-CDW transition matrix element needed is of the form $\langle \Phi^f | \hat{T} | \Phi^i \rangle$ where $|\Phi^{f,i}\rangle$ are final/initial states of the entire system e.g. $\prod_k U_k^{-1} |\psi'_k\rangle$ for the BCS system and $\hat{T} = \sum_k \hat{T}_k$ is a transition operator. In the theory of Raman Scattering from metals the dominant term in the Hamiltonian is $H_I^{(2)} = \int d\vec{r}\, \hat{\psi}^+(r) (e^2/2mc^2) A^2(r) \times \hat{\psi}(r)$. In the long wave approximation this term can be shown to give rise to tensor operator $H_{Ik}^{(2)} \sim [\hat{1} - 2\hat{J}_{3k}]$. Hence the only non-vanishing matrix elements $\langle \Phi^f | \hat{T} | \Phi^i \rangle$ are intra-series. In this theory Raman Scattering in BCS superconductors does not couple pairs and singles. The analogous selection rules for the SC-CDW system, and other results will be reported elsewhere.

4. Acknowledgement. This work was supported in part by BSC-BHE grant RF 6-64261, a grant from the Open University Research Committee, and NATO Research grant 099.82. J.L. Birman acknowledges support of his participation in the Yurmala Seminar by the USSR Academy of Sciences. The hospitality and organization by Academician M.A. Markov, and Prof. V.I. Man'ko is gratefully acknowledged.

References

1. Sooryakumar, R., Klein, M.V. (1980). Phys. Rev. Lett., 45, 660; (1981). Phys. Rev., B23, 2213; EuMo$_6$S$_8$; Huang, C.Y. (1984). CRC Critical Reviews in Solid State and Material Sciences, 12, 75.

2. Machida, K., Nokura, K., Matsubara, T. (1980). Phys. Rev., B22, 2307; Grest, G.S., Levin, K., Nass, M.T. (1982). Phys. Rev., B25, 4341, 4862; Lei, X.L., Ting, C.S., Birman, J.L. (1984). Phys. Rev., B29, 2483.

3. Birman, J.L., Solomon, A.I. (1983). In: Group Theoretical Methods in Physics (Proceedings of the International Seminar, Zvenigorod 24-26 Nov. 1982). Eds. V.I. Man'ko, A.E. Shchabad, Moscow: Nauka, 1, p. 340-345, and references therein.

4. Birman, J.L., Solomon, A.I. (1984). Progress of Theoretical Physics, Suppl., 80, p. 62. This paper contains more details and specific examples of the struc-

ture of Green Functions in the framework of the dynamical symmetry of many body problems.

5. Anderson, P.W. (1958). Phys. Rev., 112, p. 1900.
6. Birman, J.L., Solomon, A.I. (1984). In: Proceedings of the 13th Group Theory Colloquium, U. of Maryland/ Ed. by W.W. Zachary. World Scientific, p. 445-448.
7. Anderson, P.W. (1958). Phys. Rev., 112, p.1900.
8. Birman, J.L., Solomon, A.I. (1982). Phys. Rev. Lett., 49, p.230.

TRANSVECTOR ALGEBRAS IN REPRESENTATION THEORY AND DYNAMIC SYMMETRY

D.P. Zhelobenko
Patrice Lumumba University, Moscow

The problem of symmetry of an equation is classical. It is connected with the names of Galois and Sophus Lie. Unfortunately, at present this problem is as far from the solution, as it was in the past. Fortunately, however, just this problem gave rise to the remarkable apparatus of the Lie groups and Lie algebras, which is so prolific in applications in many fields of mathematics and physics.

The aim of this report is to demonstrate that for systems of equations of a special form (extremal systems) it is possible in some sense to find the full solution of the problem of describing the symmetry algebras (transvector algebras) of such systems. This solution is constructive meaning that we may give an explicit construction of generators of the transvector algebra, together with relations among them. It is remarkable that many classical "massless" equations are attached to extremal systems, e.g. Dirac's, Maxwell's, etc.

The algebraic methods exposed below originally arised in the representation theory of reductive Lie groups, but actually they have a wider field of applications in representation theory, algebraic geometry, quantum physics. In general, the existence of many profound connections among these trends is well known. It is of interest to note also some other connections between them.

1. Transvector algebras. We begin with formal definitions. Let A be an associative algebra with unit, acting by li-

near transformations in a vector space V . We fix an arbitrary set $E \subset A$, and let $V_E \subset V$ be the subspace consisting of all the solutions of the system

$$Ev = 0, \quad \text{i.e. } av = 0 \quad \text{for all } a \in E \tag{1}$$

We want to separate the "reduced" part of A whose elements transform V_E to V_E . For this define the subalgebra $A_E \subset A$ as the set of all the solutions of the system

$$Ex \equiv 0, \quad \text{i.e. } ax \equiv 0 \quad \text{for all } a \in E \tag{2}$$

where $\equiv$ is congruence modulo the left ideal $I_E = AE$. The fact that A_E is a subalgebra is evident. Actually, A_E is the normalizor of I_E , $Norm\, I_E$, i.e. the greatest subalgebra of A in which I_E is a two-sided ideal. We have

$$A_E V_E \subset V_E \tag{3}$$

i.e. A_E transforms any solution of (1) to a solution of (1).

The last assertion justifies our interest to the algebras A_E . We note that I_E annihilates V_E . Hence it is natural to consider instead of A_E the quotient algebra

$$T_E = A_E / I_E \tag{4}$$

which we call the <u>transvector algebra</u> of the system E. The passage from A_E to T_E reduces to setting $I_E = 0$. In this sense we may identify A_E and T_E .

The problem consists in studying connection between A and A_E , describing elements of A_E (or T_E) and constructing the structure theory of transvector algebras.

Clearly, knowing A_E gives us an information on the quantity of elements of V_E . For example, if A_E acts irreducibly in V_E , then any solution of (1) may be constructed by transformations of A_E from an arbitrary nonzero solution of the system (1).

Sometimes we use the following formal statement: the map $\Phi_E : (A, V) \longmapsto (A_E, V_E)$

defines a covariant functor from the category of A-modules (A-mod) to the category of A_E-modules (A_E-mod).

The last problem reduces now to the study of Φ_E : e.g. if this functor is exact, i.e. does Φ_E preserve the submodule structure?

In particular, if we know that Φ_E is exact and acts irreducibly in V then A_E and also T_E act irreducibly in V_E.

For the particular case of "extremal" systems we get a nice answer to these (and other) questions.

2. Examples.

Case U. Let $A = U(\mathfrak{g})$ be the enveloping algebra of a complex Lie algebra $\mathfrak{g}$, $\mathfrak{k}$ its subalgebra with a root system, e.g. a finite-dimensional semisimple Lie algebras. Set

$$E = \{ e_\alpha : \alpha \in \Delta^+ \} \tag{5}$$

where e_α is a root vector (generator of $\mathfrak{k}$), corresponding to the root α, Δ^+ is the system of positive roots of $\mathfrak{k}$ (with respect to a fixed Cartan subalgebra $\mathfrak{h}$). For any $\mathfrak{g}$-module V, the subspace

$$V_E = \{ v \in V : Ev = 0 \} \tag{6}$$

is the subspace of highest vectors with respect to $\mathfrak{k}$. Assume that as a $\mathfrak{k}$-module V decomposes in the direct sum of "isotypic components" V^λ (λ a highest weight). Then so does V_E :

$$V = \oplus_\lambda V^\lambda, \qquad V_E = \oplus_\lambda V_E^\lambda \tag{7}$$

Such a decomposition always takes place for finite-dimensional $\mathfrak{g}$-modules or for Harish-Chandra modules. In the latter case $\mathfrak{g} = Lie\, G_{\mathbb{C}}$, $\mathfrak{k} = Lie\, K_{\mathbb{C}}$, where G is a reductive Lie group, K is its maximal compact subgroup, and the subscript c means the complexification.

The use of decompositions (7) is the standard apparatus of the representation theory. Note, that elements of $\mathfrak{k}$ act in this scheme preserving the irreducible components, whereas other elements of $\mathfrak{g}$ mix isotypic components. Elements of A_E act in this scheme mixing highest vectors of V.

We will see that in this scheme knowledge of A_E-module V_E defines completely the initial module V up to isomorphism. In particular, it is possible in principle to solve by means of transvectors, i.e. the elements of A_E, the well-known problem of "separating multiplicities", i.e. splitting the isotypic components V^λ into irreducible $\mathfrak{k}$-components.

The simplest elements of $A_E = U(\mathfrak{g})_E$ under different names such as "step operators", "schift operators" are considered in series of works on finite-dimensional representations (cf. e.g. review [5]). The results exposed below allows us to give a uniform description of these operators and relations among them.

The scheme exposed below extends with no difficulty to the case of Lie superalgebras and infinite-dimensional Lie algebras with root systems, e.g. Kac-Moody algebras.

Case D. Let $A = \mathcal{D}(n, m)$ be the associative superalgebra over $\mathbb{C}$ generated by the creation and annihilation operators of n bosons and m fermions. A set of generators of $\mathcal{D}(n, m)$ is

$$x_i, \partial_i = \partial/\partial x_i \ (i = 1, \ldots, n); \ \xi_j, \varepsilon_j = \partial/\partial\xi_j \ (j = 1, \ldots, m) \qquad (8)$$

where x_i is the multiplication operator by the corresponding coordinate in $\mathbb{C}^n$, ξ_j is the multiplication by the odd Grassmann variable. (For $n = m$ we may put $\xi_j = dx_j$). In particular, we may consider the A-action in the subalgebra $V = \mathcal{F}(n, m)$ generated by x, ξ. (For $n = m$ this is the algebra of exterior differential forms with polynomial coefficients.)

The algebra $\mathcal{D}(n) = \mathcal{D}(n, 0)$, i.e. the algebra of differential operators with polynomial coefficients in $\mathbb{C}^n$, is called the Weyl algebra, $\mathcal{D}(0,m)$ is called the Clifford algebra. We call $\mathcal{D}(n,m)$ the <u>Weyl superalgebra</u>.

As it is well known, many operators of quantum physics are realized in Weyl (super)algebras (the simplest examples are considered in what follows). The investigation of the symmetry of the corresponding equations is the problem which in a number of cases admits the complete solution.

In earlier works on symmetries they only considered the commutant $[E,E]=E^{(1)}$ of the system E, later on they changed to the full algebra of "dynamic symmetry" DynE. Clearly, $E^{(1)} \subset A_E \subset \mathcal{D}yn\, E$. We will see later that A_E is in a sense the "right" algebra of dynamic symmetry: not too great and not too small.

<u>Case X</u>. Let X be an algebraic variety defined by the ideal $I_E \subset \mathcal{D}(n)$. This means that the algebra of regular functions $F(X)$ on X is the set of solutions of the system $Ef=0$, where $f \in F(\mathbb{C}^n)$. (As usual, the elements of E are differential operators of order ≤ 1).

The transvector algebra T_E is naturally identified in this case with the algebra $\mathcal{D}(X)$ of regular differential operators X . Accordingly, Φ_E is of the form

$$(\mathcal{D}(n), F(n)) \mapsto (\mathcal{D}(X), F(X)) \qquad (9)$$

(generalization to supervarieties is clear). Define another functor Φ_E similar to (9). Let X, Y be supervarieties such that X is distinguished by the same equations as Y plus some extra ones. Then we set $X \leq Y$ and we clearly have $F(X) = F(Y)_E$. In this case we have

$$\Phi_E \cdot (\mathcal{D}(Y), F(Y)) \mapsto (\mathcal{D}(X), F(X)) \qquad (10)$$

The problem exposed above reduces in this case to constructing the structure theory of algebras $\mathcal{D}(X)$ and investigating the functors (9), (10).

3. Extremal systems. From now on we will consider systems of the form (5), which are defined for some contragredient Lie (super)algebras, finite or infinite dimensional (Kac-Moody algebras).

Let A be an associative algebra with unit generated by elements e_i $(i \in I)$ for a finite or countable set of indices I with the relations

$$[e_i, e_j] = c_{ij}^k e_k + c_{ij}' \quad (i,j \in I) \tag{11}$$

The generators e_i here are homogeneous (odd or even) and $[a,b] = ab \pm ba$, according to the known sign rule. Now, let $\mathfrak{k} \subset A$ be a contragredient Kac-Moody (super) algebra with the standard decomposition

$$\mathfrak{k} = \mathfrak{n}^* \oplus \mathfrak{h} \oplus \mathfrak{n} \tag{12}$$

where $\mathfrak{h}$ is a Cartan subalgebra of $\mathfrak{k}$, $\mathfrak{n}$ is the maximal nilpotent subalgebra generated by e_α $(\alpha \in \Pi)$, where Π is a base, $\mathfrak{n}^*$ is the image of $\mathfrak{n}$ under involution: $(e_\alpha)^* = e_{-\alpha}$ $(\alpha \in \Pi)$.

If $\dim \mathfrak{k} < \infty$, then all the roots are multiplicity-free and the symbol e_α is well-defined up to a normalization for all $\alpha \in \pm\Delta^+$, where Δ^+ is a system of positive roots (with base Π). In this case we will use a normalization such that for $h_\alpha = [e_\alpha, e_{-\alpha}] \in \mathfrak{h}$ either $\alpha(h_\alpha) = 2$ or $\alpha(h_\alpha) = 0$. In general case, instead of e_α it suffices to consider symbols $e_{\alpha j}$ $(j = 1, \ldots, n_\alpha)$, where n_α is the multiplicity of α. In particular, $n_\alpha = 1$ for $\alpha \in \Pi$.

An injection $\mathfrak{k} \subset A$ is regular, if the following conditions hold:

(a) the linear hull A_0 of the generators e_i $(i \in I)$ is invariant under the adjoint action $\hat{a}x = [a,x]$ of $\mathfrak{k}$ and operators $\hat{e}_\alpha$ are locally nilpotent (i.e. for any x there exists n such that $(\hat{e}_\alpha)^n x = 0$);

(b) generators e_i are weight elements under Cartan subalgebra $\mathfrak{h}$: $\hat{a} e_i = \lambda_i(a) e_i$ for $a \in \mathfrak{h}$, $i \in I$;

(c) elements $k \in U(\mathfrak{h})$ act as multiplication operators in A (from left and right) without zero divisors.

The relations (11) contain the case U (for $c'_{ij} = 0$) and the case $\mathcal{D}$ (for $c^k_{ij} = 0$). If I is a finite set and $\dim \mathfrak{k} < \infty$, then the conditions (a), (b), (c) are automatically satisfied, up to a linear change of generators. The involution in $\mathfrak{k}$ is usually induced by an involution of the algebra A (for example, $x_i^* = \partial_i$ in the case $\mathcal{D}$).

We use the standard ordering of weights with respect to a Cartan subalgebra $\mathfrak{h}$: $\lambda \geq \mu$ if $\lambda - \mu$ is a sum of simple roots with nonnegative coefficients. We also assume that the elements e_i $(i \in I)$ are ordered so that their weights are nondecreasing.

The system $E = \{e_\alpha, \alpha \in \Pi\}$ or the equivalent system $E' = \{e_{\alpha_j} : \alpha \in \Delta^+, j = 1, \ldots, n_\alpha\}$ with the regular injection $\mathfrak{k} \subset A$ is called an extremal system (of kind 1).

Together with such systems, we can also consider their parabolic extensions (the systems of kind 2), their inductive limits (the systems of kind 3) and some other examples.

4. General results. Let E be an extremal system of kind 1, $\mathfrak{h}$ a Cartan subalgebra of $\mathfrak{k}$. We may assume, due to (c), that $U(\mathfrak{h}) \subset A$.

Now, we may pass from the algebra A to its extension

$$A' = A(U(\mathfrak{h}))^{-1} \tag{13}$$

(the localization with respect to $\mathfrak{h}$). The passage from A to A′ is actually a possibility to multiply in A by the elements of the field $R(\mathfrak{h}) = Q(U(\mathfrak{h}))$ (the field of quotients of the ring $U(\mathfrak{h})$). Note that in A′ there is a basis of monomials

$$e(k) = e_1^{k_1} \ldots e_n^{k_n} \quad (n = 0, 1, \ldots) \tag{14}$$

which are weight elements with respect to $\mathfrak{h}$. Multiplication of these monomials by elements of $R(\mathfrak{h})$ satisfies

$$h\,e(k) = e(k)\tilde{h} \qquad (15)$$

where $h \to \tilde{h}$ is an automorphism of $R(\mathfrak{f})$ (depending on k).

In view of (15), we may consider A' as a "pseudo-algebra" over the field $R(\mathfrak{f})$. For any subalgebra $B \subset A$ (or a quotient algebra) set

$$B' = B(U(\mathfrak{f}))^{-1} \qquad (16)$$

The following theorem describes a relation between A' and its transvector part A'_E.

<u>Theorem 1</u>. There exist linear operators $p, q : A' \to A'$ such that

(i) $A'_E \equiv pA' \equiv qA'$,

(ii) $p^2 x \equiv px$, $q^2 x \equiv qx$ for any $x \in A'$,

(iii) $\operatorname{Ker} p \equiv w^* A'$, $\operatorname{Ker} q \equiv A' w^*$,

(iv) $A' \equiv U(w^*) A'_E \equiv A'_E\, U(w^*)$

Here, as before, $\equiv$ means congruence modulo $I'_E = A'w$. Hence, modulo this ideal A'_E coincides with the images of the two "quasiprojections" p,q, kernels of which are described in (iii). Proposition (iv) means that A' is recovered from its transvector part with the help of $U(w^*)$.

A concrete form of the operators p,q is given below (for $\dim \mathfrak{k} < \infty$). We see that p annihilates I'_E, hence it defines a projection in $M' = A'/I'_E$ (such that $T'_E = pM'$). At the same time any similar interpretation for q is impossible.

The following theorem describes principal structural properties of the algebra T'_E.

<u>Theorem 2</u>. The elements $z_i = pe_i$ ($z'_i = qe_i$) generate the algebra T'_E and satisfy

$$[z_i, z_j] = \alpha_{ij}^{kl} z_k z_l + \beta_{ij}^k z_k + \gamma_{ij} \qquad (17)$$

(similarly for z'_i) with coefficients of $R(\mathfrak{f})$ and with summing over nondecreasing (similarly, nonincreasing) weights.

The relations (17) are corollaries of the fundamental relations (11). One could obtain these relations constructively using the concrete form of p,q (see below).

If $A = U(\mathfrak{g})$ (the case U) then Theorem 2 takes the following strongest form. Suppose that $\mathfrak{k}$ is reductive in $\mathfrak{g}$, i.e. the adjoint action of $\mathfrak{k}$ in $\mathfrak{g}$ is semisimple, and set $\mathfrak{g} = \mathfrak{k} \oplus \mathfrak{p}$, where $\mathfrak{p}$ is a complementary $\mathfrak{k}$-module. Note that $p\mathfrak{k} \equiv \mathfrak{f} \subset R(\mathfrak{f})$ (by (iii)). Hence, if $\{e_i\}$ form a linear basis of the space $\mathfrak{p}$, then the elements $z_i = pe_i$ are generators of T'_E. Note also that the actions of p and q preserve the weights with respect to $\mathfrak{f}$ (see sect. 5).

Theorem 3. If A is $U(\mathfrak{g})$ with reductive action of $\mathfrak{k}$ in $\mathfrak{g}$, then T'_E is generated by the elements $z_i = pe_i$ (similarly, $z'_i = qe_i$) with relations (17) complemented by (15).

Theorem 3 means that all the relations among the generators z_i follow from (15), (17). In this case for T'_E an analogue of the Poincare-Birkhoff-Witt theorem holds: the monomials

$$z(k) = z_1^{k_1} \dots z_n^{k_n} \quad (n = 0, 1, \dots) \tag{18}$$

(similarly for z'_i) form a basis of the vector space T'_E over the field $R(\mathfrak{f})$. Therefore, the algebra T'_E "remembers" its origin, the enveloping algebra, and inherits its fundamental properties.

To get the generators of T'_E belonging to T_E it suffices to multiply any element z_i (or z'_i) by an appropriate "denomerator", an element $\pi_i \in U(\mathfrak{f})$. Set

$$s_i = \pi_i z_i \quad (i \in I) \tag{19}$$

These elements, called "step operators" or "schift operators", were computed case-by-case in a number of works (e.g. in [3,4,5]). A regular method of computing these operators is exposed below.

Note that transvector algebras T'_E belong to the class of algebras with quadratic relations. These algebras became of interest in recent years in the context of mathematical physics [7].

The original definition of the algebra A may be extended by replacing (11) by quadratic conditions of type (17). This replacement does not affect Theorem 2, i.e. the functor Φ_E preserves algebras with quadratic conditions.

5. The operators p,q. We assume for simplicity that $\mathfrak{k}$ is a finite dimensional (or reductive) Lie algebra. To describe operators p,q define a normal ordering of Δ^+

The ordering of $\Delta^+ = \{\alpha_1, \ldots, \alpha_m\}$ is normal [8], if every non-simple root lies between its summands, i.e. $\alpha_j = \alpha_i + \alpha_k$ implies either $i < j < k$ or $k < j < i$. Actually, [2], a normal ordering in Δ^+ is nothing but an "optimal way" through Weyl chambers. More exactly, let C_+ be a dominant Weyl chamber, $C_- = -C_+$. An optimal way from C_+ to C_- is way through Weyl chambers with exactly m intersections with walls of these chambers. Any wall is a hyperplane with a normal vector α_i $(i = 1, \ldots, m)$.

In other words, let $\bar{w}$ be an element of the Weyl group which transforms C_+ to C_-, and let $\bar{w} = \tau_1 \ldots \tau_m$ be its reduced decomposition ($\tau_i = \tau_{\beta_i}$ is the reflection with respect to the simple root β_i). Then $\alpha_i = w_{i-1}\beta_i$ $(i = 1, \ldots, m)$, where $w_i = \tau_1 \ldots \tau_i$ and $w_0 = 1$.

Hence, there exists a bijection between the set of normal orderings of Δ^+ and the set of reduced decompositions of $\bar{w}$.

We assign to any $\alpha \in \Delta^+$ a linear operator in A' defined by the formula

$$p_\alpha = \sum_{n \geqslant 0} \frac{(-1)^n}{n!} f^{-1}_{\alpha,n} e^n_{-\alpha} \hat{e}^n_\alpha \tag{20}$$

where

$$f_{\alpha,n} = \prod_{0\le j\le n} (h_\alpha + \rho(h_\alpha) + j) \qquad (21)$$

ρ is the halfsum of roots $\alpha \in \Delta^+$. Here we consider $e_{-\alpha}, h_\alpha$ as operators of the left multiplication in A'. Set

$$p = p_{\alpha_1} \cdots p_{\alpha_m} \qquad (22)$$

with respect to a (fixed) normal ordering of Δ^+ . Similarly, to any $\alpha \in \Delta^+$ assign the linear operator

$$q_\alpha = \sum_{n\ge 0} \frac{(-1)^n}{n!} g_{\alpha,n}^{-1} e_{-\alpha}^n \hat{e}_\alpha^n \qquad (23)$$

where

$$g_{\alpha,n} = \prod_{0\le j\le n} (h_\alpha - j + 1) \qquad (24)$$

and $e_{-\alpha}, h_\alpha$ are considered as operators of the right multiplication in A'. Set

$$q = q_{\alpha_1} \cdots q_{\alpha_m} \qquad (25)$$

with respect to the normal ordering in Δ^+ . According to (a), formal series (22), (25) become finite when applied to any $x \in A'$. Therefore, the operators p,q are well-defined.

Theorem 4. The operators p,q defined by (22), (25) do not depend on the choice of normal ordering of Δ^+ and satisfy the conditions of Theorem 1.

In view of Theorem 1, any solution of (2) may be written in the form

$$x \equiv py \equiv qz \qquad (26)$$

for some $y, z \in A'$. We call the operators p,q the left and right resolvents of the system (2), respectively.

Note that any initial segment $\{\alpha_1, \ldots, \alpha_i\}$ of normal ordering in Δ^+ uniquely corresponds to some element of the Weyl group. The corresponding operator:

$$q_w = q_{\alpha_1} \cdots q_{\alpha_i} \qquad (27)$$

is the resolvents of the shortened system (2) consisting only of equations $e_\alpha x = 0$ $(\alpha = \alpha_1, \dots, \alpha_i)$ [2].

6. Extremal projections. We can expose the definition the operators p,q in a universal form considering an extension of enveloping algebra $U(\mathfrak{k})$.

Let $F(\mathfrak{k})$ be the set of all the formal series (with coefficients in $R(\mathfrak{f})$) in monomials

$$e(k,l) = e_{-\alpha_m}^{k_m} \dots e_{-\alpha_1}^{k_1} e_{\alpha_1}^{l_1} \dots e_{\alpha_m}^{l_m} \qquad (28)$$

such that $|k|+|l| \leq const$, where $|k| = k_1 + \dots + k_m$. Actually (1), $F(\mathfrak{k})$ is an algebra with respect to the multiplication. Setting

$$\mathcal{E}_\alpha = \sum_{n \geq 0} \frac{(-1)^n}{n!} f_{\alpha,n}^{-1} e_{-\alpha}^n e_\alpha^n \qquad (29)$$

we define the product

$$\mathcal{E} = \mathcal{E}_{\alpha_1} \dots \mathcal{E}_{\alpha_m} \qquad (30)$$

as an element of $F(\mathfrak{k})$.

Theorem 5. The element (30) does not depend on the choice of normal ordering in Δ^+ and is the unique solution of the system

$$e_\alpha \mathcal{E} = \mathcal{E} e_{-\alpha} = 0 \qquad \text{for all } \alpha \in \Delta^+ \qquad (31)$$

such that the constant term of $\mathcal{E}$ equals 1. This implies that $\mathcal{E}$ is a Hermitian projection in $F(\mathfrak{k})$:

$$p^2 = p, \quad p^* = p. \qquad (32)$$

We calltan extremal projection in $F(\mathfrak{k})$ (it is uniquely determined in $F(\mathfrak{k})$ after the choice of $\mathfrak{f}$ and Δ^+).

Application of extremal projections to representations of $\mathfrak{k}$ is connected with evident difficulties due to the presence of denomerators $f_{\alpha,n}$ in (29). But in the standard categories of representations these difficulties are avoided by a suitable normalization.

(1). In finite-dimensional modules, the extremal projection $\mathcal{E}$ is reduced to the known projection of Asherova, Smirnov and Tolstoy [9], by using instead of $f_{\alpha,n}$

some fraction of Γ-functions. It is the projection in a finite-dimensional k-module V onto the subspace of highest vectors (i.e. onto V_E).

(2). In Verma modules, the corresponding Γ-normalization is ineffective. In this case, after some right normalization, the extremal projections are replaced by their initial segments

$$\mathcal{E}_w = \mathcal{E}_{\alpha_1} \dots \mathcal{E}_{\alpha_i} \tag{33}$$

that enables one to compute all the singular (highest) vectors of Verma modules [1,2]. Note that all the morphisms (intertwining operators) of Verma modules may be described in similar terms.

The operators (33) may be considered as values of a cocycle on the Weyl group (or the corresponding <u>Kac semigroup</u> [20]). They also do not depend on normal ordering in (33), i.e. satisfy some "generalized Yang-Baxter equation".

(3). The extremal projection ε regularly acts in A', because in this algebra one can freely multiply by $R(\mathfrak{f})$ and operators $e_\alpha x \equiv \hat{e}_\alpha x$ are locally nilpotent (i.e. the series (29) is finite on any $x \in A'$). In this case we have $\varepsilon x \equiv px$ for all $x \in A'$ and in this case, therefore, ε coincides with the left resolvent in A.

One can get a similar interpretation for the right resolvent considering tensor products of the type $V \otimes U'(k)$ where V is a k-module.

7. <u>The functor Φ_E</u>. Let H be a category of A-modules presentable after reducing to k as sums of isotypic components V^λ. The subspace $V_E = \bigoplus_\lambda V^\lambda{}_E$ of highest vectors is invariant with respect to $\mathfrak{f}$. We will say that a module V is <u>regular</u> if all the operators $h_\alpha + \rho(h_\alpha) + j$ ($j = 1, 2, \dots$) are invertible in V_E. Let $H_0 \subset H$ be the category of regular A-modules.

<u>Theorem 6</u>. The functor $\Phi_E : (A, V) \mapsto (A_E, V_E)$ defines an injection of H_0 into A_E-mod.

Remark. The regularity condition is automatically satisfied for a finite-dimensional algebra k and a dominant λ. Actually, in this case the operators h_α are diagonalizable and $\lambda(h_\alpha)+\rho(h_\alpha)>0$ for all $\alpha\in\Delta^+$.

Theorem means that Φ_E preserves all the fundamental properties of A -modules (the structure of submodules and quotient modules, intertwining operators, etc.) and to two non-isomorphic modules V', V'' nonisomorphic modules V'_E, V''_E correspond.

In particular, if an A-module V is simple, then so is the A_E -module V_E. Moreover, V_E defines the initial module V up to isomorphism.

Theorem 6 is an evident consequence of the first structural theorem (Theorem 1). For the case $V(\dim k<\infty)$ this theorem was proved earlier [4] not using extremal projections.

General methods of computing "shift operators" were developed in [3,4]. But in any particular case these methods are reduced to solving some difficult systems of equations and therefore are of little effect, see e.g. [3,5,6].

The extremal projections are easily described for finite-dimensional Lie superalgebras (this is done independently by V.N. Tolstoy). They differ in additive normalization $(\rho(h_\alpha))$ and also the factors ε_α for odd roots are simpler than for even ones, i.e. the series (29) reduces to its first two terms. Similarly, the resolvents p,q are modified.

For infinite-dimensional Lie algebras the factorization of the operators ε, p, q is known only in some particular cases $(A_1^{(1)})$.

8. Applications (case V). In [1,2,5,6] a program is given for applying transvector algebras to representations of Lie algebras, i.e. to pairs $(\mathfrak{g}, k)$, where $\mathfrak{g}$ is a complex Lie algebra and k is its subalgebra reduct-

ive in $\mathfrak{g}$. This involves solving the following questions.

(1) Reduction $\mathfrak{g} \downarrow \mathfrak{k}$ problems for finite-dimensional $\mathfrak{g}$-modules (the questions of distinguishing multiplicities, Gelfand-Tsetlin bases, tensor products, etc.). There are numerous works dealing with solving particular problems of this kind. Using extremal projections and structural rules (17) provides us with natural apparatus for solving these problems. Some examples are considered in [1,5,6].

(2) The study of (generalized) Verma modules [1,2] and a wider category of modules with highest weights (description of singular vectors, morphisms, Jordan-Holder series, etc.) It is of interest to elucidate connection with Kazhdan-Lusztig polynomials [10].

(3) The study of (generalized) Harish-Chandra modules, i.e. modules of category H_0 (see sect.6). In accordance with [2], the use of transvector algebras allows us to get, in particular, a natural method of classifying simple Harish-Chandra modules.

(4) Theory of primitive ideals (tentatively). The extremal projections (see sect. 6) are naturally connected with the structure of flag manifolds, which, in their turn, are closely related with the structure of primitive ideals of $U(\mathfrak{g})$ [11]. It is interestly to elucidate also a connection with the theory of D-modules on the flag manifolds à la Beilinson-Bernstein [19].

Let us consider in detail the classification of simple Harish-Chandra modules, which implies, as it is known, the classification of irreducible representations of real reductive Lie groups.

The main result may be formulated as follows. First, the discrete series representations are described. These representations exist only when a Cartan subalgebra $\mathfrak{f}$ in $\mathfrak{k}$ is also a Cartan subalgebra of $\mathfrak{g}$. Now, an arbitrary irreducible representation of the group $G = \exp \mathfrak{g}$

may be recovered by induction from a representation of discrete series of a parabolic subgroup of G.

In terms of transvector algebras, any irreducible representation of G is uniquely determined by the corresponding representation of the algebra T_E in the subspace V_E (with respect to k). Actually, this representation of T_E possesses in any sense an extremal (lowest) weight and is determined by this weight up to equivalence.

The theory of representations with an extremal weight is easily constructed for T'_E (over the field $R(\mathfrak{f})$), and somewhat more complicatedly for T_E (over $\mathbb{C}$). An extremal weight is defined in regular cases by a pair (μ,ν), where μ is a highest weight for $\mathfrak{f}$ (in some sense minimal among the weights λ for which $V^\lambda \neq 0$), ν is a complex linear functional over the subspace $\mathfrak{y}_0 \subset \mathfrak{y}$, consisting of the vectors of zero weight (with respect to $\mathfrak{f}$). We denote the corresponding representation of G by $\mathcal{D}(\mu,\nu)$.

If $\mathfrak{f}$ is a Cartan subalgebra of $\mathfrak{g}$, then $\mathfrak{y}_0 = 0$ The corresponding representation $\mathcal{D}(\mu) = \mathcal{D}(\mu, 0)$ is exactly a discrete series representation of G.

Thus, using transvector algebras gives us a new approach to the whole representation theory of finite-dimensional Lie algebras.

Thus we hope that this method might be a key one in a future representations theory of infinite-dimensional Lie (super)algebras.

9. Applications (case D). In this section we only consider the simplest examples of classical "massless" equations.

(1) Wave equation $\Delta u = 0$. Let $\Delta = \partial_i \partial^i$ be the Laplace operator in $\mathbb{C}^n$. Setting $x^2 = x_i x^i$ we compute

$$[\Delta, x] = 2(\partial^i x_i + x_i \partial^i) = 4e + 2g \qquad (34)$$

where $e = x_i \partial^i$ is the Euler operator (of homogeneity), g is the trace of the metric tensor in $\mathbb{C}^n$ Setting

$$e_- = \frac{1}{2} x^2, \; e_0 = e - \frac{1}{2} g, \; e_+ = -\frac{1}{2} \Delta \qquad (35)$$

we get the basic elements of $\mathfrak{k} = \mathfrak{sl}(2)$. (The normalization is made so that Δ were colinear to $e_+ = e_\alpha$ for $\alpha > 0$).

Applying the extremal projection $\mathcal{E}$ to generators x_i, ∂_i of $\mathcal{D}(n)$ and multiplying by the corresponding denomerator, we get the generators $\tilde{x}_i, \tilde{\partial}_i$ of the transvector algebra $T = T_E$ ($E = \{\Delta\}$):

$$\tilde{\partial}_i = \partial_i, \; \tilde{x}_i = a x_i - x^2 \partial_i \; (i = 1, \ldots, n) \qquad (36)$$

where $a = -(e_0 + 2) = e + g/2 - 2$. Since $\mathcal{D}(n)$ acts irreducibly in the space of polynomials $F(n)$, we conclude (see sect.7), that T acts irreducibly in the space of harmonic polynomials V_E.

The fact that the operators $\tilde{x}_i, \tilde{\partial}_i$ are transvectors, is well known (see [14]). It is known also that they generate the Lie algebra $\mathfrak{o}(n+2)$ (recall that the commutant of E is generated by $\mathfrak{o}(n)$). We claim that in this case "there are no transvectors", i.e. any transvector is a polynomial of these operators and e_0 (which generates $R(\mathfrak{f})$).

It is clear that the operators $\tilde{x}_i, \tilde{\partial}_i$ can be used to construct the theory of spherical functions in $\mathbb{C}^n$. Actually, these operators are defined by recurrent relations among the spherical functions.

The irreducibility also implies that all the harmonic polynomials may be obtained by applying polynomials in $\tilde{x}_i$ (the latter act trivially) to the simplest harmonic polynomial $u_0(x) = 1$.

It is well known that the symmetry of hydrogen is reduced (by Fock transforms) to the symmetry of the wave equation. Thus, the operators (36) define the "maximal symmetry) of hydrogen.

(2) Dirac equation. Extending $\mathcal{D}(n)$ to the Weyl superalgebra $A=\mathcal{D}(n,m)$, we may "take the square root" $\mathcal{D}=\gamma_i\partial^i$ of the Laplace operator (here γ_i 's are analogues of the Dirac matrices), i.e. we may extend $\mathcal{H}(2)$ adding the basic elements

$$d_- = \frac{1}{\sqrt{2}}\gamma_i x^i, \quad d_+ = \frac{i}{\sqrt{2}}\gamma_i\partial^i, \quad d_\pm^2 = e_\pm$$

that close in $k=osp(1|2)$. Here $\Delta_+=\{\alpha,2\alpha\}$, where $d_+=e_\alpha$, $e_+=e_{2\alpha}$. The extremal projection may be written in the form

$$\varepsilon=(1-a^{-1}d_-d_+)\,\varepsilon_{2\alpha}$$

where $a=e_0+2$, $\varepsilon_{2\alpha}$ is an analogue of the extremal projection for $\mathcal{H}(2)$. It is easy now to compute the generators of the transvector algebra.

(3) Maxwell equations. In $A=\mathcal{D}(n,m)$ consider the following two elements:

$$d=\xi_i d^i,\quad d'=\delta_i\partial^i \tag{37}$$

Note that these elements are odd, and

$$[d,d']=dd'+d'd=\Delta. \tag{38}$$

Adding to these elements d, d' their adjoints

$$\delta=x_i\delta^i,\quad \delta'=x_i\xi^i \tag{39}$$

and computing all the commutators generated by this system, we get the Lie superalgebra $k=osp(2|2)\cong\mathcal{H}(1|2)$ of rank 2 and dimension $(4|4)$. In particular, set

$$a=[d,\delta],\ b=[d',\delta'],\ c=(a+b)/2-1 \tag{40}$$

The elements a, b form a basis of a Cartan subalgebra $\mathfrak{h}\subset k$. The full algebra k is the linear hull of $d, d', \Delta, a, b, \delta, \delta', x^2$.

The extremal system $E=\{d,d'\}$ is reduced for $n=m=4$ to the classical (homogeneous) Maxwell equations. Let V_0 be the space of all 2nd order exterior forms

$$\omega = \omega_{ij}\, dx^i \wedge dx^j \tag{41}$$

(we identify dx^i with variables ξ^i, i.e. include V_0 in the space of all the exterior forms $F(4|4)$). Here ω_{ij} is an antisymmetric tensor, components of which are interpreted as strengths of electromagnetic field. The equations

$$d\omega = d'\omega = 0 \tag{42}$$

coincide with the Maxwell equations written in the Weyl form (see e.g. [15]). We see that these equations form an extremal system (of type 1) for $k = osp(2|2)$.

The extremal projection $\mathcal{E}$ in this case is of the form $\mathcal{E} = \mathcal{E}_\alpha \mathcal{E}_\beta \mathcal{E}_\gamma$, where

$$\mathcal{E}_\alpha = 1 - a^{-1}\delta d, \quad \mathcal{E}_\beta = 1 - b^{-1}\delta' d' \tag{43}$$

$\mathcal{E}_\gamma (\gamma = \alpha + \beta)$ is an analogue of the extremal projection for $\mathcal{H}(2)$ (with modified normalization).

Computing the generators of $T = T_E$ we get:

$$\tilde{\delta}_j = a\delta_j - \delta_i (x^i \partial_j), \quad \tilde{\xi}_j = b\xi_j - \xi_i (x^i \partial_j)$$
$$\tilde{\partial}_j = \partial_j, \quad \tilde{x}_j = ab\, x_j + a_{ij} x^i + b_{ik} x^i x^k \partial_j \tag{44}$$

where

$$a_{ij} = -c\,(a\xi_i \delta_j + b\delta_i \xi_j), \quad b_{ij} = \tfrac{1}{2}(a\bar{b}\,\xi_i \delta_j + \bar{a} b\,\delta_i \xi_j), \quad \bar{a} = a-1, \ \bar{b} = b-1 \tag{45}$$

Note that T acts in the space $V = F(4|4)$. The subalgebra T_0 conserving V_0 is generated by the elements

$$\tilde{\partial}_j, \ \tilde{x}_j, \ \tilde{\xi}_i \tilde{\delta}_k \quad (i, j, k = 1,2,3,4) \tag{46}$$

hence by the elements a,b (generators of $R(f)$). Thus, the transvector algebra of the Maxwell equations coincides with T_0

The symmetry of Maxwell equations was studied in a number of works (see e.g. [16]). Our method enjoys the simplicity of algebraic approach and allows us to get automatically the maximal symmetry algebra (in the class $\mathcal{D}(n)$). This algebra is wider than the algebra of differential operators described in [16].

Therefore the full symmetry algebra $T_c \subset \mathcal{D}(4|4)$ is generated by the set of the elements (46) and the Cartan elements a,b. Moreover, this algebra acts irreducibly in the space of exterior forms V_0.

Nonlocal (integrodifferential) symmetry operators, found in [16], may also be included in a wider symmetry algebra by the general method of extremal projections.

In all the above cases (1), (2), (3) the extremal projection regularly acts in V. Therefore we have

$$V_E = \mathcal{E} V \tag{47}$$

For example, $\mathcal{E}$ coincides in case (1) with the known projection onto the space of harmonic polynomials.

Similarly, in case (3) any polynomial solution of the Maxwell equations may be constructed by a projection of V_0.

In this sense the extremal projection $\mathcal{E}$ is an analogue of the Radon-Penrouse transformation of the twistor theory [15].

It is interesting to study in detail the transvector algebra T_0 for the Maxwell equations. In particular, what Lie superalgebra is generated by (44) or (46)? What is an analogue of the theory of spherical functions for the Maxwell equations?

What shall we do if we consider nonhomogeneous equations, e.g. massive Dirac or Klein-Gordon equations? Usually, mass may be considered as a new variable and the equations become homogeneous in a space of higher dimension. But it is also interesting to find the transvector algebra for a fixed mass.

The answer to this question is obscure. In general, for a nonzero mass the transvector algebra is more complicated, but counterexamples also are known. For example, for two-dimensional Klein-Gordon or Dirac equations with nonzero mass the transvector algebra reduces to the commutant.

It is especially interesting that the classical equations (1)-(3) (and some others) are reduced to extremal systems for suitable algebras. Apparently, this fact is connected with the general principles of mechanics and with the known rules of quantization of quantum mechanical systems.

10. <u>Applications (case X)</u>. What does our theory give for the study of the ring $\mathcal{D}(X)$ of regular differential operators on an algebraic variety X ?

Of course, the principal structure theorems (sect.4) allow us to project $\mathcal{D}(n)$ onto $\mathcal{D}(X)$ (also $\mathcal{D}(Y)$ onto $\mathcal{D}(X)$ for $X \leq Y$), to compute generators of these rings and structure relations among them. Since $\mathcal{D}(n)$ irreducibly acts in $F(n)$, the $\mathcal{D}(X)$-action in $F(X)$, is also irreducible, etc. A practical use of these results is apparently connected with the consideration of some interesting examples and concrete description of corresponding transvector algebras.

One of the interesting examples is the principal affine space $X = G/N$, where G is a complex connected reductive Lie group, N its maximal unipotent subgroup (the subgroup of upper triangular matrices with units on the main diagonal for $G = SL(n)$).

Note that $F(X) = F(G)_E$, where E is the extremal system corresponding to $\mathfrak{g} = Lie\ G$ (actually, the functions on X are exactly the functions on G such that $e_\alpha f = 0$ for all $\alpha \in \Delta^+$) . For classical Lie algebras another interpretation is known; $F(X) = F(n)_{\hat{E}}$,[5] where $\hat{E}$ is the extremal system generated by the Lie algebra $\hat{\mathfrak{g}} = \mathfrak{sl}(n)$, $\mathfrak{sp}(2r)$ or $\mathfrak{o}(2r)$ for $\mathfrak{g} = \mathfrak{sl}(n)$, $\mathfrak{o}(n)$ or $\mathfrak{sp}(n)$ respectively where $r = rank\ \mathfrak{g}$ ([5]). Thus, our results are applicable in this case.

The algebra $\mathfrak{A} = F(X)$, where X is the principal affine space, is of interest because of the following. Consider $\mathfrak{A}$ as a canonical G-module on X. It is known that $\mathfrak{A}$ is the sum of all the holomorphic representations of G, occurring without multiplicities;

$$\mathfrak{A} = \bigoplus_{\lambda} \mathfrak{A}_{\lambda}$$

Moreover, $\mathfrak{A}_{\lambda} \mathfrak{A}_{\mu} = \mathfrak{A}_{\lambda+\mu}$ for all λ, μ. The space $\mathfrak{A}$ becomes a "model of representations" for G in the sense of [17].

In the theory of models some "overgroup" is usually constructed so that its action mixes all the isotypic components of G. It is not clear if such an overgroup $\widetilde{G}$ (overalgebra $\widetilde{\mathfrak{g}}$) exists, in our case, but the role of $\widetilde{G}$ is played by our transvector algebra $T_E = \mathcal{D}(X)$.

One of such models $(\widetilde{\mathfrak{g}} = \mathfrak{o}(8)$ for $\mathfrak{g} = \mathfrak{sl}(3))$ was constructed by Biedencharn and Flat, see [18], where the existence problem of more general "symmetry algebras" of given type is discussed. The answer to this problem is affirmative, but there are many such algebras. For example, the algebra $\mathcal{D}(X)$, where $X = G/N$ is a symmetry algebra for G. This algebra, as in [18], may be used for constructing the "tensor operators" for G.

This report may be considered as a plan of further investigation of transvector algebras and their applications.

References

1. Zhelobenko, D.P. (1983). Math. USSR Doklady, 273, 4, 785-788; 273, 6, 1301-1304 (Russian).

2. Zhelobenko, D.P. (1984). Funct. Anal. Appl., 18, 4, 79-80 (Russian).

3. Mickelsson, J. (1972). Rept. Math. Phys., 392; (1973). 4, 4, 307-318.

4. Hombergh, A. van den. (1975). Indag-Math., Vol.37, No.1; id. Thesis, 1976.

5. Zhelobenko, D.P. (1985). In: Proc. Soviet-Hungarian Math. School, Budapest, pp. 79-106.

6. Zhelobenko, D.P. (1985). Infinite-dimensional Lie Group Representations, Gordon and Breach, Overseas.

7. Vershik, A.M. (1984). Spectral theory of linear operators and infinite-dimensional analysis, Kiev, AN USSR, pp. 32-57 (Russian).

8. Leznov, A.N., Saveliev, M.V. (1974). Funct. Anal. Appl., 8, 4, 87-88 (Russian).

9. Asherova, R.S., Smirnov, Yu.F., Tolstoy, V.N. (1971). Teor. Math. Phys., 8, 2, 255-271 (Russian).

10. Kazhdan, D.A., Lusztig, G. (1979). Invent. Math., 53, 1.

11. Joseph, A. (1973). Lecture Notes Math., 1024, 30-77.

12. Zhelobenko, D.P. (1974). Harmonic analysis on semi-simple complex Lie groups, Nauka, Moscow (Russian).

13. Vogan, D. (1981). Representations of real reductive groups. Birkhauser, p. 754.

14. Malkin, J.A., Man'ko, V.J. (1974). Dynamic systems and coherent states of quantum systems. Nauka, Moscow (Russian).

15. Twistors and gauge fields. (1983). Mir, Moscow (Russian).

16. Fushchich, V.J., Nikitin, A.G. (1983). Physics of elementary particles and atomic nuclea, 14, 1 (Russian).

17. Gelfand, J.M., Zelevinsky, A.V. (1984). Funct. Anal. Appl., 10, 1 (Russian).

18. Flat, D.E. (1984). Bull. Amer. Math. Soc., 10

19. Bernstein, J.N. (1986). Asterisque,

20. Borovoi, M.V. (1984). Funct. Anal. Appl. 18, 2, 57-58. (Russian).

THE GALILEI GROUP IN INVESTIGATIONS OF SYMMETRY PROPERTIES OF MAXWELL EQUATIONS

G.A. Kotel'nikov,
Kurchatov Institute of Atomic Energy, Moscow

1. Introduction. Since the moment when the special relativity was created a significant attention has been concentrated on studying the symmetry of Maxwell equations. The most famous result of this study was obtained in 1904-1905 by Lorentz, Einstein and Poincaré and in 1909 by Kunningham and Bateman. The main contribution of these authors was proving the relativistic and conformal invariance of the Maxwell equations in the Minkowski space. The subsequent studies were based mainly on symmetry understood as the algebra of invariance whose elements transform a solution of the equation into a solution ([1-3] etc.). The results of the foundators of the relativity theory were enriched by numerous examples of the non-Lorentz groups of the inner symmetry, see e.g. [4-9].

The transformations of these groups, described in details in [10], are determined for the set of functions, the solutions of the Maxwell equations, do not induce transformations in the Minkowski space and therefore the corresponding groups generate the direct product with the conformal group. At the same time it is of interest to establish the existence of not only the inner symmetries of the Maxwell equations but other space-time ones, different from conformal symmetry.

Bearing in mind Einstein's remark that, "the theory of scales and clocks should be derived from the solutions of basic equations, ... and should not be considered independent of the latter" ([36], p.153) the finding of new space-time symmetries of the Maxwell equations will deepen our knowledge of the nature of space-time.

It is known (see e.g. [6], [10], [11]) that in the Minkowski space among the Lie groups generated by first order differential operators, there are no space-time groups of transformations with the conformal one as a subgroup. Therefore in the Minkowski space in this class of groups there is no symmetry group of the Maxwell equations wider than the conformal one. It means that to find other space-time symmetries it is necessary to go beyond the first order differential operators or to introduce coordinate spaces other than the Minkowski space.

In [12] transition from the 4-dimensional Minkowski space to some 5-dimensional one was carried out in which the speed of light c is considered as an independent variable much alike the radius-vector $\vec{x}$ or the time t. This 5-dimensional space includes the Minkowski space as the hyperplane c = const.

The Maxwell equations in this space turn out to be invariant with respect to some infinite group $C f_{\infty}$ which involves transformations of both the speed of light and the conformal group as a subgroup [12]. The following finite space-time transformations [13-15] may serve as an example

$$x'=\left(\gamma+\frac{a_5}{c}\right)^N\left(\alpha\,\frac{x-\beta ct}{\sqrt{1-\beta^2}}+c^N a_1\right);\ t'=\left(\gamma+\frac{a_5}{c}\right)^{N-1}\left(\alpha\,\frac{t-\beta x/c}{\sqrt{1-\beta^2}}+c^{N-1}a_0\right);$$

$$y'=\left(\gamma+\frac{a_5}{c}\right)^N\left(\alpha y+c^N a_2\right);\ z'=\left(\gamma+\frac{a_5}{c}\right)^N\left(\alpha z+c^N a_3\right);\ c'=\gamma c+a_5, \tag{1.1}$$

where $N=0,\pm 1,\ldots,\pm n$; $\alpha, \beta=V/c, \gamma, a_0-a_5$ – are the group parameters. In the case $N=0$ the homogeneous transformations (1.1) were introduced first by Romain [16] to defend the special relativity from the unjustified criticism by the President of the National Spanish Academy of Science J. Palacios; then by the author of this paper as symmetry transformations of the Maxwell equations [17], and later on by Sjödin [18]. The inhomogeneous transformations (1.1) for $N=0,\ a_5=0$ were attained by Di Jorio [19] from analysis of the symmetry properties of the D'Alambert equation in terms of the change of variables. (First this method was suggested by Voigt [20] and Umov [21] as applicable to the equation $\Box\varphi=0$). The transformations of the electromagnetic field components generated by (1.1) have the form [13-15]:

$$E'_x=\left(\gamma+\frac{a_5}{c}\right)^L\frac{E_x}{\alpha^2};\qquad H'_x=\left(\gamma+\frac{a_5}{c}\right)^L\frac{H_x}{\alpha^2};$$

$$E'_y=\left(\gamma+\frac{a_5}{c}\right)^L\frac{E_y-\beta H_z}{\alpha^2\sqrt{1-\beta^2}};\quad H'_y=\left(\gamma+\frac{a_5}{c}\right)^L\frac{H_y+\beta E_z}{\alpha^2\sqrt{1-\beta^2}};\tag{1.2}$$

$$E'_z=\left(\gamma+\frac{a_5}{c}\right)^L\frac{E_z+\beta H_y}{\alpha^2\sqrt{1-\beta^2}};\quad H'_z=\left(\gamma+\frac{a_5}{c}\right)^L\frac{H_z-\beta E_y}{\alpha^2\sqrt{1-\beta^2}};$$

$$\rho'=\left(\gamma+\frac{a_5}{c}\right)^{L-N}\rho\left(1-\beta v_x/c\right)/\alpha^3\sqrt{1-\beta^2}.$$

Here L is the number, v_x is the speed of the charge movement along the axis x and we suppose that $\dot{c}=dc/dt=0$, which may be interpreted as the condition of immobility of noninteracting electromagnetic fields in time along the axis c. If in (1.1) and (1.2) we take $\gamma=1, a_5=0$, then the obtained transformations will be apparently the symmetry transformations of the Maxwell equations from the Weyl group, the subgroup of the conformal group. Note also that the group parameters α,γ admit the kinematic realization [17,22], as a result of which the homogeneous transformations (1.1) may be written as:

$$x'=\frac{\left(1-\frac{V^2}{c^2}\right)^{(N+1)p+N(q-1)-1}}{\left(1-\frac{\vec{U}\vec{V}}{c^2}\right)^{2(N+1)p+2N(q-1)-1}}(x-Vt);\ t'=\frac{\left(1-\frac{V^2}{c^2}\right)^{Np+(N-1)(q-1)-1}}{\left(1-\frac{\vec{U}\vec{V}}{c^2}\right)^{2Np+2(N-1)(q-1)-1}}\left(t-\frac{Vx}{c^2}\right);$$

$$y',z' = \frac{\left(1-\frac{V^2}{c^2}\right)^{(N+1)p+N(q-1)-1/2}}{\left(1-\frac{\vec{U}\vec{V}}{c^2}\right)^{2(N+1)p+2N(q-1)-1}} y,z; \quad c' = \frac{\left(1-\frac{V^2}{c^2}\right)^{p+q-1}}{\left(1-\frac{\vec{U}\vec{V}}{c^2}\right)^{2(p+q-1)}} c, \quad (1.3)$$

where $-\infty < p,q < +\infty$; $\vec{U} = (U_1, U_2, U_3)$ is the velocity of some chosen reference system K_0 relative to K; $\vec{V} = (V, 0, 0)$ is the velocity of an arbitrary system K' relative to K. These transformations are not manifestly Lorentzian. However, in spite of their exotic form they (together with the relativistic theorem of composition of velocities) generate a symmetry group of Maxwell equations [17, 22]. For $N=0$, $\vec{U}=0$ formulas (1.3) include some set of transformations with invariant speed of light, namely the Voigt transformations for $p=1$, $q=0$ [20], the Ives transformations for $p=(m+1)/2, q=(-m+1)/2$ [23]; the Palakios-Gordon transformations for $p=0$, $q=1$ [24,25]; the Dewan transformations for $p=(2-n)/2$, $q=n/2$ [26]; the Podlaha transformations for $p=-\varepsilon$, $q=1+\varepsilon$ [27]. Moreover, formulas (1.3) contain also some set of transformations with noninvariant speed of light, in particular, transformations with the "common time" $t'(x')=t(x)=t_0(0)$ from our papers [17,22] and transformations of Hsu [28] for $p=1/2, q=0$. For $p=q=1/2$ formulas (1.3) transfer into the Lorentz transformations, which indicates their relativistic nature.

Another approach for seeking non Lorentz space-time symmetries of the Maxwell equations was developed by Fushchich and Nikitin [2,10]. This approach is closely connected with representations of the Lie algebra of the Poincaré group by operators which can be written in a noncovariant form involving time and space variables in an unequal way [29]. In particular, in [10] the set of operators

$$P_0 = H = \gamma_0 \gamma_k p_k; \qquad p_k = -i\partial/\partial x_k;$$
$$J_{jk} = x_j p_k - x_k p_j + S_{jk}; \quad j,k = 1,2,3, \quad x_{1,2,3} = x, y, z; \tag{1.4}$$
$$J_{0k} = t p_k - x_k H + S_{0k},$$

is constructed which on the set of 8-component functions $\Psi(x,t)$ = column $(H_1, H_2, H_3, \varphi_1, E_1, E_2, E_3, \varphi_2)$ form the sought representation of the Poincaré algebra. Here the vector-function $\Psi(x,t)$ represents a solution of the Maxwell equations in the form of the Dirac equation for the massless fields $L_1\Psi = \gamma_\mu p^\mu \Psi = 0$; $L_2\Psi = \gamma_\mu p^\mu S_{\nu\lambda} S^{\nu\lambda}$; $p_0 = i\partial/\partial t$; $\hbar = c = 1$; $\mu, \nu, \lambda = 0,1,2,3$; $g_{\mu\nu} = \mathrm{diag}(1,-1,-1,-1)$; $\gamma_\mu, S_{\nu\lambda}$ – are the 8x8 matrices [10]. As the operators (1.4) do not include any derivative with respect to time and speed of light, they appear to commute with the latter ones and hence generate the space-time transformations with invariant time and speed of light [2,10]:

$$t \to t' = t; \quad c \to c' = c; \quad \vec{x} \to \vec{x}' = f(\vec{x}, t, \lambda_1, \ldots, \lambda_{10}), \tag{1.5}$$

where λ_i are group parameters. They do not preserve the quadratic form $c^2t'^2 - \vec{x}'^2 \neq c^2t^2 - \vec{x}^2$ but nevertheless are symmetries of Maxwell equations together with the transformations

$$\vec{E}' = g(\vec{E}, \vec{H}, \partial\vec{E}/\partial x_k, \partial\vec{H}/\partial x_k, \partial^2\vec{E}/\partial x_k \partial x_j, \ldots);$$
$$\vec{H}' = f(\vec{E}, \vec{H}, \partial\vec{E}/\partial x_k, \partial\vec{H}/\partial x_k, \partial^2\vec{H}/\partial x_k, \partial^2\vec{H}/\partial x_k \partial x_j \ldots). \tag{1.6}$$

Specifically, the field transformations generated by J_{0k}, turn out to be nonlocal (integral) [10]:

$$\Psi'(\vec{x}, t) = (2\pi)^{-3/2} \int d^3p \, \Psi'(p,t) \exp(i\vec{p}\cdot\vec{x});$$
$$\Psi'(p,t) = \exp(iJ_{0k}\Theta_k)\Psi(p,t). \tag{1.7}$$

There are investigations of the symmetry of Maxwell equations in classical physics [10], where the symmetry of the first, $\nabla \times \vec{H} = i\partial_4 \vec{E} + 4\pi\vec{j}$; $\nabla\cdot\vec{E} = 4\pi\rho$, and the second, $\nabla \times \vec{E} = -i\partial_4 \vec{H}$; $\nabla\cdot\vec{H} = 0$, pairs of Maxwell equations with respect to the 20-dimensional group $IGL(4,R)$

of nonhomogeneous linear transformations of coordinates in the 4-dimensional real space are described (here $x_4 = ict$; $\vec{j} = \rho\vec{v}/c$, $\vec{v}$ is the velocity of charge). Since $IGL(4, R)$ contains the Galilei group as a subgroup, the invariance of the first and second pairs of the Maxwell equations is established also with respect to the Galilei transformations. The subtle reason of necessity to consider the subsystems of equations separately is that different linear Lie algebra representations of the Galilei group are realized on their solutions. In other words, the authors of [10] has shown that the subsystems of the Maxwell equations being invariant apart are not invariant with respect to the Galilei transformations in the class of the linear Galilei group representations if they are considered together. The situation radically changes, if one leaves the class of linear field transformations and proceeds to the nonlinear ones. As has been stated in [30-32], the Maxwell equations are invariant with respect to the Galilei transformations if the fields $\vec{E}, \vec{H}$ are transformed nonlinearly. The aim of this paper is to generalize these results and to compare them with the above results.

2. The Galilei group and its relation with Maxwell equations. The Galilei group $\mathcal{G}_{10}$ may be defined as the set of 10-dimensional linear space-time transformations

$$\vec{x}' = R\vec{x} + \vec{V}t + \vec{a}; \quad t' = t + b, \tag{2.1}$$

where R is the matrix of the 3-dimensional spatial rotation; $\vec{V} = (V_1, V_2, V_3)$ the velocity of the intertial reference frame K' relative K; $\vec{a}, b$ the parameters of space-time translations. The Lie algebra of this group is generated by the set of generators $p_0 = i\partial_t$, $p_k = -i\partial_k$ of the translation group T_4; $J_k = (\vec{x} \times \vec{p})_k$ of the 3-dimensional spatial rotation group SO_3; $\mathcal{H}_k = -t p_k$ of the pure Galilei transformation group $\mathcal{G}_3$

[33,34] (here $k=1,2,3;\ x_{1,2,3}=x,y,z$).

The relation between the Galilei group and the Maxwell equations may be obtained from understanding symmetry [30], according to which the Lie algebra of some group is the invariance algebra of an equation $A\psi=0$, if the p-fold Lie multiplication of A by any group generator q_α transforms a nonzero solution into zero: $A^p \cdot q_\alpha \psi(x) = [A[A[\cdots[A,q_\alpha]\cdots]]\psi=0$. In terms of the extended understanding of symmetry the following Theorem 1 holds: The Lie algebra of the Galilei group is the algebra of invariance of the homogeneous Maxwell equations, since [30]:

$$[\Box,q_\alpha]=0,\quad \alpha=1,2,\dots,7;\quad q_\alpha\in\{p_0,p_k,J_k\};$$
$$[\Box[\Box,\mathcal{H}_k]]=0,\quad k=1,2,3, \tag{2.5}$$

where $\Box$ is the D'Alembert operator, $[\Box,\mathcal{H}_k]=2ip_0p_k/c^2$ are the symmetry operators in Manko's sense [1]. This invariance means that there exist transformations of the electromagnetic field, which together with the Galilei group transform Maxwell equations into themselves. The problem is to find these transformations.

3. Nonlinear field transformations. The invariance of Maxwell equations with respect to the Galilei group.

Consider the main part of transformations of $\mathcal{G}_{10}$, the pure Galilei transformations generated by $\mathcal{H}_k$:

$$x_1'=x_1-V_1t;\ x_2'=x_2-V_2t;\ x_3'=x_3-V_3t;\ t'=t \tag{3.1}$$

Rewrite them, using the new variable $x_4=ict$:

$$x_1'=x_1+i\beta_1x_4;\ x_2'=x_2+i\beta_2x_4;\ x_3'=x_3+i\beta_3x_4;\ x_4'=\lambda x_4, \tag{3.2}$$

where $\beta_j=V_j/c$, $j=1,2,3$; $\lambda=c'/c=(1-2\beta\vec{\zeta}\vec{n}+\beta^2)^{1/2}$; $\beta=(\beta_1^2+\beta_2^2+\beta_3^2)^{1/2}$; $\vec{\zeta}=\vec{V}/V$ is the directional velocity vector $\vec{V}$; $\vec{n}=\vec{c}/c$ is the directional velocity vector of light propagation. The transformations of group parameters β,λ are

$$\lambda''=\lambda\cdot\lambda'; \quad \beta''_j = \beta_j + \lambda\beta'_j; \quad j=1,2,3. \tag{3.3}$$

Keeping this in mind, write Maxwell equations in the symmetric-dual form [35]:

$$\nabla\cdot\vec{E}=4\pi\varrho; \qquad \nabla\cdot\vec{H}=4\pi\rho;$$
$$\nabla\times\vec{E}=-i\partial_4\vec{H}-4\pi\vec{k}; \qquad \nabla\times\vec{H}=i\partial_4\vec{E}+4\pi\vec{j}. \tag{3.4}$$

Here ϱ and ρ are densities of the electric q and magnetic charges p; $\vec{j}=\varrho\vec{v}/c$, $\vec{k}=\rho\vec{w}/c$ are the densities of the electric and magnetic currents; $\vec{v}$ and $\vec{w}$ are velocities of the electromagnetic charges. Following [30-32], we seek the component transformation formulas of the electromagnetic field and $\varrho, \rho, \vec{j}, \vec{k}$ in the form:

$$E'_j=\lambda^n f(x,y)(aE_j+h_{jk}H_k); \qquad H'_j=\lambda^n f(x,y)(bH_j+e_{jk}E_k);$$
$$\varrho'=\lambda^n f(x,y)\varrho; \qquad \rho'=\lambda^n f(x,y)\rho; \tag{3.5}$$
$$\vec{j}'=\lambda^{n-1}f(x,y)(\vec{j}-\vec{J}); \qquad \vec{k}'=\lambda^{n-1}f(x,y)(\vec{k}-\vec{K}),$$

where $f(x,y)=\psi(x)$ is some weight function, which is not a component of the field, in general, not unity and depends on $\vec{x}$, time t and $y\in\{E_1,E_2,E_3,H_1,H_2,H_3\}$; e_{jk}, h_{jk} are numerical coefficients; $j,k=1,2,3$; the summation is performed over repeated indices; $e_{jj}-h_{jj}=0$; $\vec{J}=\varrho\vec{V}/c$; $\vec{K}=\rho\vec{V}/c$.

To find the transformation coefficients and $\psi(x)$ substitute these variables into (3.4) taking into account (3.5) and relations $\partial'_j=\partial_j$; $\partial'_4=(\partial_4-i\vec{\beta}\cdot\vec{\partial})/\lambda$; $j=1,2,3$. Let the transformed equations turn into themselves. This can be reached by imposing certain conditions on $\psi(x)$ called autotransform conditions. We have:

$$\vec{E}\cdot\nabla\psi=-\partial_j\psi h_{jk}H_k-(a-1)\partial_j\psi E_j=A_1;$$
$$\nabla\psi\times\vec{E}+i\vec{H}\partial_4\psi=-b(\vec{\beta}\cdot\vec{\partial})\psi\vec{H}-(\lambda a-1)\nabla\times\psi\vec{E}-\vec{B}+4\pi\psi\vec{K}; \tag{3.6}$$

$$\vec{H}\cdot\nabla\psi = -\partial_i\psi e_{jk}E_k-(b-1)\partial_j\psi H_j = A_2;$$
$$\nabla\psi\times\vec{H}-i\vec{E}\partial_4\psi = +a(\vec{\beta}\cdot\vec{\partial})\psi\vec{E}-(\lambda b-1)\nabla\times\psi\vec{H}+\vec{D}-4\pi\psi\vec{J}; \quad (3.7)$$

$$B_j=(i\partial_4+\vec{\beta}\cdot\vec{\partial})\psi e_{js}E_s+\lambda(\partial_k\psi h_{ls}H_s-\partial_l\psi h_{ks}H_s)+i(b-1)\partial_4\psi H_j;$$
$$D_j=(i\partial_4+\vec{\beta}\cdot\vec{\partial})\psi h_{js}H_s-\lambda(\partial_k\psi e_{ls}E_s-\partial_l\psi e_{ks}E_s)+i(a-1)\partial_4\psi E_j. \quad (3.8)$$

Consider the autotransform conditions as the system of algebraic equations for e_{jk} and h_{jk}. Let us find its solution, when $a=b=1$; $\vec{\beta}=(\beta,0,0)$; $h_{12}=h_{13}=e_{12}=e_{13}=h_{21}=h_{31}=e_{21}=e_{31}=0$, i.e. when K' moves relative K with the velocity $\vec{V}=(V,0,0)$ along x_1 [31]. Take into account, that the equations are subdivided in two subsystems (3.6) and (3.7) the determinates of which are respectively

$$\Delta_1=-\lambda(i\partial_4+\beta\partial_1)\psi H_2\cdot(i\partial_4+\beta\partial_1)\psi H_3\cdot(\partial_2\psi E_2\partial_2\psi E_3+\partial_3\psi E_2\partial_3\psi E_3),$$

$$\Delta_2=\lambda(i\partial_4+\beta\partial_1)\psi E_2\cdot(i\partial_4+\beta\partial_1)\psi E_3\cdot(\partial_2\psi H_2\partial_2\psi H_3+\partial_3\psi H_2\partial_3\psi H_3).$$

and if $\Delta_1,\Delta_2\neq 0$ each subsystem has a solution.
To distinguish the solutions we bar them. Skipping the details note that the barred and unbarred solutions are:

$$e_{23}=\bar{e}_{23};\ e_{32}=\bar{e}_{32};\ h_{23}=\bar{h}_{23};\ h_{32}=\bar{h}_{32}, \quad (3.9)$$

and $A_1, A_2, \vec{B}, \vec{D}$ from (3.6)-(3.8) at $a=b=1$ satisfy additionally

$$(i\partial_4+\beta\partial_1)^2\vec{D}/\lambda^2+\nabla\times\vec{B}/\lambda+\nabla A_1=\vec{M};\ (i\partial_4+\beta\partial_1)^2\vec{B}/\lambda^2-\nabla\times\vec{D}/\lambda+\nabla A_2=\vec{N}; \quad (3.10)$$

$$\vec{M}=\left[0;-h_{23}\left(\Delta-\frac{1}{\lambda^2}(i\partial_4+\beta\partial_1)^2\right)\psi H_3;\ -h_{32}\left(\Delta-\frac{1}{\lambda^2}(i\partial_4+\beta\partial_1)^2\right)\psi H_2\right]; \quad (3.11)$$
$$\vec{N}=\left[0;-e_{23}\left(\Delta-\frac{1}{\lambda^2}(i\partial_4+\beta\partial_1)^2\right)\psi E_3;\ -e_{32}\left(\Delta-\frac{1}{\lambda^2}(i\partial_4+\beta\partial_1)^2\right)\psi E_2\right].$$

As a result the expressions for the transformation coefficients may be written uniformly [31,32].

The weight function satisfies the equations [31,32] that in particular case $\rho = \mu = 0$ take the form:

$$\Delta\psi\varphi_\alpha - (i\partial_4 + \beta\partial_1)^2 \psi\varphi_\alpha / \lambda^2 = 0;$$
$$\alpha = 1, 2, \ldots, 6; \; \varphi \in \{E_1, E_2, E_3, H_1, H_2, H_3\}, \quad (3.12)$$

and vectors (3.11) vanish for any $e_{23}, e_{32}, h_{23}, h_{32}$. The fields $\vec{E}$ and $\vec{H}$ are supposed to be known from the solution of the Maxwell equations. The scheme presented here may be restated in another way: to determine the four unknown coefficients $e_{23}, \ldots$ we have 8 equations which are subdivided into two subsystems. Choose the weight function so that the solutions of subsystems were equal. In some cases, e.g. for constant fields, the latter scheme implying a direct reference to autotransform conditions of Maxwell equations proves to be more constructive.

Formulas (3.2) together with (3.5)-(3.7) are the sought formulas of the Galilei symmetry of Maxwell equations. They differ from the generalized relativistic transformations (1.1), (1.2) and from (1.5), (1.7). So, unlike (1.7), the formulas of transformations of fields (3.5) are not integral. The presence of the weight function $\psi(x)$ there means that in the Galilei theory transformation properties of the fields, the densities of currents and charges obviously depend in general on space-time point $(\vec{x}, t)$, unlike (1.2) where these properties obviously do not depend on $\vec{x}$ and t and have global nature. The transformation coefficients and the weight function are determined by the concrete solutions of the Maxwell equations. Therefore the transformations of the fields are nonlinear. The group multiplication of the matrices of the field transformations (3.5) for $a = b = 1$ and transformation coefficients

$$\left\| M_{\alpha\beta} \right\| = \lambda^n f(x,y) \left\| \begin{array}{ccc|ccc} 1 & 0 & 0 & 0 & 0 & 0 \\ 0 & 1 & 0 & 0 & 0 & h_{23} \\ 0 & 0 & 1 & 0 & h_{32} & 0 \\ \hline 0 & 0 & 0 & 1 & 0 & 0 \\ 0 & 0 & e_{23} & 0 & 1 & 0 \\ 0 & e_{32} & 0 & 0 & 0 & 1 \end{array} \right\| \qquad (3.13)$$

is given by the formulas [32]:

$$\psi'' = \psi \cdot \psi' ; \qquad \lambda'' = \lambda \cdot \lambda' ;$$

$$e''_{23} = e_{23} + \Lambda_1(y) e'_{23} ; \qquad h''_{23} = h_{23} + \Lambda_3(y) h'_{23} ; \qquad (3.14)$$

$$e''_{32} = e_{32} + \Lambda_2(y) e'_{32} ; \qquad h''_{32} = h_{32} + \Lambda_4(y) h'_{32} .$$

Here the transformation properties of the function $\Lambda(y)$ and the weight $\psi(x)$ under the Galilei transformations are much like those of the kinematic parameter λ: $\Lambda''(y) = \Lambda'(y') \cdot \Lambda(y)$. The transformation properties of $e_{23}, e_{32}, h_{23}, h_{32}$ coincide with those of the kinematic parameter β: $\beta'' = \beta + \lambda\beta'$. Hence, the group multiplication law of $M_{\alpha\beta}$ agrees with that of the parameters λ, β . Therefore, $M_{\alpha\beta}$ is the matrix of the Galilei group representation. Since $M_{\alpha\beta}$ depend on solutions of Maxwell equations, this representation will be called nonlinear. Some examples of calculating the weight functions $\psi(x)$ and the coefficients $\{e_{23}, e_{32}, h_{23}, h_{32}\} \subset M_{\alpha\beta}$ are presented in Table 1. The nonlinearity of the field transformations correlates with nonzero commutation relations of order $\geqslant 1$ of the generators $\mathcal{H}_k$ with D'Alembert's operator (2.5). Moreover, the following Theorem 2 holds: There are no linear transformations of fields describing in general the symmetry of Maxwell equations with respect to the Galilei transformations (cf. [10]).

Table 1. The weight functions and the coefficients of transformations of electromagnetic field.

<table>
<tr><td colspan="2">Constant homogeneous electric field of the uniformly charged plane (x_2, x_3) with surface charge density σ =const.
$\vec{E}=(0,0,E_3)$; $\vec{H}=(0,0,0)$; $E_3=2\pi\sigma, x_3>0$; $E_3=-2\pi\sigma, x_3<0$.</td></tr>
<tr><td>$\psi(x)=constant$</td><td>$e_{23}=\beta/\lambda$ [31, 32]</td></tr>
<tr><td colspan="2">Constant electric field of the point electric charge q.
$\vec{E}=E(\vec{r}/r)=(q/r^2)(\vec{r}/r)$; $\vec{H}=(0,0,0)$; $r^2=x_1^2+x_2^2+x_3^2$.</td></tr>
<tr><td>$\psi(x)=[q/(q+(B^2-1)x_1E]^{3/2}B\psi(0)$
$B^2=(1-\beta^2/\lambda^2)^{-1}$</td><td>$e_{23}=\beta/\lambda$ [31, 32]
$e_{32}=-\beta/\lambda$</td></tr>
<tr><td colspan="2">Constant magnetic field generated by constant current with density $\vec{j}=(0, I\delta(x_1)/c, 0)$ flowing on the plane (x_2, x_3) along x_2.
$\vec{E}=(0,0,0)$; $\vec{H}=(0,0,H_3)$; $H_3=-2\pi I/c, x_1>0$; $H_3=2\pi I/c, x_1<0$.</td></tr>
<tr><td>$\psi(x)=exp\left[\left(1+\frac{cH_3}{2\pi I}\right)\frac{\lambda-1+\beta h_{23}}{\lambda+\beta h_{23}}\right]$</td><td>$h_{23}=-\beta/\lambda$ [31, 32]</td></tr>
<tr><td colspan="2">Free electromagnetic field, transversal plane waves,
$\varrho(x)=\mu(x)=0$; $\vec{l}=\vec{n}\times\vec{m}$; $k=\omega/c$; $k_4=ik$; ω -frequency.
$\vec{E}=(m_1,m_2,m_3)\varphi$; $\vec{H}=(l_1,l_2,l_3)\varphi$; $\varphi=exp(-ik\cdot x)=exp[-i(k\vec{n}\cdot\vec{x}+k_4x_4)]$</td></tr>
<tr><td>$\psi(x)=(EH)^{\frac{1-\lambda}{2\lambda}}\cdot$
$\cdot exp\left[i\frac{\beta}{\lambda}(kx_1+n_1k_4x_4)\right]$</td><td>$e_{23}=\beta+n_3m_1(\lambda-1+\beta n_1)/(1-n_1^2)m_3$ [30]
$e_{32}=-\beta-n_2m_1(\lambda-1+\beta n_1)/(1-n_1^2)m_2$
$h_{23}=(\lambda-1)m_2/l_3+n_2m_1[n_1(\lambda-1)+\beta]/(1-n_1^2)l_3$
$h_{32}=(\lambda-1)m_3/l_2+n_3m_1[n_1(\lambda-1)+\beta]/(1-n_1^2)l_2$</td></tr>
<tr><td colspan="2">Massless fields of electromagnetic charges, longitudinal plane waves, $\varrho(x)=\varrho_0 exp(-ik\cdot x)$; $\mu(x)=\mu_0 exp(-ik\cdot x)$
$\vec{E}=(4\pi/\omega)\vec{n}c\varrho_0 exp-i(k\cdot x-\pi/2)$; $\vec{H}=(4\pi/\omega)\vec{n}c\mu_0 exp-i(k\cdot x-\pi/2)$</td></tr>
<tr><td>$\psi(x)=\left(-\frac{k^2EH}{16\pi^2}\right)^{\frac{n_1-\beta+n_1\lambda^2}{2n_1\lambda^2}}\cdot$
$\cdot exp\left[i\frac{\beta}{\lambda}(kx_1+n_1k_4x_4)\cdot\right.$
$\left.\cdot\frac{n_1-\beta}{n_1\lambda}\right]$</td><td>$e_{23}=(\beta n_2/n_3(n_1-\beta))(H/E)$
$e_{32}=(\beta n_3/n_2(n_1-\beta))(H/E)$
$h_{23}=(\beta n_2/n_3(n_1-\beta))(E/H)$
$h_{32}=(\beta n_3/n_2(n_1-\beta))(E/H)$</td></tr>
</table>

4. Conclusion. The obtained results allow us to formulate the following. Theorem 3: The Galilei group is the symmetry group of Maxwell equations if the fields are transformed according to a non-linear representation of this group.

The Galilei symmetry differs from the relativistic one since the latter postulates the invariance of the speed of light whereas the Galilei symmetry manifestly postulates the invariance of time. In the relativistic case the transformations of fields are linear and global, in the Galilei case they are nonlinear and generally explicitly depend on the position and time and in this sense are local.

The author is deeply grateful to Profs. D.P. Grechukhin, V.I. Manko and V.I. Fushchich for discussing various fragments of this paper and valuable comments.

References

1. Manko, V.I. (1972). Synopsis of Ph.D. thesis, Phys. Inst. of USSR Academy of Sciences, Moscow, pp. 5-6. (Russian).
2. Fushchich, V.I. (1971). Theor. Math. Phys. 7, 3-12. (Russian).
3. Malkin, I.A., Manko, V.I. (1979). Dynamic Symmetries and Coherent States of Quantum System. Nauka, Moscow, pp. 17, 566-568(Russian).
4. Danilov, Yu.A. (1967). Preprint IAE-1452, Moscow, 12 p. (Russian).
5. Ibragimov, N.Kh. (1968). Dokl. Akad. Nauk USSR 178, 566-568 (Russian).
6. Fushchich, V.I. (1974). Lett. Nuovo Cim. 11, 508-512.
7. Kotelnikov, G.A. (1982). Nuovo Cim. 72B, 68-78.
8. Strazhev, V.I., Pletyuhov, V.A. (1981). Izv. Vuzov USSR, Physics 12, 39-42. (Russian).
9. Loide, R.K. (1983). Izv. Vuzov USSR, Physics 8, 111-112.(Russian).
10. Fushchich, V.I., Nikitin, A.G. (1983). Symmetry of Maxwell Equations. Naukova Dumka, Kiev, pp.7, 17, 82-85, 116-125. (Russian).

11. Konopelchenko , B.G. (1977). Phys. Elem. Part. Atom. Nucl. (USSR), 8, 135-174. (Russian).

12. Kotelnikov, G.A. (1980). Theor. Math. Phys. 42, 139-144. (Russian).

13. Kotelnikov, G.A. (1977). Preprint IAE-2813, Moscow, 17 p. (Russian).

14. Kotelnikov, G.A. (1981). 5th Soviet Conference on Gravitation. Moscow State University, Moscow, p.99. (Russian).

15. Kotelnikov, G.A. (1981). Izv. Vuzov USSR, Physics 10, 46-51. (Russian).

16. Romain, J.E. (1963). Nuovo Cim. 30, 1254-1271.

17. Kotelnikov, G.A. (1970). Vestnik of Moscow State University, Phys.-Astron. 4, 371-374. (Russian).

18. Sjödin, T. (1979). Nuovo Cim. 51B, 229-246.

19. Di Jorio, M. (1974). Nuovo Cim. 22B, 70-78.

20. Voigt, W. (1887). Nachr. Ges. Wiss. Göttingen 2, 41-52.

21. Umov, N.A. (1950). Selected Works, Gostekhizdat, Moscow-Leningrad, pp.492-499. (Russian).

22. Kotelnikov, G.A. (1973). Preprint IAE-2306, Moscow, 8 p.; (1974). Preprint IAE-2475, Moscow, 20 p. (Russian).

23. Ives, H.E. (1937). J. Opt. Soc. Amer. 27, pp.389-392.

24. Palacios, J. (1957). Rev. Akad. Ci. Madrid 51, pp.21-101.

25. Gordon, C.N. (1962). Proc. Phys. Soc. 80, pp.569-592.

26. Dewan, E.M. (1961). Nuovo Cim. 22, pp.941-957.

27. Podlaha, M. (1969). Nuovo Cim. 64B, pp.181-187.

28. Hsu, J.P. (1976). Found. Phys. 6, pp.317-339.

29. Foldy, L.L. (1956). Phys. Rev. 102, pp.568-581.

30. Kotelnikov, G.A. (1985). In: Group Theoretical Methods in Physics. Markov, M.A., Manko, V.I., Shabad, A.E. (Ed.), 1, Harwood Acad. Publ., Chur et al, pp. 507-516, 521-535.

31. Kotelnikov, G.A. (1984). In: VANT, ser. OYaPh 4(29) CNRIAtominform, pp.17-18.

32. Kotelnikov, G.A. (1985). Izv. Vuzov USSR, Physics 8, 78-83 (Russian).

33. Hagen, C.R. (1972). Phys. Rev. D 5, 377-388.

34. Niederer, V. (1972). Helv. Phys. Acta 45, 802-810.

35. Hirata, K. (1982). Phys. Lett. 112B, 458-462.

36. Einstein, A. Physics and Reality, Nauka, Moscow (Russian).

CONFORMALLY INVARIANT NONLINEAR EQUATIONS OF ELECTROMAGNETIC FIELD

W.I.Fushchich, I.M.Zifra

The Ukrainian SSR Academy of Sciences Mathematical Institute

As it is known, Maxwell's equations alone

$$L_1 = \frac{\partial F_{\mu\nu}}{\partial x_\alpha} + \frac{\partial F_{\nu\alpha}}{\partial x_\mu} + \frac{\partial F_{\alpha\mu}}{\partial x_\nu} = 0,$$

$$L_2 = \frac{\partial \tilde{H}_{\mu\nu}}{\partial x_\alpha} + \frac{\partial \tilde{H}_{\nu\alpha}}{\partial x_\mu} + \frac{\partial \tilde{H}_{\alpha\mu}}{\partial x_\nu} = 0 \qquad (0.1)$$

are insufficient to determine electromagnetic field in various media. Beside equations (0.1) the constitute equations are imposed on D, B, E and H. We select these additional relations from symmetry's considerations.

1. Symmetry of equations (0.1). The system (0.1) is highly under-definite system of first-order partial differential equations. Therefore one should expect that (0.1) has greater symmetry than Maxwell's equations in vacuum.

Theorem 1. The system (0.1) is invariant with respect to the infinite-dimensional algebra with basis:

$$X_1 = \xi^\mu(x)\frac{\partial}{\partial x_\mu} + \eta_{F_{\mu\nu}}\frac{\partial}{\partial F_{\mu\nu}} + \eta_{\tilde{H}_{\mu\nu}}\frac{\partial}{\partial \tilde{H}_{\mu\nu}}, \qquad (1.1)$$

$$X_2 = F_{\mu\nu}\frac{\partial}{\partial F_{\mu\nu}} \equiv F_{01}\frac{\partial}{\partial F_{01}} + F_{02}\frac{\partial}{\partial F_{02}} + F_{03}\frac{\partial}{\partial F_{03}} + F_{12}\frac{\partial}{\partial F_{12}} + F_{13}\frac{\partial}{\partial F_{13}} + F_{23}\frac{\partial}{\partial F_{23}}; \qquad (1.2)$$

$$X_3=\tilde{H}_{\mu\nu}\frac{\partial}{\partial\tilde{H}_{\mu\nu}}\equiv\tilde{H}_{01}\frac{\partial}{\partial\tilde{H}_{01}}+\tilde{H}_{02}\frac{\partial}{\partial\tilde{H}_{02}}+\tilde{H}_{03}\frac{\partial}{\partial\tilde{H}_{03}}+\tilde{H}_{12}\frac{\partial}{\partial\tilde{H}_{12}}+\tilde{H}_{13}\frac{\partial}{\partial\tilde{H}_{13}}+\tilde{H}_{23}\frac{\partial}{\partial\tilde{H}_{23}}, \qquad (1.3)$$

$$X_4=F_{\mu\nu}\frac{\partial}{\partial\tilde{H}_{\mu\nu}}\equiv F_{01}\frac{\partial}{\partial\tilde{H}_{01}}+F_{02}\frac{\partial}{\partial\tilde{H}_{02}}+F_{03}\frac{\partial}{\partial\tilde{H}_{03}}+F_{12}\frac{\partial}{\partial\tilde{H}_{12}}+F_{13}\frac{\partial}{\partial\tilde{H}_{13}}+F_{23}\frac{\partial}{\partial\tilde{H}_{23}}, \qquad (1.4)$$

$$X_5=\tilde{H}_{\mu\nu}\frac{\partial}{\partial F_{\mu\nu}}\equiv\tilde{H}_{01}\frac{\partial}{\partial F_{01}}+\tilde{H}_{02}\frac{\partial}{\partial F_{02}}+\tilde{H}_{03}\frac{\partial}{\partial F_{03}}+\tilde{H}_{12}\frac{\partial}{\partial F_{12}}+\tilde{H}_{13}\frac{\partial}{\partial F_{13}}+\tilde{H}_{23}\frac{\partial}{\partial F_{23}}, \qquad (1.5)$$

where $\xi^{\mu}(x)$ are arbitrary differentiable functions

$$\eta^{F_{\mu\nu}}=-F_{\mu\alpha}\xi^{\alpha}_{\nu}-F_{\alpha\nu}\xi^{\alpha}_{\mu},\quad \eta^{\tilde{H}_{\mu\nu}}=-\tilde{H}_{\mu\alpha}\xi^{\alpha}_{\nu}-\tilde{H}_{\alpha\nu}\xi^{\alpha}_{\mu}. \qquad (1.6)$$

Proof is reduced to applying Lie's algorithm [1] to (0.1).

Theorem 2. (Corollary of Theorem 1). The system (0.1) is invariant with respect to the 20-dimensional Lie algebra of IGL(4,R) containing the Poincaré and Galilei algebras as subalgebras.

Basis elements of the Poincaré algebra P(1.3) are

$$P_{\mu}=ig_{\mu\nu}\frac{\partial}{\partial x_{\nu}},\quad J_{\mu\nu}=x_{\mu}P_{\nu}-x_{\nu}P_{\mu}+(S_{\mu\nu}\Psi)_n\frac{\partial}{\partial\Psi_n}, \qquad (1.7)$$

where n runs 1 to 12, $\Psi=$ column $(\vec{E},\vec{B},\vec{D},\vec{H})$

$$S_{ab}=\begin{pmatrix}\hat{S}_{ab} & \hat{0} & \hat{0} & \hat{0}\\ \hat{0} & \hat{S}_{ab} & \hat{0} & \hat{0}\\ \hat{0} & \hat{0} & \hat{S}_{ab} & \hat{0}\\ \hat{0} & \hat{0} & \hat{0} & \hat{S}_{ab}\end{pmatrix},\quad S_{0a}=\varepsilon_{abc}\begin{pmatrix}\hat{0} & \hat{0} & \hat{0} & \hat{S}_{bc}\\ \hat{0} & \hat{0} & -\hat{S}_{bc} & \hat{0}\\ \hat{0} & \hat{S}_{bc} & \hat{0} & \hat{0}\\ -\hat{S}_{bc} & \hat{0} & \hat{0} & \hat{0}\end{pmatrix} \qquad (1.8)$$

$$\hat{S}_{12}=\begin{pmatrix}0 & -i & 0\\ i & 0 & 0\\ 0 & 0 & 0\end{pmatrix},\quad S_{23}=\begin{pmatrix}0 & 0 & 0\\ 0 & 0 & -i\\ 0 & i & 0\end{pmatrix},\quad \hat{S}_{31}=\begin{pmatrix}0 & 0 & i\\ 0 & 0 & 0\\ -i & 0 & 0\end{pmatrix}$$

and $\hat{0}$ is the 3×3 zero matrix. The generators of the Galilei group are

$$P_0 = i\frac{\partial}{\partial x_0}, \; P_a = -i\frac{\partial}{\partial x_a}, \quad a = 1,2,3,$$

$$J_{ab} = x_b P_b - x_b P_a + (S_{ab}\Psi)_n \frac{\partial}{\partial \Psi_n}, \; G_a = tP_a + (M_a\Psi)_n \frac{\partial}{\partial \Psi_n}, \tag{1.9}$$

where

$$M_a = \frac{1}{2}\varepsilon_{abc}\begin{pmatrix} \hat{0} & \hat{S}_{bc} & \hat{0} & \hat{0} \\ \hat{0} & \hat{0} & \hat{0} & \hat{0} \\ \hat{0} & \hat{0} & \hat{0} & \hat{0} \\ \hat{0} & \hat{0} & -\hat{S}_{bc} & 0 \end{pmatrix}.$$

There are generators of the conformal group $C(1,3)$ among operators (1.1). The operators (1.7) together with

$$D = x_\nu P^\nu + 2i\Psi_n \frac{\partial}{\partial \Psi_n}, \; K_\mu = 2x_\mu D - (x_\nu x^\nu)P_\mu + 2(x^\nu S_{\mu\nu}\Psi)_n \frac{\partial}{\partial \Psi_n}, \tag{1.10}$$

form the basis of the conformal algebra $C(1.3)$.

Thus we have established that system (0.1) without constitutive equations satisfy the Lorentz-Poincaré-Einstein principle of relativity as well as the Galileian one.

2. Poincare-invariant and conformally invariant nonlinear constitutive equations. Consider constitutive equations of the following form:

$$H_{\mu\nu} = \Phi_{\mu\nu}(F_{01}, \dots, F_{23}) \equiv \Phi_{\mu\nu}(F), \tag{2.1}$$

where $\Phi_{\mu\nu}$ are arbitrary smooth (differentiable) functions $\Phi_{\mu\nu} = -\Phi_{\nu\mu}$, $\Phi_{\mu\mu} = 0$, Using Lie's method we prove the following.

Theorem 3. The system (0.1), (2.1) is invariant with respect to the $P(1.3)$ if and only if

$$H_{\mu\nu} = MF_{\mu\nu} + N\tilde{F}_{\mu\nu}, \tag{2.2}$$

where $M = M(C_1, C_2)$, $N = N(C_1, C_2)$ are arbitrary smooth functions in the invariants of electromagnetic field

$$C_1 = -\frac{1}{4} F_{\mu\nu} F^{\mu\nu} = \vec{E}^2 - \vec{B}^2, C = -\frac{1}{4} \varepsilon_{\alpha\beta\mu\nu} F^{\alpha\beta} F^{\mu\nu} = \vec{B}\vec{E},$$

where $\tilde{F}_{\mu\nu} = \varepsilon_{\mu\nu\alpha\beta} F^{\alpha\beta}$, $\tilde{H}_{\mu\nu} = \varepsilon_{\mu\nu\alpha\beta} H^{\alpha\beta}$.

In terms of D,B,E,H (2.2) takes the form:

$$\vec{D} = M\vec{E} + N\vec{B}, \vec{H} = M\vec{B} - N\vec{E}. \tag{2.3}$$

If $M = 1/L$, $N = \vec{B}\vec{E}/L$, $L = \sqrt{1 + (\vec{B}^2 - \vec{E}^2) - (\vec{B}\,\vec{E})^2}$ then (2.3) together with (0.1) coincide with the nonlinear equations of electromagnetic field proposed by Born [3] (Born-Infeld equations).

Consider a special case of constitutive equations:

$$\vec{D} = \varepsilon(\vec{E}, \vec{H}) \vec{E}, \vec{B} = \mu(\vec{E}, \vec{H}) \vec{H}. \tag{2.4}$$

<u>Corollary</u> (of Theorem 3). The system (0.1), (2.4) is Poincaré-invariant only if

$$\varepsilon(\vec{E}, \vec{H}) \cdot \mu(\vec{F}, \vec{H}) = 1$$

(we use the system of units in which the speed of light $c = 1$).

<u>Theorem</u> 4. The system (0.1), (2.2) is invariant with respect to conformal group $C(1.3)$, if

$$M = M\left(\frac{C_1}{C_2}\right), \quad N = N\left(\frac{C_1}{C_2}\right), \tag{2.5}$$

where M,N are arbitrary smooth functions depending on the ratio of C_1, C_2 only.

If we introduce the vector-potential as usually the system (0.1), (2.5) takes the form:

$$\Box A_\mu - \partial_\mu (\partial_\nu A^\nu) = \lambda \partial^\nu \left\{ \Phi\left(\frac{I_1}{I_2}\right) \right\} (\partial_\mu A_\nu - \partial_\nu A_\mu), \tag{2.6}$$

where $I_2 = -\frac{1}{4}(\partial_\mu A_\nu - \partial_\nu A_\mu)(\partial^\mu A^\nu - \partial^\nu A^\mu)$, $I_1 = -\frac{1}{4}\varepsilon_{\alpha\beta\mu\nu}(\partial^\alpha A^\beta - \partial^\beta A^\alpha)(\partial^\mu A^\nu - \partial^\nu A^\mu)$;

Φ is an arbitrary smooth functions of real variable.

The system (2.6) is invariant with respect to the conformal group and the gauge transformations. One can use conformal symmetry to find exact solutions of the nonlinear equations

$$i\{\gamma_\mu \partial^\mu + \lambda (F_{\mu\nu} F^{\mu\nu})^{-1/4} S_{\mu\nu} F^{\mu\nu}\} \psi = 0 \tag{2.7}$$

in combinations with (0.1) and (2.5), describing interaction of spinor and electromagnetic fields. The solutions of the system (0.1), (2.5), (2.7) are sought in the form

$$F_{\mu\nu} = \frac{f_{\mu\nu}}{(x^2)^2} - 2\,\frac{(f_{\mu k} x_\nu + f_{k\nu} x_\mu) x^k}{(x^2)^3}, \tag{2.8}$$

$$\tilde{H}_{\mu\nu} = \frac{\tilde{h}_{\mu\nu}}{(x^2)^2} - 2\,\frac{(\tilde{h}_{\mu k} x_\nu - \tilde{h}_{k\nu} x_\mu) x^k}{(x^2)^3}, \tag{2.9}$$

$$\psi = \frac{\gamma x}{(x^2)^2}\,\varphi(\omega), \quad \omega = \frac{\beta x}{x^2}, \tag{2.10}$$

where $f_{\mu\nu}$, $\tilde{h}_{\mu\nu}$ depend on ω, β_μ are arbitrary real constants. We get partial exact solutions of the system (0.1), (2.5), (2.7) with the help of the formulas (2.8)-(2.10):

$$F_{\mu\nu} = \frac{b_{\mu\nu}}{(x^2)^2} - 2\,\frac{(b_{\mu k} x_\nu + b_{k\nu} x_\mu) x^k}{(x^2)^3},$$

$$\tilde{H}_{\mu\nu} = \frac{\tilde{c}_{\mu\nu}}{(x^2)^2} - 2\,\frac{(\tilde{c}_{\mu k} x_\nu + \tilde{c}_{k\nu} x_\mu) x^k}{(x^2)^3}, \tag{2.11}$$

$$\psi = \frac{\gamma x}{(x^2)^2} \exp\left\{-i\,\frac{(\gamma\beta)(S_{\mu\nu} b^{\mu\nu})}{\beta^2 (b_{\mu\nu} b^{\mu\nu})^{1/4}}\right\} \chi$$

where $b_{\mu\nu}$, $\tilde{c}_{\mu\nu}$ are tensor constants, χ is a constant spinor.

References

1. Ovsyannikov,L.V. (1978). The group analysis of different equations. Nauka, Moscow (Russian).

2. Fushchich,W.I., Nikitin,A.G. (1983). Symmetry of Maxwell equations. Naukovă Dumka , Kiev (Russian).

3. Born,M., Infeld,L. (1934). Roy. Soc. A 144, 852, 4225-51.

A NONASSOCIATIVE EXTENSION OF THE DIRAC MATRIX ALGEBRA

J.Lõhmus, L.Sorgsepp[†]

Institute of Physics, Academy of Sciences of the Estonian SSR, Tartu

[†]Institute of Astrophysics and Atmospheric Physics, Academy of Sciences of the Estonian SSR, Tartu

1. In the treatments of relativistic-invariant equations usually the algebras representable by matrices (i.e. the associative and Lie algebras) are exploited. However, several attempts have been made [1-9] to use the nonassociative algebra of octonions for these purposes (see e.g. [10, 11]). The preference of octonions is comprehensible from the standpoint of both mathematics and physics. In mathematics, it is well-known that the four algebras: of real and complex numbers, quaternions, and octonions are the only division algebras over the real numbers. In physics, the majority of relativistic-invariant equations may be formulated in terms of quaternions (see e.g. [1]). The use of quaternions does not lead to any new result except for an elegant formulation of equations. While the octonions are the direct generalization of quaternions, there is a hope to get some new results by formulating relativistic-invariant equations in terms of octonions.

So far [1-9] these attempts stemed from some presupposed correspondence between octonion components and physical quantities (fields). However, there has been no systematic use of the theory of bimodule representation of nonassociative algebras [12]. In some papers (not dealing with relativistic-invariant equations), the Hilbert spaces with octonionic scalars [13] and the Clifford algebra C_7

(corresponding to octonions) [14, 15] have been investigated.

2. For an arbitrary linear finite dimensional algebra A the following simple construction may be carried through. Let us introduce the left and right multiplications and the corresponding matrices L_a and R_a:

$$L_a x = ax \text{ for any } a,\ x \in A,$$
$$R_a x = xa \text{ for any } a,\ x \in A. \tag{1}$$

The matrices L_a, R_a act upon the elements of A as vectors of the linear space of the algebra. They form the associative multiplication algebra of A. This algebra may be considered as the simplest, regular bimodule representation of the (nonassociative) algebra A (for our purposes it suffices to confine to this representation and we do not give any general definition [12] (see also [10, 11]). In such representation the nonassociativity of A leads to the noncommutativity of L- and R-matrices:

$$\{a,\ x,\ b\} = (ax)b - a(xb) = [R_b,\ L_a]x \text{ for any } a,\ b,\ x \in A. \tag{2}$$

The representation of this type is not a homomorphism: if we take the correspondences $a \to L_a$, R_a, $b \to L_b$, R_b we generally do not get $ab \to L_{ab} = L_a L_b$, $ab \to R_{ab} = R_a R_b$. In the case of associative algebras $[L,\ R] = 0$, and we have representation in the usual sense. For the alternative algebras (including octonions) we get the following correspondences:

$$ab \to L_{ab} = L_a L_b + [L_a,\ R_b],$$
$$ab \to R_{ab} = R_a R_b + [R_a,\ L_b].$$

3. Let us proceed from the Dirac equation with the electromagnetic interaction term and γ-matrices in the representation where the helicity operator γ^5 is diagonal (see e.g. [16]):

$$\left[\frac{1}{c}\frac{\partial}{\partial t} - \frac{ie}{\hbar c} A_o + \alpha^k\left(\frac{\partial}{\partial x_k} + \frac{ie}{\hbar c} A_k\right) + \frac{imc}{\hbar}\beta\right]\Psi = 0, \tag{3}$$

or

$$-i\hbar\gamma\mu \frac{\partial\Psi}{\partial x\mu} + \frac{e}{c}\gamma\mu A_\mu\Psi - imc\Psi = 0, \qquad (3')$$

where $x_4 = ix_o = ict$, $iA_o = A_4$; $\gamma^4 = \beta$, $\gamma^k = -i\beta\alpha^k$, $\alpha^k = i\gamma^4\gamma^k$, $k = 1, 2, 3$.

The Dirac matrices $\gamma\mu$ satisfy the relations

$$\{\gamma^\mu, \gamma^\nu\} = \gamma^\mu\gamma^\nu + \gamma^\nu\gamma^\mu = 2\delta_{\mu\nu}I; \quad \mu,\nu = 1,2,3,4.$$

For the extension of the matrix structure of the Dirac equation we use the R-matrices of the regular bimodule representation of the octonion algebra, defined as

$$R_i\underline{x} = \underline{xe_i}, \quad i = 1,2, \ldots, 7, \ (R_o = I), \qquad (4)$$

where $\underline{x}$ and $\underline{xe_i}$ are columns (8-vectors) corresponding to the octonions x and xe_i, respectively, e_i are octonionic units, for these we use the multiplication table with positive cycles 123, 154, 176, 246, 275, 347, 356 (this is the table $\tilde{5}$ in [10]).

R-matrices satisfy the following equations

$$\{R_i, R_j\} = -2\delta_{ij}I; \quad i,j = 1,2, \ldots, 7; \quad R_1R_2R_3R_4R_5R_6R_7 = 1$$

Now for the role of Dirac matrices we can choose the following triple products of R-matrices

$$"\gamma^1" = R_3R_2R_6, \quad "\gamma^2" = R_3R_2R_7, \quad "\gamma^3" = R_3R_2R_4, \quad "\gamma^4" = R_3R_2R_5$$

(in the sequel we omit the quotation marks). To motivate such a choice note that the matrices "$\gamma\mu$", $\mu = 1, 2, 3, 4$, give us the 8-dimensional real representation which coincides with the real representation obtained with the help of the usual Dirac matrices by replacing 1 and i by $\begin{pmatrix}1 & 0\\ 0 & 1\end{pmatrix}$ and $\begin{pmatrix}0 & -1\\ 1 & 0\end{pmatrix}$ respectively. The choice of the particular multiplication table is due to the (perhaps not obligatory) requirement that in the equation only single R-matrices and not their products will enter coefficients of the derivatives.

In our formalism the imaginary unit is: $i = R_2R_3$ (except the i that enters the time components of relativistic 4-vectors) The matrix $\gamma^5 = \gamma^1\gamma^2\gamma^3\gamma^4$ is $R_1R_2R_3$. Also

$\alpha^1=R_1R_4R_7$, $\alpha^2=R_1R_4R_6$, $\alpha^3=R_1R_6R_7$.

Let us write now Eq. (3'), using the new matrices:

$$\hbar\left(R_6\frac{\partial\psi}{\partial x_1}+R_7\frac{\partial\psi}{\partial x_2}+R_4\frac{\partial\psi}{\partial x_3}+R_5\frac{\partial\psi}{\partial x_4}\right)$$
$$+\frac{e}{c}R_2R_3\left(R_6A_1+R_7A_2+R_4A_3+R_5A_4\right)\psi+mcR_2R_3\psi=0. \tag{5}$$

The wave function here is an 8-component column consisting of coefficients of the real and imaginary parts of the components of the Dirac bispinor.

The information about reflection operators P, C, T in terms of new matrices may be resumed as follows:

$$P=\gamma^4=R_3R_2R_5,\quad C=C'f=R_1R_2R_5,\quad C'=-\gamma^2=R_2R_3R_7,$$
$$f=R_2R_6R_4,\quad T=T'f=-R_2,\quad T'=\gamma^3\gamma^1=R_6R_4, \tag{6}$$

where f is the 8x8-matrix obtained from a 4x4 unit matrix by replacing $1\to\begin{pmatrix}1&0\\0&-1\end{pmatrix}$ on the main diagonal. This matrix represents the operation of complex conjugation included in the operators C, T of charge conjugation and time reversal. It is useful also to consider the following products

$$PC'T'=\gamma^5=R_1R_2R_3,\quad PCT=-i\gamma^5=\alpha^1\alpha^2\alpha^3=R_1.$$

4. Let us now manipulate with Eq. (5). At first, we may symmetrize the electromagnetic interaction term, taking for each triple product of R-matrices the two more products obtained from the initial one by the cyclic permutation of indices. All these three terms are equal. To preserve the equation we multiply the electromagnetic interaction term by $^1/_3$:

$$\ldots+\frac{1}{3}\frac{e}{c}\big[(R_2R_3R_6+R_3R_6R_2+R_6R_2R_3)A_1$$
$$+(R_2R_3R_7+R_3R_7R_2+R_7R_2R_3)A_2+$$
$$+(R_2R_3R_4+R_3R_4R_2+R_4R_2R_3)A_3+$$
$$+(R_2R_3R_5+R_3R_5R_2+R_5R_2R_3)A_4\big]\psi+\ldots \tag{7}$$

Eq. (5) with the symmetrized electromagnetic interaction term (7) may be reformulated into an octonionic equation with the help of Eq. (4):

$$\hbar\left(\frac{\partial\psi}{\partial x_1}e_6+\frac{\partial\psi}{\partial x_2}e_7+\frac{\partial\psi}{\partial x_3}e_4+\frac{\partial\psi}{\partial x_4}e_5\right)+$$
$$+\frac{1}{3}\cdot\frac{e}{c}\Big[\big(\sum_{cycl}((\psi e_6)e_3)e_2\big)A_1+\big(\sum_{cycl}((\psi e_7)e_3)e_2\big)A_2+ \tag{8}$$
$$+\big(\sum_{cycl}((\psi e_4)e_3)e_2\big)A_3+\big(\sum_{cycl}((\psi e_5)e_3)e_2A_4\Big]+$$
$$+mc(\psi e_3)e_2=0.$$

Here ψ is the octonion wave function with components equal the coordinates of the 8-vector of the wave function in Eq. (5).

Consider now all possible placings of parentheses in Eq. (8). It gives Eq. (8) the maximal generality with respect to the rearrangements of parentheses but threatens with serious changes of the equation. A more detailed analysis, however, shows that only the last term changes.

Let us make all possible rearrangements of parentheses in the expression $((\psi e_6)e_3)e_2$ at the field component A_1:

$$((\psi e_6)e_3)e_2+((\psi e_6)e_3)e_2+(\psi(e_6e_3))e_2+(\psi e_6)(e_3e_2)+ \tag{9}$$
$$+\psi((e_6e_3)e_2)+\psi(e_6(e_3e_2)).$$

Obviously, the two last terms cancel out. Considering the whole sum over the cycles (3 terms), we see that all the terms with replaced parentheses cancel out and there remains only the initial one. Thus the electromagnetic interaction term remains unchanged.

The mass term in Eq. (8) turns into

$$mc\,[(\psi e_3)e_2+\psi(e_3e_2)]=mc\,[(\psi e_3)e_2-\psi e_1]. \tag{10}$$

After all possible rearrangements of parentheses we can write Eq. (8) with mass term (10) again in terms of R-matrices, it will be now "more octonionic" than the initial one (Eq. (5)):

$$\hbar\left(R_6\frac{\partial\psi}{\partial x_1}+R_7\frac{\partial\Psi}{\partial x_2}+R_4\frac{\partial\Psi}{\partial x_3}+R_5\frac{\partial\psi}{\partial x_4}\right)+$$
$$+\frac{1}{3}\cdot\frac{e}{c}\{\text{Eq. }(7)\}\psi+mcR_2R_3(1+R_1R_2R_3)\psi=0 \tag{11}$$

Then we multiply (from the left) Eq.(11) by $1-\gamma^5=1-R_1R_2R_3$. The matrix $R_1R_2R_3$ anticommutes with the matrices R_6, R_7, R_4, R_5 and commutes with R_2R_3; now, in the first two terms, the multiple $1-R_1R_2R_3$ may be transferred from the left to the right to the wave function ψ; the multiple itself changes into $1+R_1R_2R_3$. Now the expression $(1+\gamma^5)\Psi$ may be considered as a new wave function. The mass term of Eq.(11) multiplied by $1-\gamma^5$ changes as follows:

$$mc(1-R_1R_2R_3)R_2R_3(1+R_1R_2R_3)\psi=$$

$$=mcR_2R_3(1-R_1R_2R_3)(1+R_1R_2R_3)\psi=0$$

Thus, the requirement of all possible rearrangements of parentheses in the octonionic equation (8) leads to the vanishing of the mass term for one chiral projection, the other remaining intact.

In formula (7), there is a hint at a possibility to introduce a fractional charge e/3. It is possible only if we consider it, not t, as a true time variable. The quantity it does not change under any reflection (6), it is the irreversible time, the time variable t only characterizes the direction of motion.

5. Concluding remarks. We start with the usual Dirac equation with Dirac 4x4-matrices generating the Clifford algebra C_4 (with 16 basis elements). A passing over to R-matrices extends the Clifford algebra to C_6 (with 64 basis elements). For example, here opens the possibility to introduce the complex conjugation operator. If we take also L-matrices (1) (analoguously to Eq.(4)) the Clifford algebra extends to the C_7 [13-15] (see also [10]). The R- and L-matrices complement each other in the bimodule representation of octonions. The L-matrices could have appeared in the additional terms in Eqs.(8), (9), which originated from the rearrangement of parentheses. These terms cancelled out and we did not pay attention to them any more. We guess that the L-matrices are required if we

want to describe some internal degrees of freedom (e.g. the colour). In the full bimodule representation R-matrices are connected with the space-time, L-matrices, with an internal space. Nonzero commutators of type (2) (measuring nonassociativity!) are connecting these spaces.

References

1. Bourret,R. (1959). Canad. J. Phys. 37, 183.
2. Penney,R. (1968). Amer. J. Phys. 36,871; (1971). Nuoovo Cim., 3B, 95.
3. Buoncristiani,A.M.-J. (1973). J. Math. Phys. 14,849; (1974). Util. Math., 6, 23.
4, Nahm,W. (1978). Preprint TH 2489, CERN, Geneva.
5. Konstein,S.E. (1978). Short reports on physics, 1,12 (Russian).
6. Okubo,S., Tosa,Y. (1979). Phys. Rev. D20, 462.
7. Oliveira,C.G., Maia,M.D. (1979). J. Math. Phys. 20, 923.
8. Petri,J. (1979). Bull. cl.sci.Acad. roy. Belg.65,6
9. Bogush,A.A., Kurochkin,Yu.A. (1980). In: Group-theoretical methods in physics. Nauka, Moscow, 2, 42.
10. Sorgsepp,L., Lohmus,J. (1978). Preprint F-7, Tartu; (1979). Hadronic J., 2, 1388.
11. Lohmus,J.,Sorgsepp,L. (1985). Preprint F-24,25, Tartu.
12. Schafer,R.D. (1966). An introduction to nonassociative algebras. N.Y.-London, Academic Press.
13. Goldstine,H.H., Horwitz,L.P. (1962). Proc. Nat. Acad. Sci. (USA), 48, 1134; (1964). Math. Ann. 154,1; (1966), 164, 291.
14. Basri,S.A., Horwitz,L.P. (1975). Phys. Rev. D11,572.
15. Biedenharn,L.C., Horwitz,L.P. (1979). In: Quantum Theory and the Structure of Time and Space, vol. 3 (ed. by L.Castell et al.); München: Carl Hanser Verlag, p. 61.
16. Pauli,W. (1933). In: Handbuch der Physik 2-te Aufl. Hrsg. von Geiger,H.,Scheel,K. Bd. 24,t. 1. Springer, Berlin.

Part II
REPRESENTATION THEORY

CANONICAL BASIS IN IRREDUCIBLE REPRESENTATIONS OF gl_3 AND ITS APPLICATIONS.

I.Gelfand, A.Zelevinsky
Moscow State University, USSR

In this paper we define and study the new basis in irreducible finite dimensional representations of the Lie algebra gl_3, which we call canonical. This basis has some advantages against the commonly used Gelfand-Tsetlin basis. The Gelfand-Tsetlin basis is defined by means of the restriction of representations of gl_3 to the chain of subalgebras $gl_1 \subset gl_2 \subset gl_3$. But this chain depends on the arbitrary choice of ordering of the basis vectors e_1, e_2, e_3 in the standard representation of gl_3; another choice leads to another Gelfand-Tsetlin basis whose relation to the first basis is rather complicated (see [6]). Meantime, the canonical basis does not depend on this arbitrary choice. (In more invariant terms, the canonical basis depends only on the choice of Cartan and Borel subalgebras in gl_3 whereas to define the Gelfand-Tsetlin basis we need in addition an ordering of simple roots). We will see that the canonical basis is related to the standard Gelfand-Tsetlin basis as well as to the opposite Gelfand-Tsetlin basis by a simple triangular transformation; so in some sense it lies "in the middle" between these two bases.

Another advantage of the canonical basis is that the action of the generators of gl_3 in it is given by integral matrices, and the matrix elements have no denominators. Besides, the canonical basis has a beautiful polynomial realization.

The canonical basis has an unexpected application to

the classical Racah and Clebsch-Gordan coefficients (related to $\mathfrak{sl}_2$). Namely, as indicated above the canonical basis gives the factorization of the transition matrix between two Gelfand-Tsetlin bases for gl_3 into the product of two triangular matrices. But it is shown in [6], [7] that elements of this matrix are closely related to Racah coefficients (or Wigner's 6j-symbols). As a consequence we get a simple triangular factorization of the matrix of 6j-symbols and of the Clebsch-Gordan matrix. We are grateful to V.N.Tolstoy who called our attention to the papers [6],[7].

The paper is organized as follows. In § 1 the canonical basis is defined. The transition matrix between the canonical basis and the Gelfand-Tsetlin basis and the action of generators of gl_3 in the canonical basis are computed in §2. In §3 we evaluate the canonical basis in the realization of irreducible representations in the space of polynomials on the base affine space of the group GL_3 .

In §4 the action of the subgroup of permutations $S_3 \subset GL_3$ in the canonical basis is computed. Since a permutation sends the Gelfand-Tsetlin basis into another Gelfand-Tsetlin basis, this enables one to compute the transition matrices between the canonical basis and all Gelfand-Tsetlin bases.

Finally, the applications of the results of §4 to Clebsch-Gordan coefficients and 6j-symbols are given in §5.

The main results of this paper were announced in [3]. A possibility of generalization of the results to gl_n is discussed in [2].

1. <u>Definition of the canonical basis</u>. We recall that a weight for gl_n is an integral vector $\gamma = (\gamma_1, \ldots, \gamma_n)$; the weight γ is called highest if $\gamma_1 \geq \gamma_2 \geq \ldots \geq \gamma_n$. Let V_λ denote the irreducible finite dimensional representation of gl_n with highest weight λ , and

$V_\lambda(\gamma)$ be the weight subspace of weight γ in V_λ ; thus, $V_\lambda(\gamma) = \{ v \in V_\lambda : E_{ii}v = \gamma_i v$ for $1 \leq i \leq n \}$. If $\nu =$ $= (\nu_1, \ldots, \nu_n)$ is one more highest weight for gl_n then we put $V_\lambda(\gamma,\nu) = \{ v \in V_\lambda(\gamma) : E_{i,i+1}^{\nu_i - \nu_{i+1}+1} v = 0$ for $1 \leq i \leq n-1\}$.

A basis in V_λ will be called proper if each of the subspaces $V_\lambda(\gamma,\nu)$ (for all possible γ and ν) is spanned by its subset.

Theorem 1. Each irreducible finite dimensional representation of gl_n has a proper basis.

Theorem is proved in [2] .

For general n and λ a proper basis in V_λ is not unique. It turns out, however that it is essentially unique for $n=3$.

Theorem 2. In each irreducible finite dimensional representation of gl_3 there is only one proper basis up to scalar multiples. We call this basis (normalized as indicated in §2 below) canonical.

This theorem is also proved in [2] . Another proof is outlined in §2.

Hereafter we deal only with gl_3 . For each vector f belonging to an irreducible representation of gl_3 we define the exponents $d_1 = d_1(f)$ and $d_2 = d_2(f)$ to be the least nonnegative integers such that $E_{12}^{d_1+1} f =$ $= E_{23}^{d_2+1} f = 0$.

Proposition 1. Let $\lambda = (\lambda_1, \lambda_2, \lambda_3)$ be a highest weight for gl_3 and $\gamma = (\gamma_1, \gamma_2, \gamma_3)$ a fixed weight in V_λ . Then each vector f of the canonical basis in $V_\lambda(\gamma)$ is uniquely determined by $d_1(f)$ (as well as by $d_2(f)$). When f varies, each of the exponents of f runs some interval in Z ; the sum $d_1(f) + d_2(f)$ is independent of f and equals $\max (\lambda_1 - \gamma_1, \gamma_3 - \lambda_3)$.

This Proposition follows from the results in [2]; another proof is outlined in § 2.

By Proposition 1, a vector f of the canonical basis in V_λ may be parametrized by its weight and any one of the exponents. However, another parametrization will be more convenient for our purposes viz., the parametrization by Gelfand-Tsetlin patterns. It will be introduced in the next section.

2. Canonical basis and Gelfand-Tsetlin basis. In this section we compute the transition matrix between the canonical basis and the Gelfand-Tsetlin basis, and the action of the generators of gl_3 on the canonical basis. We begin with some well-known facts about Gelfand-Tsetlin patterns.

Let V_λ be an irreducible representation of gl_3 with highest weight $\lambda = (\lambda_1, \lambda_2, \lambda_3)$. Recall that a Gelfand-Tsetlin pattern (G-pattern) of type λ is a triangular array

$$\Lambda = \begin{matrix} \lambda_1 & & \lambda_2 & & \lambda_3 \\ & \alpha_1 & & \alpha_2 & \\ & & \beta & & \end{matrix}$$

of integers satisfying the betweenness conditions: $\lambda_1 \geqslant \alpha_1 \geqslant \lambda_2 \geqslant \alpha_2 \geqslant \lambda_3$, $\alpha_1 \geqslant \beta \geqslant \alpha_2$. The weight $\gamma = (\gamma_1, \gamma_2, \gamma_3)$ of a G-pattern Λ is defined from the equations

$$\gamma_1 = \beta, \quad \gamma_1 + \gamma_2 = \alpha_1 + \alpha_2, \quad \gamma_1 + \gamma_2 + \gamma_3 = \lambda_1 + \lambda_2 + \lambda_3 \quad (1)$$

It is known that the Gelfand-Tsetlin basis (G-basis) in $V_\lambda(\gamma)$ is parametrized by G-patterns of type λ and weight γ ; let $|\Lambda\rangle$ denote the normalized vector of the basis corresponding to Λ. It will be more convenient for us to consider $e_\Lambda = d(\Lambda)\,|\Lambda\rangle$ instead of $|\Lambda\rangle$, where the normalization constant $d(\Lambda)$ is defined to be

$$d(\Lambda) = \{(\lambda_1-\alpha_1)!\,(\lambda_1-\alpha_2+1)!\,(\lambda_2-\alpha_2)!\,(\alpha_1-\beta)!\,(\alpha_1-\lambda_2)! \cdot (\alpha_1-\lambda_3+1)!\,(\alpha_2-\lambda_3)!\,(\beta-\alpha_2)!/(\alpha_1-\alpha_2+1)!\,(\alpha_1-\alpha_2)!\}^{\frac{1}{2}} \quad (2)$$

When the type λ of Λ is fixed we write simply $\Lambda =$ $= \begin{smallmatrix}\alpha_1 & & \alpha_2\\ & \beta & \end{smallmatrix}$ and $e_\Lambda = e^{\begin{smallmatrix}\alpha_1 & & \alpha_2\\ & \beta & \end{smallmatrix}}$. Rewriting the well-known Gelfand-Tsetlin formulas [4], [5] with the account of the normalization (2) we see that the action of the generators of gl_3 on the basis (e_Λ) is given by

$$E_{12}e^{\begin{smallmatrix}\alpha_1 & & \alpha_2\\ & \beta & \end{smallmatrix}} = (\alpha_1 - \beta)e^{\begin{smallmatrix}\alpha_1 & & \alpha_2\\ & \beta+1 & \end{smallmatrix}}, \quad E_{21}e^{\begin{smallmatrix}\alpha_1 & & \alpha_2\\ & \beta & \end{smallmatrix}} = (\beta - \alpha_2)\, e^{\begin{smallmatrix}\alpha_1 & & \alpha_2\\ & \beta-1 & \end{smallmatrix}},$$

$$E_{23}e^{\begin{smallmatrix}\alpha_1 & & \alpha_2\\ & \beta & \end{smallmatrix}} = (\lambda_1 - \alpha_1)e^{\begin{smallmatrix}\alpha_1+1 & & \alpha_2\\ & \beta & \end{smallmatrix}} + \frac{(\lambda_2-\alpha_2)(\beta-\alpha_2)(\lambda_1-\alpha_2+1)}{(\alpha_1-\alpha_2+1)(\alpha_1-\alpha_2)} \cdot e^{\begin{smallmatrix}\alpha_1 & & \alpha_2+1\\ & \beta & \end{smallmatrix}}, \tag{3}$$

$$E_{32}e^{\begin{smallmatrix}\alpha_1 & & \alpha_2\\ & \beta & \end{smallmatrix}} = \frac{(\alpha_1-\beta)(\alpha_1-\lambda_2)(\alpha_1-\lambda_3+1)}{(\alpha_1-\alpha_2+1)(\alpha_1-\alpha_2)} e^{\begin{smallmatrix}\alpha_1-1 & & \alpha_2\\ & \beta & \end{smallmatrix}} + (\alpha_2-\lambda_3)\, e^{\begin{smallmatrix}\alpha_1 & & \alpha_2-1\\ & \beta & \end{smallmatrix}}.$$

Now we return to the canonical basis. We reformulate the definition in terms of the exponents $d_1(f)$ and $d_2(f)$ introduced in §1. Namely, the canonical basis in V_λ is determined up to normalization by the condition that for any $d_1, d_2 \geq 0$ and any weight γ the subspace of $f \in V_\lambda(\gamma)$ such that $d_1(f) \leq d_1$, $d_2(f) \leq d_2$ is spanned by a subset of this basis. Our next goal is to define the normalization of the vectors of the canonical basis and their parametrization by G-patterns. We need some more definitions.

Let $\Lambda = \begin{smallmatrix}\lambda_1 & & \lambda_2 & & \lambda_3\\ & \alpha_1 & & \alpha_2 & \\ & & \beta & & \end{smallmatrix}$ and $\Lambda' = \begin{smallmatrix}\lambda_1' & & \lambda_2' & & \lambda_3'\\ & \alpha_1' & & \alpha_2' & \\ & & \beta' & & \end{smallmatrix}$ be two G-patterns.

We write $\Lambda' < \Lambda$ if Λ and Λ' have the same type and weight and $\alpha_1' < \alpha_1$. By (1), a G-pattern of fixed type and weight is determined by one parameter, e.g. by α_1; therefore, the relation $\Lambda' < \Lambda$ is a linear order on such patterns. For each integer k we put $\Lambda[k] = \begin{smallmatrix} \lambda_1 & & \lambda_2 & & \lambda_3 \\ & \alpha_1+k & & \alpha_2-k & \\ & & \beta & & \end{smallmatrix}$; thus $\Lambda' < \Lambda$ if and only if $\Lambda' = \Lambda[-k]$ for some $k > 0$. Let $V(\Lambda)$ denote the subspace of V_λ (where λ is the type of a G-pattern Λ) spanned by $e_{\Lambda'}$'s with $\Lambda' \leqslant \Lambda$.

<u>Proposition 2</u>. Let Λ be a G-pattern of type λ and weight γ. Then $V(\Lambda) = \{f \in V_\lambda(\gamma) : d_1(f) \leqslant d_1(e_\Lambda)\}$.

<u>Proof</u>. By (3) we have

$$d_1(e^{\alpha_1 \ \alpha_2}_{\ \ \beta}) = \alpha_1 - \beta \tag{4}$$

Hence $d_1(e_{\Lambda'}) < d_1(e_\Lambda)$ if $\Lambda' < \Lambda$. This implies our proposition.

<u>Proposition 3</u>. For any G-pattern Λ there is the unique vector of the canonical basis belonging to $V(\Lambda) - \sum_{\Lambda'<\Lambda} V(\Lambda')$. We normalize it so that the coefficient of e_Λ in its expansion in the G-basis were 1; denote this vector by f_Λ.

Thus, this proposition gives the normalization of the canonical basis and its parametrization by G-patterns.

<u>Proof</u>. By Proposition 2, each of the subspaces $V(\Lambda)$ is spanned by a subset of the canonical basis. This implies our proposition by induction in Λ.

The vector f_Λ may be written in the form

$$f_\Lambda = \sum_{k \geqslant 0} (-1)^k b_k(\Lambda) e_{\Lambda[-k]} ; \tag{5}$$

it is also clear that e_Λ is of the form

$$e_\Lambda = \sum_{k \geqslant 0} a_k(\Lambda) f_{\Lambda[-k]} \tag{6}$$

We compute $a_k(\Lambda)$ and $b_k(\Lambda)$, and the action of generators of gl_3 on the (f_Λ). It turns out that all formulas are piecewise polynomial with singularities on a certain hyperplane in the space of G-patterns. This hyperplane is defined by the equation $\lambda_2 = \gamma_2$, where $\lambda = (\lambda_1, \lambda_2, \lambda_3)$ is the type of a G-pattern and $\gamma = (\gamma_1, \gamma_2, \gamma_3)$ is its weight.

A G-pattern Λ of type λ and weight γ is of class I if $\gamma_2 > \lambda_2$, of class II if $\gamma_2 < \lambda_2$ and of class 0 if $\gamma_2 = \lambda_2$.

<u>Theorem 3.</u> Let $\Lambda = \begin{smallmatrix}\alpha_1 & & \alpha_2\\ & \beta & \end{smallmatrix}$ be a G-pattern of type λ. Then

$$a_k(\Lambda) = \binom{\beta-\alpha_2}{k}\binom{\lambda_2-\alpha_2}{k}\Big/\binom{\alpha_1-\alpha_2}{k}, \qquad (7^{I})$$

$$b_k(\Lambda) = \binom{\beta-\alpha_2}{k}\binom{\lambda_2-\alpha_2}{k}\Big/\binom{\alpha_1-\alpha_2+1-k}{k} \qquad (8^{I})$$

if Λ is of class I or 0;

$$a_k(\Lambda) = \binom{\alpha_1-\beta}{k}\binom{\alpha_1-\lambda_2}{k}\Big/\binom{\alpha_1-\alpha_2}{k}, \qquad (7^{II})$$

$$b_k(\Lambda) = \binom{\alpha_1-\beta}{k}\binom{\alpha_1-\lambda_2}{k}\Big/\binom{\alpha_1-\alpha_2+1-k}{k} \qquad (8^{II})$$

if Λ is of class II or 0.

<u>Theorem 4.</u> The action of generators of gl_3 on the (f_Λ) is given by the formulas

$$E_{12} f^{\,\alpha_1\ \alpha_2}_{\ \ \beta} = (\alpha_1-\beta) f^{\,\alpha_1\ \alpha_2}_{\ \ \beta+1} + (\lambda_2-\alpha_2)\chi(I) f^{\,\alpha_1-1\ \alpha_2+1}_{\ \ \beta+1},$$

$$E_{21} f^{\,\alpha_1\ \alpha_2}_{\ \ \beta} = (\beta-\alpha_2) f^{\,\alpha_1\ \alpha_2}_{\ \ \beta-1} + (\alpha_1-\lambda_2)\chi(II) f^{\,\alpha_1-1\ \alpha_2+1}_{\ \ \beta-1},$$

$$E_{23} f^{\,\alpha_1\ \alpha_2}_{\ \ \beta} = (\lambda_1-\alpha_1) f^{\,\alpha_1+1\ \alpha_2}_{\ \ \beta} + (\lambda_1+\lambda_2+\beta-2\alpha_1-\alpha_2)\cdot$$
$$\cdot\ \chi(II) f^{\,\alpha_1\ \alpha_2+1}_{\ \ \beta}, \qquad (9)$$

$$E_{32} f^{\,\alpha_1\ \alpha_2}_{\ \ \beta} = (\alpha_2-\lambda_3) f^{\,\alpha_1\ \alpha_2-1}_{\ \ \beta} + (\alpha_1+2\alpha_2-\lambda_2-\lambda_3-\beta)\cdot$$
$$\cdot\ \chi(I) f^{\,\alpha_1-1\ \alpha_2}_{\ \ \beta}.$$

where the symbol $\varkappa(J)$ equals 1 if $\begin{smallmatrix}\lambda_1 & & \lambda_2 & & \lambda_3\\ & \alpha_1 & & \alpha_2 & \\ & & \beta & & \end{smallmatrix}$ is of class J, and equals 0, otherwise.

Now we give a sketch of the proof of Theorems 2,3,4 and Proposition 1. Denote temporarily the right-hand side of (5) by $\tilde{f}_\Lambda$, where the $b_k(\Lambda)$ are defined by (8^I) (8^{II}). The formulas (6) and (9) with f replaced by $\tilde{f}$ and $a_k(\Lambda)$ defined by (7^I), (7^{II}) are directly verified (to verify (9) one only needs to use (5) and (3)). It easily follows from (9) that

$$d_1(\tilde{f}^{\,\alpha_1\,\alpha_2}_{\;\;\beta}) = d_1(e^{\,\alpha_1\,\alpha_2}_{\;\;\beta}) = \alpha_1 - \beta \text{ , and that}$$

$$d_2(\tilde{f}^{\,\alpha_1\,\alpha_2}_{\;\;\beta}) = \lambda_1 - \alpha_1 + \max(0,\lambda_2-\gamma_2) \text{ . Hence for}$$

$\Lambda' < \Lambda$ we have $d_1(\tilde{f}_{\Lambda'}) < d_1(\tilde{f}_\Lambda)$ and $d_2(\tilde{f}_{\Lambda'}) > d_2(\tilde{f}_\Lambda)$; moreover, $d_1(\tilde{f}_\Lambda) + d_2(\tilde{f}_\Lambda)$ depends only on the type and weight of Λ, and has the value given by Proposition 1.

Let λ be a highest weight for gl_3, γ a weight, and d_1, d_2 nonnegative integers. The just proven inequalities between the exponents of $\tilde{f}_\Lambda$ easily imply that the subspace $\{f \in V_\lambda(\gamma) : d_1(f) \leq d_1 , d_2(f) \leq d_2\}$ is spaned by the vectors $\tilde{f}_\Lambda$ contained in it. So $\tilde{f}_\Lambda$ form a proper basis. Furthermore, if Λ is a G-pattern of type λ and weight γ, and $d_1 = d_1(\tilde{f}_\Lambda)$, $d_2 = d_2(\tilde{f}_\Lambda)$ then the corresponding subspace is spanned by $\tilde{f}_\Lambda$. This proves uniqueness of a proper basis for gl_3. Therefore, the basis $(\tilde{f}_\Lambda)$ is canonical, which proves all our assertions.

The formulas (7)-(9) have interesting symmetry properties. For each G-pattern

$$\Lambda = \begin{matrix}\lambda_1 & & \lambda_2 & & \lambda_3\\ & \alpha_1 & & \alpha_2 & \\ & & \beta & & \end{matrix} \qquad \text{we put} \qquad \tilde{\Lambda} = \begin{matrix}-\lambda_3 & & -\lambda_2 & & -\lambda_1\\ & -\alpha_2 & & -\alpha_1 & \\ & & -\beta & & \end{matrix}$$

The following properties of the mapping $\Lambda \to \tilde{\Lambda}$ follow at once from definitions.

Proposition 4. The mapping $\Lambda \to \tilde{\Lambda}$ is an involution, i.e. $\tilde{\tilde{\Lambda}} = \Lambda$. If Λ is of type λ and weight γ then $\tilde{\Lambda}$ is of type $\tilde{\lambda} = (-\lambda_3, -\lambda_2, -\lambda_1)$ and weight $-\gamma$. If Λ is of class I (resp. II, O) then $\tilde{\Lambda}$ is of class II (resp. I, O). We have $\widetilde{\Lambda[k]} = \tilde{\Lambda}[k]$; in particular, if $\Lambda' < \Lambda$ then $\tilde{\Lambda}' < \tilde{\Lambda}$.

Let $T: V_\lambda \to V_{\tilde{\lambda}}$ denote the linear operator such that $T(e_\Lambda) = e_{\tilde{\Lambda}}$ for any G-pattern Λ of type λ. Let ω be the automorphism of gl_3 such that $\omega(E_{ii}) = -E_{ii}$, $\omega(E_{i,i\pm 1}) = E_{i\pm 1,i}$.

Proposition 5. (a) For any G-pattern Λ of type λ we have $T(f_\Lambda) = f_{\tilde{\Lambda}}$.

(b) We have $T \circ \omega(X) = X \circ T$ for any $X \in gl_3$.

Proof. By (7), (8) and Proposition 4, for each G-pattern Λ we have

$$a_k(\tilde{\Lambda}) = a_k(\Lambda),\ b_k(\tilde{\Lambda}) = b_k(\Lambda) \tag{10}$$

This proves (a). Part (b) follows at once from (3) or (9).

3. Geometric realization of the canonical basis. In this section we compute the canonical basis in the realization of irreducible finite dimensional representations of $\mathfrak{gl}_3$ in the space of functions on the base affine space A of GL_3. By definition, $A = GL_3/N$, where N is the subgroup of unipotent upper triangular matrices in GL_3. We realize A as a cone in the seven-dimensional affine space. Namely, let $V = V_{(1,0,0)}$ be the standard three-dimensional representation of gl_3 and (e_1, e_2, e_3) the standard basis in V. We let $W = V \oplus \Lambda^2 V \oplus \Lambda^3 V$; then the vectors e_i, $e_i \wedge e_j$ for $i<j$, and $e_1 \wedge e_2 \wedge e_3$ form a basis in W. Let (x_i, x_{ij}, x_{123}) be the corresponding coordinates in W. Then A may be realized as the subset of W determined by the inequalities

$(x_1,x_2,x_3) \neq 0$, $(x_{12},x_{13},x_{23}) \neq 0$, $x_{123} \neq 0$, and the equation

$$x_1x_{23} - x_2x_{13} + x_3x_{12} = 0 \tag{11}$$

It is known that the natural representation of GL_3 (or gl_3) in the space $C[A]$ of regular (algebraic) functions on A is the sum of all the irreducible representations V_λ each occuring exactly once.

A monomial $\prod_i x_i^{m_i} \cdot \prod_{i<j} x_{ij}^{m_{ij}} \cdot x_{123}^{m_{123}}$ is said to be proper if all its exponents are integers, all except m_{123} are nonnegative, and $m_2m_{13} = 0$. It follows easily from (11) that proper monomials form a basis in $C[A]$. It turns out that union of canonical bases in all V_λ is just this basis. To be more precise we assign to each G-pattern $\Lambda = \begin{smallmatrix} \alpha_1 & & \alpha_2 \\ & \beta & \end{smallmatrix}$ of type λ and weight γ the proper monomial $x^{m(\Lambda)}$ whose exponents are defined as follows: if Λ is of class I or O then

$$m_1 = \beta - \alpha_2,\ m_3 = \lambda_1 - \alpha_1,\ m_{12} = \alpha_2 - \lambda_3,\ m_{23} = \lambda_2 - \alpha_2,$$
$$m_{123} = \lambda_3,\ m_2 = \gamma_2 - \lambda_2,\ m_{13} = 0; \tag{12^{I}}$$

if Λ is of class II or O then

$$m_1 = \alpha_1 - \lambda_2,\ m_3 = \lambda_1 - \alpha_1,\ m_{12} = \alpha_2 - \lambda_3,\ m_{23} = \alpha_1 - \beta,$$
$$m_{123} = \lambda_3,\ m_2 = 0,\ m_{13} = \lambda_2 - \gamma_2 . \tag{12^{II}}$$

<u>Theorem</u> 5. The linear mapping $\bigoplus_\lambda V_\lambda \longrightarrow C[A]$ sending each f_Λ to the proper monomial $x^{m(\Lambda)}$ is an isomorphism of representations of GL_3 (and gl_3).

<u>Proof</u>. It suffices to compute the action of generators of gl_3 in the basis of proper monomials (this is done

in a straightforward way with the help of (11)) and then to note that the replacement of monomials by G-patterns according to (12) transforms this action into the action given by (9).

Remarks. 1) Theorem 5 and the formulas (6) and (7) give the realization of the G-basis in $C[A]$. Similar formulas were obtained in [5], p. 303.

2) The basis of proper monomials in $C[A]$ was mentioned in [5], p. 341 but its canonical definition was not given.

4. The action of permutations in GL_3 on the canonical basis. In this section we compute the action of the permutation subgroup $S_3 \subset GL_3$ on the canonical basis in each V_λ. It is convenient for us to choose two permutations

$$s_1 = \begin{matrix} 0 & 1 & 0 \\ 1 & 0 & 0 \\ 0 & 0 & 1 \end{matrix} \quad \text{and} \quad s = \begin{matrix} 0 & 0 & 1 \\ 0 & 1 & 0 \\ 1 & 0 & 0 \end{matrix}$$

as generators of S_3.

Proposition 6. The action of s_1 in the canonical basis in V_λ is given by

$$s_1 f^{\alpha_1 \alpha_2}_{\beta} = (-1)^{\alpha_2} \sum_{\alpha_1' \leq \alpha_1} \binom{d}{\alpha_1 - \alpha_1'} f^{\alpha_1' \quad \alpha_1+\alpha_2-\alpha_1'}_{\alpha_1+\alpha_2-\beta} ,$$

where $d = \min(\alpha_1-\lambda_2, \alpha_1-\beta, \lambda_2-\alpha_2, \beta-\alpha_2)$.

Sketch of the proof. We realize f_Λ's as proper monomials on the base affine space A (see Theorem 5). Now s_1 acts on $C[A]$ by an automorphism, and we have $s_1x_1 = x_2$, $s_1x_2 = x_1$, $s_1x_3 = x_3$, $s_1x_{12} = -x_{12}$, $s_1x_{13} = x_{23}$, $s_1x_{23} = x_{13}$, $s_1x_{123} = -x_{123}$. Acting by s_1 on the proper monomial $x^{m(\Lambda)}$ corresponding to f_Λ we obtain a monomial not proper in general. But it is decomposed in a linear combination of proper monomials with the help of (11). It remains to use (12) for translating the results back into the language of G-patterns.

To describe the action of s we introduce some notation. For each weight $\gamma = (\gamma_1,\gamma_2,\gamma_3)$ we let $s\gamma = (\gamma_3,\gamma_2,\gamma_1)$. Now let $\Lambda = \begin{smallmatrix}\alpha_1 & & \alpha_2\\ & \beta & \end{smallmatrix}$ be a G-pattern of type λ and weight γ. Define the G-pattern $s\Lambda$ of type λ and weight $s\gamma$ to be $s\Lambda = \begin{smallmatrix}\lambda_1+\alpha_2-\beta, & \lambda_2+\lambda_3-\alpha_2\\ \gamma_3 & \end{smallmatrix}$ if Λ is of class I or 0, and $s\Lambda = \begin{smallmatrix}\lambda_1+\lambda_2-\alpha_1, & \lambda_3+\alpha_1-\beta\\ \gamma_3 & \end{smallmatrix}$ if Λ is of class II. It is directly verified that the mapping $\Lambda \longrightarrow s\Lambda$ is an involution, i.e. $s(s\Lambda) = \Lambda$ and that $s(\Lambda[k]) = (s\Lambda)[-k]$ for all integer k; in particular, if $\Lambda' < \Lambda$ then $s\Lambda' > s\Lambda$.

<u>Theorem</u> 6. The action of $s \in GL_3$ on the canonical basis in V_λ is given by $sf_\Lambda = (-1)^{\lambda_2} f_{s\Lambda}$.

Proof is similar to that of Proposition 6.

Theorems 3 and 6 allow us to decompose the matrix of the operator $s : V_\lambda \longrightarrow V_\lambda$ in the G-basis (e_Λ) into the product of two triangular matrices. Put $g_\Lambda = (-1)^{\lambda_2} s\, e_{s\Lambda}$. By (5), (6) and Theorem 6 we have:

$$f_\Lambda = \sum_{k\geq 0} (-1)^k b_k(s\Lambda)\, g_\Lambda[k] \tag{13}$$

$$g_\Lambda = \sum_{k\geq 0} a_k(s\Lambda)\, f_{\Lambda[k]} \tag{14}$$

(the coefficients in (13) and (14) are given by (7) and (8)).

The elements g_Λ form one more basis in $V_\lambda(\gamma)$ parametrized by G-patterns of type λ and weight γ. Up to parametrization (2), this is the G-basis corrresponding to the chain $gl_1 \subset gl_2 \subset gl_3$ which is obtained from the standard chain by conjugation by $s \in GL_3$. Define the transition matrix $(c(\Lambda,\Lambda'))$ by

$$g_\Lambda = \sum_{\Lambda'} c(\Lambda,\Lambda')\, e_{\Lambda'} \; ; \tag{15}$$

by (5) and (14) $c(\Lambda,\Lambda') = 0$ unless $\Lambda' = \Lambda\,[r]$ for some $r \in Z$, and

$$c(\Lambda,\Lambda\,[r]) = \sum_{k,\ell\geq 0,\,k-\ell=r} (-1)^{\ell} a_k(s\Lambda)\, b_\ell(\Lambda\,[k]) \tag{16}$$

5. <u>Triangular factorization for 6j-symbols and Clebsch-Gordan coefficients.</u> It is shown in [6], [7] that the transition matrix between two G-bases for gl_3 may be expressed in terms of Racah coefficients (or Wigner's 6j-symbols) corresponding to sl_2. So there is an interpretation of (16) in terms of 6j-symbols. Recall the definitions of the Clebsch-Gordan coefficients and 6j-symbols.

As is common in physical papers we parametrize irreducible (finite dimensional) representations of sl_2 by a half-integer $j \geq 0$; the representation V_j corresponding to j has dimension $2j+1$. Choose an invariant scalar product on V_j and an orthonormal basis of weight vectors; these vectors are denoted by $|jm\rangle$ ($m = -j, -j+1,\dots,j-1,j$). Sometimes another basis (e^j_m) in V_j is more convenient, namely

$$e^j_m = \{(j+m)!\,(j-m)!\,\}^{\frac{1}{2}}\,|j\,m\rangle \tag{17}$$

Let E_+, E_- and H be standard generators of sl_2 satisfying the commutation relations $[H,E_\pm] = \pm 2E_\pm$, $[E_+,E_-] = H$; then their action on the e^j_m is given by

$$E_+e^j_m = (j-m)e^j_{m+1}\,,\quad E_-e^j_m = (j+m)e^j_{m-1}\,,\quad He^j_m = 2m\,e^j_m \tag{18}$$

It is known that the tensor product $V_{j_1} \otimes V_{j_2}$ is multiplicity free. Furthermore, V_j occurs in $V_{j_1} \otimes V_{j_2}$ if and only if j_1+j_2+j is an integer and each of the numbers j_1, j_2 and j does not exceed the sum of two others (these two conditions will be referred to as the triangle condition). Thus there is the unique (up to scalar multiple) operator $T(j_1 j_2 j) : V_j \longrightarrow V_{j_1} \otimes V_{j_2}$ commuting with the action of sl_2. Its matrix elements, $C^{j_1 j_2 j}_{m_1 m_2 m}$, defined by

$$T(j_1 j_2 j)\ |jm\rangle = \sum_{m_1, m_2} C^{j_1 j_2 j}_{m_1 m_2 m}\ |j_1 m_1\rangle \otimes |\ j_2 m_2\rangle \qquad (19)$$

are called the Clebsch-Gordan coefficients; therefore, $C^{j_1 j_2 j}_{m_1 m_2 m} = 0$ unless j_1, j_2 and j satisfy the triangle condition and $m_1 + m_2 = m$. The normalization of $T(j_1 j_2 j)$ is uniquely determined by the condition that this is an isometry and the phase condition $C^{j_1 j_2 j}_{j_1, j-j_1, j} > 0$.

Now consider the tensor product of three irreducible representations $V_{j_1} \otimes V_{j_2} \otimes V_{j_3}$. We may decompose it into irreducible components in two different ways. The first is to decompose $V_{j_1} \otimes V_{j_2}$ into $\oplus V_{j_{12}}$ and then to decompose each $V_{j_{12}} \otimes V_{j_3}$; alternatively, we may decompose $V_{j_2} \otimes V_{j_3}$ into $\oplus V_{j_{23}}$ and then decompose each $V_{j_1} \otimes V_{j_{23}}$. Hence for any V_j there are two different bases in the space of operators $V_j \longrightarrow V_{j_1} \otimes V_{j_2} \otimes V_{j_3}$ commuting with the action of sl_2. The first is parametrized by j_{12}'s such that each of the triples (j_1, j_2, j_{12}) and (j_{12}, j_3, j) satisfies the triangle condition; the corresponding operator is the composition of $T(j_{12} j_3 j) : V_j \longrightarrow V_{j_{12}} \otimes V_{j_3}$ and $T(j_1 j_2 j_{12}) \otimes \mathbb{1} : V_{j_{12}} \otimes V_{j_3} \longrightarrow V_{j_1} \otimes V_{j_2} \otimes V_{j_3}$. Denote this operator by $L(j_1 j_2 j_3 j j_{12})$. Similarly, for each

j_{23} such that (j_2, j_3, j_{23}) and (j_1, j_{23}, j) satisfy the triangle condition define $R(j_1 j_2 j_3 j j_{23})$ to be the composition of $T(j_1 j_{23} j) : V_j \longrightarrow V_{j_1} \otimes V_{j_{23}}$ and $1 \otimes T(j_2 j_3 j_{23}) : V_{j_1} \otimes V_{j_{23}} \longrightarrow V_{j_1} \otimes V_{j_2} \otimes V_{j_3}$; then the operators $R(j_1 j_2 j_3 j j_{23})$ form another basis in the same vector space. Let $\left(\left\langle\begin{matrix} j_1 & j_2 & j_{12} \\ j_3 & j & j_{23}\end{matrix}\right\rangle\right)$ denote the transition matrix between these bases, i.e.

$$R(j_1 j_2 j_3 j j_{23}) = \sum_{j_{12}} \left\langle\begin{matrix} j_1 & j_2 & j_{12} \\ j_3 & j & j_{23}\end{matrix}\right\rangle L(j_1 j_2 j_3 j j_{12}) \qquad (20)$$

This matrix is expressed in terms of 6j-symbols as follows:

$$\left\langle\begin{matrix} j_1 & j_2 & j_{12} \\ j_3 & j & j_{23}\end{matrix}\right\rangle = (-1)^{j_1+j_2+j_3+j} \{(2j_{12}+1)(2j_{23}+1)\}^{\frac{1}{2}} \left\{\begin{matrix} j_1 & j_2 & j_{12} \\ j_3 & j & j_{23}\end{matrix}\right\} \qquad (21)$$

(see e.g. [6] , (1.18b)).

Now we return to gl_3 . For any G-pattern $\Lambda = \begin{matrix}\alpha_1 & & \alpha_2 \\ & \beta & \end{matrix}$ we let $j(\Lambda) = (\alpha_1 - \alpha_2)/2$; if sl_2 is embedded in gl_3 by $E_+ = E_{12}$, $E_- = E_{21}$, $H = E_{11} - E_{22}$ then e_Λ generates under the action of this subalgebra an irreducible representation isomorphic to $V_{j(\Lambda)}$.

Theorem 7. Let Λ and Λ' be G-patterns of type λ and weight γ . Then the coefficient $c(\Lambda, \Lambda')$ from (15) is

$$c(\Lambda, \Lambda') = \frac{d(s\Lambda)}{d(\Lambda')} \left\langle\begin{matrix} \frac{\lambda_1-\gamma_2}{2} & \frac{\lambda_1-\gamma_1}{2} & j(\Lambda') \\ \frac{\lambda_2-\lambda_3}{2} & \frac{\lambda_1-\gamma_3}{2} & j(s\Lambda)\end{matrix}\right\rangle \qquad (22)$$

where $d(\Lambda)$ is defined by (2) and $s\Lambda$ is defined in §4.

This Theorem is a reformulation of (2.4b) from [6] (see also [7]).

Using (21) and (22) we see that (16) gives the "triangular factorization" for 6j-symbols.

It is well-known that Clebsch-Gordan coefficients may be considered as limits of Racah coefficients (see e.g. [1], (3.300)). In our notation this result has the form

$$C^{j_1 j_2 j}_{m_1, m-m_1, m} = \lim_{j_3 \to \infty} \left\langle \begin{matrix} j_3 & j_1 & j_3+m_1 \\ j_2 & j_3+m & j \end{matrix} \right\rangle \tag{23}$$

By (23), Theorem 7 implies the triangular factorization of the Clebsch-Gordan coefficients. For this we assign to each j_1, j_2, m_1, m_2 as above and any integer $t \geq 0$ the G-pattern

$$\Lambda_t \begin{pmatrix} j_1 & j_2 \\ m_1 & m_2 \end{pmatrix} = \begin{matrix} 2j_1 + 2j_2 + t \ , & 2j_2 \ , & 0 \\ m_1 + j_1 + 2j_2 + t \ , & m_2 + j_2 & \\ 2j_2 + t & & \end{matrix} \tag{24}$$

Theorem 8. Fix j_1, j_2, m_1, m_2, and j. For any integer $t \geq 0$ put $\Lambda'_t = \Lambda_t \begin{pmatrix} j_1 & j_2 \\ m_1 & m_2 \end{pmatrix}$, $r = j(s\Lambda'_t) - j$ and $\Lambda_t = \Lambda'_t [r]$. Then

$$C^{j_1 j_2 j}_{m_1, m_2, m_1+m_2} = \lim_{t \to \infty} \frac{c(\Lambda_t, \Lambda'_t)\, d(\Lambda'_t)}{d(s\,\Lambda_t)} \tag{25}$$

Proof follows at once from (22)-(24).

Theorem 8 has an interpretation and proof not using 6j-symbols. Fix j_1 and j_2 and for each integer $t \geq 0$ consider the subspace $V(t) = \bigoplus_{\gamma} V_\lambda(\gamma) \subset V_\lambda$, where

$\lambda = (2j_1 + 2j_2 + t, 2j_2, 0)$ and γ runs all the weights such that $\gamma_1 = 2j_2 + t$. If $\Lambda = \Lambda_t\begin{pmatrix} j_1 & j_2 \\ m_1 & m_2 \end{pmatrix}$ we write $e_{m_1m_2}(t)$, $f_{m_1m_2}(t)$ and $g_{m_1m_2}(t)$ instead of e_Λ, f_Λ, and g_Λ. Clearly, each of the families $(e_{m_1m_2}(t))$, $(f_{m_1m_2}(t))$ and $(g_{m_1m_2}(t))$, where $m_i = -j_i, -j_i+1, \ldots, j_i$ $(i = 1,2)$, is a basis in $V(t)$. We consider sl_2 as the subalgebra in gl_3 generated by $E_+ = E_{23}$, $E_- = E_{32}$, $H = E_{22} - E_{33}$; clearly, $V(t)$ is invariant under this subalgebra.

Assign to each pair (m_1,m_2) as above the pair (j,m) as follows: $m = m_1+m_2$, and $j = j_1+m_2$ (resp. $j = j_2 - m_1$) if $m \geqslant j_2-j_1$ (resp. $m \leqslant j_2-j_1$). Geometrically, $2j$ is the length of the hook shown on Fig. 1, which contains the point (m_1,m_2). It is easy to see that this construction gives a one-to-one correspondence between the pairs (m_1,m_2) and pairs (j,m) such that j_1,j_2 and j satisfy the triangle condition and $m \in \{-j, -j+1,\ldots, j\}$. In all subsequent formulas we suppose that (m_1,m_2) and (j,m) are related by this correspondence. In terms of G-patterns this means the following: if $\Lambda = \Lambda_t\begin{pmatrix} j_1 & j_2 \\ m_1 & m_2 \end{pmatrix}$ then $j = j(s\Lambda)$; furthermore, Λ is of class I (resp. II, 0) if $m > j_2-j_1$ (resp. $m < j_2-j_1$, $m = j_2-j_1$), see Fig. 1.

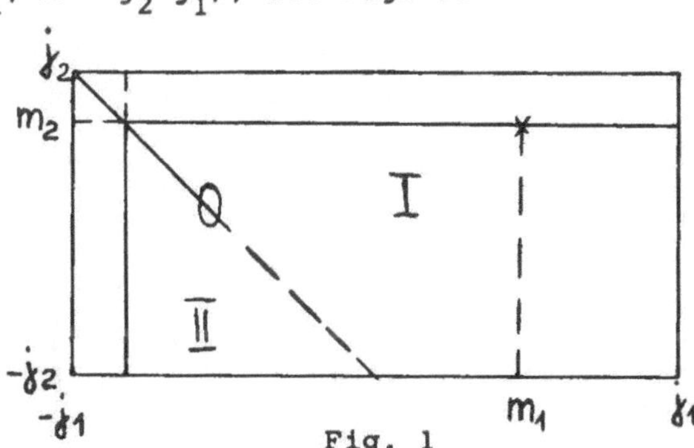

Fig. 1

The action of sl_2 on each of the bases $(e_{m_1m_2}(t))$, $(f_{m_1m_2}(t))$ and $(g_{m_1m_2}(t))$ is computed easily, using (3) and (9). The result for $g_{m_1m_2}(t)$ may be formulated as follows:

Proposition 7. For each integer $t \geqslant 0$ the linear mapping $V_j \longrightarrow V(t)$ sending e_m^j to $g_{m_1m_2}(t)$ (where (m_1,m_2) and (j,m) are related as above) commutes with the sl_2- action (see (17), (18)).

By Proposition 7, we may identify each $V(t)$ as the representation of sl_2 with $V_{j_1} \otimes V_{j_2}$ by means of the isomorphism sending $g_{m_1m_2}(t)$ to $T(j_1j_2j)\, e_m^j$ (see (19)). Using this identification we obtain the basis $(g_{m_1m_2})$ in $V_{j_1} \otimes V_{j_2}$ and two one-parametric families of bases $(e_{m_1m_2}(t))$ and $(f_{m_1m_2}(t))$.

Proposition 8.

$$g_{m_1m_2} = \sum_{k \geqslant 0} \binom{j_1-m_1}{k}\binom{j_2+m_2}{k} \Big/ \binom{2j}{k} f_{m_1+k,m_2-k}(t) \tag{26}$$

$$f_{m_1m_2}(t) = \sum_{k \geqslant 0} (-1)^k \frac{\binom{j_1+j_2-j}{k}\binom{j_1+j_2-j+t}{k}}{\binom{j_1+j_2+m_1-m_2+1-k+t}{k}} e_{m_1-k,m_2+k}(t) \tag{27}$$

$$e_{m_1m_2}(t) = \sum_{k \geqslant 0} \frac{\binom{j_1+j_2-j}{k}\binom{j_1+j_2-j+t}{k}}{\binom{j_1+j_2+m_1-m_2+t}{k}} f_{m_1-k,m_2+k}(t) \tag{28}$$

$$f_{m_1m_2}(t) = \sum_{k \geqslant 0} (-1)^k \binom{j_1-m_1}{k}\binom{j_2+m_2}{k} \Big/ \binom{2j+1-k}{k} g_{m_1+k,m_2-k} \tag{29}$$

This follows at once from (5)-(8), (13) and (14).

In particular, we see that the basis $(f_{m_1m_2}(t))$ is independent on t (so we may write simply $(f_{m_1m_2})$) and is connected with each of the bases $(g_{m_1m_2})$ and $(e_{m_1m_2}(t))$ by a triangular transformation. Now let $t \to \infty$ in (27) and (28); clearly, the limit vectors $e_{m_1m_2} = \lim_{t\to\infty} e_{m_1m_2}(t)$ exist and form a basis in $V_{j_1} \otimes V_{j_2}$, and we have

$$f_{m_1m_2} = \sum_{k\geq 0} (-1)^k \binom{j_1+j_2-j}{k} e_{m_1-k,m_2+k} \tag{27'}$$

$$e_{m_1m_2} = \sum_{k\geq 0} \binom{j_1+j_2-j}{k} f_{m_1-k,m_2+k} \tag{28'}$$

<u>Proposition</u> 9. The vector $e_{m_1m_2}$ in $V_{j_1} \otimes V_{j_2}$ equals the tensor product $e^{j_1}_{m_1} \otimes e^{j_2}_{m_2}$. So up to normalization factors of type (17) the transition matrix between the bases $(g_{m_1m_2})$ and $(e_{m_1m_2})$ is the Clebsch-Gordan matrix.

<u>Proof</u>. Letting $t \to \infty$ in the formulas expressing the action of sl_2 on the $e_{m_1m_2}(t)$ we obtain that the limit basis $(e_{m_1m_2})$ is transformaed under the action of sl_2 in the same way as the basis $(e^{j_1}_{m_1} \otimes e^{j_2}_{m_2})$. It follows that these two bases differ only by normalization factors; the fact that all the factors are equal to 1 is proved directly.

Now we see that (26) and (27') give the triangular factorization of the Clebsch-Gordan matrix. This is our interpretation of Theorem 8.

References

1. Biedenharn,L.C., Louck,J.D. (1981). Angular momentum in quantum physics. Addison-Wesley, Reading, Massachusetts.

2. Gelfand,I.M., Zelevinsky,A.V. Multiplicities and proper bases for gl_n - in this volume.

3. Gelfand,I.M., Zelevinsky,A.V. (1985). Funct. Anal. and Appl. 19, No. 2, 72-75 (Russian).

4. Gelfand,I.M., Tsetlin,M.L. (1950). Doklady Akad. Nauk SSSR, 71, No. 5, 825-828 (Russian).

5. Zhelobenko,D.P. (1970). Compact Lie groups and their representations. Nauka, Moscow (Russian).

6. Pluhař,Z., Smirnov,Yu.F., Tolstoy,V.N. (1982). A novel approach to the SU(3) symmetry calculus. Preprint, Karlov University, Praha.

7. Moshinsky,M., Chacón,E. (1968). Racah coefficients and states with permutational symmetry. In: Spectroscopic and group theoretical methods in physics, North-Holland, Amsterdam, pp. 99-117; Gal,A., Lipkin,H.J., U(3) transpositions and the isospin-U-spin transformation, ibid., pp. 119-124.

MULTIPLICITIES AND PROPER BASES FOR gl_n

I.Gelfand, A.Zelevinsky
Moscow State University

1. Formulations of results. The first main result of this paper is a new expression for multiplicities of irreducible components in the tensor product of two irreducible finite-dimensional representations of the Lie algebra gl_n . The answer is the number of integral points in a certain convex polytope in the space of Gelfand-Tsetlin patterns (G-patterns). The Kostant function and the weight multiplicity for gl_n may also acquire a similar interpretation. Thus for gl_n all these three fundamental functions of the representation theory may be treated in the same geometric way: there is a convex polytope in the space of large dimension with a linear projection on the domain of our function and the function value at each point x equals the number of integral points in the fiber of the projection over x (clearly this fiber is also a convex polytope). This approach gives a natural explanation of the piece-wise polynomial behaviour of these functions.

To formulate the result we need some notations. Recall that the weight lattice for gl_n consists of integral vectors $\gamma = (\gamma_1,\ldots,\gamma_n)$. The weight $\lambda = (\lambda_1,\ldots,\lambda_n)$ is called highest if $\lambda_1 \geq \lambda_2 \geq \ldots \geq \lambda_n$. Irreducible finite-dimensional representations of gl_n are parametrized by highest weights; denote the representation with highest weight λ by V_λ .

A G-pattern for gl_n is defined as a sequence $(\lambda^{(1)}, \lambda^{(2)},\ldots, \lambda^{(n)})$, where $\lambda^{(i)}=(\lambda_1^{(i)}, \lambda_2^{(i)},\ldots,$

$\lambda_i^{(i)}$) is a highest weight for gl_i such that

$$\lambda_j^{(i+1)} \geqslant \lambda_j^{(i)} \geqslant \lambda_{j+1}^{(i+1)} \qquad \text{for } 1 \leqslant j \leqslant i \leqslant n-1 \tag{1}$$

A G-pattern is usually represented as a triangular array

$$\Lambda = \begin{matrix} \lambda_1^{(n)} & \lambda_2^{(n)} \dots & \lambda_n^{(n)} \\ & \lambda_1^{(n-1)} \dots & \lambda_{n-1}^{(n-1)} \\ & \lambda_1^{(1)} & \end{matrix}$$

We call $\lambda^{(n)}$ the type of Λ and define the weight $\gamma = (\gamma_1, \dots, \gamma_n)$ of Λ by

$$\gamma_1 + \dots + \gamma_i = \lambda_1^{(i)} + \lambda_2^{(i)} + \dots + \lambda_i^{(i)} \qquad (1 \leqslant i \leqslant n) \tag{2}$$

For $1 \leqslant j \leqslant i \leqslant n-1$ we put

$$d_j^{(i)}(\Lambda) = \sum_{1 \leqslant h < j} (\lambda_h^{(i+1)} - 2\lambda_h^{(i)} + \lambda_h^{(i-1)}) + $$

$$+ (\lambda_j^{(i+1)} - \lambda_j^{(i)}) \tag{3}$$

We call the $d_j^{(i)}(\Lambda)$ the exponents of a G-pattern Λ.

Theorem 1. For any three highest weights λ, μ, ν for gl_n the multiplicity of the irreducible representation V_μ in the tensor product $V_\lambda \otimes V_\nu$ equals the number of G-patterns Λ of type λ and weight $\mu - \nu$ such that the exponents of Λ satisfy

$$d_j^{(i)}(\Lambda) \leqslant \nu_i - \nu_{i+1} \qquad (1 \leqslant j \leqslant i \leqslant n-1) \tag{4}$$

This theorem as well as all the other results will be proved in § 2.

Now we give a similar expression for the weight multiplicities for gl_n. Let $V_\lambda(\gamma)$ denote the weight subspace of weight γ in V_λ, i.e. $V_\lambda(\gamma) = \{ v \in V_\lambda : E_{ii}v = \gamma_i v \text{ for } 1 \leqslant i \leqslant n \}$; by definition the multiplicity of γ in V_λ is $\dim V_\lambda(\gamma)$.

Proposition 1. The multiplicity of a weight γ in V_λ equals the number of G-patterns of type λ and weight γ.

This proposition in an equivalent form with Young tableaux instead of G-patterns goes back to A.Young (cf. e.g. [6], Ch. 1, (6.4)).

Let us give a geometric interpretation of Theorem 1 and Proposition 1. Consider the vector space $R^{n(n+1)/2}$ with coordinates $(\lambda_j^{(i)})$ $(1\leq j \leq i \leq n)$ and the (unbounded) convex polytope Γ in it defined by inequalities (1); then G-patterns for gl_n may be viewed as integral points in Γ. The mapping assigning to each G-pattern Λ its type and weight extends to the linear mapping $p: \Gamma \to R^{2n}$. Proposition 1 means that the multiplicity of γ in V_λ equals the number of integral points in $p^{-1}(\lambda,\gamma)$. Furthermore, for any highest weight ν let Γ_ν denote the subset of Γ determined by inequalities (4); clearly Γ_ν is also a convex polytope. Let $p_\nu : \Gamma_\nu \to R^{2n}$ be the restriction of p to Γ_ν. Theorem 1 asserts that the multiplicity of V_μ in $V_\lambda \otimes V_\nu$ equals the number of untegral points in $p_\nu^{-1}(\lambda, \mu - \nu)$.

Proposition 1 has natural representation-theoretic interpretation. Namely there is the Gelfand-Tsetlin basis in V_λ parametrized by G-patterns of type λ, and each weight subspace $V_\lambda(\gamma)$ is spanned by its subset corresponding to G-patterns of weight γ. It turns out that Theorem 1 may be similarly interpreted.

Definition. (a) Let $\lambda = (\lambda_1,\ldots,\lambda_n)$ and $\nu = (\nu_1,\ldots,\nu_n)$ be two highest weights for gl_n, and γ a weight. Set

$$V_\lambda(\gamma,\nu) = \{v\in V_\lambda(\gamma):\ E_{i,i+1}^{\nu_i-\nu_{i+1}+1}v = 0 \ \text{for}\ 1\leq i \leq n-1\}$$

(b) For any $v \in V_\lambda$ and $i = 1,2,\ldots,n-1$ define the exponent $d_i = d_i(v)$ to be the smallest nonnegative integer such that $E_{i,i+1}^{d_i+1} v = 0$.

(c) A basis in V_λ is proper if each of the subspaces

$V_\lambda(\gamma,\nu)$ for all possible γ and ν is spanned by its subset.

(d) Let B be a basis in V_λ. By a proper parametrization of B we mean its parametrization by G-patterns of type λ $(\Lambda \longrightarrow f_\Lambda)$ such that:

(1) if the weight of Λ is γ then $f_\Lambda \in V_\lambda(\gamma)$;

(2) $d_i(f_\Lambda) = \max\limits_{1\leqslant j\leqslant i} d_j^{(i)}(\Lambda)$, where $d_j^{(i)}(\Lambda)$ is defined by (3), for each $i = 1,\ldots, n-1$.

Proposition 2. For any three highest weights λ, μ, ν for gl_n the multiplicity of V_μ in $V_\lambda \otimes V_\nu$ equals $\dim V_\lambda(\mu-\nu,\nu)$.

This fact is well-known (see e.g. [5]).

Proposition 3. (a) The subspace $V_\lambda(\gamma,\nu)$ consists of vectors $v \in V_\lambda(\gamma)$ such that $d_i(v) \leqslant \nu_i - \nu_{i+1}$ for $1 \leqslant i \leqslant n-1$.

(b) A properly parametrized basis in V_λ is proper. Conversely, each proper basis in V_λ has a proper parametrization.

Theorem 2. In each irreducible finite dimensional representation V_λ of gl_n it is possible to choose a properly parametrized basis.

Now let (f_Λ) be a properly parametrized basis in V_λ. By Proposition 3, each of the subspaces $V_\lambda(\gamma,\nu)$ in V_λ has as a basis the subset of (f_Λ) corresponding to G-patterns Λ of type λ and weight γ satisfying (4); in particular, $\dim V_\lambda(\gamma,\nu)$ is the number of such patterns. By Proposition 2 this implies Theorem 1. So, a properly parametrized basis "explains" Theorem 1 in the same way as the Gelfand-Tsetlin basis explains Proposition 1.

Theorem 3. In the case of gl_3 a properly parametrized basis in each irreducible finite dimensional representation is unique up to scalar factors.

The proper basis for gl_3 will be called canonical. Its properties are studied in [1].

Note that our proof of Theorem 2 uses the result of the very interesting paper [7]. Namely, in [7] for each irreducible finite dimensional representation of gl_n there is constructed a special basis with remarkable branching properties; we show that these properties imply that the basis is proper.

Proper bases may be used for explicit decomposition into irreducible components of the tensor product of two irreducible finite dimensional representation of gl_n. Namely, a choice of a basis in $V_\lambda(\mu-\nu,\nu)$ allows one to construct the complete family of irreducible components of $V_\lambda \otimes V_\nu$ which are isomorphic to V_μ. More precisely, note that Proposition 2 may be refined as follows: for each λ,μ,ν there is a natural isomorphism between $V_\lambda(\mu-\nu,\nu)$ and the space $\mathrm{Hom}(V_\mu, V_\lambda \otimes V_\nu)$ of linear operators $T: V_\mu \longrightarrow V_\lambda \otimes V_\nu$ commuting with the gl_n-action. A proper basis in V_λ by definition gives the basis in each $V_\lambda(\mu-\nu,\nu)$ and so makes it possible to choose the basis $T_1,\dots,T_N$ in $\mathrm{Hom}(V_\mu, V_\lambda \otimes V_\nu)$. The subspaces $T_i(V_\mu)$ are irreducible components of $V_\lambda \otimes V_\nu$ isomorphic to V_μ; taking all these subspaces for all possible μ we obtain a decomposition of $V_\lambda \otimes V_\nu$ into irreducible components. This application of proper bases stimulates their investigation.

The main results of this paper are announced in [2]. Another approach to the decomposition problem of tensor products based on the study of representation models is worked out by the authors in [3].

2. Proofs. Proof of Theorem 1. We call a weight $\gamma=(\gamma_1,\dots,\gamma_n)$ for gl_n polynomial if each γ_i is nonnegative. The first step in the proof is reduction to the case of polynomial λ,μ,ν. For this we put $\delta = (1,1,\dots,1)$ and note that the multiplicity of V_μ in $V_\lambda \otimes V_\nu$ as well as the number of G-patterns from Theorem 1 do not change under the replacement $(\lambda \longrightarrow \lambda + k\delta, \nu \longrightarrow \nu + \ell\delta, \mu \longrightarrow \mu + (k+\ell)\delta)$ for arbitrary in-

tegers k and ℓ; for k and ℓ large enough λ , μ and ν are polynomial. In this case the multiplicity of V_μ in $V_\lambda \otimes V_\nu$ is given by the classical Littlewood-Richardson rule (see e.g. [6] , Ch. 1, §9). Recall its formulation.

First of all, the multiplicity of V_μ in $V_\lambda \otimes V_\nu$ may be nonzero only when $\mu - \nu$ is polynomial, i.e., $\mu_i \geqslant \nu_i$ for $1 \leqslant i \leqslant n$. Let $D(\mu - \nu)$ denote the set $\{(i,j) \in Z^2 : 1 \leqslant i \leqslant n,\ \nu_i < j \leqslant \mu_i\}$; this set is called a skew diagram. By a Young tableau of shape $\mu - \nu$ we mean the mapping $T : D(\mu - \nu) \longrightarrow \{1,2,\ldots,n\}$ satisfying the following conditions:

$$T(i,j+1) \geqslant T(i,j) \tag{5}$$

$$T(i+1,j) > T(i,j) \tag{6}$$

A vector $\lambda = (\lambda_1,\ldots,\lambda_n)$, where λ_k is the number of points $(i,j) \in D(\mu - \nu)$ such that $T(i,j) = k$, is called the weight of T.

In drawing, the nodes of a skew diagram are usually replaced by squares as in Fig. 1. A Young tableau T is described graphically by numbering each square (i,j)

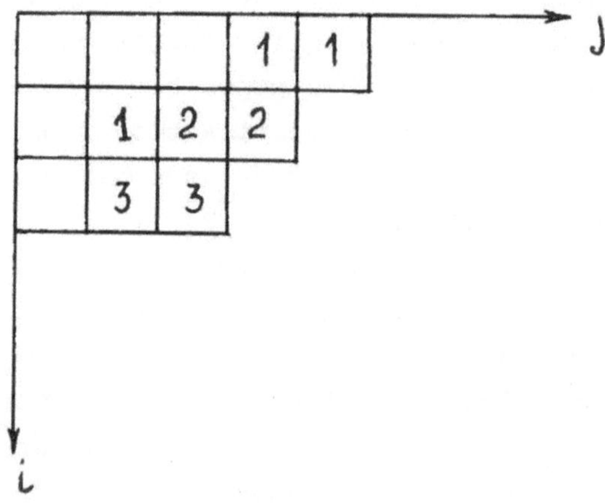

Fig. 1

with the number $T(i,j)$; the conditions (5) and (6) mean that the numbers must increase strictly down each column, and weakly from left to right along each row. The component λ_k of the weight of T is the number of squares numbered with k. For example, the Young tableau in Fig.1 is of shape $(5,4,3) - (3,1,1)$ and weight $(3,2,2)$. Let T be a Young tableau. Let $w(T)$ denote the word obtained by reading the symbols in T from right to left in successive rows, starting with the top row; for example, for T in Fig. 1 $w(T) = 1122133$. A word $w = a_1a_2...a_N$ in the symbols $1, 2, ..., n$ is said to be a lattice permutation if for $1 \leq r \leq N$ and $1 \leq i \leq n-1$ the number of occurences of the symbol i in $a_1a_2...a_r$ is not less than the number of occurences of $i+1$.

Littlewood-Richardson rule. Let λ, μ, ν be polynomial highest weights for gl_n. Then the multiplicity of V_μ in $V_\lambda \otimes V_\nu$ equals the number of tableaux T of shape $\mu - \nu$ and weight λ such that $w(T)$ is a lattice permutation.

For the proof of Theorem 1 it remains to give a one-to-one correspondence between G-patterns described in Theorem 1 and Young tableaux described in the Littlewood-Richardson rule. Let T be a Young tableau of shape $\mu-\nu$ and $1 \leq i, j \leq n$; denote by $\lambda_j^{(i)}$ the number of occurences of j in the first i rows of T (to be more formal, $\lambda_j^{(i)}$ is the number of points $(i',j') \in D(\mu - \nu)$ such that $i' \leq i$ and $T(i', j') = j$). Let $\Lambda = \Lambda(T)$ denote the pattern formed by the numbers $\lambda_j^{(i)}$ for $1 \leq j \leq i \leq n$.

Proposition 4. The mapping $T \longrightarrow \Lambda(T)$ gives a one-to-one correspondence between Young tableaux T of shape $\mu - \nu$ and weight λ such that $w(T)$ is a lattice permutation, and G-patterns Λ of type λ and weight $\mu - \nu$ satisfying (4).

Proof. It is verified directly that $w(T)$ is a lattice permutation if and only if $\lambda_j^{(i)} = 0$ for $j > i$ and

$\lambda_j^{(i)} \geqslant \lambda_{j+1}^{(i+1)}$ for $1 \leqslant j \leqslant i \leqslant n-1$; it follows that the pattern $\Lambda(T)$ is triangular and satisfies half of the conditions (1). Another half, namely, $\lambda_j^{(i)} \geqslant \lambda_j^{(i-1)}$, follows from the fact that $\lambda_j^{(i)} - \lambda_j^{(i-1)}$ is the number of squares in the i-th row of T occupied by j. By (5) T is uniquely recovered from $\Lambda = \Lambda(T)$ i.e. the i-th row of T consists of $(\lambda_1^{(i)} - \lambda_1^{(i-1)})$ 1's, $(\lambda_2^{(i)} - \lambda_2^{(i-1)})$ 2's,..., $\lambda_i^{(i)}$ i's . This description of T readily implies that the conditions (6) for T are equivalent to the inequalities (4) for $\Lambda(T)$. Finally, the fact that $\Lambda(T)$ is of type λ and weight $\mu - \nu$ follows at once from definitions.

Proposition 4 and Theorem 1 are proved.

Proof of Proposition 3. Part (a) follows at once from definitions. To prove (b) we use the following.

Proposition 5. Let λ, ν be highest weights for $\mathfrak{gl}_n$ and γ a weight. Then $\dim V_\lambda(\gamma, \nu)$ equals the number of G-patterns Λ of type λ and weight γ satisfying (4).

Proof. If $\gamma+\nu$ is a highest weight then our statement follows from Theorem 1 and Proposition 2. We show that if it is not the case then $V_\lambda(\gamma, \nu) = 0$ and there are no G-patterns of weight γ satisfying (4).

Lemma 1. If v is a nonzero vector of V_λ of weight $\gamma = (\gamma_1, \ldots, \gamma_n)$ and $1 \leqslant i \leqslant n-1$ then $d_i(v) \geqslant \gamma_{i+1} - \gamma_i$.

Proof. Restrict V_λ to the subalgebra with the basis $\{E_- = E_{i+1,i}, H = E_{ii} - E_{i+1,i+1}, E_+ = E_{i,i+1}\}$; this subalgebra is isomorphic to $\mathfrak{sl}_2$, and $Hv = (\gamma_i - \gamma_{i+1})v$. It remains to use the following well-known property of representations of $\mathfrak{sl}_2$: if v is a nonzero vector in a finite dimensional representation of $\mathfrak{sl}_2$ such that $Hv = -hv$ for some integer $h \geqslant 0$ then $E_+^h v \neq 0$.

By this lemma and part (a) of proposition 3, if $V_\lambda(\gamma, \nu) \neq 0$ then $\gamma_{i+1} - \gamma_i \leqslant \nu_i - \nu_{i+1}$ for $1 \leqslant i \leqslant n-1$ and hence $\gamma + \nu$ is a highest weight.

Now let Λ be a G-pattern of weight γ satisfying (4).

Using (1) to (4) we deduce

$$\gamma_{i+1} - \gamma_i = \sum_{j\geq 1} (\lambda_j^{(i+1)} - 2\lambda_j^{(i)} + \lambda_j^{(i-1)}) = d_i^{(i)}(\Lambda) -$$

$$- \lambda_i^{(i)} + \lambda_{i+1}^{(i+1)} \leq d_i^{(i)}(\Lambda) \leq \nu_i - \nu_{i+1} ,$$

which again implies that $\gamma+\nu$ is a highest weight. This proves Proposition 5.

Now let (f_Λ) be a properly parametrized basis in V_λ. By definition and part (a) of Proposition 3 $f_\Lambda \in V_\lambda(\gamma,\nu)$ if and only if a G-pattern Λ is of type λ, weight γ and satisfies (4). Since by Proposition 5 $\dim V_\lambda(\gamma,\nu)$ is the number of such patterns, $V_\lambda(\gamma,\nu)$ is spanned by the vectors f_Λ contained in it. Therefore, the basis (f_Λ) is proper.

Conversely, let B be a proper basis in V_λ. We fix a weight γ. For each family $(d_1,\ldots,d_{n-1})$ of nonnegative integers let $\varphi_1(d_1,\ldots,d_{n-1})$ denote the number of vectors $f\in B$ with weight γ and such that $d_i(f) = d_i$ for $1\leq i\leq n-1$; denote also by $\varphi_2(d_1,\ldots,d_{n-1})$ the number of G-patterns Λ of type λ and weight γ such that $\max_j d_j^{(i)}(\Lambda) = d_i$ for $1 \leq i \leq n-1$. By Proposition 3 (a) and Proposition 5, for all $d_1,\ldots,d_{n-1}$ we have

$$\sum_{0\leq d_i'\leq d_i} \varphi_1(d_1', \ldots, d_{n-1}') =$$

$$\sum_{0\leq d_i'\leq d_i} \varphi_2(d_1', \ldots, d_{n-1}') ;$$

indeed, both parts equal $\dim V_\lambda(\gamma,\nu)$, where ν is such that $\nu_i - \nu_{i+1} = d_i$ for $1\leq i\leq n-1$. It follows that $\varphi_1(d_1,\ldots,d_{n-1}) = \varphi_2(d_1,\ldots,d_{n-1})$ for all $d_1,\ldots,d_{n-1}$. Therefore, B admits a proper parametrization. Proposition 3 is proved.

Proof of Theorem 2. We fix an irreducible finite dimensional representation V of gl_n. By Proposition 3 it suffices to prove that V has a proper basis. In [7] a special basis in V is constructed satisfying some branching properties; we show that these properties imply that the basis is proper. We give the result of [7] in a convenient form.

For each family $\underline{s} = (s_1,\dots,s_k)$ of positive integers with sum n denote by $gl_{\underline{s}}$ the product $gl_{s_1} \times \dots \times gl_{s_k}$; $gl_{\underline{s}}$ is viewed as a Lie subalgebra of gl_n. Irreducible finite dimensional representations of $gl_{\underline{s}}$ are parametrized by families $\underline{\lambda} = (\lambda^{(1)},\dots,\lambda^{(k)})$, where $\lambda^{(i)}$ is a highest weight for gl_{s_i}; the corresponding representation of $gl_{\underline{s}}$ is $V_{\underline{\lambda}} = V_{\lambda^{(1)}} \otimes \dots \otimes V_{\lambda^{(k)}}$. Let $\lambda = (\lambda_1,\dots,\lambda_m)$ and $\lambda' = (\lambda'_1,\dots,\lambda'_m)$ be highest weights for gl_m; we shall write $\lambda \geqslant \lambda'$ if

$$\lambda_1 \geqslant \lambda'_1,\ \lambda_1+\lambda_2 \geqslant \lambda'_1+\lambda'_2,\ \dots,\ \lambda_1+\dots+\lambda_{m-1} \geqslant \lambda'_1+\dots+\lambda'_{m-1},\ \lambda_1+\dots+\lambda_m = \lambda'_1+\dots+\lambda'_m .$$

If $\underline{\lambda} = (\lambda^{(1)},\dots,\lambda^{(k)})$ and $\underline{\lambda}' = (\lambda^{(1)'},\dots,\lambda^{(k)'})$ are two families of highest weights for the same $gl_{\underline{s}}$ then we shall write $\underline{\lambda} \geqslant \underline{\lambda}'$ if $\lambda^{(i)} \geqslant \lambda^{(i)'}$ for $1 \leqslant i \leqslant k$.

Consider the restriction of V to $gl_{\underline{s}}$. For each $\underline{\lambda} = (\lambda^{(1)},\dots,\lambda^{(k)})$ corresponding to $gl_{\underline{s}}$ let $V(\underline{s},\underline{\lambda})$ denote the sum of all irreducible components of this restriction which are isomorphic to $V_{\underline{\lambda}}$. We put

$$\overline{V(\underline{s},\underline{\lambda})} = \bigoplus_{\underline{\lambda}' \leqslant \underline{\lambda}} V(\underline{s},\underline{\lambda}'), \quad V^{o}(\underline{s},\underline{\lambda}) = \bigoplus_{\underline{\lambda}' < \underline{\lambda}} V(\underline{s},\underline{\lambda}')$$

so that $V(\underline{s},\underline{\lambda})$ is identified with $\overline{V(\underline{s},\underline{\lambda})}/V^{o}(\underline{s},\underline{\lambda})$.

Theorem 4. [7]. In each irreducible finite dimensional representation V of gl_n there is a basis B such that for all $\underline{s}$ and $\underline{\lambda}$ the following conditions hold:

(a) The subspace $V(\underline{s},\underline{\lambda})$ is spanned by a subset of B;

(b) The vectors of B belonging to $\overline{V(\underline{s},\underline{\lambda})} - V^{o}(\underline{s},\underline{\lambda})$ are separated into some groups so that the images of vectors of each group in $\overline{V(\underline{s},\underline{\lambda})}/V^{o}(\underline{s},\underline{\lambda})$ form a basis in some irreducible component under the action of $gl_{\underline{s}}$.

The conditions (a) and (b) are very strong; it is even possible that B is uniquely determined by them up to scalar factors. The fact that B is proper turns out to be equivalent to a small part of these conditions.

Proposition 6. A basis B in irreducible representation V of gl_n is proper if and only if it satisfies the condition (a) of Theorem 4 for all $\underline{s} = (s_1,\dots,s_k)$ such that $s_i \leq 2$ for all i (and all $\underline{\lambda}$).

Proof. Recall that a family of subspaces in V is called a lattice if it is closed under sums and intersections. Let B be a basis in V; it is clear that the subspaces spanned by the subsets of B form a lattice. Therefore, to prove Proposition it suffices to show that the families of subspaces $V(\gamma,\nu)$ (for all possible γ and ν) and $V(\underline{s},\underline{\lambda})$ (for $\underline{s} = (s_1,\dots,s_k)$ such that $s_i \leq 2$ and all possible $\underline{\lambda}$) generate the same lattice. We prove this statement for an arbitrary finite dimensional representation of gl_n (not nesseccarily irreducible).

If $\underline{s} = (1,1,\dots,1)$ then the subspaces $V(\underline{s}, \underline{\lambda})$ are just the weight subspaces in V. Now we use a lemma about representations of gl_2 .

Lemma 2. Let V be a finite dimensional representation of gl_2 , $\gamma=(\gamma_1,\gamma_2)$ a weight and $d \geq \max(0,\gamma_2 - \gamma_1)$ an integer, Let v be a nonzero vector in V of weight γ. Then the following conditions are equivalent: (1) $d_1(v) \leq d$; (2) $v \in \overline{V(\underline{s},\underline{\lambda})}$, where $\underline{s} = (2)$, $\underline{\lambda} = (\gamma_1 + d, \gamma_2 - d)$.

This lemma follows at once from definitions and the well-known description of irreducible representation of gl_2 .

Now for each $i = 1,\dots,n-1$ consider the subalgebra of gl_n isomorphic to gl_2 with the basis (E_{ii}, $E_{i,i+1}$, $E_{i+1,i}$, $E_{i+1,i+1}$). Using Proposition 3 (a) and Lemma 1 it is easy to see that our statement that two lattices coincide follows from Lemma 2 applied to the restrictions of V to these subalgebras. This proves

Proposition 6 and Theorem 2.

Proof of Theorem 3. We use the following lemma.

Lemma 3. Let Λ be a G-pattern for gl_3 of type $\lambda = (\lambda_1, \lambda_2, \lambda_3)$ and weight $\gamma = (\gamma_1, \gamma_2, \gamma_3)$. Then $d_1^{(1)}(\Lambda) + \max(d_1^{(2)}(\Lambda), d_2^{(2)}(\Lambda))$ depends only on λ and γ and equals $\max(\lambda_1 - \gamma_1, \gamma_3 - \lambda_3)$.

Proof. Follows at once from (2) and (3) in §1.

Now let (f_Λ) be a properly parametrized basis in the irreducible representation V_λ of gl_3. Let Λ be a G-pattern of type λ. Let γ denote the weight of Λ, and $\nu = (\nu_1, \nu_2, \nu_3)$ be a highest weight for gl_3 such that $\nu_1 - \nu_2 = d_1^{(1)}(\Lambda)$, $\nu_2 - \nu_3 = \max(d_1^{(2)}(\Lambda), d_2^{(2)}(\Lambda))$.

Using Lemma 3, Proposition 5 and the definition of a proper parametrization we see that the subspace $V_\lambda(\gamma, \nu)$ is one dimensional and is spanned by f_Λ. Thus, f_Λ is uniquely determined by Λ up to a multiple. This proves Theorem 3.

Remark. Let B be a basis in the irreducible finite dimensional representation V_λ of gl_n satisfying (a) and (b) from Theorem 4. For each $i = 1,\dots,n$ consider the restriction of V_λ to the subalgebra gl_i embedded in gl_n in a standard way (i.e. gl_i has a basis (E_{jk}) with $1 \leq j, k \leq i$). Let $f \in B$; it is easy to see that there is the unique highest weight $\lambda^{(i)}$ for gl_i such that $f \in V_\lambda((i), \lambda^{(i)}) - V^o((i), \lambda^{(i)})$. Furthermore, the highest weights $\lambda^{(1)}, \lambda^{(2)}, \dots, \lambda^{(n)}$ form a G-pattern of type λ. Therefore, we obtain a parametrization of B by G-patterns of type λ. Apparently, for the basis constructed in [7] this parametrization is proper.

References

1. Gelfand, I.M., Zelevinsky, A.V. Canonical basis in irreducible representations of gl_3 and its applications. - in this volume.

2. Gelfand, I.M., Zelevinsky, A.V. (1985). Funct. Anal.

and its Appl., 19, No. 2, 72-75 (Russian).

3. Gelfand,I.M., Zelevinsky,A.V. (1984). Funct. Anal. and its Appl., 18, No. 3, 14-31 (Russian).
4. Gelfand,I.M., Tsetlin,M.L. (1950). Doklady Akad. Nauk SSSR, 71, No. 5, 825-828 (Russian).
5. Zhelobenko,D.P. (1970). Compact Lie groups and their representations. Nauka, Moscow (Russian).
6. Macdonald,I.G. (1979). Symmetric functions and Hall polynomials. - Clarendon Press, Oxford.
7. Concini,C.De, Kazhdan,D. (1981). Israel J. Math., 40, Nos. 3-4, 275-290.

ON R-MATRIX QUANTIZATION OF FORMAL LOOP GROUPS

I.V.Cherednik
Moscow State University

1. <u>The general plan of quantization.</u> We choose a basis $\{I_\alpha\}$ in the associative algebra M_N of NxN - matrices with unit I_o and for any $X \in M_N$ define $X^\alpha \in \mathbb{C}$ from the expansion $X = \sum_\alpha X^\alpha I_\alpha$ in terms of I_α

$$^1I_\alpha = I_\alpha \otimes I_o \otimes \dots \otimes I_o, \quad ^2I_\alpha = I_o \otimes I_\alpha \otimes I_o \otimes \dots \otimes I_o, \text{ etc. },$$

$$^{ij}X = \sum_{\alpha,\beta} X^{\alpha\beta}\, {}^iI_\alpha\, {}^jI_\beta \quad \text{for } X = \sum_{\alpha,\beta} X^{\alpha\beta}\, I_\alpha \otimes I_\beta \in \in M_N \otimes M_N$$

etc. Given $X(u) \in M_N$ or $X(u,v) \subset M_N^{\otimes 2}$ for $u,v \in \mathbb{C}$ we write $^iX = {}^i(X(u_i))$, $^{ij}X = {}^{ij}(X(u_i,u_j))$, where $\{u_j\}$ are independent parameters, connected with each component of tensor products under consideration. In the similar way for $X(u) \in M_N^{\otimes 2}$ we set $^{ij}X = {}^{ij}(X(u_{ij}))$, where $u_{ij} = u_i - u_j$.

The infinitely-dimensional group $\tilde{GL}_N$ of invertible formal series $L = \sum_{i=-k}^{\infty} L_i u^i$, $k \in \mathbb{Z}$, in $u \in \mathbb{C}$ with coefficients $L_i \in M_N$ is called a loop group. For simplicity we will further deal only with the variety $\tilde{M}_N$ of all such L (not only invertible) instead of $\tilde{GL}_N$ and will quantize its multiplicative structure; $\tilde{GL}_N$ is open and dense in $\tilde{M}_N$. Formal power series of the coefficients L_i^α as variables, which have for every $k \in \mathbb{Z}$ only a finite number of terms (monomials) without any

L_i^α , $i \leq -k$, form the ring $\mathbb{C}[\tilde{M}_N]$ of regular functions on $\tilde{M}_N$.

The series $(LL')_i^\alpha$ in L_j^β , L'^γ_k can be considered as elements of $\mathbb{C}[\tilde{M}_N] \otimes \mathbb{C}[\tilde{M}'_N]$ after the substitution $L_j^\beta L'^\gamma_k \longmapsto L_j^\beta \otimes L'^\gamma_k$. Here LL' denotes the usual product of power series. Let $\delta : \mathbb{C}[\tilde{M}_N] \longrightarrow \mathbb{C}[\tilde{M}_N] \otimes \mathbb{C}[\tilde{M}'_N]$ be a homomorphic extension of the mapping $L_i^\alpha \to (LL')_i^\alpha$.

A quantization of $\tilde{M}_N$ is a $\mathbb{C}$-vector space isomorphism ρ of $\mathbb{C}[\tilde{M}_N]$ and a $\mathbb{C}$ - algebra A , generated by monomials in $\rho(L_i^\alpha)$. A quantization is called consistent with the multiplication, when there exists a rings homomorphism $\Delta : A \to A \otimes A'$ such that $\Delta(\rho(L_i^\alpha)) = \rho(\delta(L_i^\alpha))$. Here we suppose that the elements of A commute with elements of A' in $A \otimes A'$. This definition is a matrix version of Drinfeld's definition of a quantum group [6] .

Following L.D.Faddeev and the others (see e.g. [1]) a quantization is constructed by means of some initial analitical (or meromorphic) matrix function $R(u) \in M_N^{\otimes 2}$, $u \in \mathbb{C}$. The equation

$$ {}^{12}R \; {}^{1}L \; {}^{2}L = {}^{2}L \; {}^{1}L \; {}^{12}R \tag{1} $$

becomes equivalent to some relations in terms of L_i^α after expanding R at $u = 0$. All functions that vanish when (1) holds form a two-sided ideal K in the ring $\mathcal{T}$ of all "non-commutative" power series in L_i^α with a finiteness condition like that for $\mathbb{C}[\tilde{M}_N]$ ($\mathcal{T}$ is the projective limit of the tensor algebras of the spaces $V_k = \bigoplus_{\alpha, i \geq -k} \mathbb{C}L_i^\alpha$). Denote $R_N = \mathcal{T}/\mathcal{K}$ Later on we suppose that $R(0)$ is the permutation matrix: $(R(0)(I_\alpha \otimes I_\beta) = (I_\beta \otimes I_\alpha) R(0))$ and let R depend on a parameter $\eta \in \mathbb{C}$ so that $\lim_{\eta \to 0} R = I_0 \otimes I_0$ after a normalization.

Reordering L_i^α in monomials of $\mathbb{C}[\tilde{M}_N]$ exicographically for the standard ordering on i's and some fixed ordering on $\{ \alpha \}$. Next, let us construct the corresponding injection $\rho : C[\tilde{M}_N] \longrightarrow \mathcal{T}$. This ρ induces an isomorphism $\rho : C|\tilde{M}_N| \xrightarrow{\sim} \mathcal{R}_N$ if R satisfies the factorization equation

$$ {}^{12}R \; {}^{13}R \; {}^{23}R = {}^{23}R \; {}^{13}R \; {}^{12}R \tag{2} $$

The inverse is true when (2) is treated up to multiplication by a scalar matrix. This observation has been actually made by A.B.Zamolodchikov in the context of indeterminate dependence on u (u is considered as some continuous index of L^{α}) and for the "continuous" variants of $\mathbb{C}\,[\tilde{M}_N]$ and $\mathcal{R}_N$. The case of formal series L has no new essential features.

The existence of Δ is a direct consequence of (1). Namely,we should prove that if both L and L' with pairwise commuting coefficients are solutions of (1) then so is LL' . For factorizable R the quotient algebra $\mathcal{R}_N$ is called an R-algebra. Its continuous (see above) analogue was first introduced by L.D.Faddeev, L.A.Takhtadzhan and E.K.Sklyanin and plays the main role in the quantum method for the inverse scattering problem.

The general solution of (2) is Baxter-Belavin's R-matrix in terms of elliptic functions of u and η (see e.g. [2,3]). We can obtain finite-dimensional subvarieties $Z_N^k \subset \tilde{M}_N$, confining ourselves to a rational elliptic $L(u)$ (of the same type as R) with poles only at $u = 0$ (modulo symmetries) of order $\leq k$ ($L_i^{\alpha} = 0$ for $i < -k$). Futher, we show that the image $\rho\,(J^k)$ of the ideal $J^k \subset \mathbb{C}\,[\tilde{M}_N]$ of all functions that vanish at two-sided ideal in $\mathcal{R}_N$. Hence, the restriction

$$ \rho : \mathbb{C}\,[Z_N^k] \overset{\mathrm{def}}{=} \mathbb{C}\,[\tilde{M}_N]\,/J^k \longrightarrow \mathcal{R}_N^k \overset{\mathrm{def}}{=} \mathcal{R}_N/\rho\,(J^k) $$

may be considered as a quantization of Z_N^k. Then, we construct some (probably all) finite-dimensional irreducible representations of $\mathcal{R}^k_N$.

When η tends to zero (in the quasi-classical limit) we obtain the Poisson bracket { , } on $\mathbb{C}[\tilde{M}_N]$. It was verified for Z^1_2 in [4] and for arbitrary Z^k_N in [5], that $\{ J^k , \mathbb{C}[\tilde{M}_N] \} \subset J^k$. This result is a preliminary version of the quantization of Z^k_N mentioned above. We will transfer to $\mathcal{R}^k_N$ other results of [5] (concerning the centre and the maximal commutative subalgebra of $\mathbb{C}[Z^k_N]$ with { ,}).

The algebra $\mathcal{R}^1_2$ was introduced by E.K.Sklyanin [4]. V.G.Drinfeld constructed $\mathcal{R}^\infty_N$ and its variants for other simple Lie algebras, when the R-matrix is Yang's one (a degeneration of Belavin's R-matrix). Here $\mathcal{R}^\infty_N = \varprojlim \mathcal{R}^k_N$ as $k \longrightarrow \infty$ corresponds to $Z^\infty_N = \bigcup_k Z^k_N$.

Some irreducible representations of $\mathcal{R}^1_2$ of arbitrary dimension ≥ 2 are found in [7]. The algebra $\mathcal{R}^1_N$ and a generalization of the symmetric and exterior powers of the fundamental N-dimensional representation of $\mathcal{R}^1_N$ are defined in [8]. The technique of [8] is based on the approach of [9] (for Belavin's R-matrix developed in [2]). The relation between representations of $\mathcal{R}^\infty_N$ for Yang's matrix and the degenerate affine Hecke algebra $\mathcal{H}_N$ (q=1) is the main result of [11]. The paper [12] is devoted to irreducible finite-dimensional representations of $\mathcal{H}_N$ (q=1) and their applications. Some representations of Yang's $\mathcal{R}^\infty_N$ were obtained in [9] (in an R-matrix form).

2. <u>Generalized Young projections</u>. Let S_n be the symmetric group acting on any n-element set, $s_i = (i,i+1)$ the permutation which interchanges elements in places i and i+1, s_o the unit of S_n, $\mathbb{C}[S_n] = \bigoplus_{w \in S_n} \mathbb{C}w$ the group algebra of S_n. Permutations $w \in S_n$ act on any function f of $u_1,\dots,u_n$ by the formula $({}^w f)(u_1,\dots,u_n) = f(w^{-1}(u_1,\dots,u_n))$, Let ε be an associative $\mathbb{C}$-algeb-

ra. The minimal number of s_i's in a representation of $w \in S_n$ as a product of s_i's is the length of w; the element w_o of maximal length transforms $(1,\dots,n)$ into $(n,\dots,1)$.

We start with a collection of functions $\{\varphi_x(u_1,\dots,u_n)\}$, $x \in S_n$, with values in ε, which is a homogeneous 1 - cocycle, i.e. $\varphi_{xy} = \varphi_y^{\,y^{-1}} \varphi_x$ if $l(xy) = l(x) + l(y)$. We should check the relations

$$\varphi_{s_i}^{\;s_i} \varphi_{s_{i+1}}^{\;s_i s_{i+1}} \varphi_{s_i} = \varphi_{s_{i+1}}^{\;s_{i+1}} \varphi_{s_i}^{\;s_{i+1} s_i} \varphi_{s_{i+1}} \tag{3}$$

Later we suppose that $\varphi_{s_i} = \varphi_i(u_{i,i+1})$ for some analitical functions $\varphi_i(u)$, $1 \le i \le n$, $u \in \mathbb{C}$. Let us assume that both sides of (3) identically vanish when $u_{i,i+2} = 0 = u_{i,i+1} - \eta$ for some nonzero $\eta \in \mathbb{C}$ $(i = 1,\dots,n-1)$.

A generalized Young diagramm μ of degree n is given by two sequences of integers $\{m_i' \le m_i\ ,\ 1 \le i \le r\}$ such that $\sum_{1 \le i \le r} (m_i - m_i') = n - r$ and either $m_i \ge m_{i+1}$ or $m_{i+1} - m_i = 1 \ge m_{i+1}' - m_i'$ for each $1 \le i \le r$. It can be visualized by nodes placed in points $(i,j) \in \mathbb{Z}^2$ such that $m_i' \le j \le m_i$. One obtains a usual Young diagram if $m_i' = 1$, $m_i \ge m_{i+1}$. The diagram is skew when $m_i \ge m_{i+1}$ and $m_i' \ge m_{i+1}'$. Suppose that μ is represented as a disjoint union of subdiagrams $\mu = \bigcup_p \mu_p$, where $p = p_1,\dots,p_r$, $\mu_p = \{(i,j),\ p_i = p\}$, $1 \le p_1 \le \dots \le p_r \le r$ and $m_i' \ge m_{p_k}$ if $p_i < p_k$. Any different $\mu_p, \mu_{p'}$ can be separated by diagonals $i - j = \text{const}$. Let us fix a sequence $\{d_i \in \mathbb{C}\ ,\ 1 \le i \le r\}$, in which $d_i = d_k \iff p_i = p_k$. Further, let $d_i - d_k \notin \mathbb{Z} \setminus 0$ for $1 \le i, k \le r$.

One can construct a natural Young tableau by replacing the nodes of μ by $1,\dots,n$ in order to the right successive rows: $k > l \Leftrightarrow$ either $i_k > i_l$ or $i_k = i_l$ and $j_k > j_l$, where (i_k, j_k) are coordinates of k. Let us consider φ_{w_o} as a function of $v \in \mathbb{C}$ setting

$$u_k = (i-p_i)v + u_k^o\,\eta\ ,\quad u_k^o = d_i + i - p_i - j_k + m'_{p_i}\ . \tag{4}$$

We need p_i, m'_{p_i} in (4) to get $u_k = d_k\eta$ for the first (minimal) k in any μ_p.

Theorem 1 (cf. [10, 12, 13]). The function $\varphi_{w_o}(v)$ has zero at $v = 0$ of order $\geq \varkappa - \gamma$, where $\varkappa$ is the number (#) of pairs $\{k,l\}$, $1 \leq k < l \leq n$, for which $i_k - j_k = i_l - j_l$, $\gamma = \#\,\{1 \leq i < t \leq r \mid m_i - m_t = t - i = m'_i - m'_t\}$.

The Young projectors for φ, μ are some $\hat{\Phi}_\mu$, $\check{\Phi}_\mu \in \varepsilon$ such that $v^{-\varkappa+\gamma}\varphi_{w_o}(v{=}0) = \hat{\Phi}_\mu \hat{F}_\mu = \check{F}_\mu \check{\Phi}_\mu$, $\Phi_\mu^2 = \Phi_\mu$, where F_μ are invertible.

Theorem 2 (cf. [8,10]). For μ_1 to be the only node (i.e. for $p_2 = 2$) and $u = d_1\eta$ the function $\varphi_{w_o}(v,u)v^{-\varkappa+\gamma}\prod (u-a\eta)^{-\varkappa_a+\gamma_a}\big|_{v=0}$ is analitical in a neighbourhood of $v = 0 = u - a\eta$, where $\varkappa_a = \#\,\{k > 1 \mid u_k^o = a\}$, $\#\{i\}_a - \gamma_a = \max\{r \mid \exists\, i_1 < i_2 < \ldots < i_r,\ t_1 < \ldots < t_r,\ i_s \in \{i\}_a,\ 0 \leq m'_{i_s} - m'_{t_s} = i_s - t_s - 1 \geq m_{i_s} - m_{t_s}\}$, $\{i\}_a = \{i = i_k > 1 \mid j_k = m'_i,\ u_k = a\}$.

To introduce the R-matrix language one needs the assumption $\mathbb{C}[S_n] \subset \varepsilon$. We also require φ to be "local": ${}^{ij}\varphi = w({}^{w}\varphi_{s_1})w^{-1}$ is not to be dependent on the choise of the permutation $w \in S_n$ translating (1,2) to places (i,j). Setting $R_w = \varphi_w w$, ${}^{ij}R = {}^{ij}\varphi\,(i,j)$ makes (3) and (2) equivalent.

Untill the end of this section ${}^{i\hat{3}}R = {}^{i3}R \ldots {}^{in+2}R$ for $i = 1,2$. One can easily show that ${}^{2\hat{3}}R = R_\pi$, ${}^{1\hat{3}}R = s_1({}^{s_1}{}^{1}R_\pi)s_1$ for $\pi = s_{n+1} \ldots s_2 : (1,\ldots,n+2) \to (1,3,\ldots,n+2,2)$ in S_{n+2}. Then ${}^{12}R\,{}^{1\hat{3}}R\,{}^{2\hat{3}}R = {}^{2\hat{3}}R\,{}^{1\hat{3}}R\,{}^{12}R$.

Exactly the same is true for $^{1\check{3}}R = {}^{1\,n+2}R \ldots {}^{13}R$ (we use the notations of § 1 with $\hat{3} = (3,\ldots,n+2)$, $\check{3} = (n+2,\ldots,3)$). Starting with μ, w_o such that $\deg w_o = n$ and the corresponding Φ_μ let us substitute $v=0$ in (4) (after the shift of indices by 2) for $u_3,\ldots, u_{n+2}$ and construct the functions

$$^{i\hat{}}L = {}^{1\hat{3}}R \; {}^{3\hat{}}\Phi_\mu = {}^{\hat{3}}\hat{\Phi}_\mu \; {}^{1\hat{3}}R \; {}^{\hat{3}}\hat{\Phi}_\mu$$

$$^{i\check{}}L = {}^{\hat{3}}w_o(\check{\Phi}_\mu) \; {}^{1\check{3}}R = {}^{\hat{3}}w_o(\check{\Phi}_\mu) \; {}^{1\check{3}}R \; {}^{\hat{3}}w_o(\check{\Phi}_\mu)$$

of u_i, $i=1,2$, where $w_o(\check{\Phi}_\mu) \overset{\text{def}}{=} w_o \check{\Phi}_\mu w_o$. Then (1) is valid both for $R, \hat{L}$ and $R, \check{L}$.

We will consider bel-ow two important examples of cocycles φ: intertwining operators of irreducible finite-dimensional modules over the Hecke algebra and the Baxter-Belavin R-matrix.

3. <u>Representations of Hecke algebras</u>. The Hecke algebra H_n^q, $q \in \mathbb{C}^*$ is generated over $\mathbb{C}$ by $1, T_1,\ldots,T_{n-1}$ with the relations $[T_i, T_j] = 0$ whenever $i \neq j \pm 1$, $T_i T_{i+1} T_i = T_{i+1} T_i T_{i+1}$ and $(T_i - q)(T_i + 1) = 0$. The algebra H_n^1 becomes $\mathbb{C}[S_n]$ after replacing T_i by s_i.

<u>Proposition 3</u> [13] . The set of functions $\varphi_i^q(u) = 1 + \dfrac{1-e^u}{1-e^\eta} T_i$, where $e^\eta = q \neq 1$, extends to a homogeneous 1-cocycle $\{\varphi_x^q\}$ with η satisfying the conditions of §2.

When we substitute $u \mapsto \delta u$, $q \mapsto e^{\delta\eta}$ and make η tend to zero we obtain "local" (§2) $\varphi_i^1 = 1 + (u/\eta) s_i$ after the identification of H_n^1 with $\mathbb{C}[S_n]$. The corresponding $R^1 = \varphi_1^1 s_1 = s_1 + u/\eta$ is called the Yang R-matrix [14] , if s_1 is interpreted as the permutation matrix for some $\mathbb{C}^N \otimes \mathbb{C}^N$.

By adding new pairwise commuting independent elements

$x_1,\ldots,x_n$ with the relations

$$x_iT_i - T_ix_{i+1} = (q-1)x_{i+1} = T_ix_i - x_{i+1}T_i \qquad (5)$$

to $\{T_i\}$ for $q \neq 1$ we get the affine Hecke algebra $\mathcal{H}_n^q$ (see e.g. [13]). We omit x_i^{-1} met in the standard definition of $\mathcal{H}_n$. After substituting $q \mapsto e^{\delta\eta}$, $x_j \mapsto e^{\delta x_j}$, then reducing by δ and making $\delta \to 0$ we obtain some algebra $\mathcal{H}_n^1$ with relations

$$x_iT_i - T_ix_{i+1} = \eta = T_ix_i - x_{i+1}T_i \qquad (5')$$

instead of (5) (see [11]).

Given a diagram μ we construct the element $\tilde{\varphi}_{w_o}^q = v^{-\kappa+\gamma}\varphi_{w_o}^q(v=0) \in H_n^q$ (see §2). Let $x_1,\ldots,x_n$ act on $1 \in H_n^q$ by the formulas $x_k(1) = e^{-u_k^o\eta}$ for $q \neq 1$ and $x_k(1) = -u_k^o\eta$ for $q = 1$. Here u_k^o are from (4). These relations produce the only H_n^q action on H_n^q , extending the right regular action of H_n^q on itself. The latter is introduced by formulas $T_i(a) = aT_i$, $T_iT_j(a) = aT_jT_i$, etc.

<u>Theorem 4</u> (cf. [18, 13, 12]). The subspace $V_\mu = \tilde{\varphi}_{w_o}^q H_n^q$ is an irreducible $\mathcal{H}_n^q$ - submodule of the $\mathcal{H}_n^q$ - module H_n^q (see above) for q from some open dense subset of $\mathbb{C}^*$, containing 1. Furthermore, every finitedimensional irreducible $\mathcal{H}_n^q$ - module is of the form V_μ for a proper μ . The non-ordered collections of pairs $\{(\tilde{\mu}_p, d_p)\}$ are in one-to-one correspondence with the isomorphism classes of V_μ , where $\tilde{\mu}_p$ is the class of μ_p modulo a shift in $\mathbb{Z}^2$.

<u>Theorem 5</u> [12] . For every $q \in \mathbb{C}^*$ an arbitrary finite-dimensional irreducible $\mathcal{H}_n^q$ - module with a semi-simple operation of the subalgebra $\mathbb{C}[x_1,\ldots,x_n]$ is isomorphic to V_μ for some skew diagram μ (see § 2).

Given a skew μ let us define the permutation ω transforming $(1,\dots,n)$ into the sequence of integers written out of the tableau μ in the following order: k presedes l $\Leftrightarrow$ either $j_k > j_l$ or $i_k > i_l$ if $j_k = j_l$. We write $w' \geq w \Leftrightarrow l(w'w^{-1}) + l(w) = l(w')$ for w, $w' \in S_n$.

Theorem 6 [12]. For every $w \geq \omega$ and v from (4) we have $\varphi^q_w(v) = E^\mu_w v^\kappa + o(v^{\kappa+1})$, $V_\mu = E_w H^q_n$. The elements $E_w = E^\mu_w$, $w \geq \omega$, form a $\mathbb{C}$ - basis in V_μ. Set $z_i = \eta(u^o_{w^*_i} - u^o_{w^*_{i+1}})$, where w transfers an element from the place w^*_i to the place i, $1 \leq i \leq n$. Then the expression $A = E_w(1 + \frac{1-e^{z_i}}{1-e^{\eta}} T_i)$ for $q \neq 1$ and $A = E_w(1 + \frac{z_i}{\eta} T_i)$ for $q = 1$ vanishes if and only if $\sigma_i w \not\geq \omega$ or, which is the same, $z_i = \pm \eta$. In other cases $A = E_{\sigma_i w}$ for any q if $\sigma_i w \geq w$, $A = (1 + \frac{2-e^{z_i}-e^{-z_i}}{(1-e^{\eta})^2} e^{\eta}) E_{\sigma_i w}$ for $q \neq 1$ and $A = (1 - z_i^2) E_{\sigma_i w}$ for $q = 1$ whenever $\omega \leq \sigma_i w \leq w$. Furthermore, $x_i(E_w) = \exp(-u^o_k \eta) E_w$ for $q \neq 1$ and $x_i(E_w) = -u^o_k \eta E_w$ for $q = 1$, where $k = w^*_i$.

Next, we state an analogue of Young's branching property for $\{E_w\}$. Let $j_s = m_{i_s}$ for some s $(1 \leq s \leq n)$ and the complement $\mu \setminus (i_s, j_s) = \mu^s$ be again a skew diagram of degree $n-1$. Set $V^s_\mu = \bigoplus_{w'} \mathbb{C} E_{w'}$, where $w'(1,\dots,n) = (s,\dots)$. Ommiting s one constructs the permutation "w withouts" $\hat{w}'$: $(1,\dots,\hat{s},\dots,n) \mapsto (\hat{s},\dots)$. We map $\mathcal{H}^q_{n-1}$ into $\mathcal{H}^q_n$ setting $T_i \mapsto T_{i+1}$, $x_i \mapsto x_{i+1}$ for $1 \leq i < n-1, n$. The image will be denoted by $\check{\mathcal{H}}^q_{n-1}$.

Theorem 7 [12]. Every V^s_μ is a $\check{\mathcal{H}}^q_{n-1}$ -submodule of $V_\mu = \bigoplus_s V^s_\mu$. The mapping $E_{w'} \mapsto E_{\hat{w}'}$ of V^s_μ to the $\mathcal{H}^q_{n-1}$ - module V_{μ^s} is an $\mathcal{H}^q_{n-1}$ - module isomorphism. The same is true, when $j_s = m'_{i_s}$, $w'(1,\dots,n) = (\dots,s)$ and $\mathcal{H}^q_n$ contains $\mathcal{H}^q_{n-1}$ in a natural way.

Theorem 8. Under the assumptions of Theorem 4 V_μ is irreducible as an H^q_n - module if and only if μ is skew and either $m'_1 = m'_r$ or $m_1 = m_r$. In the case $q = 1$ if $m'_1 = m'_r$ the action of $x_i + d_1\eta$ and $\chi_i = \sum_{l=i+1}^{n} (i,l) \in \mathbb{C}[S_n] = H^1_n$ is the same. When $m_1 = m_r$ this is true for $x_i + u^o_n\eta$ and $\chi_i = -\sum_{l=i}^{i-1} (i,l)$ $(i = 1,\dots,n)$.

The map $T_i \to s_i$, $x_i \to \chi_i$ is a homomorphism $\mathcal{H}^1_n \to H^1_n$, as was shown in [11] (see also [9] for some preliminary results). It follows from Theorem 8 that the action of x_i cannot be expressed in terms of elements of H^1_n for any skew μ with $m'_1 \neq m'_r$ and $m_1 \neq m_r$.

The module V_μ for $q = 1$ and a skew μ coincides as an S_n - module with the standard S_n - module corresponding to μ (see [15]). It follows from Theorem 6,7 or

Proposition 9. $E_\omega = c_\mu P_\mu Q_\mu$ for a skew μ and $q = 1$, where P_μ is the symmetrizer of the horizontal group (the group of permutations which preserve the number of elements in the rows) and Q_μ is the anti-symmetrizer of the vertical group. If either $m'_1 = m'_r$ or $m_1 = m_r$, then $E_{w_o} = c'_\mu P_\mu Q_\mu P_\mu$ $(c_\mu, c'_\mu \in \mathbb{C}^*)$.

Some formulas of a multiplicative type for Young projections follow from Proposition 9. For example, if μ is a row or a column then

$$\sum_{w \in S_n} (\pm 1)^{\mathrm{sgn}\, w}\, w = \mathrm{const} \prod_{i>j} (i-j \pm (i,j)) ,$$

where (i,j) precedes (i',j') whenever $j > j'$ or $j = j'$

and $i > i'$ (see [9] and [2,3]). The author does not know whether such formulas have ever occured before[9].

One can construct for a given skew μ two skew diagrams μ^{v}, μ^{*} by reflections in a point of $\mathbb{Z}^2$ and a line $i-j$ = const respectively. We briefly note that the automorphism of $\mathcal{H}_n^1$, transfering s_i to s_{n-i} and x_i to $-x_{n+1-i}$, maps V_μ onto V_μ and transforms E_w^μ into $E^{\mu^v}_{w_o w w_o}$. The involution $T_i \to -T_{n-i}$, $x_i \to x_{n+1-i}$ transfers E^μ_ω to $E^{\mu^*}_{\omega^*}$ ($\omega^* = w_o \omega^{-1} w_o$) and maps V_μ onto the left analogue of V_{μ^*}. The latter one is related with the Zelevinsky involution [20].

In conclusion we mention that there are analogoues of $\mathcal{H}_n^q$ and $\mathcal{Y}^q$ for arbitrary Weyl groups (see [13,16]). We hope that after some degeneration the cocycles $\mathcal{Y}$ can be transformed into the cocycles by Asherova, Smirnov, Tolstoy, Zhelobenko used for a constructive definition of the extremal projection in the enveloping algebras of simple Lie algebras [19] .

4. Representations of R-algebras. Denote by $\theta^o_\alpha(u) = \theta^o_\alpha(u;\tau)$ the theta-function for some fixed $\tau \in \mathbb{C}$ ($\operatorname{Im}\tau > 0$) with the characteristics $\alpha_1/N + 1/2$, $\alpha_2/N + 1/2$, where $\alpha = (\alpha_1, \alpha_2) \in \mathbb{Z}_N^2$ ($\mathbb{Z}_N = \{0,1,\dots,N-1\}$). Set

$$\rho_\alpha(u) = \theta^o_\alpha(u+\eta/N) \,/\, \theta^o_\alpha(\eta/N), \qquad (6)$$

where $\eta \in \mathbb{C}^*$, $\alpha \in \mathbb{Z}_N^2$ (see, e.g. [2,3]). Then ρ_α/ρ_o are doubly periodic functions both of u and η for the lattice of periods $N\Lambda = N\mathbb{Z} + N\mathbb{Z}\tau$ (i.e. ρ_α/ρ_o are rational functions on the elliptic curve $E = \mathbb{C}/N\Lambda$). We also have $\rho_\beta/\rho_o(u+\alpha_1+\alpha_2\tau) = \varepsilon^{\langle\alpha,\beta\rangle}\rho_\beta/\rho_o(u)$, where $\varepsilon = \exp(2\pi i/N)$, $\langle\alpha,\beta\rangle = \alpha_1\beta_2 - \beta_1\alpha_2$, $\alpha,\beta \in \mathbb{Z}_N^2$.

Let matrices $I_\alpha = g^{\alpha_1} h^{\alpha_2}$ be defined for $\alpha \in \mathbb{Z}_N^2$, some fixed $\sigma \in \mathbb{Z}_N^*$ and fixed matrices g,h such that $g^N = h^N = I_o$ and $hg = \varepsilon gh$. The matrix function $R(u) =$

$= \frac{1}{N} \sum_{\alpha \in \mathbb{Z}_N^2} \rho_\alpha(u)\, I_\alpha \otimes I_\alpha^{-1}$ of $u \in \mathbb{C}$ is called the Baxter-Belavin R-matrix. It satisfies all the assumptions of §2 (the locality, the relation (2), the condition for η) for $\varepsilon = M_N^{\otimes n}$ and S_n, operates by interchanging components (upper-left indices) in tensor products. Hence, one can introduce R_w, $\varphi_w = R_w w^{-1}$ and define $\varphi_{w_o}(\eta) = v^{-\kappa+\gamma}\ {}_{w_o}(v{=}0)$ for a diagram μ (and numbers $\alpha_1, \ldots, \alpha_r$). The matrix $\tilde{\varphi}_{w_o}$ up to a scalar matrix is a rational function of η on E.

Let us calculate $\tilde{\varphi}_{w_o}(\eta = 0)$. After rescaling $u, \eta \mapsto \delta u, \delta\eta$ in (6) we make δ tend to zero. The result is $I_o \otimes I_o \frac{u}{\eta} + P$, where $P = \frac{1}{N} \sum_\alpha I_\alpha \otimes I_\alpha^{-1}$ is the permutation matrix. After identification the elements of $H_n^1 = \mathbb{C}[S_n]$ with the corresponding matrices in $M_N^{\otimes n}$ we obtain Yang's matrix R^1. Hence, $\tilde{\varphi}_{w_o}(\eta = 0) = \varphi^1_{w_o}$ ($\varphi^1_w = R^1_w w$). There is some degeneration procedure $\varphi \to \varphi^q$, $q \neq 1$, too, but we do not consider it here.

Proposition 11. For η of a neighborhood of zero $\mathrm{rk}\,\tilde{\varphi}_{w_o}(\eta) = \mathrm{rk}\,\tilde{\varphi}^1_{w_o}$.

To prove it we use

<u>Lemma 12</u> (cf. [13]). There exist elements $\{z_i\} \subset S_n$ with the following properties. Set $z_i^{-1} \varphi_{z_i}(v) = v^{\lambda_i} \hat{\varphi}_{z_i} + O(v^{\lambda_i+1})$, $\hat{\varphi}_{z_i} \neq 0$ for an appropriate $\{\lambda_i\} \subset \mathbb{Z}_+$, v from (4) and η of a punctured neighborhood of zero. Then $\{c \in \mathbb{C}[S_n] \mid \hat{\varphi}^1_{z_i} c = 0$ for each $i\} = V_\mu = \tilde{\varphi}^1_{w_o} \mathbb{C}[S_n]$ (may be after some rearranging of the rows of),

When η tends to zero the ranks of $\tilde{\varphi}_{w_o}$, $\{\hat{\varphi}_{z_i}\}$ might decrease and the kernels of the operators with these matrices might get in a less general position. This proves Proposition 11.

Starting with $\tilde{\varphi}_{w_o}$ let us construct left and right projections $\hat{\Phi}_\mu$, $\check{\Phi}_\mu$ (onto the image of $\tilde{\varphi}_{w_o}$ and parallel to the kernel of $\tilde{\varphi}_{w_o}$). Since $(I_\alpha \otimes {}^oI_\alpha)R = R(I_\alpha \otimes I_\alpha)$ we may suppose that $(I_\alpha^{\otimes n})\,\Phi_\mu(I_\alpha^{\otimes n})^{-1} = \Phi_\mu$, where $I_\alpha^{\otimes n} = {}^1I_\alpha \ldots {}^nI_\alpha$, $\alpha \in \mathbb{Z}_N^2$. Using the rationality of $\tilde{\varphi}_{w_o}$ we choose $\hat{\Phi}_\mu$, $\check{\Phi}_\mu$ to be rational (in η) with $\Phi_\mu^1 \overset{o}{=} \Phi_\mu$ $(\eta = 0)$ being projections for $\tilde{\varphi}^1_{w_o}$.

After adding to μ one more component with only one node as μ_1 we can use notations of Theorem 2 for the resulting diagram. Put $\sum_a \gamma_a = g_\mu$ ($g_\mu = g$ is a positive integer), $\sum_a \gamma_a \eta a = \xi_\mu = \xi \in \mathbb{C}$ (a runs all the complex numbers, but $\gamma_a = 0$ for almost all a). Let us introduce the doubly periodic mod Λ (rational on $\mathbb{C}/\Lambda$) function of u

$$f_\mu(u) = \theta_o^o\left(u + (n\eta/N - \xi)g^{-1}\right)^g \theta_o^o(\eta/N)^{-n} \cdot \prod_a \theta_o^o(u - a\eta)^{\kappa\, a^-\gamma_a}$$

and the matrices ${}^i\hat{L}' = f_\mu^{-1}(u)\,{}^i\hat{L}(u)$, ${}^i\check{L}'(u) = f_\mu^{-1}(u)\,{}^i\check{L}(u)$ satisfying (1) (see the end of § 2). From Theorem 2 and the symmetry properties of R, Φ one can deduce

<u>Proposition 10</u> (see [8,10]). ${}^iL' = {}^i\hat{L}'$ (or ${}^i\check{L}'$) =

$$= \sum_{\substack{0\le k\le g-1\\ \alpha\in\mathbb{Z}_N^2}} \rho_\alpha^{(k)}\left(u + \left(\frac{(n-g)\eta}{N} - \xi\right)g^{-1}\right) {}^iI_\alpha\, {}^{\hat{3}}J_{k\alpha},$$

where $\hat{3} = (3,\ldots,n+2)$, $J_{k\alpha} = \hat{J}_{k\alpha}, \check{J}_{k\alpha} \in M_N^{\otimes n}$ depend only on η, $(I_\alpha^{\otimes n})\,J_{k\beta}\,(I_\alpha^{\otimes n})^{-1} = \varepsilon^{\sigma<\alpha,\beta>}$ for $0 \le k \le g-1$, $\alpha,\beta \in \mathbb{Z}_N^2$. (Here after $\rho_\alpha^{(k)} = \rho_{\alpha_o}^k\, \rho_o^{-k-1} \rho_{\alpha - k\alpha_o}$ for some fixed $0 \ne \alpha_o \in \mathbb{Z}_N^2$).

Let us consider

$$L = \sum_{\substack{0\le k\le g-1\\ \alpha\in \mathbb{Z}_N^2}} \rho_\alpha^{(k)}(u)\, I_\alpha\, A_{k\alpha}\ , \tag{7}$$

where $A_{k\alpha}$ are independent (pair-wise non-commuting) variables. All polynomials in $A_{k\alpha}$, that vanish when (1) holds for such L and arbitrary u_1, u_2, form a two-sided ideal $\mathcal{K}_\eta$ in the tensor algebra $\mathcal{T}$ of $\bigoplus_{k,\alpha} \mathbb{C}\, A_{k\alpha}$. Setting $\tau(\alpha) : A_{k\beta} \to \omega^{\sigma<\alpha,\beta>} A_{k\beta}$ we define an action τ of $\mathbb{Z}_N^2 \ni \alpha$ on $\mathcal{T}$. The degree of polinomials gives us a $\mathbb{Z}_+$ - grading on $\mathcal{T}$ (deg 1 =0). The ideal $\mathcal{K}_\eta$ is homogeneous (it is generated by polynomials of deg=2) and τ- invariant. Hence, $\mathcal{R}^g_\eta = \mathcal{T}/\mathcal{K}_\eta$ inherits the action τ and grading.

When $\eta \bmod N\Lambda$ belongs to some open subset $U \subset E$ one can similarly define for $\mathcal{K}_U$ the algebra $\mathcal{R}^g_U$ that consists of all polynomials with functions rational on E and regular on U as coefficients. By the standard construction we get from $\{\mathcal{R}^g_U\}$ a locally free sheaf $\mathcal{R}^g$ of algebras on E called an R-algebra. For each η the (closed) fiber $\mathcal{R}_{\bar\eta}{}^g$ of $\mathcal{R}^g$ is "less than or equal to" $\mathcal{R}^g_\eta$ defined above for fixed η, because $\mathcal{K}_\eta$ is generated by some elements of an appropriate $\mathcal{K}_U$. A priori, it is possible that $\mathcal{R}^g_\eta \ne \mathcal{R}^g_{\bar\eta}$ for some (necessarily finite) set of points E. The common fiber $\mathcal{R}^g_?$ of $\mathcal{R}^g$ is an algebra over the field $\mathbb{C}(E)$ of all rational functions on E. Note that $A_{k\alpha}$ become global sections of $\mathcal{R}^g$.

By changing the fiber only at $\bar\eta = 0$ define two other locally free sheaves of algebras $\tilde{\mathcal{R}}^g$ and $\bar{\mathcal{R}}^g$. After substituting $\rho_\alpha^{(k)}(u-\frac{\eta}{N})$ for $\rho_\alpha^{(k)}(u)$ and $\bar{A}_{k\alpha}$ for $A_{k\alpha}$ we obtain a sheaf of algebras with global sections $\bar{A}_{k\alpha}$. If we rescale $u \mapsto u\eta$ in (7), (1) in a neighborhood of $\eta = 0$ and redenote $A_{k\alpha} \mapsto \tilde{A}_{k\alpha}$ we get the sheaf $\tilde{\mathcal{R}}^g$ with sections $\tilde{A}_{k\alpha}$. Expanding L at $\eta = 0$

and then comparing the coefficients of u^{-i} one can linearly express $\{\bar{A}_{k\alpha}\}$ in terms of $\{A_{k\alpha}\}$ and $\{A_{k\alpha}\}$ in terms of $\{\tilde{A}_{k\alpha}\}$ with power series in η as coefficients. Hence, one has homomorphisms $\tilde{\mathcal{R}}^g \to \mathcal{R}^g \to \bar{\mathcal{R}}^g$ in a neighborhood of $\eta = 0$, which are isomorphisms outside $\eta = 0$.

Theorem 13 (cf. [8,10]), The map $A_{k\alpha} \to J_{k\alpha}$ extends to homomorphisms $\psi_{\mu_g} : \mathcal{R}^g_\eta \to M_d$ for a sufficiently general η and $\psi_\mu : \mathcal{R}^g_{\tilde\eta} \to M_\alpha \otimes_{\mathbf{C}} \mathbb{C}(E)$ for the common $\tilde{\eta}$

after restricting $J_{k\alpha}$ to the d-dimensional subspace Im (Φ_μ), where $\psi = \hat{\psi}, \check{\psi}$, $J = \hat{J}, \check{J}$, $d = \mathrm{rk}\,\tilde{\varphi}^1_{w_0}$ (see proposition 11), $g = g_\mu$.

Note that (for general η) the isomorphism class of ψ_μ does not depend on the choice of Φ_μ. Furthermore, $\hat{\psi}_\mu \simeq \check{\psi}_\mu$ if Im $(\tilde{\varphi}^1_{w_0}) \cap \ker (\tilde{\varphi}^1_{w_0}) = \{0\}$. The latter

holds for a skew μ with either $m'_1 = m'_r$ or $m_1 = m_r$. I do not know if this is always true.

5. The structure of $\mathcal{R}^g$ at generic point. Our first aim is to calculate the ranks of the sheaves $\mathcal{R}^g(p)$, generated by homogeneous polynomials in $A_{k\alpha}$ of deg = p. It suffices to define the fiber at $\eta = 0$ of $\mathcal{R}^g$ (or of $\tilde{\mathcal{R}}^g$, $\bar{\mathcal{R}}^g$).

Following [3,17] we can obtain some relations in $\mathcal{R}^g_0$:

$$[A_{k\alpha}, A_{k\beta}] = (\varepsilon^{-\sigma\alpha_1\beta_2} - \varepsilon^{-\sigma\alpha_2\beta_1}) A_{k+1\,\alpha+\beta} A_{oo},$$

$$[A_{oo}, A_{k\alpha}] = 0 \tag{8}$$

Set $U^g = \mathcal{T}/(8)$, $U^g_o = U^g/\{A_{oo} = 1\}$. Then U^g_o is isomorphic to the universal enveloping algebra of the Lie algebra $\underline{gl}_N[t]/(t^g) = \underline{gl}_N \otimes \mathbb{C}[t] \bmod t^g$. The isomorphism is given by the formulas $A_{k\alpha} \to t^k I_\alpha^{-1}$. A priori, U^g is maped onto $\mathcal{R}^g_0$ but perhaps, is not equal to $\mathcal{R}^{g}_{\bar{o}}$.

Similarly, the equation (1) for $\tilde{A}_{k\alpha}$ produces some relations for $\tilde{\mathcal{R}}^{g}_{\bar{o}}$ that may be written in the form (1) for Yang's R-matrix $R^1 = u/\eta + P$ at $\eta = 1$. One must substitute $\rho_o \mapsto Nu+1$, $\rho_\alpha \mapsto 1$ for $\alpha \neq 0$ in (1) to get the corresponding equation. We do not describe here the resulting algebra $\tilde{U}^g$ because its relations are considered in [6] (in a more general situation). Finally, for $\bar{A}_{k\alpha}$ we obtain the algebra U^g , which is isomorphic to the ring $\mathbb{C}[\bar{A}_{k\alpha}]$ of polynomials in commuting $\bar{A}_{k\alpha}$ ($\bar{U}^g$ majorizes $\bar{\mathcal{R}}^{g}_{\bar{o}}$).

Now we can formulate and sketch the proof of the main

<u>Theorem 14</u> (cf. [10]). $U^g = \mathcal{R}^{g}_{\bar{o}}$, $\tilde{U}^g = \tilde{\mathcal{R}}^{g}_{\bar{o}}$, $\bar{U}^g = \bar{\mathcal{R}}^{g}_{\bar{o}}$. The rank of $\mathcal{R}^{g}(p)$ (as a vector bundle) equals the dimension of the space of homogenous polynomials of deg = = p in (commuting) $A_{k\alpha}$, $0 \le k \le g-1$, $\alpha \in \mathbb{Z}_N^2$.

To prove it we find for every p the diagram μ with $g_\mu = g$, for which the corresponding homomorphism $\tilde{U}^g \to M_\alpha$ (we apply Theorem 13 to Yang's matrix) is injective on the component $\tilde{U}^g$ of deg = p. Introduce the sheaf $\tilde{\tilde{\mathcal{R}}}^{g}$)may be not locally free) by imposing only (1) for R^1 ($\eta = 0$) at $\eta = 0$. Then there exists a homomorphism $\tilde{\tilde{\mathcal{R}}}^{g}_{\eta} \to M_\alpha$ lifting $\tilde{U}^g \to M_\alpha$ in a neighborhood of $\eta = 0$. This homomorphism factors through $\tilde{\mathcal{R}}^{g}_{\eta}$ ($\tilde{\mathcal{R}}^g$ is the universal locally free quotient-sheaf of $\tilde{\tilde{\mathcal{R}}}^g$) and has no kernal on the p-component of $\tilde{\mathcal{R}}^g$. As p is arbitrary, we are done.

V.G.Drinfel'd proved that irreducible finite-dimensional representations of $\tilde{U}^g$ (for an indeterminate g) and of $\mathcal{H}^1_n$ (for n = 1,2,...) are in some correspondence. Hence, we proved incidentally.

<u>Theorem 14A</u>. Each irreducible d-dimensional representation of $\tilde{U}^g$ is lifted to a representation of $\mathcal{R}^{g}_{\tilde{\eta}}$ in terms of matrices of order d with coefficients in $\mathbb{C}$ (E) (and no poles at $\bar{o}$).

For $\{d_i\}$ depending on η the construction of Theorem 13 is the only such lifting. For $N = 2$, $g = 1$ the dimension of $\mathcal{R}^g_\eta(p)$ (see Theorem 14) and the structure of the centre of $\mathcal{R}^g_\eta$ (see below) were conjectured in [4].

Corollary 15 (cf. [8]). For a sufficiently general η (including $\eta = 0$) relations in $^g_\eta$ are generated by the linearly independent relations, which are the coefficients of $\hat{u}_1^{-i}\,\hat{u}_2^{-j} I_\alpha \otimes I_\beta$ in (1), where $\hat{u} = u + \eta/N$, $0 \leq i, j < g$, $(i, \alpha) \prec (j, \beta)$ for some ordering on $\mathbb{Z}_N^2 \times \mathbb{Z}_g \ni (i, \alpha)$. When $g = 1$ we have the following defining relations:

$$\sum_{\substack{a+b=c \\ a,b}} (A^d_{a,b} - A^d_{b,a})(\beta_{b-d} + \beta_a - \beta_{a-d} - \beta_b) = 0,$$

where $A^d_{a,b} \stackrel{\mathrm{def}}{=} \varepsilon^{(a_1-d_1)(d_2-b_2)} A_{ao} A_{bo}$, $a,b,c,d \in \mathbb{Z}_N^2$,

$c-d \neq \alpha$, $\rho_\alpha / \rho_o(u) = 1 + \beta_\alpha u + O(u^2)$

in a neighbourhood of $u = 0$.

The maximal commutative subalgebra. For $1 < r < s-1 \in \mathbb{Z}_+$ denote by $\tilde{Q}_1$ "the anti-symmetrizer" $\hat{\Phi}_\mu \in M_N^{\otimes r}$ for μ consisting only one column (i.e. for $m'_i = m_j = 1$). Similarly, $\tilde{Q}_2 \in {}^{r+1}M_N \ldots {}^{s}M_N$ for the set of indicies $(r+1, \ldots, s)$. Note, that for $\sigma = -1$ the standard anti-symmetrizers can be used as $\tilde{Q}_{1,2}$ (see [2,3]). Starting with L in the form (7) (where $A_{k\alpha}$ are elements of $\mathcal{R}^g_\eta$) construct

$$\tilde{}^1L(\tilde{u}_1) = \tilde{Q}_1 \left(\prod_{i=r}^{1} {}^iL(\tilde{u}_1 + (i-1)\eta) \right) \tilde{Q}_1 = \tilde{Q}_1\, {}^rL(\tilde{u}_1 + (r-1)\eta) \ldots {}^1L(\tilde{u}_1) \tilde{Q}_1,$$

$$\tilde{}^2L(\tilde{u}_2) = \tilde{Q}_2 \left(\prod_{i=s-r}^{1} {}^{i+r}L(\tilde{u}_2 + (i-1)\eta) \right) \tilde{Q}_2$$

(pay attention to the order of multiple). For uniformity set $\tilde{Q}_1 = I_o$, ${}^{1}L = {}^{1}L(\tilde{u}_1)$ for $r = 1$, ${}^{2}L = {}^{s}L(\tilde{u}_2)$ for $r = s-1$. Let ${}^{\tilde{1}\tilde{2}}R = \tilde{Q}_1\tilde{Q}_2 \prod_{j=r}^{1} (\prod_{i=1}^{s-r} {}^{j(i+r)}R(\tilde{u}_{12} + (j-i)\eta)) \tilde{Q}_1\tilde{Q}_2$.

<u>Proposition 16</u> ([2]). ${}^{\tilde{1}\tilde{2}}R\, {}^{\tilde{1}}L\, {}^{\tilde{2}}L = {}^{\tilde{2}}L\, {}^{\tilde{1}}L\, {}^{\tilde{1}\tilde{2}}R$,

$[H_r(\tilde{u}_1), H_{s-r}(\tilde{u}_2)] = 0$, where $H_r(\tilde{u}_1) = Sp\, {}^{\tilde{1}}L$, Sp is the matrix trace, $1 \leq r < s-1$.

Denote by $H_r^{k,i} \in \mathcal{R}_\eta^g$ the coefficient of $(u+k\eta)^{-i}$ in the expansion of $H_r(u)$ at $u = -k\eta$, $1 \leq r \leq N$, $0 \leq k \leq r-1$, $1 \leq i \leq g$. The degenerate algebra $\bar{\mathcal{R}}_{\,o}^{\,g} = \bar{U}^g = \mathbb{C}[\bar{A}_{k\alpha}]$ of $\bar{\mathcal{R}}_\eta^g$ is commutative and it also has the induced Poisson bracket $\{\, ,\, \}$. For $N=2$, $g=1$ this bracket was obtained in [4] and for general N,g it was constructed and studied in [5] (see also [3]). From [5] we can easily deduce that $\mathbb{C}[H_r^{k,i}]$ is a maximal commutative subalgebra of $\bar{U}^g$ with respect to $\{\, ,\, \}$. Therefore we get

<u>Theorem 17</u>. At generic point $\tilde{\eta}$ the elements $H_r^{k,i}$ generate a maximal commutative subalgebra in $\mathcal{R}_\eta^g$ and are all algebraically independent.

The centre. By Proposition 11 we may deduce that for $r=N$, $s=r+1$, ${}^{\tilde{1}\tilde{2}}R = \tilde{Q}_1\, c(\eta)\, \theta_o^o(\eta/N)^{-N}\, \nu(\tilde{u}_{12})$, where

$$\nu(u) = \theta_o^0 (u+N\eta) \prod_{k=1}^{N-1} \theta_o^o (u+(k-1)\eta), \quad c \text{ is a scalar}$$

function of η. Hence, $H_N^{k,i}$ $(0 \leq k \leq N-1,\ 1 \leq i \leq g)$ lie in the centre of $\mathcal{R}_\eta^g$. With the aid of [5] and by direct calculating with tensor products we obtain

<u>Theorem 18.</u> For a sufficiently general η (including $\eta = 0$) elements $H_N^{k,i}$ generate the centre of $\mathcal{R}_\eta^g$.

For a diagram μ let us substitute $J_{k\alpha}$ for $A_{k\alpha}$ with $g_\mu = g$. Then $H_N(u+\zeta-\eta/N) = c^n(\eta)\, \varepsilon(u)\, \hat{\Phi}_\mu$, where

$$\zeta = \left(\frac{n\eta}{N} - \xi\right) g^{-1},$$

$$\varepsilon(u) = \frac{\prod_a \theta_o^o(u-a\eta + N\eta)^{\kappa_a}}{\prod_a \theta_o^o(u-a\eta + (N-1)\eta)^{\kappa_a}} \prod_{i=0}^{N-1} \frac{\prod_a \theta_o^o(u-a\eta+i\eta)^{\gamma_a}}{\theta_o^o(u+i\eta+\zeta)^g}$$

<u>Corollary 19</u> [10] . For a (usual) Young diagram $g=1$, $\xi=0$ and

$$\varepsilon(u) = \prod_{i=1}^{N} (\theta_o^o(u+(m_i+N-i)\eta)/\theta_o^o(u+\frac{n}{N}\eta+(N-i)\eta)) \tag{8}$$

where for $N \leq i < r$ the numbers m_i are defined by zeroes.

In (8) substitute $\lambda\eta$ for u and make η tend to zero. Then $C(\eta) \to 1$.

$$\varepsilon(u) \mapsto \prod_{i=1}^{N} ((\lambda+m_i+N-i)/(\lambda+\frac{n}{N}+N-i)) ,$$

and we get, in particular, a well-known formula for the action of Casimir operators on irreducible finite-dimensional representations. The analogue of (8) for $N=2$ see in [7] .

In conclusion, let us resume that $\mathcal{R}_\eta^g$ provides us with an interesting example of a deformation of the universal enveloping algebra of $gl_N[t]/(t^g)$ for which the centre, maximal commutative subalgebras and irreducible finite-dimensional representations can be explicitly described. Note, that for some reasons the maximal commutative subalgebra is the canonical one (unlike the cases of rational and trigonometric degenerations of R and the usual enveloping algebras). But we have no natual Cartan subalgebras and root space decompositions for $\mathcal{R}_\eta^g$ and have difficulties calculating the spectrum of the maximal commutative subalgebra in the adjoint and other representations.

<u>Aknowledgements.</u>The author is grateful to A.B.Zelevinsky for fruitful discussions and to V.G.Drinfel'd for his kind attention to this work.

References

1. Faddeev,L.D. (1981). Quantum scattering transformation. Proc. of Frieburg Summer Inst.

2. Cherednik,I.V. (1982). Yadern. Phys. (Nucl. Phys.) 36, N 2, 549 (Russian).

3. Cherednik,I.V. (1984). Itogi Nauki, VINITI AN SSSR. Algebra, Geometry, Topology, 22, 205 (Russian).

4. Sklyanin,E.K. (1982). Funct. anal. appl., 16, N 4, 27 (Russian).

5. Cherednik,I.V. (1983). DAN SSSR, 271, N 1, 51(Russian).

6. Drinfel'd,V.G. (1985). DAN SSSR, 283, N 2, 90 (Russian).

7. Sklyanin,E.K. (1983). Funct. anal. appl., 17, N 4, 34 (Russian).

8. Cherednik,I.V. (1985). Funct. anal. appl., 19, N 1, 89 (Russian).

9. Kulish,P.P., Reshetikhin,N.Yu., Sklyanin,E.K. (1981). Lett. Math. Phys., 5, N 5, 393.

10. Cheresnik,I.V. (1985). DAN SSSR, 283, N 1, 41 (Russian).

11. Drinfel'd,V.G. (1985). Fuct. anal. appl., 19, N 4, 52 (Russian).

12. Cherednik,I.V. (1985). Funct. anal. appl., 19, N 4, 83 (Russian).

13. Rogawski,J.D. (1984). On modules over the Hecke algebra of a p-adic group. Princeton, Preprint.

14. Yang,C.N. (1968). Phys. Rev., 168, N 5, 1920.

15. James,G., Kerber,A. (1981). The representation theory of the symmetric group. Addison-Vesley P.G.

16. Cherednik,I.V. (1984). Theor. mat. phys., 61, N 1, 35 (Russian).

17. Gel'fand,I.M., Cherednik,I.V. (1983). Uspekhi Mat. Nauk (Russian Math. Surveys), 38, N 3, 3 (Russian).

18. Zelevinsky,A. (1980). Ann. Sci. Ec. Norm. Supp. 4-e serie, 13, N 2, 165.

19. Zhelobenko,D.P. (in this volume).

20. Zelevinsky,A. (1981). Representations of finite classical groups, Lect. Notes. Math. 869.

SYMMETRIC BASIS IN LIE GROUP REPRESENTATION THEORY AND CONSTRUCTING CLEBSCH-GORDON COEFFICIENTS

A.L.Shelepin, L.A.Shelepin

P.N.Lebedev Physics Institute, the USSR Academy of Sciences, Moscow

In contemplating physical aspects of the LIE group representation theory (including the calculation of Clebsch-Gordon(GG) coefficients) possibility of developing a unified approach to various given groups and specific problems should not be underestimated. But on the way to such approach (similar to the theory of angular momemta for SU (2)) one meets a number of difficulties. One of them is e.g. the existence of multiple weights and representations if CG series. Thus in order to distinguish basic vectors for multiple weights one has to be given additional conditions bearing no direct connection with the group. Frequently in this capacity various reduction schemes are used. They correspond to manifold bases of representations and, consequently, different CG coefficients. Distinguishing multiple weights in a CG series invariably suggests an additional rule or operation to be set. CG coefficient formulas for $SU(n)$ are frequently clumsy, and virtually boundless. Therefore the interrelations among CG coefficients pertaining to different groups and arrived at through different methods are not easy to establish. As for non-compact pseudo-unitary groups $SU(m,n)$, the programme of constructing irreducible representations and CG series is yet to be completed.

The aim of this paper is to emphasize the feasibility of the aforementioned approach, founded on the application of symmetric basis. Among its features are simplicity, universality, basic invariability in application to all classical groups (in the sense of both unitary system of symbolic expression and sequence of operations based thereon). Until recently the significance of symmetric basis for the Lie group theory was gravely underestimated. And this is attributable to its profusion; to the fact that not all of its components are linearly independent (although the expansion with respect to this basis in single-valued).

Introducing symmetric basis [1] is connected to the theory of classical groups invariants [2]. Ref. [1,3] give a construction mode for the symmetric basis in its application to groups $SU(n)$ and provides a method of calculation of CG coefficients, based on the expansion of groups $SU(n)$ generating invariants in terms of polynomial symmetric basis. The initial invariant for groups $SU(n)$ is the $n \times n$ determinant. The basis vectors are introduced as independent minors of U_{ik} determinant, i, $k=1, \ldots, n-1$ [1].

For the group $SU(2)$, representation $D(P)$, $P=2j$, we have:

$$|n_1 n_2\rangle = \left(C_P^{n_1 n_2}\right)^{1/2} U_1^{n_1} U_2^{n_2} \tag{1}$$

where $P=n_1+n_2$, $C_P^{n_1 n_2}$ a binomial coefficient. For the group $SU(3)$, representation $D(PQ)$, we have:

$$\left|\begin{matrix} P_1 & P_2 & P_3 \\ q_1 & q_2 & q_3 \end{matrix}\right\rangle = \left(C_P^{P_1 P_2 P_3} C_Q^{q_1 q_2 q_3}\right)^{1/2} U_1^{P_1} U_2^{P_2} U_3^{P_3} \xi_1^{q_1} \xi_2^{q_2} \xi_3^{q_3} \tag{2}$$

where $P=P_1+P_2+P_3$, $Q=q_1+q_2+q_3$, $C_P^{P_1 P_2 P_3} = \frac{P!}{P_1! P_2! P_3!}$ a threenomial coefficient, $\xi = \varepsilon_{ikl} U_k V_l$. Between quantities U_i and ξ_i there exists an evident relation $U_1 \xi_1 = 0$. The basis of representation $D(P,R,Q)$ of $SU(4)$ is constructed, therefore, of quantities:

$$U_i^{P_i}, \lambda_{ik}^{2ik} = (\varepsilon_{ikem} U_e V_m)^{2ik}, \xi_i^{q_i} = (\varepsilon_{ikem} U_k V_e W_m)^{q_i}$$

On the whole the following features may be listed as characteristic of the symmetric basis: 1) constructed as a polynom of basis of fundamental representations; 2) factorization: consists of different factors, all being bases of representations $D(P,0,0)$, $D(0,P,0)$, $D(0,0,P)$ etc.; 3) profusion: not all of its components are linearly independent, although the expansion in terms of this basis is single-valued; 4) maximum symmetry of CG coefficients corresponding to the basis. In application to orthogonal and sympletic groups the symmetric basis holds similar properties.

The symmetric basis for Lie groups representations also stands markedly aloof in infinitesimal aspect. Within the Lie algebra the set of commutative operators determines their eigenvalues - weights of representations. Raising and lowering the operators rule the transformation between them. The basis of representations - the eigenfunctions of commutative operators - is dependent on their specific realization. We will study the weights and bases of representations of compact and noncompact unitary groups $U(m+n)$, $U(m,n)$ with regard to the elementary case $U(2)$ and $U(1,1)$. The Lie algebra commutation relations for them are identical with those for the group GL (2,R) [4]:

$$[\hat{n}_1, \hat{n}_2] = 0 \; , [\hat{n}_1, \hat{E}_\pm] = \pm \hat{E}_\pm \; , [\hat{n}_2, \hat{E}_\pm] = \mp \hat{E}_\pm \; , [\hat{E}_+, \hat{E}_-] = \hat{n}_1 - \hat{n}_2 \tag{3}$$

This algebra splits into two. The operator $\hat{n}_1 + \hat{n}_2$ commutates with every other;

$$[\hat{M},\hat{E}_\pm]=\pm\hat{E}_\pm,\quad [\hat{E}_+,\hat{E}_-]=2M \text{ where } \hat{M}=(\hat{n}_1-\hat{n}_2)/2 \tag{4}$$

The Casimir operator for the last system is

$$J^2=\frac{1}{4}[2\hat{E}_+\hat{E}_-+2\hat{E}_-\hat{E}_++(\hat{n}_1-\hat{n}_2)^2] \tag{5}$$

Transformations for the groups $SU(2)$, $SU(1,1)$ are ruled by the operators e^{iL} , where respectively:

$$SU(2):\ \hat{L}=ia_1\hat{M}+ia_2\hat{L}_2+ia_3\hat{L}_3$$

$$SU(1,1):\ \hat{L}=ia_1\hat{M}+a_2\hat{L}_2+a_3\hat{L}_3 \tag{6}$$

Here $\hat{L}_2=\frac{\hat{E}_++\hat{E}_-}{2}$, $\hat{L}_3=i\frac{\hat{E}_--\hat{E}_+}{2}$, $a_i\in R$.

Group transformations and unitarity condition yield for the operators:

$$SU(2):\ \hat{M}^+=\hat{M},\ \hat{L}_2^+=\hat{L}_2,\ \hat{L}_3^+=\hat{L}_3 \text{ or } \hat{M}^+=M,\ \hat{E}_+^+=\hat{E}_-$$

$$SU(1,1):\ \hat{M}^+=\hat{M},\ \hat{L}_2^+=-\hat{L}_2,\ \hat{L}_3^+=-\hat{L}_3 \text{ or } \hat{M}^+=M,\ \hat{E}_-^+=-\hat{E}_- \tag{7}$$

The weights and the basis in representation slaces are defined as eigenvalues and eigenfunctions of commutative operators

$$\hat{n}_1|n_1n_2\rangle=n_1|n_1n_2\rangle,\quad \hat{n}_2|n_1n_2\rangle=n_2|n_1n_2\rangle \tag{8}$$

From the commutation relations (3) we obtain

$$\hat{E}_+|n_1n_2\rangle=C(n_1n_2)|n_1+1,n_2-1\rangle \tag{9}$$

$$\hat{E}_-|n_1n_2\rangle=C(n_2n_1)|n_1-1,n_2+1\rangle \tag{10}$$

$$n_1-n_2=C(n_1-1\ n_2+1)C(n_2n_1)-C(n_1n_2)C(n_2-1\ n_1+1) \tag{11}$$

where $C(n_1n_2)$ is a function of the variables n_1, n_2 . Irreducible representations are characterised by eigenvalues of the operators commutating with all group operators. With the help of the Casimir operator eigenvalues (5) we get:

$$\hat{E}_+\hat{E}_-|n_1 n_2\rangle = n_1(n_2+1)|n_1 n_2\rangle, \quad \hat{E}_-\hat{E}_+|n_1 n_2\rangle = n_2(n_1+1)|n_1 n_2\rangle \tag{12}$$

Relations (9)-(12) explicitly define the set of weights for the irreducible groups $SU(2)$ and $SU(1,1)$. Thus, the condition of unitarity:

$$\hat{M}^+ = \hat{M} \Rightarrow Im\, n_1 = Im\, n_2$$

$$\begin{aligned} SU(2):\ & \hat{E}_+^+ = E_-^+ \Rightarrow C^*(n_2 n_1) = C(n_1 - 1\ \ n_2+1) \\ & \hat{E}_+\hat{E}_-|n_1 n_2\rangle = |C(n_2 n_1)|^2 |n_1 n_2\rangle = n_1(n_2+1)|n_1 n_2\rangle \\ & \hat{E}_-\hat{E}_+|n_1 n_2\rangle = |C(n_1 n_2)|^2 |n_1 n_2\rangle = n_2(n_1+1)|n_1 n_2\rangle . \end{aligned} \tag{13}$$

$$\begin{aligned} SU(1,1):\ & \hat{E}_+^+ = -\hat{E}_- \Rightarrow C^*(n_2 n_1) = -C(n_1 - 1\ \ n_2+1) \\ & \hat{E}_+\hat{E}_-|n_1 n_2\rangle = -|C(n_2 n_1)|^2 |n_1 n_2\rangle = n_1(n_2+1)|n_1 n_2\rangle \\ & \hat{E}_-\hat{E}_+|n_1 n_2\rangle = -|C(n_1 n_2)|^2 |n_1 n_2\rangle = n_2(n_1+1)|n_1 n_2\rangle \end{aligned} \tag{14}$$

Unitarity regions are shown on Fig.1, when there is a plane of all admissible real parts n_1 and n_2 of all irreducible representations. The unitarity region for $SU(2)$, shaded with crossed lines, is determined by the relations

$$n_2(n_1+1) \geq 0 \qquad n_1(n_2+1) \geq 0 \tag{15}$$

For $SU(1,1)$ the unitarity region (Shaded with lines) is determined by the relations

$$n_2(n_1+1) \leq 0 \qquad n_1(n_2+1) \leq 0 \tag{16}$$

From (13), (14) we also get (excluding an arbitrary phase factor):

$$\begin{aligned} \hat{E}_-|n_1 n_2\rangle &= \sqrt{n_1(n_2+1)}\,|n_1 - 1\ \ n_2+1\rangle \\ \hat{E}_+|n_1 n_2\rangle &= \sqrt{(n_1+1)n_2}\,|n_1 + 1\ \ n_2-1\rangle \end{aligned} \tag{17}$$

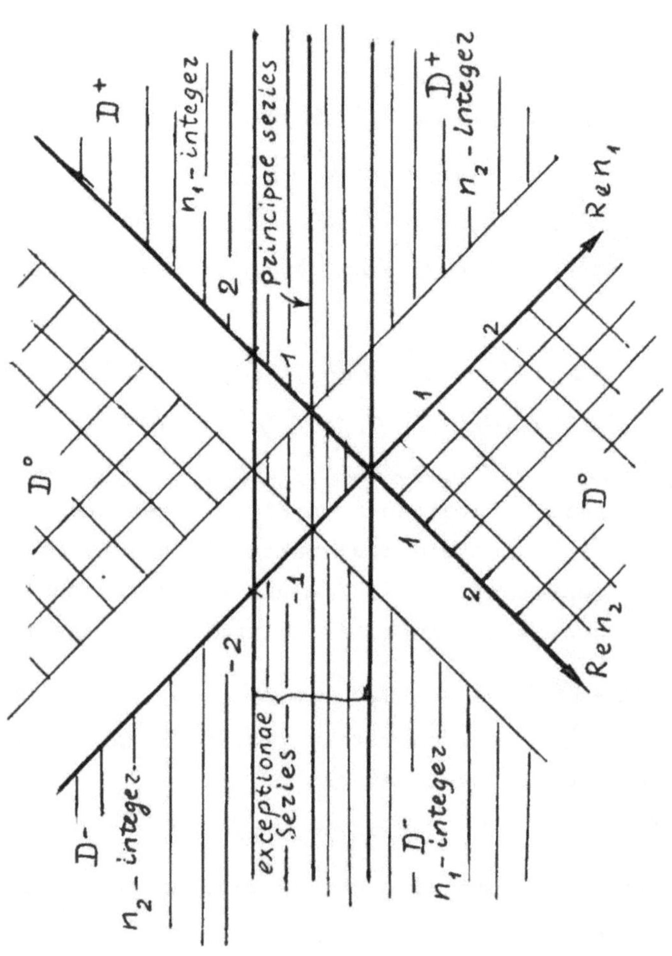

Fig. 1. The unitarity regions for representations of SU(2) and SU(1,1)

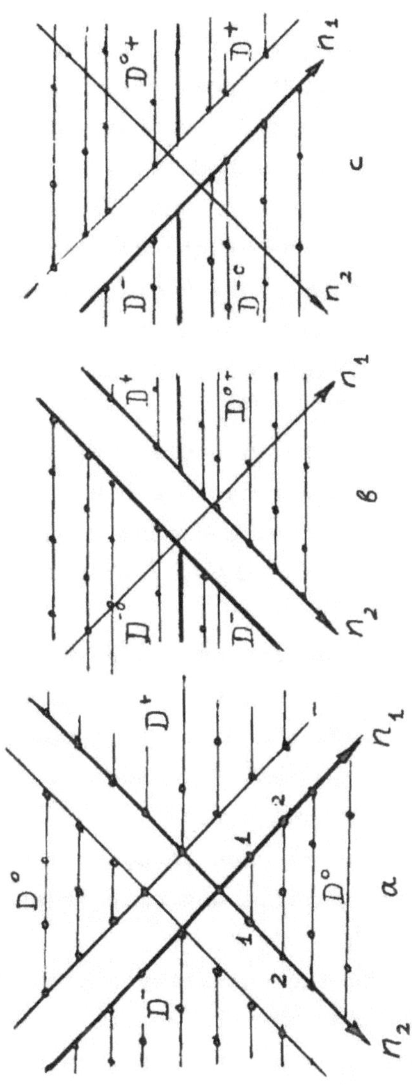

Fig. 2. The set of weights for irreducible representations

Since for irreducible representations $n_1+n_2=2j=p$, their weights should lie on the lines $n_1+n_2=$ const, which run parallel to the X-axis. For a representation to be unitary such a line (rather, the weights it bears at 2 linear measure units intervals) should never extend into the non-unitary region. By means of $\hat{E}_\pm$ one can reach any arbitrary weight starting from any one given. At the point where the numerical factor vanishes, the representation breaks off, indicating that the highest (lowest) weight is reached. It is noteworthy that for the imaginary part of n_1 and n_2 operatos $\hat{E}_\pm$ prove ineffective, i.e. the weight diagram of a representation in the plane $Im n_1$, $Im n_2$ is but a point.

From relations (17) we get the following classification and the set of weights for the irreducible representations of $SU(2)$ and $SU(1,1)$:

1. n_1 and n_2 are integers. The representation $D(p)$ splits into three irreducible ones: $D^-(p)$ with the highest wwight, a finite-dimensional $D^0(p)$, $\dim D^0(p)=|p+1|$, $D^+(p)$ with the lowest weight (Fig. 2a).

2. n_1 (or n_2) is integer, $n_2(n_1)$ - non-integer (Fig.2b and 2c respectively). We have two types of representations with the highest (lowest) weight: $D^-(p)$ and $D^{-0}(p)$ ($D^+(p)$ and $D^{0+}(p)$) that differ in the set of weights.

3. n_1 and n_2 are non-integer. The representation $D(p)$ is irreducible and prossesses neither highest nor lowest weight.

Let us examine the unitarity region for $SU(2)$.

1. Let n_1 and n_2 be reals. Therefore, in order to stay within the unitarity region both the highest and lowest weights should be integer. In this case the repre-

sentations are finite-dimensional. There are two series, symmetric with respect to the line $n_1 + n_2 = 2j = -1$ (the so-called mirror symmetry representations [8]) with equal eigenvalues of the operator $J^2 = 1/4\,(n_1+n_2+2)(n_1+n_2)$.

2. Assume that n_1 and n_2 have the imaginary part. Then, to satisfy (15) one needs $Re\,n_1 - Re\,n_2 = 1 = 2Re\,m$ The region of unitarity will be inevitably. These representations are non-unitary.

Let us consider unitary representations of $SU(1,1)$

1. Assuming that n_1 and n_2 are real numbers, it is easy to notice that the weights of all presentations $D^+(p)$ and $D^-(p)$, $-\infty < p < +\infty$, Lie within the unitarity region. This constitutes the so-called discrete series of unitary representations.

Within the range $-2 < p < 0$, along with the discrete series, there appears an exceptional series - due to the fact that the non-unitary region could well be crossed over. To characterise representations of the exceptional series one needs an extra number E , which would define the closest to $m=0$ value of , $2m = n_1 - n_2; -1 + |p+1| \leq E \leq 1 - |p+1|$. In the latter relation the equation is satisfied by the representations $D^{0+}(p)$ and $D^{-0}(p)$ with the lowest (highest) weight, the strict inequality - by the irreducible $D(p)$.

2. If n_1 and n_2 have the imaginary part, (14) yields $Re\,n_1 + Re\,n_2 = -1$, $Im\,n_1 = Im\,n_2$. Therefore the representations of the so-called principal series $D(-1+i\rho)$, where ρ is real, $-1 \leq E \leq 1$.

As we see, the considered approach to enumerating the irreducible representations of $SU(2)$ and $SU(1,1)$ and to establishing their properties simple and demonstrative. It can be explicitly generalised to cover the groups $SU(3)$ and $SU(2,1)$. There are three operators for this: $\hat{n}_1$, $\hat{n}_2$, $\hat{n}_3$. The space of weights, comprising all representations, is three-dimensional. The eigenvalues n_1, n_2 and n_3, serving as coordinates in that space, lie on the edges of trihedral angle (Fig.3a). Weights, corresponding to an irreducible representation, spread not along the line (or parts of it) but over the planes $p = n_1 + n_2 + n_3$ = const (Fig.3 shows such regions for an integer n). For the group $SU(3)$ the bounded shaded region denotes highest and lowest weights. A unitary representation is finite-dimensional. If for the groups $SU(2)$ and $SU(3)$ the representations $D(p)$ and $D(-p-1)$, symmetric with respect to the line $p = n_1 + n_2 = -1$ are equivalent, in this case they are not. Shaded region for $Re\, p > -1$ agree with the representation $D^0(PO)$, $p_i = n_i$, for $Re\, p < -1$ the representation $D^0(OQ)$, $q_i = -n_i - 1$, $Q = q_1 + q_2 + q_3$. As for $SU(2)$ and $SU(1,1)$ the representations, symmetric with respect to the point $Re\, n = -1/2$ are conjugate.

A method of determining irreducible unitary representations for $SU(2,1)$ is analogous to that for $SU(1,1)$. At this point we shall only enumerate unitary representations with non-multiple weights. Translating the condition for hermiticity (antihermiticity) generators to their eigenvalues, for $SU(2,1)$ we get (compact $SU(2)$ subgroup agrees with n_1 and n_2):

$$
\begin{array}{lll}
n_1(n_2+1) \geqslant 0 & n_2(n_3+1) \leq 0 & n_1(n_3+1) \leq 0 \\
n_2(n_1+1) \geqslant 0 & n_3(n_2+1) \leq 0 & n_3(n_1+1) \leq 0
\end{array}
\qquad (18)
$$

For the group $SU(3)$ the sign is always $\geqslant 0$, for $SL(3,R)$ it is always ≤ 0. We have the following unitary series: $D^-(PO)$: $p_1 \geq 0$, $p_2 \geqslant 0$, $p_3 \leq 1$; $p \geqslant -1$, p_1, p_2, p_3 integers, $D^+(PO)$: $p_1 \leq -1$, $p_2 \leq -1$, $p_3 \geq p_1 + 2$; $p \geqslant -1$, p_1 and p_2 integer, p_3 real,

$D^{-0}(P0): P_1 \leq 0, P_2 \geq 0, P_3 \leq P; -1 \leq P < 0, P_1$ and P_2 integer, P_3 real.

Weight diagrams for D^+ and D^- representations are given on Fig.3, D^{-0} covers the regions of D^0 and D^-; weight diagrams for the representarions $D^+(0P), D^-(0P)$, $D^{0+}(0P)$ may be obtained by inversion of the P_i-axis.

The weight diagrams help to establish the following formulas for reducing representations $SU(2,1)$ to compact $SU(2)$

$$D^+(P0) = \sum_{\alpha=0}^{\infty} D^0\alpha \qquad D^-(P0) = \sum_{\alpha=1}^{\infty} D^0(P+\alpha)$$
$$D^0(P0) = \sum_{\alpha=0}^{P} D^0(\alpha) \qquad D^{0-}(P0) = \sum_{\alpha=0}^{\infty} D^0(\alpha) \tag{19}$$

and non-compact $SU(1,1)$ subgroups:

$$D^+(P0) = \sum_{\alpha=1}^{\infty} D^+(P+\alpha) \quad D^-(P0) = \sum_{\alpha=0}^{\infty} D^-(P-\alpha)$$
$$D^{-0}(P0) = \sum_{\alpha=0}^{\alpha=[P]+1} D^{-0}(P-\alpha) + \sum_{\alpha=[P]+2}^{\infty} D^-(-\alpha) \tag{20}$$

It is noteworthy that in (20) the expansion of $D^-(P0)$ comprises certain representations twice because of the equivalence of $D^-(-P-1)$ and $D^-(P)$; [ ] here stand for the integer part.

The general procedure fits higher rank groups. Thus, for $SU(4)$, $SU(3,1)$, $SU(2,2)$ thw weights of irreducible representations fall within regions of the volume $n_1 + n_2 + n_3 + n_4 =$ const.

From infinitesimal point of view the symmetric basis is directly connected with the method considered above. The basis includes the components of fundamental representation vector in eigenvalues n_i of Casimir operators; it is directly related to the weight space. The symmetric basis for finite-dimensional representations of compact groups ((1), (2)), is immediately generalized to the non-compact one. In this case n_1 and n_2 in relation (1) may take any values not only non-negative integer one; the binomial coefficient are substituted by the B-function:

$$(B(n_1+1,n_2+1))^{-1/2} u_1^{n_1} u_2^{n_2} = \left(\frac{\Gamma(P+1)}{\Gamma(n_1+1)\Gamma(n_2+1)}\right)^{1/2} u_1^{n_1} u_2^{n_2} \quad (21)$$

The normalizing coefficient in (21) can be obtained in a way similar to that applicable to finite-dimensional case 5 . Assuming that

$$|n_1 n_2\rangle = \rho_{n_1 n_2} u_1^{n_1} u_2^{n_2}, \quad u_1 = |10\rangle, \quad u_2 = |01\rangle \quad (22)$$

apply $\hat{E}$ to (22):

$$\sqrt{(n_2+1)n_1}\,|n_1+1, n_2-1\rangle = \rho_{n_1 n_2} n_1 u_1^{n_1-1} u_2^{n_2} \hat{E}|1,0\rangle =$$

$$= \rho_{n_1,n_2} n_1 u_1^{n_1-1} u_2^{n_2+1} = \frac{\rho_{n_1,n_2}}{\rho_{n_1-1,n_2+1}} n_1 |n_1-1, n_2+1\rangle$$

Whence $\rho_{n_1,n_2}/\rho_{n_1-1,n_2+1} = \sqrt{(n_2+1)/n_1}$ and (up to the last factor) $\rho_{n_1,n_2} = (B(n_1+1, n_2+1))^{-1/2}$.

Let us get a closer look at (21). For non-negative integer n_1 and n_2 , we have (1), a basis of the finite-dimensional representation $D^0(P)$. With $-1 \leq P < \infty$ formula (21) yields bases for representation $D^{0+}(P)$, $D^{-0}(P)$, $D(P)$. Indeed, let n_1 be an integer; then, with negative n_1 the normalizing factor vanishes (the denominator contains the factorial of a negative integer). Therefore, the basis vectors of the representation $D^{0+}(P)$, $P \geq -1$ take the form (we have made use of the formula $\Gamma(z)\Gamma(1-z)\sin\pi z = \pi$):

$$|n_1\, p-n_1\rangle = \left(\frac{\Gamma(p+1)}{n_1!\,\Gamma(p-n_1+1)}\right)^{1/2} u_1^{n_1} u_2^{p-n_1} =$$
$$= \left(\frac{\sin\pi(n_1-p)}{\pi}\cdot\frac{\Gamma(p+1)\Gamma(n_1-p)}{n_1!}\right)^{1/2} u_1^{n_1} u_2^{p-n}, \quad (23)$$
$$n_1 = 0, 1, 2 \ldots$$

The notation (23) fits for positive arguments $\Gamma(Z)$ in cases $n_2 = p - n_1 > 0$ and $n_2 < 0$ respectively. It is evident that the closer n_2 to an integer, the closer the basis of the representation to that of a finite-dimensional one and the more trivial is the role of the "tail", lying outside D^0 region and proportional to $\sin E_2$, $E_2 = n_2 - [n_2]$. For an irreducible $D(p)$ the extension beyond D^0 region is characterised by two parameters: $E_1 = n_1 - [n_1]$ and $E_2 = n_2 - [n_2]$ the left-hand side and right-hand side in Fig.1 respectively). With $E_1 = 0$ and $E_2 = 0$ we get the basis for the representation $D^0(p)$, $E_2 = 0 - D^0(p)$.

With $p \leq -1$ and integer $n_1\,(n_2)$ (21) yields the bases for representations D^+ and D^- respectively:

$$|n_1, p-n_1\rangle = i^{n_1}\left(\frac{\Gamma(n_1-p)}{\Gamma(-p)\,n_1!}\right)^{1/2} u_1^{n_1} u_2^{p-n_1}, \quad n_1 = 0,1,2\ldots$$
$$|p-n_2, n_2\rangle = i^{n_2}\left(\frac{\Gamma(n_2-p)}{\Gamma(-p)\,n_2!}\right)^{1/2} u_1^{p-n_2} u_2^{n_2}, \quad n = 0,1,2\ldots \quad (24)$$

Note that for the representation $D(-1)$ the absolute value of the normalizaing factor equals 1 (n_1 and n_2 are real).

The construction of symmetric bases makes it possible to establish series and CG coefficients by means of the standard procedure of generating invariants, developed formerly for compact groups [1,3]. For $SU(2)$ and $SU(1,1)$

the generating invariants are of the form:

$$\begin{vmatrix} U_1 & V_1 \\ U_2 & V_2 \end{vmatrix}^{A_1} \begin{vmatrix} U_1 & W_1 \\ U_2 & W_2 \end{vmatrix}^{A_2} \begin{vmatrix} V_1 & W_1 \\ V_2 & W_2 \end{vmatrix}^{A_3} \tag{25}$$

Unlike the unitary finite-dimensional representations of $SU(2)$ where the powers of determinants A_i are integer non-negative numbers, A_i and n_1 and n_2 may take arbitrary values. If the generating invariant is a single determinant, it rules the contraction of a representation with a conjugate. For the contraction of finite-dimensional representations we have (D is a non-negative integer):

$$(U_1V_2 - V_1U_2)^{D} = \sum_{n=0}^{D} (-1)^n \frac{D!}{n!(D-n)!} [U_1^{D-n} U_2^{n}][V_1^{n} V_2^{D-n}] \tag{26}$$

for the contraction of $D^+(-B)$ and $D^-(-B)$, $B \geq 1$,

$$(U_1V_1 - U_2V_2)^{-B} = \sum_{\ell=0}^{\infty} (-1)^{\ell} \frac{\Gamma(B+\ell)}{\Gamma(B)\ell!} [U_1^{-B-\ell} U_2^{\ell}][V_1^{\ell} V_2^{-B-\ell}] \tag{27}$$

for the contraction of $D^{0+}(C)$ and $D^{0}(C)$, $C \geq 1$,

$$(U_1V_1 - U_2V_2)^{C} = \sum_{m=0}^{\infty} (-1)^{m} \frac{\Gamma(C+1)}{\Gamma(-m+1)m!} [U_1^{C-m} U_2^{m}][V_1^{m} V_2^{C-m}] \tag{28}$$

for the contraction of $D(A)$, $A \geq -1$,

$$(U_1V_1 - U_2V_2)^{A} = \sum_{K=-\infty}^{\infty} (-1)^{K} \frac{\Gamma(A+1)}{\Gamma(A-K-E+1)\Gamma(K+E+1)} \left[U_1^{A-K-E} U_2^{K+E}\right]\left[V_2^{A-K-E} V_1^{K+E}\right] \tag{29}$$

Series (26)-(28) are the binomial expansion in terms of the Newton formula, series (29) is formal. In general case (25), expanding every determinant with respect to formulas (26)-(29) we get (up to a multiple) the expression valid for an arbitrary component:

$$U_1^{A_2-d_2+d_1} U_2^{A_1-d_1+d_2} V_1^{A_1-d_1+d_3} V_2^{A_3-d_3+d_1} W_1^{A_3-d_3+d_2} W_2^{A_2-d_2+d_3} \tag{30}$$

CG series can be obtained comparing (30) products of the basis elements of representations. The sence of the fundamental representations is that multiplying $P_1 = A_1 + A_2$ and $P_2 = A_1 + A_3$ we get $P = A_2 + A_3 = P_1 + P_2 - 2A_1$. For $D^+(P_1) \otimes D(P_2)$ we compare (30) with

$$(U_1^{K_1} U_2^{P_1 - K_1})(V_1^{K_2} V_2^{P_2 - K_2}), \; P_1, P_2 \leq -1, \; K_1, K_2 \geq 0, \quad k_1, k_2 \text{ integer} \tag{31}$$

and see that A_1 is a non-negative integer (otherwise expansion (25) would involve terms with negative K_1 and K_2, absent from the set of basis vectors (31)). Thus, the CG series has the form:

$$D^+(P_1) \otimes D^+(P_2) = \bigoplus_{A_1 \geq 0} D^+(P_1 + P_2 - 2A_1) \tag{32}$$

This expression is also valid for $D^+(P_1) \otimes D^{0+}(P_2)$, $D^{0+}(P_1) \otimes D^{0+}(P_2)$ (here after for definiteness of the pair of equivalent mirror-symmetrical representations we pick out $D^+(P)$ for $P \leq 1$ and $D^{0+}(P)$ for $P \geq 1$). From (32) it follows that the expansion of the direct product of unitary representations, each one possessing a highest (lowest) weight (i.e. $D^+(P_1)$ and $D^{0+}(P_2)$, $D^-(P_1)$ and $D^{-0}(P_2)$, $0 < P_2 < -1$), contains only unitary represenatations.

In the basis $(U_1^{K_1} U_2^{P_1 - K})(V_1^{K_2} V_2^{P_2 - K_2})$ of $D^+(P_1) \otimes D^0(P_2)$ where $K_1, K_2, P_2 \geq 0$ are integers, V_1 and V_2 are to finite powers; therefore, P_1 and P_2 are non-negative integers. Besides, $A_1 \leq P_2$ due to $P_2 = A_1 + A_3$. The corresponding series

$$D^+(P_1) \otimes D^0(P_2) = \bigoplus_{A_1 = 0}^{P_2} D(P_1 + P_2 - 2A_1) \tag{33}$$

after substituting D^{0+} for D^+ retains its form.

In case of $D^+(P_1) \otimes D^-(P_2)$ the expansion would comprise not only the series $\bigoplus_{A \geq 0} D^+(P_1 - P_2 - 2A) \oplus \bigoplus_{A \geq 0} D^-(P_2 - P_1 - 2A)$

related to (32) by the Wigner symmetry of representation permutation, but also the direct integral of $D(P)$ for A is not restricted to be integer. For an integer $P_1 - P_2$ the expansion would comprise $\bigoplus_{A \geq 0} D^0(|P_1 - P_2| + 2A$ related to (33) by the Wigner symmetry. Notations for all CG series of $SU(1,1)$, including discrete ones, as well as the principal and the exceptional series, could be similarly constructed.

The expansion (30) also yields the normalized Wigner coefficient for the representations of $SU(1,1)$. It is defined as the coefficient preceding the normalized product of bases in (30). As is clear from the general calculation procedure, there exists a close interrelation between CG coefficients of finite- and infinite- dimensional representations. This interrelation justifies the application of results, previously acquired for finite-dimensional representations (see also [6]).

The scheme for constructing the symmetric basis for higher-rank non-compact groups is similar to that pertaining to compact group. Thus, the symmetric basis for $SU(2,1)$ has the form (2), although P_1, P_2, P_3, q_1, q_2, q_3 may take non integer values (the factorial in normalization in this case is replaced by the Γ-function). The method of constructing CG series and CG coefficients within the symmetric basis, developed for compact groups, is immediately generalized to noncompact groups. In this case one should consider complex rank tensors [7].

The peculiarity of symmetric basis is its adequacy to the group itself. Group symmetry, distorted when bases with requirement surpassing those of a group are used, is preserved here. The transition from symmetric basis to other bases is accomplished by means of CG coefficients of a particular kind (for example, the transition to the canonical basis in $SU(3)$ is ruled by the coefficient $\langle PO \| CQ \| PQ \rangle$). For the symmetric basis there is no multiplicity problem. The CG coefficient is uniquely defined by its eigenvalues $n_1, n_2 \ldots$.

The applications of a symmetric basis are determined by its simple, universal and standard approach to compact and noncompact groups. It can be applyed to quantum mechanics calculations, the theory of complex physics systems, physical symmetries studies. In terms of symmetric basis the symmetry of values, defining a system, is never violated. This can be manifestly exemplified by the symmetry of CG coefficients of the group $SU(n)$. CG coefficients of these groups have, in principle, rather high symmetric properties. The symmetry consists of the conventional Wigner symmetry, related to the permutations of angular momenta and the weight diagrams symmetry (of rotation, reflection, inversion). Thus, for $SU(2)$ there are six Wigner symmetry operations and two weight diagram operations (identity and $m \to -m$ operation), in all, 12 symmetries. For $SU(3)$ there are already 12 weight diagram symmetries. Those include rotations and reflections ($n!$ permutations of symmetric indices) along with inversion (a permutation of covariant and contravariant indices). There are 6.12 CG coefficient symmetries for $SU(3)$. The number of symmetries steeply increases as n grows. But this is true only within the framework of symmetric basis. For the canonic basis, based on reduction schemes, thbs is no longer here and symmetries are broken (for CG coefficients of $SU(3)$ only 4 weight diagram symmetries stay intact).

In general, the application of a symmetric basis as a cornerstone of the physical aspect of the Lie group theory opnes entirely new perspectives due to its simplicity, universality and physics-oriented approach.

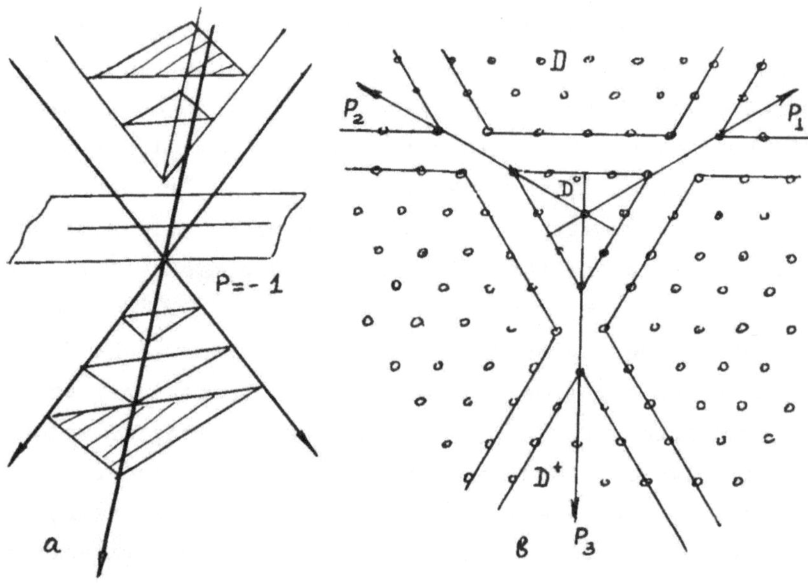

Fig. 3. The space of weights for representations of SU(3) and SU(2,1)

References

1. Shelepin, L.A.(1973). In: TrudyFIAN (Proceedings of P.N.Lebedev Institute) 70, pp.3-119. (Russian).

2. Weyl, H. (1939). The Classical Groups, their Invariants and Representations.

3. Karassev, V.P. (1973). In: Trudy FIAN (Proceedings of P.N.Lebedev Institute) 70, pp.147-220 (Russian).

4. Barut, O., Raczka, R. (1977). Theory of Group Representations and Applications. Warszawa.

5. Lubarsky, G.Ya. (1957). Groups theory and its application in physics. - Moscow (Russian).

6. Holman, W.J. III, Biedenharn, L.C. Jr. (1966). Annals Phys. 39, 1

7. Fronsdal, C. (1965). In: High-Energy Physics and Elementary Particles IAEA Vienna, 585

8. Jueys, A.P., Bandzaitis, A.A. (1977). Theory of Angular Momentum in Quantum Mechanics, Mosklas, Vilnius.

INVARIANT ALGEBRAIC METHODS AND SYMMETRIC ANALYSIS OF COOPERATIVE PHENOMENA

V.P. Karassiov

P.N. Lebedev Physical Institute of the USSR Academy of Sciences, Moscow

1. Cooperative phenomena and processes in various many-body physical systems (atoms, molecules, nuclei etc.) are now the topic of great interest (see e.g. [1-6] and references therein). A natural and efficient tool for studying cooperative phenomena in quantum systems is the second quantization method [7], which is closely connected with the representation theory of Lie groups and supergroups [8-12]. Therefore, group-theoretical methods are very adequate to an analysis of different cooperative effects within the general approach of dynamic symmetries [13].

For this purpose we need an adequate mathematical technique for different groups of dynamic symmetries. Such a technique should contain the construction of (1) bases of irreducible representations (IR), (2) Wigner-Racah algebras [10, 14], (3) generalized coherent states and (4) methods for calculating various characteristics of physical systems (that are based on resolution of the above mentioned questions [1]).

The method of generating invariants (GI) in the form [15,9] appears to be an efficient tool for achieving this purpose in the case of classical Lie groups, particularly, SU(n) (that are groups of dynamic symmetries of multilevel quantum systems) [16,17]. Here a resolution of the Wigner-Biedenharn problem [18,17] (an explicit constructing orthogonal systems of GI) for SU(n) Wigner coefficients is of the key role. Below we consider some possibilities for solving this problem within general programme [16]. In addition, we give an abstract of the paper [19] on applications of the finite difference calculus in group-theoretical formalism and discuss some physical applications of obtained results.

2. In [16,17] we have established an equivalence of the problems of constructing GI for SU(n) 3rd rank Wigner coefficients (corresponding to the decomposition $\prod_{1\le i\le 3} D(P^i) \to D(\dot{O}_{n-1})$, $P^i \equiv (p^i_1, \ldots, p^i_{n-1})$, $\dot{O}_k = \overbrace{0, \ldots, 0}^{k}$) and $S(\prod_{1\le i\le 3} U_i(n-1))$ -highest vectors of a noncanonical polynomial $S[U(n-1)\otimes 3]$- bases of $SU(3n-3)$ IR (corresponding to the reduction $SU(3n-3) \supset \prod_{i=1} \otimes SU_i(n-1)$). Hence we have proposed algorithms for constructing these GI as linear combinations of vectors e^J_{μ} of the Gelfand-Tsetlin (GT) basis of SU(3n-3) IR $D(\dot{O}_{n-1} J \dot{O}_{2n-4})$ realized in terms of $[x_{i}, \ldots x_{in}] \equiv \varepsilon_{\alpha_1 \ldots \alpha_n\, n+1 \ldots 3n-3}\, x^{\alpha_1}_{i_1} \ldots x^{\alpha_n}_{i_n}$ (that are simultaneously elementary SU(n)-invariants and minors of SU(3n-3) component determinant). In addition, in [17] for the case n = 3 we have given similar procedures for constructing such SU(3) GI in terms of vectors $f^J_{a,\nu}$ of the Littlewood basis of SU(6) IR $D(\dot{O}_2 J \dot{O}_2)$ corresponding to Littlewood's rules of constructing GI for SU(3) Wigner coefficients [20]. This basis is generated by a nonorthogonal set of vectors $f^J_a \equiv f^J_{a,\nu(a)}$, that are highest vectors of $S[U(2)\otimes 3]$ IR $(S[U(2)\otimes 3] \subset SU(6))$

$$
\begin{aligned}
f^J_a([ik\ell]) &= \varrho(J;a)(s_2!)^{-3}[156]^{p_1-s_1-u} \times \\
&\times [356]^{p_2-s_1-v}[134]^{u}[123]^{v}[125]^{q_1-s_2-v} \times \\
&\times [345]^{q_2-s_2-u}[135]^{s_1}[\bar{a}_1\bar{a}_2\bar{a}_3]^{s_2} \times \qquad (1) \\
&\times ([a_1 12][a_2 34][a_3 56])^{s_2}, \quad s_1 s_2 = 0 \\
3J &= \sum_{i=1}^{3}(p_i + 2q_i), \quad [ik\ell] \equiv [x_i x_k x_\ell].
\end{aligned}
$$

Here $\varrho(\dots)$ is a normalization factor; $x_i = (x_i^{\alpha})^3_{\alpha=1}$ are vectors of the fundamental IR D(10) of SU(3), $\bar{x}_i = (\bar{x}_i^{\alpha}) \equiv (\partial/\partial x_i^{\alpha})$, $P_i = (p_i, q_i)$ is simultaneously a signature of SU(3) IR $D(P_i)$ in the decomposition $D(P_1)\times D(P_2)\times D(P_3) \to D(00)$ and a highest weight of $SU_i(2)$ IR in the reduction $\prod_i \otimes SU_i(2) \subset SU(6)$; $a = (\{p_i q_i\}; u, v, s_1, s_2)$. For $S_2 = 0$ vectors f^J_a from (1) equal constituents of symmetric basis $6^{\gamma}_{\{a_{ik\ell}\}}([ik\ell]) = \prod_{1\leq i<k<\ell} [ik\ell]^{a_{ik\ell}}$ of IR $D(\dot{O}_2 J \dot{O}_2)$, and for $S_1 = 0$ vectors f^J_a are sums of two vectors $6\{a_{ik\ell}\}(\dots)$.

We have found that the procedure for constructing $S[U(2)^{\otimes 3}]$-highest vectors π^J_a of an orthonormal $S[U(2)^{\otimes 3}]$-basis of IR $D(\dot{O}_2 J \dot{O}_2)$ (and corresponding SU(3) GI system) in terms of f_a^J is simpler than that in terms e^J_{μ}. Indeed, for vectors f^J_a the conditions

$$\begin{cases} E_{i\,i+1}\, \pi^J_a = 0, \quad E_{ik} = (x_i \bar{x}_k) \equiv \sum_{\alpha=1}^{3} x_i^{\alpha} \bar{x}_k^{\alpha}, \\ (E_{ii} - E_{i+1\,i+1})\pi^J_a = p_i \pi^J_a,\ E_{i+1\,i+1}\pi^J_a = q_i \pi^J_a,\ i = 1,3\ 5, \end{cases} \tag{2}$$

defining $S[U(2)^{\otimes 3}]$-highest vectors, are satisfied automatically. But for vectors $e^J_{\mu}([ik\ell])$ this is not the case. Specifically, the conditions (2) with i=1,5 fix parameters m_{ik} in a GT-pattern $\mu = [m_{ik}]$:

$$\mu = \begin{vmatrix} J\ J\ J\ 0\ 0\ 0 \\ J\ J\ m_{35}\ 0\ 0 \\ J\ m_{24}\ m_{34}\ 0 \\ m_{13}\ m_{23}\ m_{33} \\ m_{12}\ m_{22} \\ m_{11} \end{vmatrix} \quad \begin{aligned} & m_{12} = m_{11} = p_1 + q_1,\ m_{22} = q_1, \\ & m_{24} = m_{35} = J - q_3, \\ & m_{34} = J - p_3 - q_3, \\ & m_{13} + m_{23} + m_{33} = p_1 + 2q_1 + p_2 + q_2, \end{aligned} \tag{3}$$

but condition $E_{34}\, e^J_{\mu} = 0$ is not satisfied. We can prove this fact immediately if we make use of quasi-monomial realization of vectors $e^J_{\mu}([ik\ell])$ with m_{ik} from (3) that results from [11],

$$e^{\xi}_{\mu}([ik\ell]) = [\bar{a}_1\bar{a}_2\bar{z}_1]^{m_{13}-m_{23}}[\bar{a}_1\bar{z}_1\bar{z}_2]^{m_{23}-m_{33}} \times$$
$$[3a_1a_2]^{m_{13}-m_{12}}[1a_1a_2]^{m_{12}-m_{23}}[12a_1]^{m_{22}-m_{33}} \times$$
$$[123]^{m_{33}}[13a_1]^{m_{23}-m_{22}}[z_1 56]^{m_{13}-m_{24}} \times \tag{4}$$
$$\times[456]^{J-m_{13}}[z_1z_2 5]^{m_{23}-m_{34}}[z_1 45]^{m_{24}-m_{23}} \times$$
$$\times[z_1z_2 4]^{m_{34}-m_{33}}, [ika] \equiv [x_i x_k a]$$

Note that the expanded form of $e^{J}_{\mu}([ik\ell])$ in (4) (obtained by differentiating) is a multifold sum of constituents of the symmetric basis [17]. Therefore, the matrix of the transformation from $\{f^{J}_{a}\}$ to $\{\pi^{J}_{a}\}$ is simpler than a similar matrix for transformation from $\{e^{J}_{\mu}\}$ to $\{\pi^{J}_{a}\}$. It is a consequence of a close inter-relation between the bases $\{f^{J}_{a,\nu}\}$ and $\{\sigma\{a_{ik\ell}\}\}$. This situation is typical for IR $(\dot{O}_{n-1} J \dot{O}_{2n-4})$ of SU(m), and therefore, we only demonstrate it in the case n=2, m=4, where we can calculate corresponding matrices explicitly.

In this case vectors of SU(4) IR D(OJO) are identical with GI for SU(2) 4th rank Wigner coefficients, and, therefore, we proceed in what follows with these GI. Vectors $e^{J}_{\mu}([ik])$ of GT-basis coincide with GI $F^{j_{12}}_{\{j_i\}}([ik])$ that are given up to an inessential normalization factor by the expression

$$F^{j_{12}}_{\{j_i\}}([ik]) = e^{J}_{\mu}([ik]) \propto [(2j_{12})!]^{-1}[\bar{a}\bar{c}]^{2j_{12}} \times$$
$$\times[12]^{J_1-2j_{12}}[1a]^{J_1-2j_2}[2a]^{J_1-2j_1}[34]^{J_2-2j_{12}}[3c]^{J_2-2j_4} \times$$
$$\times[4c]^{J_2-2j_3} = \prod_{i=1,2}(J_1-2j_i)!\prod_{i=3,4}(J_2-2j_i)![12]^{J_1-2j_{12}}[34]^{J_2-2j_{12}} \times \tag{5}$$
$$\times \sum_{\alpha+\beta=J_2-2j_3} \frac{[14]^{\alpha}[24]^{\beta}[23]^{J_1-2j_1-\beta}[13]^{\xi_1-2j_2-\alpha}}{\alpha!\beta!(J_1-2j_1-\beta)!(J_1-2j_2-\alpha)!} \propto$$

$$\propto [12]^{J_1-2j_{12}}[34]^{J_2-2j_{12}} \sum_{\nu_1,\nu_2} \begin{pmatrix} L_1 & L_2 & j_{12} \\ \nu_1 & \nu_2 & j_3-j_3 \end{pmatrix} \times$$

$$\times \frac{[14]^{L_1-\nu_1}[13]^{L_1+\nu_1}[23]^{L_2+\nu_2}[24]^{L_2-\nu_2}}{[(L_1+\nu_1)!(L_2+\nu_2)!(L_1-\nu_1)!(L_2-\nu_2)!]^{1/2}},$$

$J_1=j_1+j_2+j_{12}$, $J_2=j_{12}+j_3+j_4$, $L_1=J_1/2-j_2$, $L_2=J_1/2-j_1$, $J=\sum_{i=1}^{4} j_i$, $2j_1=m_{11}$, $2j_4=J-m_{23}$, $2j_2=m_{12}+m_{22}-m_{11}$, $2j_3=J+m_{23}-m_{12}-m_{22}$, $2j_{12}=m_{12}-m_{22}$,

where $\begin{pmatrix} L_1 & L_2 & j_{12} \\ \nu_1 & \nu_2 & j_3-j_4 \end{pmatrix}$ are general (2d rank) SU(2) Clebsh-Gordan coefficients corresponding to the reduction $D(2j_1)\times D(2j_2)\to D(2j_1+2j_2)$. From (5) it follows that the matrix $C=\left\| C\begin{pmatrix}\{a_{ik}\}\\ j_{12}\end{pmatrix}\right\|$ of the transformation from the symmetric basis $\{\sigma_{\{a_ik\}}\}$ to the orthonormal basis $\{e^J_\mu\}$ is given in terms of Clebsh-Gordan coefficients rather than $6j$-symbols in which a transformation matrix between two orthonormal bases is expressed [10]. Note that the matrix C^{-1} of the transformation inverse to (5) is also expressed in terms of SU(2) Clebsh-Gordan coefficients,

$$\sigma^J_{\{a_{ik}\}} = \mathfrak{F}^J_{\{a_ik\}}([ij]) = \sum_{j_{12}} C^{-1}\begin{pmatrix}\{a_{ik}\}\\ j_{12}\end{pmatrix} \mathfrak{F}^{j_{12}}_{\{j_i\}}([ik]),$$

$$C^{-1}\begin{pmatrix}\{a_{ik}\}\\ j_{12}\end{pmatrix} \equiv \left[\prod_{i=1}^{4}(2j_i)!\right]^{-1} \mathfrak{F}^{j_{12}}_{\{j_i\}}([\bar{x}_i,\bar{x}_k]) \times$$

$$\mathfrak{F}^J_{\{a_ik\}}([x_i x_k]) \simeq \frac{A_{13}!A_{23}!A_{14}!A_{24}!}{(J_1-A_{12}+1)} \times \tag{6}$$

$$\times \sum_\beta (-1)^\beta [\beta!(-J_1+2j_1+A_{23}+\beta)!(A_{24}-\beta)!(J_1-2j_1-\beta)!(\beta+A_{13}-A_{24})!]^{-1} \times$$

$$\times [(J_1-2j_1+A_{14}-A_{23}-\beta)!]^{-1} \propto \begin{pmatrix} \frac{1}{2}(J_1-2j_1+A_{14}) & \frac{1}{2}(J_1-2j_2+A_{14}) & j_3-\frac{A_{34}}{2} \\ \frac{1}{2}(A_{14}-J_1+2j_1 & \frac{1}{2}(A_{13}+A_{23}-J_1)+A_{24} & -A_{34}/2+j_4-j_{12} \end{pmatrix}$$

Hence we obtain an expression for $6j$-symbols in the form of a sum of only two Clebsh-Gordan coefficients. Indeed, performing a change $x_1 \leftrightarrow x_3$ and $j_1 \leftrightarrow j_3, j_{12} \rightarrow j_{23}$ in (5), we obtain the GI:

$$\mathcal{F}^{j_{23}}_{\{j_i\}}([ik]) \simeq [23]^{J_3-2j_{23}}[14]^{J_4-2j_{23}}(-1)^{j_1+j_2+j_3-j_4} \times$$

$$\times \sum_{\nu_3,\nu_4} \begin{pmatrix} L_3 & L_4 & j_{23} \\ \nu_3 & \nu_4 & j_5-j_4 \end{pmatrix} \frac{[34]^{L_3-\nu_3}[13]^{L_3+\nu_3}[24]^{L_4-\nu_4}}{[(L_3-\nu_3)!\,(L_3+\nu_3)!\,(L_4-\nu_4)!]^{1/2}} \times \qquad (7)$$

$$\times \frac{[12]^{L_4+\nu_4}}{[(L_4+\nu_4)!]^{1/2}}, \quad J_3 = j_2+j_3+j_{23},\ J_4 = j_{32}+j_1+j_4,\ j_{32} = j_{23}, \quad L_3 = J_3/2 - j_2,\ L_4 = J_3/2 - j_3$$

Hence the above statement follows immediately if we take into account formula (6). Note that we can interpret the substitution $x_1 \leftrightarrow x_3$ as inproper (det s=-1) discrete "rotation" $s: x_i \rightarrow s x_i,\ s = (13) \in S_4 \subset U(4) \supset SU(4)$- in the space of SU(4) IR $D(1\dot{0}_2)$. Therefore, SU(2) $6j$ -symbols (and arbitrary transformation matrices of SU(n) in general can be interpreted as values of D-function of SU(4) (or SU(m)) corresponding to discrete "rotation angles" (cf. [21]).

Returning now to the Wigner-Biedenharn problem for SU(n) note that similar transformations appear to be useful for constructing SU(n) Gi by discrete "rotation" of the GT-basis vectors. For example, it is the case for the SU(3) GI corresponding to the decomposition $D(p_1 \cdot q_1) \times D(p_2 0) \times D(p_3 q_3) \rightarrow D(0\,0)$. But in general we need to make use of more complex transformations of variables in vectors $e^J_\mu([i_1 \dots i_n])$. The work along this line is in progress.

3. Now we discuss some applications of the finite difference calculus methods in the group-theoretical formalism following [19]. These methods appear naturally when we construct the Wigner-Racah algebras of various groups within the infinitesimal approach. Then different elements

of the Wigner-Racah algebras are defined by difference (recurrence) equations [10,20]. These equations have a most simple form if we use nonnormalized bases of IR because matrix elements of group generators in these bases are expressed by rational function of integers (cf. [22]). Therefore, it is natural to look for solutions of such equations in a set of generalized hypergeometric functions ${}_pF_q\left(\begin{smallmatrix}(a)_p\\(b)_q\end{smallmatrix}, z\right)$ [23]. Indeed, this conjecture is true for SU(2) Clebsch-Gordan coefficients and $6j$-symbols. In [19] we have investigated ${}_pF_q$ -functions along this line. Our main result is representating ${}_pF_q$ -functions in the form

$$
\begin{gathered}
{}_{(p+1)}F_q\left(\begin{matrix}(a(x))_p, -s\\(b(x))_q\end{matrix}; z\right)=\left[\Psi(x;(a)_p,(b)_q;z)\right]^{-1}\times \\
\times\Delta_x^s\Psi(x;(a)_p,(b)_q;z),\ (a)_p=(a_1,\dots,a_p),\ \Delta_x f(x)=f(x+1)-f(x),
\end{gathered}
\tag{8}
$$

where generating functions $\Psi(\dots)$ are products of symbolic powers $A^{(B)} \equiv A!\,(A-B)!$ of some expressions A(x) depending on discrete variables "x". For example, setting in (8) $\Psi=\prod\limits_{1\leqslant i\leqslant p}(a_i+x)^{(\alpha_i)}\prod\limits_j(b_j-x)^{(\beta_j)}$ we get

$$
\begin{gathered}
{}_{1+u}F_u\left(\begin{matrix}-s, (a_i+x+1)_p, (x-b_j+\beta_j)_q\\(a_i+x-\alpha_i+1)_p, (x-b_j)_q\end{matrix}; z\right)=(-1)^s\times \\
\times\left[z^x\prod_i(a_i+x)^{(\alpha_i)}\prod_j(b_j-x)^{(\beta_j)}\right]^{-1}\Delta_x^s\left[z^x\prod_i(a_i+x)^{(\alpha_i)}\prod_j(b_j-x)^{(\beta_j)}\right].
\end{gathered}
\tag{9}
$$

Hence we have obtained different "sum rules" for ${}_pF_q$ -functions using formulas (8) and

$$
\Delta_x^s\left[u(x)\cdot v(x)\right]=\sum_{k,n}\frac{s!\left[\Delta_x^k u(x)\right]\cdot\left[\Delta_x^n v(x)\right]}{(n+k-s)!\,(s-k)!\,(s-n)!}\,. \tag{10}
$$

Using the representation (8) we also have established interrelations between theories of angular momenta and orthogonal polynomials in discrete variables [24]. Particularly, we have identified the SU(2) Clebsch-Gordan coefficients with the Hahn polynomials, and, in addition, we conjectured [19] that a new class of orthogonal polynomials may be associated with $6j$-symbols (cf. [24,25]) and other 3nj-symbols. We also expect that the representation (8) and its generalizations may be useful in other group-theoretical topics.

4. In conclusion, we briefly consider some physical applications of developed techniques.

In [1,16] we have demonstrated the fruitfulness of the GI-technique for investigating various coherent cooperative processes within algebraic models of matter-radiation interaction. In addition, we point out that formula (8) and its analogues appear to be useful for solving discrete kinetics equations (cf. [19] and references therein). Another region of the application of the SU(n) GI formalism is calculating the group-theoretical models in nuclear and particle physics (cf. [3,4]).

The SU(n) GI-technique has also applications in analysis of phase transitions and of associated problem of the symmetry breaking [26], specifically, in minimizing the Higgs-Landau potentials in the SU(n)-symmetric field-theoretic models [2,26,27].

References

1. Karassiov, V.P., Shelepin, L.A. (1984). Trudy PhIAN, 144, 124 (Russian).

2. Umezawa, H., Matsumoto, H., Tachiki, M. (1982). Thermofield dynamics and condensed states. North Holland, Amsterdam et al.

3. Neudatchin, V.G. Obukhovsky, I.T., Smirnov, Yu.F.(1984). Phys. Elementary Par. and Atomic Nuclei, 15, 1165 (Russian).

4. Filippov, G.F., Vasilevsky, V.S., Chopovsky, L.L.(1984). Phys. Elementary Part. and Atomic Nuclei, 15, 1338 (Russian).

5. Haake, F., Reibold, R. (1984). Phys. Rev., A29,3208.

6. Manykin, E.A., Samartsev, V.V. (1984). The optical echo-spectroscopy. Nauka, Moscow (Russian).

7. Berezin, F.A. (1965). The second quantization method. Nauka, Mosoow (Russian).

8. Jordan, P. (1934). Z.Phys., v.94, p.531.

9. Weyl, H. (1931). Gruppentheorie und Quantenmechanik, Hizzel, Leipzig.

10. Biedenharn, L.C., Louck, J.D. (1981). Angular momentum in Quantum Physics. Theory and Applications. Addison Wesley, Massachusetts.

11. Karassiov, V.P., Karassiov, P.P., Sanko, V.A. et al. (1979). Trudy PhIAN, v.106, p.119 (Russian).

12. Bars, I., Gunaydin, M. (1983). Commun. Math. Phys., v.91, p.31.

13. Malkin, I.A., Manko, V.I. (1979). Dynamic Symmetries and Coherent States of Quantum Systems, Nauka, Moscow (Russian).

14. Butler, P.M. (1975). Phil. Trans. Roy. Soc. (London), A277, 545.

15. Karassiov, V.P. (1976). Kratkie soobshcheniya po fiz. PhIAN, 4, 21 (Russian).

16. Karassiov, V.P. (1985). In: Group-theoretical Methods in Physics. (Markov, N.A., Manko, V.I., Shabad, A.E. eds), Harwood Academic publishers, New York et al., 2, pp.107,661.

17. Karassiov, V.P., Shchelock, N.P. (1986). Trudy PhIAN, 173, 115 (Russian).

18. Biedenharn, L.C. (1962). Phys. Lett., v.3, p.254.

19. Karassiov, V.P., Shelepin, L.A. (1976). Trudy PhIAN, 87, 55. (Russian).

20. Karassiov, V.P. (1973). Trudy PhIAN, 70, 147. (Russian).

21. Fujiwara, Y., Horiuchi, H. (1982). Mem. fac. Sci. Kyoto Univ. Press, ser. Phys. et al., 36, 197.

22. Zhelobenko, D.P. (1982). Compact Lie groups and their representations. Transl. AMS, Providence.

23. Bateman, H., Erdelyi, A. (1953). Higher transcendental functions. McGraw-Hill, New York et al., vv1,2.

24. Nikiforov, A.F., Suslov, S.K., Uvarov, V.B. (1985). Classical orthogonal polynomials in a discrete variable. Nauka, Moscow (Russian).

25. Askey, R., Wilson, J.A. (1979). SIAM F. Math. Anal., 10, 1008.

26. Jaric, N.V., Michel, L., Sharp, R.T. (1984). J. Physique, 45, 1.

27. Hubsch, T., Pallua, S., Meljanac, S. (1985). Phys. Rev., D32, 1021.

MULTIPLICATIVE INTEGRALS OF REPRESENTATIONS OF PSU(1.1), SU(2) AND VERTEX OPERATORS

Yi.A.Verdiev

Physical Institute of the Azerbaijan SSR Academy of Sciences, Baku

A multiplicative integral of group representations (also called continuous tensor product of representations [2 , 3]) is in essence a nonlocal construction of representations [1].

This report is based on the results of Vershik-Gelfand-Grayev [1] ans Ismagilov [4] who studied PSU(1.1) and SU (2) respectively. Since the construction of [1] also enters the general scheme of the "factorizable" representations, we briefly recapitulate it, cf.[5].

<u>1. The space exp K.</u> Let K be a Hilbert space (complex or real). Denote by $\hat{K}$ the complex linear space with a basis consisting of the symbols exp x, $x \in K$. We introduce the scalar product $[\ ,\ ]$ setting $[\exp x, \exp y] = e^{(x,y)}$, where $(\ ,\ )$ is the scalar product in K. The completion of $\hat{K}$ with respect to the scalar product $[\ ,\ ]$ is denoted by exp K.

The expansion $e^z = 1 + \frac{z}{1} + \frac{z^2}{2} + \dots$ defines a natural isomorphism $\exp K \simeq \sum_{n=0} \oplus K^n_{symm}$, where $K^{(n)}_{symm}$ for $n \geq 1$ is the completion of the symmetrized tensor product of n copies of K, $K^0 = C$.

Let D(K) be the group of all motions $x \to Ux + h$, where U is a unitary operator in K, $h \in K$. Define the operators P(U,h): $\exp K \to \exp K$ setting

$$B(U,h): \exp x \to e^{-(Ux,h)-\frac{1}{2}|h|^2} \exp(Ux+h). \qquad (1)$$

When K is real these operators define a unitary representation of D(K). For the complex K the representation (U,h) $\longrightarrow$ B(U,h) is projective.

2: Representations in exp K. Let $\tilde{G}$ be a group, $\tilde{\varphi}$ its homomorphism in D(K). Clearly $\tilde{\varphi}(g)x = \tilde{V}(\tilde{g})x + \tilde{\beta}(\tilde{g})$, where $x \in K$, $\tilde{V}$ is a unitary representation of $\tilde{G}$ in K, the mapping $\tilde{\beta} : \tilde{G} \to K$ satisfies $\tilde{\beta}(\tilde{g}_1\tilde{g}_2) = \tilde{V}(\tilde{g}_1)\tilde{\beta}(\tilde{g}_2) + \tilde{\beta}(\tilde{g}_1)$. The function $\tilde{\beta}$ is called 1-cocycle.

Using the representation (U,h) $\longrightarrow$ B(U,h) we obtain the representation $T(\tilde{g}) = B(\tilde{\varphi}(g)) = B(\tilde{V}(\tilde{g}), \tilde{\beta}(\tilde{g}))$, $\tilde{g} \in \tilde{G}$, in the space exp K. We have:

$$T(\tilde{g})\exp(x) = e^{-(\tilde{V}(\tilde{g})x, \tilde{\beta}(\tilde{g})) - \frac{1}{2}(\tilde{\beta}(\tilde{g}), \tilde{\beta}(\tilde{g}))} \exp(\tilde{V}(\tilde{g})x + \tilde{\beta}(\tilde{g})). \qquad (2)$$

Usually $\tilde{G} \equiv G^X$ is a group of measurable G-valued functions on X, K is the Hilbert space of measurable maps of X to the Hilbert space H_o, in which the representation $g \to V(g)$, $g \in G$, acts. The representation $\tilde{V}$, and β are defined by the formula:

$$(\tilde{V}(\tilde{g})f)(x) = V(g(x))f(x), \; \tilde{\beta}(\tilde{g})(x) = \beta(g(x)),$$

where $\beta : G \to H_o$ is a 1-cocycle. This is the Araki construction. Thus, our problem is to construct β for each partucular case. First, consider the noncompact group PSU (1.1) [1].

3. The discreat representation of PSU (1.1). The group $G \equiv$ PSU (1.1) consists of complex-valued matrices g =

$= \begin{pmatrix} \alpha & \beta \\ \bar{\beta} & \bar{\alpha} \end{pmatrix}$, where $|\alpha|^2 - |\beta|^2 = 1$, g and -g are identified. It is well known that PSU (1.1) and $SO_0(1,2)$ are isomorphic.

Let H be the space of smooth functions $f(\xi)$ on the circle $|\xi| = 1$ in which V(g) acts by the formula:

$$(V(g)f)(\xi) = f\left(\frac{\alpha\xi + \bar{\beta}}{\beta\xi + \bar{\alpha}}\right)|\beta\xi + \bar{\alpha}|^{-2}. \tag{3}$$

There is an invariant linear functional $l(f) = \frac{1}{2}\int_0^{2\pi} f(e^{i\varphi})\, d\varphi$ in the space H. Define the scalar product

$$(f_1, f_2) = c\int_0^{2\pi}\int_0^{2\pi} \ln\left|1 - e^{-i(\varphi_1 - \varphi_2)}\right| f_1(e^{i\varphi_1})\overline{f_2(e^{i\varphi_2})}\, d\varphi_1\, d\varphi_2 \tag{4}$$

in the subspace H_0 of the functions $f(\xi)$ for which $l(f)=0$. Since $\int_0^{2\pi} f(e^{i\varphi})\, d\varphi = 0$, this scalar product is invariant under the action (3).

Using the expansions:

$$f(e^{i\varphi}) = \sum_{n\neq 0} a_n e^{in\varphi}, \quad \ln|1 - e^{i\varphi}|^{-2} = \sum_{n\neq 0} \frac{1}{|n|} e^{in\varphi}$$

we obtain from (4)

$$\|f\|^2 = c' \sum_{n\neq 0} \frac{1}{|n|}|a_n|^2 \geqslant 0, \quad c' = -2\pi^2 c.$$

The quotient H/H_0 is one-dimensional. Instead of V we introduce now the operator V': $V'(g) = V(g) - \beta(g)$, where $\beta(g) = V(g)f_0 - f_0$ and $f_0 = a_0$ is an element of H/H_0. We have: $V'(g)f_0 = f_0$. Consequently, H/H_0 is the invariant subspace under action of the operator V'. The invariance of l implies: $l(\beta(g)) = l(V(g)f_0) - l(f_0) = 0$, i.e. $\beta(g) \in H_0$. It is easy to verify that $\beta(g_1 g_2) =$

$= V(g_1)\ \beta(_2) + \beta(g_1)$. Therefore β is a 1-cocycle with values in H_o. It follows from (3) that

$$\beta(g.\xi) = |\beta\xi + \bar{\alpha}|^{-2} - 1, \text{ where } f_o \equiv 1. \tag{5}$$

The function $\beta(g,\xi)$ can be represented in the form:

$$\beta(g) = \sum_{n\neq 0} (-1)^n \beta_n e^{in\varphi} \tag{6}$$

where $\beta_n = \left(\frac{\beta}{\bar{\alpha}}\right)^n$ for $n>0$, $\beta_n = \left(\frac{\bar{\beta}}{\alpha}\right)^{|n|}$ for $n<0$.

<u>4.Vershik-Gelfand-Graev's construction of the representation of the group G^X, $G = PSU(1,1)$.</u> Denote by H^X the space of the smooth functions $f(x;\xi): X \to H$ for which $1(f(x,\xi)) = 1$, $x \in X$. The scalar product in $\exp H_o^X \equiv \exp K$ is defined by the formula:

$$[\exp\tilde{f}'_1, \exp\tilde{f}'_2] = \exp\Big(c\int_0^{2\pi}\int_0^{2\pi} \ln\left|1-e^{-i(\varphi_1-\varphi_2)}\right| \cdot f'_1(x,e^{i\varphi_1})\, f'_2(x,e^{i\varphi_2})\, d\varphi_1 d\varphi_2 dx\Big) \tag{7}$$

where $f'_i = f_i - f_o = f_i - 1$, $i = 1,2$, dx is the measure in X.

The representation is of the form:

$$T(\tilde{g})\exp f'(x,\xi) = \lambda(\tilde{g},\tilde{f})\exp\big(V(g(x))f'(x,\xi) + \beta(g(x),\xi)\big),$$

where

$$V(g(x))f(x,\xi) = f\Big(x, \frac{\alpha(x)\xi + \overline{\beta(x)}}{\beta(x)\xi + \overline{\alpha(x)}}\Big)\,|\beta(x)\xi + \overline{\alpha(x)}|^{-2},$$

$$\lambda(\tilde{g},\tilde{f}) = \exp(-c\int \ln|1-e^{-i(\varphi_1-\varphi_2)}|((V(g(x))f'(x,e^{i\varphi_1})\cdot$$

$$\cdot\beta(g(x),e^{i\varphi_2}) + \frac{1}{2}\beta(g(x),e^{i\varphi_1})\beta(g(x),e^{i\varphi_2}))\cdot$$

$$\cdot d\varphi_1 d\varphi_2 dx). \tag{8}$$

The operators $T(\tilde{g})$ form a representation, if

$$\lambda(\tilde{g}_1,\tilde{f}_1)\ \lambda(\tilde{g}_2,\tilde{f}) = \lambda(\tilde{g}_1\tilde{g}_2,\tilde{f}) \text{ for any } \tilde{g}_1,\tilde{g}_2 \in G \text{ and } \tilde{f} \in H^X, \tag{9}$$

where $f_1(x,\xi) = T(g_2(x))\ f(x,\xi)$.

It is easy to verify that $T(\tilde{g})$ are unitary. The definition of the scalar product implies that $[1,1] = 1$. On the other hand, from (8) we have:

$$T(\tilde{g})\cdot 1 = \exp(-\frac{c}{2}\int \ln|1-e^{-i(\varphi_1-\varphi_2)}|\beta(g(x),e^{i\varphi_1})\beta(g(x),e^{i\varphi_2})$$

$$d\varphi_1 d\varphi_2 dx)\cdot \exp\beta(g(x),\xi).$$

If $\tilde{g} = g(x)$ is an element of the maximal compact subgroup consisting of the matrices $k(x) = \begin{pmatrix} e^{i\varphi(x)/2} & 0 \\ 0 & e^{-i\varphi(x)/2}\end{pmatrix}$ then $\beta(k(x)) = 0$ and $T(k(x)).1 = 1$. A vector of H with the norm invariant with respect to K^X is called a vacuum vector. The spherical function of the representation $T(g)$ defined by the formula $\psi(\tilde{g}) = [T(\tilde{g})\ 1,1]$ is of the form:

$$\psi(\tilde{g}) = \exp(-\frac{c}{2}\int \ln|1-e^{i(\varphi_1-\varphi_2)}|\beta(g(x),e^{i\varphi_1})\cdot$$

$$(g(x), e^{i\varphi_2})d\varphi_1 \, d\varphi_2 \, dx).$$

Using the expansion (6) the function Ψ can be rewritten in the convenient form:

$$\Psi(\tilde{g}) = \exp(2c' \int \ln \Psi(g(x)dx), \text{ where}$$

$$\Psi(g(x)) = \frac{1}{|\alpha(x)|}, \quad c' = -2\pi^2 c.$$

The spherical function Ψ uniquely determines the representation.

The representation space exp H_0^X can be also realized as the space of sequences:

$$F = (1, a_n(x), \ldots, a_n(x), \ldots, a_{n_k}(x), \ldots)$$

with the norm

$$\| F \|^2 = \exp(c' \sum_{n \neq 0} \frac{1}{|n|} \int |a_n(x)|^2 dx).$$

The vector $\xi_0 = (1, 0, \ldots)$ is a vacuum vector. The representation in the space of sequences is given by the formula:

$$T(\tilde{g})F = \lambda(\tilde{z}, a) \; (1, a'_n(x), \ldots, a'_n(x) \ldots a'_{n_k}(x), \ldots)$$

where

$$a'_n(x) = \sum_{m=-\infty}^{+\infty} P_{mn}(g(x)) a_m(x),$$

$$P_{mn}(g(x)) \frac{1}{2\pi} \int_0^{2\pi} \left(\frac{\alpha(x)e^{-i\varphi} - \overline{\beta(x)}}{-\beta(x)e^{-i} + \overline{\alpha(x)}} \right)^n e^{-im\varphi} d\varphi$$

$$\lambda(\tilde{z}, a) = \Psi(z) \exp(c' \sum_{n=1}^{\infty} \frac{1}{n} \int (z(x))^n a_{-n}(x) a_0(x) dx + \qquad (10)$$

$$+\sum_{n=1}^{\infty}\frac{1}{n}\int(\overline{z(x)})^n a_n(x)a_o(x)dx)$$

$$\tilde{z}=z(x)=\overline{\beta(x)}\,/\overline{\alpha(x)},$$

$$\Psi(z)=[T(\tilde{g})\xi_o,\xi_o]=\exp(c'\int \ln(1-|z(x)|^2)dx.$$

5. Fubini-Veneziano vertex operator. From formula (10) it is easy to see a connection between the representation operator $T(\tilde{g})$ and the vertex operator in the dual-resonance model. Indeed, let the space X be the sphere S^3 in the momentum space R^4:

$$S^3=\{p=(p_1,p_2,p_3,p_4),\ [p,p]=p_1^2+p_2^2+p_3^2+p_4^2=1\}$$

Now, we suppose that the functions $a_n(p)$ $(a_{-n}(p))$, $n>0$, are of the form: $a_n(p)=a_{n\nu}p_\nu$ and $z(p)=z$ is independent of p, $|z|\leqslant 1$. Then, since $a_o(p)=1$ and, consequently, $a_{o\nu}=p_\nu$, we have:

$$\lambda(z,a)=\Psi(z)\exp(c'(\sum_{n=1}^{\infty}\frac{1}{n}z^{n}a_{-n\nu}p_\nu+\sum_{n=1}^{\infty}\frac{1}{n}\bar{z}^{-n}a_{n\nu}p_\nu)). \tag{11}$$

As is well known, the infinitesimal operators of representation of $PSU(1,1)\equiv SO_o(1.2)$ can be realized by the creation and annihilation operators of linear ossilator. Suppose that in formula (11) $a_{n\nu}$ for $n>0$ are the annihilation operators and $a_{-n\nu}\equiv\overline{a_{n\nu}}\equiv a^+_{n\nu}$ ate the creation operators with the commutation relations:

$$[a_{n\nu}, a_{m\nu'}] = [a^+_{n\nu}, a^+_{m\nu'}] = 0, [a_{n\nu}, a^+_{m\nu'}] = n\delta_{mn}\delta_{\nu\nu'}.$$

Then formula (11) can be rewritten in the form:

$$\lambda(z,a) = \Psi(z)\exp\left(c'\sum_{n=1} \frac{1}{n} z^n a^+_{n\nu} p_\nu\right)\exp\left(c'\sum_{n=1}^{\infty} \frac{1}{n}\cdot\right. \tag{12}$$
$$\left.\cdot (z)^{-n} a_{n\nu} p_\nu\right).$$

Here the order of the operators is chosen so that

$$\langle 0|\lambda(z,a)|0\rangle = [T(\tilde{g})\xi_0, \xi_0] = \Psi(z)$$

where $|0\rangle$ is the vacuum state of oscilator: $a_{n\nu}|0\rangle =$ $= 0$. The operator $\lambda(z,a)\Psi^{-1}(z) = V(z,Q)$ for $c' = -1$ coincides with the reggeon-particle-reggeon vertex operator in the dual-resonance model.

As a consequence of (9) we have:

$$\lambda(z,a')\,\lambda(g,a) = \lambda(z',a)\,|\alpha|^{-2c'}\,|\beta z + \bar{\alpha}|^{2c'} \tag{13}$$

or $V(z, a')\ V(g, a) = V(z', a)$

where

$$a' \equiv a'_{n\nu} = \sum_{m'} P_{mn}(g)\, a_{n\nu} + a_{o\nu}, P_{mo}(g) = 0 \text{ for } m \neq 0,$$

$$z' = \frac{\alpha z + \bar{\beta}}{\beta z + \bar{\alpha}}, \quad g = \begin{pmatrix} \alpha & \bar{\beta} \\ \beta & \bar{\alpha} \end{pmatrix} \in SU(1.1).$$

This formula gives the symmetry properties of the operator $\lambda(z, a)$ with respect to the projective transformations, in other words, it defines the duality property of the scattering amplitudes calculated by means of these operators.

6. Three reggeon vertex operator. In order to construct

three reggeon vertex operator it is necessary to consider an invariant 3-linear functional.

We know that the invariant 3-linear functionals in the space H_0 of the functions $f(\xi)$ for which $l(f) = 0$ is of the form:

$$(f_1, f_2, f_3) = c \int K^{(3)}(\varphi_1, \varphi_2, \varphi_3) \prod_{i=1}^{3} f_i(e^{i\varphi_i}) d\varphi_i \tag{14}$$

where

$$K^{(3)}(\varphi_1, \varphi_2, \varphi_3) = \ln|1-e^{-i(\varphi_1-\varphi_2)}| \ln|1-e^{-i(\varphi_2-\varphi_3)}| \cdot$$
$$\cdot \ln|1-e^{-i(\varphi_1-\varphi_3)}|$$

Using this formula one can define the 3-linear functional in the space $\exp H_0^X$. In realising this space as the space of sequences F we have:

$$[\exp a^{(1)}, \exp a^{(2)}, \exp a^{(3)}] = \exp(c \sum_{n_1, n_2, n_3} W_{n_1 n_2 n_3} \int a^{(1)}_{n_1}(x)$$
$$\cdot a^{(2)}_{n_2}(x) a^{(3)}_{n_3}(x) dx)$$

where $W_{n_1 n_2 n_3}$ are the Wigner coefficients of SO(1.2)

$$W_{n_1 n_2 n_3} = \sum_{n \neq 0} \frac{1}{|n|\,|n+n_1|\,|n+n_1+n_2|} \delta_{n_1+n_2+n_3, 0}$$

The representation formula takes the form:

$$T(\tilde{g}) F^{(i)} = \lambda(\tilde{z}, a^{(1)}, a^{(2)}, a^{(3)}) (1, a'^{(i)}_{n}(x), \dots a'^{(i)}_{n1}(x) \dots$$
$$a'^{(i)}_{n_k}(x) \dots)$$

where

$$\lambda(\tilde{z}^{(i)}, a^{(1)}, a^{(2)}, a^{(3)}) = \exp(\frac{c}{3} \sum_{n_1, n_2, n_3} W_{n_1 n_2 n_3} \cdot$$

$$\int [a^{(1)}_{n_1}(x) a^{(2)}_{n_2}(x) z^{(3)}_{n_3}(x) + a^{(1)}_{n_1}(x) z^{(2)}_{n_2}(x) a^{(3)}_{n_3}(x) +$$

$$+ z^{(1)}_{n_1}(x) a^{(2)}_{n_2}(x) a^{(3)}_{n_3}(x) + a^{(1)}_{n_1}(x) z^{(2)}_{n_2} z^{(3)}_{n_3}(x) +$$

$$+ z^{(1)}_{n_1}(x) a^{(2)}_{n_2}(x) z^{(3)}_{n_3}(x) + z^{(1)}_{n_1}(x) z^{(2)}_{n_2}(x) a^{(3)}_{n_3}(x) +$$

$$+ z^{(1)}_{n_1}(x) z^{(2)}_{n_2}(x) z^{(3)}_{n_3}(x)]\, dx,$$

$$\tilde{z}^{(i)} \equiv z^{(i)}_n = \left(\frac{\overline{\beta(x)}}{\bar{\alpha}(x)}\right)^n a^{(i)}_0(x), \text{if } n>0, \; z^{(i)}(x) = \left(\frac{\beta(x)}{\alpha(x)}\right)^{|n|} a^{(i)}_0(x)$$

if $n<0$, $i=1,2,3$.

Introduce the commuting operators $a_n(p) = a^{(1)}_n(p)$, $b_n(p) = a^{(2)}_n(p)$, $c_n(p) = a^{(3)}_n(p)$. Then the operators $:\lambda(z^i, a, b, c):$ (where $::$ denotes the normal product of the operators) the three reggeon vertex operators.

<u>7. Ismagilov's construction on a representation of the group G^X, where $G = SU(2)$.</u> Note, that in Vershik-Gelfand-Graev's construction of representations of G^X the fact that the trivial representation is not an isolated point in the space of irreducible unitary representations is crucial. In case of compact Lie group G the trivial representation is isolated and it is imposible to follow Vershik-Gelfand-Graev's scheme. For the group G^X of continously differentiable mappings instead of the group of

measurable mappings it is possible, however, to construct an irreducable nonlocal representation. The main idea of this construction is to consider first the semidirect product G·L, where L is Lie algebra of G. Therefore, the group $(G.L)^X$ is isomorphic to a skew-product of G^X and $\mathfrak{R}(X, L)$ (the group of L-valued forms on X).

Thus, consider the group $\theta(X,G) = G^X \mathfrak{R}(X,L)$. The action V of G^X on $K \equiv \mathfrak{R}(X,L)$ is the adjoint yction:

$$(V(g)\omega)(x) = \mathrm{Ad}g(x)\cdot\omega(x), \quad \omega(x) \in K.$$

The scalar product in K is defined by the formula:

$$(\omega_1, \omega_2)_P = \int (\omega_1(x)\,\omega_2(x))_{P(x)}\, dx$$

where (,) is the Cartan scalar product in L, P(x) is given by a positive definite matrix $\{P_{ik}(x)\}_1^m$. For an arbitrary $u \equiv g(x) \in G^X$ consider $\dot{u} = u^{-1}du \in \mathfrak{R}(X, L)$. It is clear that $(uv)^{\cdot} = \mathrm{ad}v\cdot\dot{u} + \dot{v}$, i.e. $\dot{u}$ is a 1-cocycle.

According to the general scheme we obtain a representation in the space exp K:

$$T_P(u)\colon \exp\omega \to e^{-(\mathrm{Ad}u\,\omega,\,\dot{u})_P - \frac{1}{2}(\dot{u},\dot{u})_P} \exp(\mathrm{Ad}u\,\omega + \dot{u}).$$

Application of this scheme to the gauge field theory is straightforward. Consider gauge transformations:

$$\hat{A}_\mu(x) \to u^{-1}(x)\,\hat{A}_\mu(x)\,u(x) + u^{-1}(x)\,\partial_\mu u(x)$$

where $\hat{A}_\mu(x) = \sum_i A_\mu^i(x)\,\tau^i$, τ^i are the generators of SU(2), i = 1,2,3. The pure gauge term $u^{-1}\partial_\mu u$ is a 1-cocycle. Therefore we have:

$$T(u): \exp \hat{A}_\mu \to e^{-\int (u^{-1}\hat{A}_\mu u, u^{-1}\partial_\mu u)dx - \frac{1}{2}(u^{-1}\partial_\mu u, u^{-1}\partial_\mu u)dx}$$

$$\exp(u^{-1}\hat{A}_\mu u + u^{-1}\partial_\mu u). \tag{15}$$

It is well known that the pure gauge term is also a vacuum solution. Ir we take an instanton solution as a vacuum solution, formula (15) gives the vertex operator "gluon-instanton-gluon".

References

1. Vershik, A.M., Gelfand, I.M., Graev, M.I. (1973). Representations of the group SL(2,R), where R is a ring of function. Russ.Math.Sorv. 28, 5, 83.

2. Streater, R.F. (1969). Current commutation relations, continuous tensor product and infinitly divisible group representations. Rendiconti di Se Int. di Fisica E.Fermi, 11, 247.

3. Partasarathy, K.R., Shmit, K. (1972). Rositive definite kernal, continuous tensor product and central theorems of probability theory. Lect.Notes in Math. 272, Springer, Berlin.

4. Ismagilov, R.S. (1976). On unitary representations of the group $C_0(X,G)$, G = SU(2). Mat.Sbornik, 100,1, 117 (Russian).

5. Araki, H. (1970). Factorisable representations of current algebra. Publ. of RIMS, Kyoto Univ.Ser.A. 5, 3, 361.

ON STRUCTURE OF THE REPRESENTATIONS OF U(2,1) WITH EXTREMAL VECTOR

Yu.F.Smirnov, V.N.Tolstoy, V.A.Knyr[+] , L.Ya.Stotland[+]
Institute of Nuclear Physics, Moscow State University,
[+] Khabarovsk Polytechnical Institute, Khabarovsk, 680035

1. Introduction. During the recent decade different noncompact Lie algebras and their infinite-dimensional unitary irreducible representations (UIR) have found applications in physics. These algebras mainly serve as dynamical symmetry algebras for various quantum mechanical systems [1] . It is well known that one (or several) UIR of the dynamical symmetry algebra contains the total energy spectrum of the system. Since the spectrum of each physical system has the ground state energy as the lower boundary, only the UIRs of noncompact algebras with containing some extremal weight (EW) are perfinent to physical applications. The type of dynamical symmetry algebra (i.e. the properties of the quantum mechanical system) is determined by quantum numbers of the (usually nondegenerate) ground state, whose wave function can be considered as in extremal vector (EV) of UIR corresponding to the EW. Therefore from the physical viewpoint it is interesting to consider the structure of the UIR containing a unique EV for various noncompact Lie algebras. Below it is done for U(2,1). The UIRs for the pseudo-unitary Lie algebras U(p,q) are discussed in [2-7] . The comparison of our results with those of [2-7] shows that all the types of UIRs of U(2,1) belonging to discrete or continuous series can be characterized by specific EW. (The same is true for U(p,1). Therefore, the class of UIR's with single EV is rather broad and ade-

quate for the study of UIR for other noncompact Lie algebras. The main tool in our approach is the Mickelsson-Zhelobenko algebra [8] based on the projection operator theory for simple Lie algebras [9].

2. <u>The classification of UIR's for U(2,1)</u>. The generators A_{ik} (i,k = 1,2,3) of U(2,1) acting in the space R of the weight vectors satisfy the usual commutation relations

$$[A_{ik}, A_{lm}] = \delta_{kl} A_{im} - \delta_{im} A_{lm} \tag{1}$$

We shall use the canonical reduction

$$U(2,1) \supset U(2) \supset U(1) \tag{2}$$

In this case every basic vector of a UIR is indexed by a Gelfand-Graev scheme

$$\left| \begin{matrix} f_{13} & & f_{23} & & f_{33} \\ & f_{12} & & f_{22} & \\ & & f_{11} & & \end{matrix} \right\rangle . \tag{3}$$

It means that this vector belongs to a definite UIR of each algebra entering the chain (2). The signatures of UIRs of these algebras are indicated in corresponding rows of the Gelfand-Graev schemes. Since the structure of the UIR $[f_{12} f_{22}]$ of U(2) is well known, it suffices to consider the highest weight vector with respect to this subalgebra. These vectors will be called rank-one vectors and denoted by the shortened scheme

$$\left| \begin{matrix} f_{13} & & f_{23} & & f_{33} \\ & f_{12} & & f_{22} & \end{matrix} \right\rangle ,$$

where $f_{12}-f_{22} \geq 0$, $(f_{12}-f_{22})$ is a non-negative integer. The subspace of all rank-one vectors in R will be labelled as R_H. For the rank-one vectors we clearly have

$$A_{12}\left|\begin{matrix} f_{13} & f_{23} & f_{33} \\ & f_{12} & f_{22} \end{matrix}\right\rangle = 0 \tag{4}$$

Let us introduce raising (lowering) operators Z_i, Z_{-i} $(i = 1,2)$ of the Mickelsson-Zhelobenko algebra acting in R_H according to [8] by the formulas

$$Z_i = PA_{i3}P, \quad Z_{-i} = PA_{3i}P, \quad i=1,2, \quad Z_i^{+} = -Z_{-i}. \tag{5}$$

where P projects any vector of R into R_H and is of the form [9]

$$P = \sum_{r=0}^{\infty} (-1)^r \frac{1}{r!} \prod_{s=1}^{r} \frac{1}{(A_{11}-A_{22}+S+1)} A_{21}^r A_{12}^r . \tag{6}$$

Using (5) and (6) we get

$$Z_{-1} Z_{+2} = Z_2 Z_{-1}, \qquad Z_{-2} Z_1 = Z_1 Z_{-2},$$

$$Z_2 Z_{-2} = Z_{-2} Z_2 \frac{(A_{11}-A_{22}+1)^2}{(A_{11}-A_{22})(A_{11}-A_{22}+2)} +$$

$$+ Z_1 Z_{-1} \frac{(A_{11}-A_{22}+1)}{(A_{11}-A_{22})(A_{11}-A_{22}+2)} - \frac{(A_{11}-A_{22}+1)(A_{33}-A_{22}+1)}{(A_{11}-A_{22}+2)}, \tag{7}$$

$$Z_{-1} Z_1 = Z_1 Z_{-1} \frac{(A_{11}-A_{22}+1)^2}{(A_{11}-A_{22})(A_{11}-A_{22}+2)} +$$

$$+ Z_{-2} Z_2 \frac{(A_{11}-A_{22}+1)}{(A_{11}-A_{22})(A_{11}-A_{22}+2)} + \frac{(A_{11}-A_{22}+1)(A_{33}-A_{11}-1)}{(A_{11}-A_{22}+2)}.$$

The classification of UIRs of $U(2,1)$ can be based on the fact that in every UIR there is a single extremal vector $|f\rangle$ with the extremal weight $[f]=[f_{13},f_{23},f_{33}]$ such that

$$\mathcal{Z}_{-1}|f\rangle=\mathcal{Z}_{2}|f\rangle=0. \tag{8}$$

Different extremal weights generate nonequivalent UIRs. The UIR of the discrete series are characterized by an additional index indicating the type of series. In the case of the continuous series it is also necessary to introduce additional parameters, the eigenvalues of the Casimir operators G_2, G_3 (see below).

Because of cyclicity of the EV the arbitrary rank-one vector takes the form

$$\left|\begin{matrix} f_{13} & f_{23} & f_{33} \\ f_{12} & f_{22} & \end{matrix}\right\rangle=\frac{1}{N(a,b)}\mathcal{Z}_{-2}^{a}\,\mathcal{Z}_{1}^{b}\,|f\rangle, \tag{9}$$

where $N(a,b)$ is the normalization factor, $a=f_{23}-f_{22}$, $b=f_{12}-f_{13}$. Using (7) we get the recurrences

$$|N(a,b)|^2=|N(a,b-1)|^2\frac{(f_{13}-f_{23}+b)}{(f_{13}-f_{23}+a+b+1)}\cdot$$

$$\cdot(b+\delta_{f_{13}f_{23}}\lambda)\cdot(f_{13}-f_{33}-\delta_{f_{13}f_{23}}\lambda+b),$$

$$\tag{10}$$

$$|N(a,0)|^2=|N(a-1,0)|^2\frac{(f_{12}-f_{23}+a)}{(f_{13}-f_{23}+a+1)}\cdot$$

$$\cdot(a-\delta_{f_{13}f_{23}}\lambda)\cdot(f_{33}-f_{23}+a+\delta_{f_{13}f_{23}}\lambda).$$

Here δ_{ij} is the Kronecker symbol; λ is the complex parameter, appearing for $f_{13}=f_{23}$ when denominators in (7) vanish. The values of the extremal weight $[f]$ and (for $f_{13}=f_{23}$) the additional parameter λ allow one to define the normalization factors (10) unambiguously and to describe completely the structure of UIRs. Thus it suffices to enumerate the extremal weights $[f]$ and to find the values λ compatible with the unitarity condition. The unitarity condition is satisfied if the squared normalization factor (10) is positive. Therefore λ must be real or $(f_{13}-f_{23})/2+i\nu$, where $\nu>0$. The real values of λ may be discrete or conditions. The real discrete values of λ correspond to UIR of the discrete series. The real continuous values of λ and $\lambda=(f_{13}-f_{33})/2+i\nu$ correspond to UIR of the continuous series. The total list of the UIRs of U(2,1) obtained in this way is given below.

1. <u>Intermediate discrete series D^{int}</u>

a) If $f_{13}\neq f_{23}$, $f_{13}\geq f_{23}$, $f_{33}\geq f_{23}$ we have a discrete series with unbounded a and b. It will be referred to as an intermediate DS. In the notations of [2] the signature of this DS is $[m_{13},m_{23},m_{33}]$, where $m_{13}=f_{13}-1$, $m_{23}=f_{33}$, $m_{33}=f_{23}+1$. Then for $m_{12}=f_{12}$ and $m_{22}=f_{22}$ we obtain well-known inequalities [2,4]: $m_{12}\geq m_{13}+1$, $m_{22}\leq m_{33}-1$. In [2] only standard DS satisfying Gelfand-Graev's inequalities: $m_{13}\geq m_{23}\geq m_{33}$ are considered. A nonstandard Δ^{int} with $m_{13}=m_{23}-1$ or $m_{23}=m_{33}-1$ also exists. The possibility of violation of the Gelfand-Graev inequalities was noted in [10]. The structute of the D^{int} and Δ^{int} series is shown an Fig. 1a and 1b, where EV's are indicated by circles.

b) If $f_{13}=f_{23}$, $f_{13}-f_{33}\leq -4$, $\lambda=-2,-3,\dots,$ $f_{13}-f_{33}+2$ we have the one-side bounded discrete series D^- . It is the so-called DS with highest weight. In notation of [2] ($m_{13} = f_{33} + \lambda - 1$, $m_{23} = f_{13} - \lambda$, $m_{33} = f_{13} + 1$) we get $m_{13} + 1 \geq m_{23} \geq m_{33} + 1$, $m_{23} - 1 \geq m_{12} \geq m_{33} - 1 \geq m_{22}$. In addition to the results of [2,4] the nonstandard DS Δ^- exists for $m_{13} = m_{23} - 1$. The structure of the series D^- and Δ^- is shown on Fig. 1a and 1b. Noted that at $\lambda = -1$ ($m_{23} = m_{33}$) the diagram of D^- degenerates into a line. Such series will be refered to as degenerate DS (DDS).

c) If $f_{13} = f_{23}$, $f_{13} - f_{33} \geq 2$, $\lambda = 2,3,\ldots$ $\ldots, (f_{13} - f_{33})$ we have another bounded discrete series D^+ from below (DS with the lowest weight). In notation of [2] we get the structure of D^+ in the form : $m_{13} + 1 \geq m_{23} \geq m_{33} - 1$, $m_{12} \geq m_{13} + 1 \geq m_{22} \geq m_{23} + 1$. The structure of D^+ is shown on Fig 1a. The violation of the Gelfand-Graev inequalities produces two degenerate series: Δ_0^+ with $m_{13} = m_{23} - 1$ (Fig. 1d) and Δ_1^+ with $m_{23} = m_{33} - 1$ (Fig. 1e). If $\lambda = 1$ ($m_{13} = m_{33} - 1$) we get a nonstandard generate series Δ^+ (Fig. 1c).

Each DDS is characterized by an EV, which is annihilated by the three operators $Z_{\pm i}$ simultaneously. Besides the above-mentioned series , two nonstandard DDS with the signatures $[m_{13}, m_{23} - 1, m_{23} + 1]$ and $[m_{23} - 2, m_{23}, m_{33}]$ exist.

II. <u>Continuous series (CS) (Fig. 1f)</u>

a) C^+ corresponds to

$$f_{13} = f_{23}, \quad 1 < \lambda < f_{13} - f_{33} - 1; \quad n < f_{13} - f_{33} - \lambda < \lambda < n+1,$$

$$n \in Z; \quad f_{12} \geq f_{13}, \quad f_{22} \leq f_{23}$$

b) C^- corresponds to

$$f_{13} = f_{23}, \quad f_{13} - f_{33} + 1 < \lambda < -1; \quad n < f_{33} - f_{13} + \lambda < -\lambda < n+1,$$

$$n \in Z; \quad f_{12} \geq f_{13}, \quad f_{22} \leq f_{23}$$

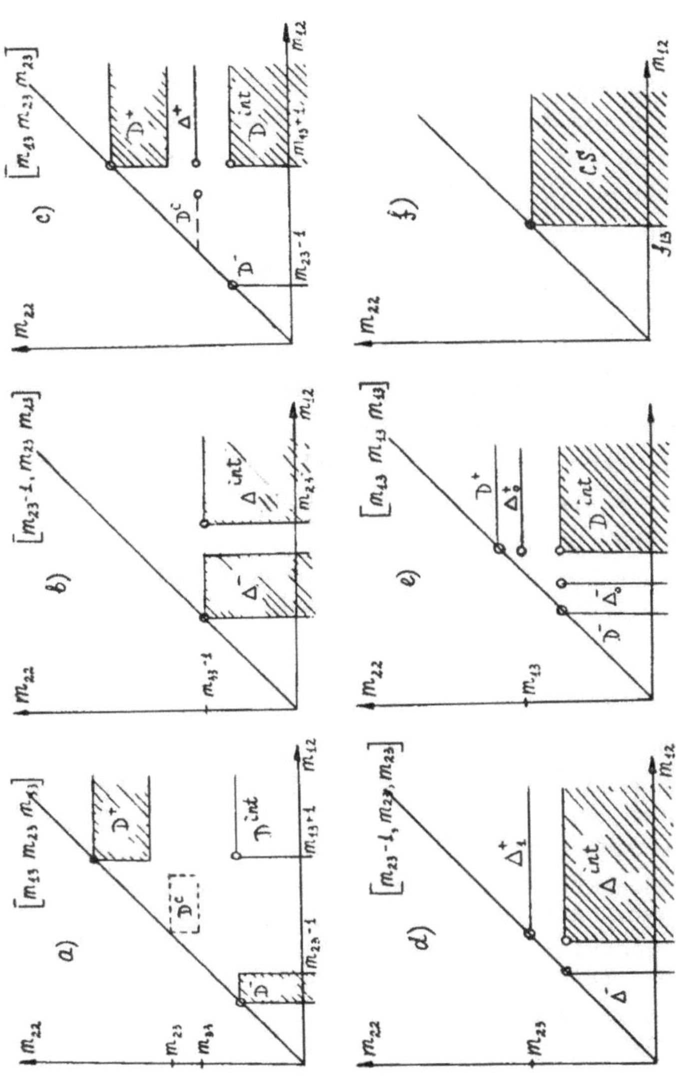

Fig. 1. The structure of the various series of UIRs of U(2,1)

c) C^{o} corresponds to

$$f_{13} = f_{23}, \quad \lambda = \frac{f_{13} - f_{23}}{2} + i\nu, \quad \nu > 0,$$

$$f_{12} \geq f_{13}, \quad f_{22} \leq f_{23}$$

Since an explicit form of the orthonormalized basic vectors is known, it is easy to calculate the eigenvalues of the Casimir operators and the matrix elements of the generators of U(2,1). The obtained result agrees with the theorem [11] that the eigenvalues of the Casimir operators for all the types of DS with the same signature $[m_{13}\ m_{23}\ m_{33}]$ coinside. Matrix elements of the generators for all DS are described by Gelfand-Graev's formulas [2-7]. Comparison of our results with [3,4] allows us to conclude that our list contains all the types of IR belonging to the discrete and continuous series of U(2,1). Thus it is established, that in every IR of U(2,1) a single EV exists which determines the structure of the corresponding UIR.

The authors are thankful to R.M.Asherova and D.P.Zhelobenko for valuable discussions.

References

1. Malkin,I.A., Manko,V.I. (1979). Dynamical Symmetries and Coherent States of the Quantum Systems, Nauka, Moscow (Russian).

2. Gelfand,I.M., Graev,M.I. (1965). Izv. Acad. Nauk SSSR, ser. math., 29, 1329 (Russian).

3. Ottoson,U. (1968). Comm. Math. Phys., 10, 144.

4. Bidenharn,L.C., Nuyts,J., Strautmann,N. (1965). Ann. Inst. Henri Poincaré, 3, 13.

5. Bidenharn,L.C., Gruber,B., Weber,H.J. (1967). Proc. Roy. Irish. Acad., 67A, 1.

6. Olshansky,G.I. (1980). Funk. Anal. Appl., 14, 32 (Russian).

7. Barut,A.O., Raczka,R. (1977). The Theory of Group Representations and Applications. Warszawa .

8. Zhelobenko,D.P. (1984). Dokl. Akad. Nauk SSSR, 4, 1317 (Russian).

9. Asherova,R.M., Smirnov,Yu.F., Tolstoy,V.N. (1979). Matem. zam. 26, 15 (Russian).

10. Todorov,I.T.(1966). Discrete series of hermitian representations of the Lie algebra U(p,q). ICTP, Trieste, Preprint lC/66/71.

11. Klimyk,A.U. (1979). Lett. Math. Phys., 3, 315.

RECURRENT CONSTRUCTING SU(4) SUPERMULTIPLET BASIS

D.M. Vladimirov, Yu.V. Gaponov

I.V. Kurchatov Institute of Atomic Energy, Moscow

The main well-known difficulty of constructing of general analytic form of a supermultiplet basis for an arbitrary irreducible representation $(pp'p'')$ of SU(4) is connected with the degeneracy of the most spin-isospin multiplets of an arbitrary representation. This problem is solved only for some multiplets of degeneracy $\leqslant 2$ [1]. Here we give a general recurrent procedure allowing us to successively construct the basis functions of degenerate spin-isospin multiplets of the $(ts)=(tt_z=tss_z=s)$ -type of an arbitrary representation (pp'p'') proceeding from the highest weight (pp'). Basis functions of the two series of multiplets with the degeneracy two: $t+s=$ $=p+p'-1$, $t-s=p-|p''|-1$ and three: $t=p-1$, $|p''|+1\leqslant s\leqslant p'-1$ (see Fig.1) are constructed in an

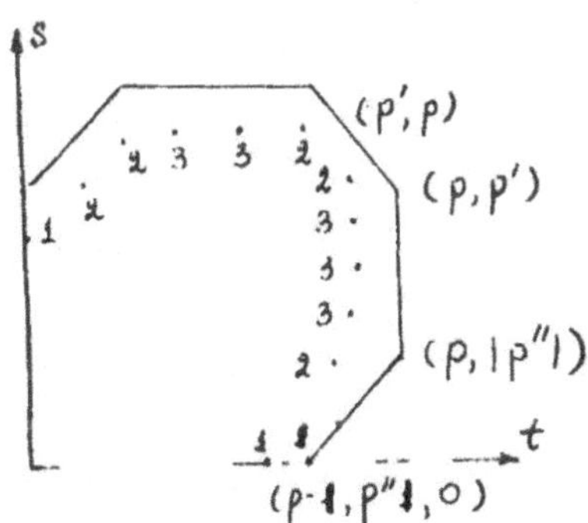

Fig. 1

explicit analytic form. Also, the problem of the Clebsch-Gordan series is solved for the tensor product of the representation (110) and an arbitrary representation $(f)=(pp'p'')$ in the expansion of which (f) enters with degeneracy three.

1. Recurrent procedure of supermultiplet basis construction. Let us present SU(4) in the $SU_T(2)\times SU_S(2)$ reduction by 15 generators with the relations

$$[T_\alpha,T_\beta]=\epsilon_{\alpha\beta\gamma}T_{-\gamma},\ [S_\alpha,S_\beta]=\epsilon_{\alpha\beta\gamma}S_{-\gamma},\ [T_\alpha,S_\beta]=0,$$
$$[T_\alpha,E_{\beta\gamma}]=\epsilon_{\alpha\beta\gamma}E_{-\delta\gamma},\ [S_\alpha,E_{\beta\gamma}]=\epsilon_{\alpha\beta\gamma}E_{\beta-\delta},$$
$$[E_{\alpha\beta},E_{\gamma\delta}]=\delta_{\alpha-\gamma}\epsilon_{\beta\delta\epsilon}S_{-\epsilon}+\delta_{\beta-\delta}\epsilon_{\alpha\gamma\epsilon}T_{-\epsilon},\epsilon_{+-0}=1\ . \quad (1)$$

From these generators we construct the operator functions introduced by Hecht and Pang [2] and connecting neighbouring spin-isospin multiplets:

$$O_{\alpha\beta}(ts)|(ts)\rangle=N_{\alpha\beta}(ts)|(t+\alpha\ s+\beta)\rangle\,, \quad (2)$$

where $|(ts)\rangle$ denotes the basis function of the state $(ts)=(tt_z=ts s_z=s)$, $N_{\alpha\beta}(ts)$ is a normalization coefficient:

$$O_{\alpha\beta}(ts)=\sum_{\gamma\delta=0\pm1}B_{\alpha\gamma}(t)B_{\beta\delta}(s)(T_-)^{\gamma-\alpha}(S_-)^{\delta-\beta}E_{\gamma\delta},$$
$$E_{\alpha\beta}=\sum_{\gamma\delta}(B_{\alpha\gamma}(t))^{-1}(B_{\beta\delta}(s))^{-1}(T_-)^{\gamma-\alpha}(S_-)^{\delta-\beta}O_{\gamma\delta}(ts), \quad (3)$$
$$B(t)=\begin{pmatrix}1&0&0\\(t+1)^{-1}&1&0\\-t(2t+1))^{-1}-t^{-1}&&1\end{pmatrix},\quad B(s)=\begin{pmatrix}1&0&0\\(s+1)^{-1}&1&0\\-(s(2s+1))^{-1}-s^{-1}&&1\end{pmatrix}.$$

Hecht and Pang used $O_{\alpha\beta}(ts)$ to construct the basis of a number of special supermultiplets of SU(4) containing only non-degenerate states. Let us generalize their method for degenerate states. We separate degenerate spin-isospin multiplets entering an SU(4)-multiplet by an additional index. Then $O_{\alpha\beta}(ts)$ acting on some basis function $|(ts)_i\rangle$ of the degenerate state (ts) in the general case changes it into a linear superposition:

$$O_{\alpha\beta}(ts)|(ts)i\rangle = \sum_k N^k_{\alpha\beta}(ts_i)\ |(t+\alpha\ s+\beta)k\rangle\ . \tag{4}$$

For convenience we separate the operator functions into step-down $O_{-11}, O_{-10}, O_{0-1}, O_{-1-1}$ and step-up O_{1-1}, O_{10}, O_{01}, O_{11} ones and the function which does not change spin-isospin variables O_{00}. In the general case various degenerate basis functions of the multiplet $(ts) = (p+b-a, p'-b)$ ($b \leqslant a$ are integers, $0 \leqslant a \leqslant$ $\leqslant p+p', -(p+p') \leqslant b \leqslant p'$) can be obtained applying step-down operators to the highest weight state

$$N_{cd}(ts)|(ts)cd\rangle = (O_{-11})^{c+d-b}(O_{-10})^{a-c-2d}(O_{0-1})^{c}(O_{-1-1})^{d}|(pp')\rangle \tag{5}$$

where c, d are nonnegative integers such that $c+2d \leqslant a, c+d \geqslant b$. It is significant that any other position of the step-down operators with respect to each other yields the basis function of (tS) linearly connected with (5). At the same time the number of various functions of the type (5) exactly coincides with the multiplet degeneracy of (tS). The first statement can be proved with the help of the linear relations (9) among matrix elements for quadratic products of the step-down operators and the second one, by calculating the Racah degeneracy N_{ts} of (tS) and the general number of allowed pairs c, d. Thus e.g. the degeneracy of states lying on the straight line $t+s = p+p'-a$, $t \geqslant s$, $0 \leqslant a \leqslant p' - |p''|$ is calculated with the help of the Maguin and Partensky recurrences [4] and equals:

$$N_{p+b-a,p'-b} = \begin{cases} N_{max}\ , b \leqslant 0 \\ N_{max}\ \ b(b+1)/2, 0 < b \leqslant [a/2]+1 \\ (a-b+1)(a-b+2)/2\ , [a/2] \leqslant b \leqslant a\ , \end{cases} \tag{6}$$

where $N_{max} = 1/4(a+2)^2$ for even a and $N_{max} = 1/4(a+1)(a+3)$ for odd a, [ ] is the integer part.

Let us state the problem of constructing a recurrence in which proceeding from the highest weight state (pp') of the supermultiplet (pp'p'') and using step-down operators one could successively obtain basis functions of degenerate multiplets (tS). The peculiarity of the degenerate state case as compared with that of non-degenerate multiplets is that in going over to one and the same state (tS) with the help of different types of step-down operators we obtain different basis functions

$$O_{\alpha\beta}|(t-\alpha s-\beta)i\rangle = N_{\alpha\beta}(t-\alpha s-\beta i)|(ts)_{(\alpha\beta)i}\rangle \quad (7)$$

Here the degenerate state index $(\alpha\beta)i$ indicates the step-down operator and the function from which the basis function $|(ts)(\alpha\beta)\rangle$ is obtained. Generally, these functions may turn out to be too many and a method of selecting an orthonormal basis from them is needed. Usually, for this purpose two commuting operators are fixed in the SU(4)-symmetry scheme (e.g. Ω and $\Phi[1]$) whose eigenfunctions form an orthonormal set. However, physical meaning of these operators is not clear. We apply to (7) another orthonormalization procedure, most convenient for physical applications e.g. for calculating the Clebsch-Gordan coefficients for SU(4). This procedure includes the study of the matrix

$$K_{(\gamma\delta)j,(\alpha\beta)i}(ts) = \langle (ts)_{(\gamma\delta)j}|(ts)_{(\alpha\beta)i}\rangle \quad (8)$$

$(K_{(\alpha\beta)i,(\alpha\beta)i} = 1)$ determining its rank, (tS) degeneracy, and diagonalizing it for constructing an orthonormal basis.

In the recurrent procedure the step-up operators applied to the highest weight state (pp') yield zero and O_{00} yields p''. Let us assume that in compliance with the recurrent procedure the inductive character at a certain step of the procedure a basis for each of the states

$(t-\alpha\ S-\beta)$ ($\alpha\beta$ runs through the indices of the step-down operators) preceding the multiplet (tS) is built, the matrices $K(t-\alpha\ S-\beta)$ are calculated which become unitary after the orthogonalization and the action of the step-up operators and O_{00} on the basis functions is known. Let us show that in this case the normalization coefficients of the (tS) multiplet basis functions (7) and the matrix (8) are expressed through known values. For this consider the matrix elements of $O_{-\gamma-\delta}\,O_{\alpha\beta}$ ($\alpha\beta$ runs all the indices, $-\gamma-\delta$ runs the indices of step-up operators and 00). Using the coupling of $O_{\alpha\beta}$ with generators of the group (3) and (1) we obtain:

$$\langle (t-\gamma\ S-\delta)j|O_{-\gamma-\delta}\,O_{\alpha\beta}|(t-\alpha\ S-\beta)i\rangle =$$

$$=\delta_{\gamma\alpha}\delta_{\delta\beta}A_{\alpha\beta}(t-\alpha S-\beta)\langle (t-\alpha S-\beta)j|(t-\alpha S-\beta)i\rangle + \tag{9}$$

$$+\sum_{-\mu-\nu} D^{-\mu-\nu}_{-\gamma-\delta\alpha\beta}(t-\alpha\ S-\beta)\langle (t-\delta\ S-\delta)j|O_{-\gamma+\alpha+\mu,-\delta+\beta+\nu}O_{-\mu-\nu}|(t-\alpha S-\beta)i\rangle$$

where the sum runs the indices of the step-up operators. The values of $A_{\alpha\beta}(tS)$ are

$$A_{00}=p(p+4)+p'(p'+2)+(p'')^2-t(t+4)-S(S+2),$$

$$A_{-11}=t-S,\ A_{-10}=(t+1)S/(S+1)-A_{00}/t, \tag{10}$$

$$A_{0-1}=S-A_{00}/S,\ A_{-1-1}=(t+S+1)(2S-1)/(2S+1)-A_{0-1}/t$$

and $D^{\mu\nu}_{\gamma\delta\alpha\beta}$ (tS) are coupled with the matrices B(t) and B(S) as follows:

$$D^{\mu\nu}_{\gamma\delta\alpha\beta}(tS)=\sum_{(\varkappa x)(\omega\rho)}\delta_{-\gamma-\omega,\alpha-\varkappa}\delta_{\delta-\rho,\beta-x}\big(B_{\alpha\varkappa}(t)B_{\beta x}(S)C_{\gamma\omega}C_{\delta\rho}+$$

$$+(1-\delta_{0\varkappa}\delta_{0x})B_{\alpha\omega}(t)\cdot B_{\beta\rho}(S)C_{\gamma\varkappa}C_{\delta x}\big)\big)\cdot \tag{11}$$

$$\cdot\Big(\sum_{(\varepsilon\xi)}\delta_{\mu-\varkappa,\omega-\varepsilon}\delta_{\nu-x,\rho-\xi}(B_{\varkappa\mu}(t))^{-1}(B_{x\nu}(S))^{-1}C_{\omega\varepsilon}C_{\rho\xi},$$

where $\varkappa x = 00, 1\text{-}1, 10, 01, 11,$

$$C_{\alpha\beta} = 0\ (\alpha > \beta),\ C_{\alpha\alpha} = 1,\ C_{\alpha\beta} = (-1)^{\beta}\ (\alpha < \beta)$$

Since

$$(N_{\alpha\beta}(t\text{-}\alpha\ s\text{-}\beta)i))^2 = \langle(t\text{-}\alpha S\text{-}\beta)i|O_{-\alpha-\beta}O_{\alpha\beta}|(t\text{-}\alpha S\text{-}\beta)i\rangle,$$

$$K_{(\gamma\delta)j\,(\alpha\beta)i}(tS) = \langle(t\text{-}\gamma S\text{-}\delta)j|O_{-\gamma-\delta}O_{\alpha\beta}|(t\text{-}\alpha S\text{-}\beta)i\rangle\cdot$$

$$\cdot(N_{\gamma\delta}(t\text{-}\gamma S\text{-}\delta)j)^{-1}(N_{\alpha\beta}(t\text{-}\alpha S\text{-}\beta)i))^{-1}, \qquad (12)$$

where $\alpha\beta, \gamma\delta = -11, -10, 0\text{-}1, -1\text{-}1$, then (9) is a recurrence for these values. Some of the basis functions are linearly dependent, which can be established with the help of (9) for six pairs of step-up operators $O_{-\gamma-\delta}O_{\alpha\beta}(-\gamma\text{-}\delta \neq \alpha\beta = 1\text{-}1,\ 10,\ 01, 11)$. For the next step of the recurrent procedure it is necessary to find the action of the step-up operators and O_{00} on the constructed or the orthonormal basis. Normalization coefficients for these operators satisfy the system of linear equations whose rank equals the degeneracy of the state (tS):

$$\sum_m N^m_{\alpha-\beta}(tS_{(\gamma\delta)j})K_{mi}(t\text{-}\alpha S\text{-}\beta) = N_{\alpha\beta}(t\text{-}\alpha S\text{-}\beta i)K_{(\alpha\beta)i\,(\gamma\delta)j}(tS) \qquad (13)$$

$$\sum_m N^m_{00}(tS_{(\gamma\delta)j})K_{(\alpha\beta)i,m}(tS)N_{\alpha\beta}(t\text{-}\alpha S\text{-}\beta\ i) = \qquad (14)$$

$$= \sum_{-n-\nu}^{-\mu-\nu} D_{-\alpha\beta\ 00}(tS)\langle(t\text{-}\alpha S\text{-}\beta)i|O_{-\alpha+\mu,-\beta+\nu}O_{-\mu-\nu}|(tS)_{(\gamma\delta)j}\rangle.$$

Thus, the next step of the recurrent procedure has been prepared. The application of the described procedure allowed us to construct basis functions for non-degenerate states (tS) lying on the borderline of a polygon and containing all the spin-isospin multiplets of the given supermultiplet (pp'p'') and their neighbouring two-fold and three-fold degenerate states shown on Fig.1. Thus, for non-degenerate states $t+S=p'+p$

$$N_{10}(tS)=N_{01}(tS)=N_{11}(+S)=0,\ N_{00}(tS)=p'',$$

$$N_{-11}(tS)=N_{1-1}(t-1S+1)=[(t-1)S-p(p'-1)]^{1/2}. \tag{15}$$

For two-fold states $t+S=p+p'-1$

$$N_{-10}(t+1S)=[(t+1)-(N_{00}(t+1S))^2/(t+1)-(N_{1-1}(t+1S))^2/(S+1)]^{1/2},$$

$$N_{0-1}(tS+1)=[(S+1)-(N_{00}(tS+1))^2/(S+1)-(N_{1-1}(tS+1))^2/(t+1)]^{1/2},$$

$$K_{(-10),(0-1)}(tS)=[(t+1)^{-1}+(S+1)^{-1}][N_{00}(tS+1)/N_{0-1}(tS+1)][N_{-11}(t+1S))/$$

$$/N_{-10}(t+1S))], \tag{16}$$

$$\det K=\frac{[(p+1)^2-(p'')^2][(p')^2-(p'')^2]}{(t+1)(S+1)[N_{0-1}(tS+1)N_{-10}(t+(S))]^2}.$$

For non-degenerate states $t=p, |p''|\leqslant S\leqslant p'$

$$N_{1-1}(pS)=N_{10}(pS)=N_{11}(pS)=0,\ N_{00}(pS)=(p'+1)p''/(S+1),$$

$$N_{0-1}(pS)=N_{01}(pS-1)=[((p'+1)^2-S^2)(S^2-(p'')^2)/S(2S+1)]^{1/2}. \tag{17}$$

For three-fold states $t=p-1, |p''|+1\leqslant S\leqslant p'-1$

$$N_{-11}(pS-1)=[p-S+1-(N_{0-1}(pS))^2/p]^{1/2},$$

$$N_{-10}(pS)=[(p+1)S/(S+1)-(N_{00}(pS))^2/p]^{1/2},$$

$$N_{-1-1}(pS+1)=[(p+S+2)(2S+1)/(2S+3)-(N_{01}(pS))^2/p]^{1/2}, \tag{18}$$

$$K_{(-1\alpha),(-1\beta)}(p-1S)=-1/p[N_{0-\alpha}(pS)/N_{-1\alpha}(pS-\alpha)]\cdot[N_{0-\beta}(pS)/N_{-1\beta}(pS-\beta)],$$

$$\det K=\frac{S(2S+1)[(p+1)^2-(p'+1)^2][(p+1)^2-(p'')^2]}{(S+1)(2S+3)p[N_{-11}(pS-1)N_{-10}(pS)\cdot N_{-1-1}(pS+1)]^2}.$$

For non-degenerate states $t-S=p-|p''|$

$$N_{1-1}(tS)=N_{10}(tS)=N_{0-1}(tS)=0,\; N_{00}(tS)=(p'+1)S/(S+1),$$

$$N_{11}(tS)=N_{-1-1}(t+1\,S+1)=\left[(p-t)(t+|p''|+2)(2S+1)/(2S+3)\right]^{1/2} \tag{19}$$

For two-fold degenerate states $t-S=p-|p''|-1$

$$N_{01}(tS-1)=\left[((p'+1)^2-S^2)(2S-1)/S(2S+1)-(N_{11}(tS-1))^2/(t+1)\right]^{1/2},$$

$$N_{-10}(t+1S)=\left[(p+1)(|p''|+1)S/(S+1)^2-(N_{00}(t+(S))^2/(t+1)\right]^{1/2},$$

$$K_{(01),(-10)}(tS)=\left[(p-|p''|)/(t+1)S\right]\cdot\left[N_{00}(t+1S|)/N_{10}(t+1S|)\right]\cdot\left[N_{11}(tS-1)/N_{01}(tS-1)\right],$$

$$\det K=\frac{S(2S-1)\left[(p-|p''|)^2(p'+1)^2-(p+1)(|p''|+1)-(p'+1)^2)^2\right]}{(S+1)^2(2S+1)(t+1)\left[N_{-10}(t+1S)\cdot N_{01}(tS-1)\right]^2}\,. \tag{20}$$

2. <u>Constructing Clebsch-Gordan coefficients for representations with external degeneracy</u>. Investigating the nuclear phenomena in the SU(4)-symmetry scheme leads to the problem of constructing Clebsch-Gordan coefficients (CGC) of SU(4). In [3] recurrent expressions are obtained for CGC. The first step begins from the highest weight state and requires to find CGC of the type

$$\left\langle \begin{matrix} (f_1) & (f_2) \\ (t_1S_1)_i & (t_2S_2)_j \end{matrix} \right\| \left. \begin{matrix} (f)=(pp'p'') \\ (pp') \end{matrix} \right\rangle \tag{21}$$

If the representation (f) enters $(f_1)\otimes(f_2)$ multiplicity-free, constructing these coefficients is simple enough. At the same time it is generally unknown how to compute CGC if (f) enters $(f_1)\otimes(f_2)$ with the degeneracy $\geqslant 1$.

One of the problems important for applications is that when $(f_1)=(f)$ is an arbitrary representation, $(f_2)=(110)$ and the external degeneracy is three. In order to solve the problem construct from the generators of the group (1) three tensor operators $R^{(k)}_{\tau\tau_z\sigma\sigma_z}$ $(k=1,2,3)$ which are transformed according to the representation (110) containing three spin-isospin multiplets (tS)=(10), (01), (11). The simplest one is:

$$R^{(1)}_{1\alpha 00}=(-1)^{\eta}T_\alpha,\ R^{(1)}_{0\alpha}=(-1)^{\eta}S_\alpha,\ R^{(1)}_{1\gamma 1\beta}=(-1)^{\eta}E_{\gamma\beta}, \qquad (22)$$

where $\eta=1$ for $\alpha=1$ and $(\gamma\beta)=(00),(-10),(0-1),(-1-1)$, (11) and $\eta=0$ in the other cases. Calculating matrix elements for the components of this tensor operator between the orthonormal states of the supermultiplet (pp'p'') and employing the Wigner-Eckart theorem, we get a set of CGC:

$$\langle (t'S')_j(10)\|(tS)i\rangle_1=\delta_{tt'}\delta_{SS'}\delta_{ij}\,(t(t+1))^{1/2}/\langle R^1\rangle,$$

$$\langle (t'S')_j(01)\|(tS)i\rangle_1=\delta_{tt'}\delta_{SS'}\delta_{ij}\,(S(S+1))^{1/2}/\langle R^1\rangle,$$

$$\langle (t+\alpha\ S+\beta)_j(11)\|(tS)i\rangle=(-1)^{\eta(\alpha\beta)}N^{j}_{\alpha\beta}(tSi)/\langle R^1\rangle\cdot$$

$$\cdot(\langle t+\alpha t+\alpha 1-\alpha|tt\rangle\langle S+\beta\,S+\beta\,1-\beta|SS\rangle)^{-1}, \qquad (23)$$

where $\langle R^1\rangle=\sqrt{C_2}=[p(p+4)+p'(p'+2)+(p'')^2]^{1/2}$ is a reduced matrix element of the tensor operator $R^{(1)}$. One can show that these CGC are are normed and orthogonalized. Other tensor operators are conducted recursively:

$$R^{(K)}_{\tau\tau_z\sigma\sigma_z}=\sum_{\substack{(\tau_1\tau_{1z}\sigma_1\sigma_{1z})\\(\tau_2\tau_{2z}\sigma_2\sigma_{2z})}}\left\langle\begin{matrix}(110) & (110)\\(\tau_1,\sigma_1) & (\tau_2\sigma_2)\end{matrix}\right\|\left.\begin{matrix}(110)\\(t\sigma)\end{matrix}\right\rangle\cdot \qquad (24)$$

$$\langle\tau_1\tau_{1z}\tau_2\tau_{2z}|\tau\tau_z\rangle\langle\sigma_1\sigma_{1z}\sigma_2\sigma_{2z}|\sigma\sigma_z\rangle R^{(1)}_{\tau_1\tau_{1z}\sigma_1\sigma_{1z}}R^{(K-1)}_{\tau_2\tau_{2z}\sigma_2\sigma_{2z}}$$

$$\langle(01)(11)\|(10)\rangle\doteq\langle(11)(01)\|(10)\rangle=\langle(10)(11)\|(01)\rangle=\langle(11)(10)\|(01)\rangle=-1/\sqrt{2}$$

$$\langle(10)(01)\|(11)\rangle=\langle(01)(10)\|(11)\rangle=1/2\,\langle(11)\|(11)\rangle=1/\sqrt{6}. \qquad (25)$$

The CGC of the (21)-type corresponding to the tensor

operators $R^{(K)}$ can be obtained using the formula

$$\langle (tS)_i(\tau\sigma)\|(pp')\rangle_K = \langle R^1\rangle\langle R^{K-1}\rangle/\langle R^K\rangle \cdot$$

$$\cdot \sum_{(\tau_1\sigma_1)(\tau_2\sigma_2)(t's')_j} \langle (\tau_1\sigma_1)(\tau_2\sigma_2)\|(\tau\sigma)\rangle\langle (tS)_i(\tau_1\sigma_1)\|(t'S')_j\rangle_1 \cdot \quad (26)$$

$$\cdot\langle (t'S')_j(\tau_2\sigma_2)\|(pp')\rangle_{K-1} U\langle t\tau_1 p\tau_2; t'\tau\rangle U(S\sigma_1 p'\sigma_2; S'\sigma),$$

where U is the Racah coefficient of the SU(2) group.

To calculate the reduced matrix elements $\langle R^K\rangle$, let us find the scalar products of the tensor operators

$$\langle R^i R^k\rangle \equiv \left\langle \begin{matrix}(pp'p'')\\(tS)\end{matrix}\right| \left[R^{(i)}_{\tau\tau_z\sigma\sigma_z} \times R^{(K)}_{\tau-\tau_z\sigma-\sigma_z}\right]^{(j)=(000)} \left|\begin{matrix}(pp'p'')\\(tS)\end{matrix}\right\rangle . \quad (27)$$

These are expressed via Casimir operators C_2, C_3, C_4 of SU(4) as follows

$$\langle R^i R^k\rangle = \begin{pmatrix} C_2 & -\sqrt{6}C_3 & 2/3\, C_4 \\ -\sqrt{6}C_3 & 2/3 C_4 & -\sqrt{8/27}\, C_3(C_2+3) \\ 2/3 C_4 & -\sqrt{8/27}\,C_3(C_2+3) & 1/9\left[C_4(C_2+4)+3(C_3)^2\right] \end{pmatrix} \quad (28)$$

where $C_3=(p+2)(p'+1)p''$, $C_4=(p+2)^2(p'+1)^2+(p+2)^2\cdot (p'')^2+(p'+1)^2(p'')^2-C_2-4$. Since the determinant of this matrix is non-zero in general, the three sets of CGC are linearly independent. The orthonormality of these sets can be obtained due to a linear transformation of $R^{(K)}$ which reduces (28) to a unitary matrix.

References

1. Jeugt, J. Van der, et al. (1983). Ann. of Phys. 147, 85-139.
2. Hecht, K.T., Pang, S.C. (1969). J. Math. Phys. 10, 1571.
3. Vladimirov, D.M. (1984). Preprint IAE-3949/1, Moscow.
4. Partensky, A., Maguin, C.(1978).J.Math.Phys. 19, 511.

THE ELEMENTS OF THE SU(3) WIGNER-RACAH ALGEBRA

V.N. Tolstoy, Yu.F. Smirnov, Z. Pluhař[+]
Institute of Nuclear Physics, Moscow State University, Moscow
+ Charles University, Praha

By full analogy with the quantum theory of angular momenta, the $3n\Lambda$ symbols (i.e. $3\Lambda x$-, 6Λ-, 9Λ -coefficients) are introduced. The properties of these quantities are described in detail. The extension of the graphic technique elaborated in the angular momentum theory to the SU(3) group is briefly discussed.

1. Introduction. At present SU(3) is extensively used in the nuclear and particle theory. The Elliott SU(3) scheme, the SU(3) limit of the interacting bosons model, the symplectic collective model in the nuclear theory, and the flavour and colour SU(3) symmetry in the particle physics can be indicated as examples of the SU(3) group applications. The SU(3) symmetry is approximate in the nuclear physics, while the colour SU(3) symmetry is considered to be exact in the elementary particle physics. In any case it is clear that at present the SU(3) symmetry begins playing the same fundamental role in quantum physics as the angular momentum theory.

In this connection the development of the SU(3) Wigner-Racah algebra (WRA) including the algebra of the irreducible SU(3) tensors is urgent because this formalism is necessary for practical calculations. In essence, the WRA is a theory of the recoupling coefficients connecting various sets of basis vectors for the tensor product of the irreducible representations (IR).

In spite of the great efforts aimed at elaborating

the SU(3) WRA, the situation with this problem is not satisfactory nowadays compared with the quantum angular momentum theory (i.e. the SU(2) WRA) [1-2]. The main difficulties arise from the fact that the SU(3) group is not multiplicity-free. Although some elements of the SU(3) WRA have been developed, the systematization of these results was not made as yet. In particular, the problem of classification of the multiple IR's in the tensor product of two IR's was solved differently by different authors. Therefore, no unique symmetry relations for the Clebsch-Gordan coefficients (CGC) of SU(3) group are available. Besides, the theory of $3n\Lambda$-symbols is in embryonal state.

Recently in Refs [3-5] the general approach to the theory of the SU(3) IR was proposed on the basis of a combination of the projection operator method developed in Ref. [6] and the SU(2) WRA. In this way it was shown that the copious results on the SU(3) IR theory can be reduced to the same simple and compact form as in case of the SU(2) group. Among these results, the general formula expressing the SU(3) CGC's in terms of the SU(2) recoupling coefficients should be mentioned. The second important point is the use in Refs. [3-5] of the special classification operator for labelling the multiple IRs in the tensor product of two IR's which makes it possible to obtain the CGC's with the same simple symmetry properties as in the angular momentum theory. We shall develop the SU(3) WRA on the basis of these results.

In this paper we introduce the $3\Lambda\varkappa$-, 6Λ-, 9Λ-symbols for the SU(3) group (similar to $3j$-, $6j$-, $9j$-symbols in the angular momentum theory); and describe their classical non Regge-like) symmetry properties. The extension of the 3nj-symbol graphic technique to the SU(3) group case is also discussed.

2. The Wigner coefficients for the SU(3) group. Let us denote by symbols

$$|(\lambda\mu)\,y\,t\,t_z\rangle \quad \text{or} \quad |(\lambda\mu)\,j\,t\,t_z\rangle \tag{1}$$

the orthonormalized basis vectors of the arbitrary IR$(\lambda\mu)$ of the SU(3) group corresponding to the canonical reduction

$$SU(3) \supset U_y(1) \otimes SU_T(2). \tag{2}$$

Here y, t and t_z are the quantum numbers of the hypercharge, the isospin, and the isospin projection, respectively (see [3-5]). Using the quantum number j instead of y $(= -\frac{1}{3}(2\lambda+\mu)+2j)$ is more convenient because the SU(2)-content of the SU(3) IR can be expressed easily in terms of this quantum number. Namely, in the IR $(\lambda\mu)$ all IR's of the $U_y(1)\otimes SU_T(2)$ subgroup are present that satisfy the following conditions [3-5]:

$$\begin{aligned} &\tfrac{1}{2}\vec{\mu}+\vec{j}+\vec{t}=\vec{0}\,, \\ &\tfrac{1}{2}\mu+j+t \le \lambda+\mu\,. \end{aligned} \tag{3}$$

The first equation means the "triangle rule" for the vector coupling of three angular momenta. Henceforth, we shall use the abridged notation $\Lambda := (\lambda\mu)$, $\varkappa := (j,t,t_z)$, $\tilde{\Lambda} := (\mu\lambda)$, $\tilde{\varkappa} := (\frac{1}{2}(\lambda+\mu)-j,\, t,\, t-t_z)$. The transition from the $|\Lambda\varkappa\rangle$ to $|\tilde{\Lambda}\tilde{\varkappa}\rangle$ vectors means the turnover from the IR Λ to the contragredient (conjugate) IR $\tilde{\Lambda}$. Therefore the vectors (1) are characterized by quantum numbers $\tilde{\Lambda}$ and $\tilde{\varkappa}$ with respect to the contragredient representation [7].

For brevity we shall write "the SU(3) spin Λ" instead of "the SU(3) IR Λ" and a set of quantum numbers $\varkappa$ will be called "the SU(3) spin projection". If the IR Λ_{12} is constructed out of the tensor product of two IRs Λ_1 and Λ_2, this procedure of "the vector coupling of the SU(3) spins" will be denoted as

$$\vec{\Lambda}_{12} = \vec{\Lambda}_1 + \vec{\Lambda}_2 \ .$$

The SU(3) CGC's connect two orthonormalized basis sets, namely, the simple tensor basis $|\Lambda_1 \varkappa_1\rangle |\Lambda_2 \varkappa_2\rangle$ and the reduced basis $|\Lambda_1 \Lambda_2 : S \Lambda_{12} \varkappa_{12}\rangle$:

$$|\Lambda_1 \Lambda_2 : S \Lambda_{12} \varkappa_{12}\rangle = \\ = \sum_{\varkappa_1, \varkappa_2} (\Lambda_1 \varkappa_1 \Lambda_2 \varkappa_2 | S \Lambda_{12} \varkappa_{12}) |\Lambda_1 \varkappa_1\rangle |\Lambda_2 \varkappa_2\rangle , \tag{4}$$

where an eigenvalue S of the classification operator $\hat{S}$ [3-5] is used as an index labelling multiple IR's $\vec{\Lambda}_{12}$ in the tensor product $D^{\Lambda_1} \otimes D^{\Lambda_2}$. The general formula to calculate the SU(3) CGC's is given in Refs. [3-5]. For our purposes it is sufficient to lead the explicit expression only for the CGCs with $\Lambda_{12} = (00)$. In this case the index S can be omitted and we have (see [3-5])

$$(\Lambda \varkappa \Lambda' \varkappa' | 00) = (-1)^{\lambda + j - \frac{1}{2}\mu + t_z} \cdot \frac{\delta_{\Lambda', \bar{\Lambda}} \cdot \delta_{\varkappa', \tilde{\varkappa}}}{\sqrt{\dim \Lambda}} \tag{5}$$

where $\dim \Lambda = \frac{1}{2}(\lambda + 1)(\mu + 1)(\lambda + \mu + 2)$ is the dimension of the IR Λ.

By full analogy with the quantum theory of angular momenta, the Wigner coefficients for the SU(3) group (or $3\Lambda\varkappa$ -symbols) may be introduced:

$$\begin{pmatrix} \Lambda_1 & \Lambda_2 & \Lambda_3 \\ \varkappa_1 & \varkappa_2 & \varkappa_3 \end{pmatrix} S \Bigg) = \eta \sum_{\Lambda', \varkappa'} (\Lambda_1 \varkappa_1 \Lambda_2 \varkappa_2 | S \Lambda' \varkappa')(\Lambda' \varkappa' \Lambda_3 \varkappa_3 | 00). \tag{6}$$

Thus, the $3\Lambda\varkappa$ - symbol is the amplitude of the probability for three SU(3)-spins $\vec{\Lambda}_1$, $\vec{\Lambda}_2$, $\vec{\Lambda}_3$ with "projections" $\varkappa_1$, $\varkappa_2$ and $\varkappa_3$ respectively to form the vanishing total SU(3)-spin ($\vec{\Lambda}_1 + \vec{\Lambda}_2 + \vec{\Lambda}_3 = \vec{0}$). The phase factor $\eta = (-1)^{\lambda_2 + \mu_2 + \lambda_3}$ is so chosen that the $3\Lambda\varkappa$ -symbol is invariant with respect to cyclic permutations of SU(3)-spins Λ_1, Λ_2 and Λ_3 (see below).

Substituting the Eq.(5) into Eq.(6), we obtain the following relation between the $3\Lambda\varkappa$ - symbol and the CGC:

$$\begin{pmatrix} \Lambda_1 & \Lambda_2 & \Lambda_3 \\ \varkappa_1 & \varkappa_2 & \varkappa_3 \end{pmatrix} S \Big) = \frac{(-1)^{\lambda_1+\mu_2+\lambda_3+\mu_3+j_3-\frac{1}{2}\mu_3+t_{3z}}}{\sqrt{\dim \Lambda_3}} \cdot (\Lambda_1 \varkappa_1 \Lambda_2 \varkappa_2 | S \tilde{\Lambda}_3 \tilde{\varkappa}_3) . \tag{7}$$

Let now the main properties of the $3\Lambda\varkappa$-symbols be described.

1°. <u>Nonvanishing conditions</u>. It is clear that the symbols are nonvanishing only when the conditions (3) are satisfied for every set of quantum numbers $\Lambda_i, \varkappa_i$, $i = 1,2,3$.

These coefficients must satisfy the vector coupling rule for the SU(3) spins

$$\vec{\Lambda}_1 + \vec{\Lambda}_2 + \vec{\Lambda}_3 = \vec{0} , \tag{8}$$

which means

$$y_1 + y_2 + y_3 = 0, \qquad \vec{t}_1 + \vec{t}_2 + \vec{t}_3 = 0 \tag{9}$$

and the "triangle" condition for SU(3) spins (an analogue of the "triangle" condition for three angular momenta) must be satisfied. In the notations

$$A_i := \frac{1}{3}(2\lambda_i + \mu_i), \quad B_i := \frac{1}{3}(\lambda_i + 2\mu_i) \tag{10}$$

the "triangle" condition for the SU(3) spins can be written as follows [8].

$$\begin{aligned} &A_i + A_l - B_k \geq 0, \qquad A_i - B_i + A_l - B_l + B_k \geq 0, \\ &B_i + B_l - A_k \geq 0, \qquad B_i - A_i + B_l - A_l + A_k \geq 0. \end{aligned} \tag{11}$$

Here $i \neq l \neq k$ and the numbers in the left-hand sides of these inequalities must be integers. The multiplicity $\mathrm{Mult}(\Lambda_1\Lambda_2\Lambda_3)$ of the vanishing total SU(3)-spin in a system of three spins Λ_1, Λ_2 and Λ_3 (i.e. the number of various S eigenvalues) is equal to 1+ K, where

K is the minimum of the left sides of the inequalities (11) and numbers λ_i, μ_i, $i=1,2,3$.

2°. Unitarity relations. The unitarity property of the SU(3) CGCs gives the following relations for the $3\Lambda\varkappa$-symbols

$$\sum_{\varkappa_1,\varkappa_2} \begin{pmatrix} \Lambda_1 & \Lambda_2 & \Lambda_3 \\ \varkappa_1 & \varkappa_2 & \varkappa_3 \end{pmatrix} S \Bigg) \begin{pmatrix} \Lambda_1 & \Lambda_2 & \Lambda'_3 \\ \varkappa_1 & \varkappa_2 & \varkappa'_3 \end{pmatrix} S' \Bigg) = \delta_{\varkappa_3\varkappa'_3}\delta_{\Lambda_3\Lambda'_3}\delta_{S,S'} \frac{\{\Lambda_1\Lambda_2\Lambda_3|S\}}{\sqrt{\dim\Lambda_3}} \tag{12}$$

$$\sum_{\varkappa_3,\Lambda_3,S} \dim\Lambda_3 \left(\begin{matrix} \Lambda_1 & \Lambda_2 & \Lambda_3 \\ \varkappa_1 & \varkappa_2 & \varkappa_3 \end{matrix}\middle| S\right) \left(\begin{matrix} \Lambda_1 & \Lambda_2 & \Lambda_3 \\ \varkappa'_1 & \varkappa'_2 & \varkappa_3 \end{matrix}\middle| S\right) = \delta_{\varkappa_1,\varkappa'_1}\delta_{\varkappa_2,\varkappa'_2} \tag{13}$$

where the "3Λ-symbol" $\{\Lambda_1\Lambda_2\Lambda_3|S\}=1$ if the SU(3) spins $\bar{\Lambda}_1, \bar{\Lambda}_2, \bar{\Lambda}_3$ satisfy the "triangle condition" and S belongs to the spectrum of the classification operator; in the rest cases we get $\{\Lambda_1\Lambda_2\Lambda_3|S\}=0$.

3°. Symmetry relations. The $3\Lambda\varkappa$-symbols have the following symmetry properties [5]

$$\left(\begin{matrix} \Lambda_1 & \Lambda_2 & \Lambda_3 \\ \varkappa_1 & \varkappa_2 & \varkappa_3 \end{matrix}\middle| S\right) = (-1)^{p_{ilk}\Phi(\Lambda_1\Lambda_2\Lambda_3)} \left(\begin{matrix} \Lambda_i & \Lambda_l & \Lambda_k \\ \varkappa_i & \varkappa_l & \varkappa_k \end{matrix}\middle| (-1)^{p_{ilk}} S\right) \tag{14}$$

$$\left(\begin{matrix} \Lambda_1 & \Lambda_2 & \Lambda_3 \\ \varkappa_1 & \varkappa_2 & \varkappa_3 \end{matrix}\middle| S\right) = (-1)^{\Phi(\Lambda_1\Lambda_2\Lambda_3)} \left(\begin{matrix} \tilde{\Lambda}_1 & \tilde{\Lambda}_2 & \tilde{\Lambda}_3 \\ \tilde{\varkappa}_1 & \tilde{\varkappa}_2 & \tilde{\varkappa}_3 \end{matrix}\middle| -S\right) \tag{15}$$

where p_{ilk} is a parity of the permutation $\begin{pmatrix} 1 & 2 & 3 \\ i & l & k \end{pmatrix}$;

$\Phi(\Lambda_1\Lambda_2\Lambda_3) = \max\{\lambda_1+\lambda_2+\lambda_3, \mu_1+\mu_2+\mu_3\} + \mathrm{Mult}(\Lambda_1\Lambda_2\Lambda_3) - 1.$

Thus, the $3\Lambda\varkappa$-symbols are invariant with respect to the even permutation of columns. In case of the odd permutation of columns, the $3\Lambda\varkappa$-symbol is given the phase factor $(-1)^{\Phi(\Lambda_1\Lambda_2\Lambda_3)}$ and sign inversion of the eigenvalue S takes place. Similar replacements must be made when all parameters $\Lambda_i, \varkappa_i$, $i=1,2,3$ are substituted by conjugated ones $\tilde{\Lambda}_i, \tilde{\varkappa}_i$ in the $3\Lambda\varkappa$-symbol.

3. <u>6Λ -symbols</u>. The 6Λ-symbols are related to the recoupling coefficients for three SU(3) spins. Let us form the total SU(3)-spin $\vec{\Lambda}$ with the projection $\varkappa$ out of three SU(3) spins $\vec{\Lambda}_1, \vec{\Lambda}_2, \vec{\Lambda}_3$. This may be made in two ways: 1) $\vec{\Lambda}_1 + \vec{\Lambda}_2 = \vec{\Lambda}_{12}$, $\vec{\Lambda}_{12} + \vec{\Lambda}_3 = \vec{\Lambda}$ and 2) $\vec{\Lambda}_2 + \vec{\Lambda}_3 = \vec{\Lambda}_{23}$, $\vec{\Lambda}_1 + \vec{\Lambda}_{23} = \vec{\Lambda}$. The states corresponding to these coupling schemes are of the form

$$|\Lambda_1\Lambda_2(S_{12}\Lambda_{12}), \Lambda_3 : S_{12,3}\,\Lambda\varkappa\rangle = \sum_{\varkappa_1,\varkappa_2,\varkappa_3} (\Lambda_1\varkappa_1\Lambda_2\varkappa_2|S_{12}\Lambda_{12}\varkappa_{12}) \times$$
$$\times (\Lambda_{12}\varkappa_{12}\Lambda_3\varkappa_3|S_{12,3}\Lambda\varkappa)\,|\Lambda_1\varkappa_1\rangle|\Lambda_2\varkappa_2\rangle|\Lambda_3\varkappa_3\rangle, \quad (16)$$

$$|\Lambda_1, \Lambda_2\Lambda_3(S_{23}\Lambda_{23}) : S_{1,23}\,\Lambda\varkappa\rangle = \sum_{\varkappa_1,\varkappa_2,\varkappa_3} (\Lambda_2\varkappa_2\Lambda_3\varkappa_3|S_{23}\Lambda_{23}\varkappa_{23}) \times$$
$$\times (\Lambda_1\varkappa_1\Lambda_{23}\varkappa_{23}|S_{1,23}\Lambda\varkappa)\,|\Lambda_1\varkappa_1\rangle|\Lambda_2\varkappa_2\rangle|\Lambda_3\varkappa_3\rangle. \quad (17)$$

The 6Λ-symbol $\{ \because \because | \ldots \}$ can be defined by the relation

$$\langle\Lambda_1\Lambda_2(S_{12}\Lambda_{12}), \Lambda_3 : S_{12,3}\,\Lambda\varkappa|\Lambda_1, \Lambda_2\Lambda_3(S_{23}\Lambda_{23}) : S_{1,23}\,\Lambda'\varkappa'\rangle =$$

$$= \delta_{\varkappa,\varkappa'}\,\delta_{\Lambda,\Lambda'}(-1)^{\varphi}\sqrt{\dim\Lambda_{12}\cdot\dim\Lambda_{23}}\begin{Bmatrix}\Lambda_1 & \Lambda_2 & \Lambda_{12} \\ \Lambda_3 & \Lambda & \Lambda_{23}\end{Bmatrix}\Gamma\Bigg\}, \quad (18)$$

where

$$\varphi := \lambda_1 + \mu_1 + \lambda_3 + \mu_3 + \lambda + \mu + \lambda_{23} + \mu_{23},$$
$$\Gamma := (S_{12}, S_{12,3}, S_{23}, S_{1,23})$$

From the definition (18) the following expression of the 6Λ-symbol in terms of $3\Lambda\varkappa$-symbols can be obtained

$$\left\{\begin{matrix}\Lambda_1 & \Lambda_2 & \Lambda_3 \\ \Lambda_4 & \Lambda_5 & \Lambda_6\end{matrix}\middle|\Gamma\right\} = \sum_{\varkappa_i}(-1)^{\sum_{i=1}^{6}(j_i - \frac{1}{2}\mu_i + t_{i\varkappa})}\begin{pmatrix}\Lambda_1 & \Lambda_2 & \Lambda_3 \\ \varkappa_1 & \varkappa_2 & \varkappa_3\end{pmatrix}S_{123}\Bigg)\times$$
$$\times\left(\begin{matrix}\tilde{\Lambda}_1 & \Lambda_5 & \tilde{\Lambda}_6 \\ \tilde{\varkappa}_1 & \varkappa_5 & \tilde{\varkappa}_6\end{matrix}\middle|S_{\tilde{1}5\tilde{6}}\right)\left(\begin{matrix}\tilde{\Lambda}_4 & \tilde{\Lambda}_2 & \Lambda_6 \\ \tilde{\varkappa}_4 & \tilde{\varkappa}_2 & \varkappa_6\end{matrix}\middle|S_{\tilde{4}\tilde{2}6}\right)\left(\begin{matrix}\Lambda_4 & \tilde{\Lambda}_5 & \tilde{\Lambda}_3 \\ \varkappa_4 & \tilde{\varkappa}_5 & \tilde{\varkappa}_3\end{matrix}\middle|S_{4\tilde{5}\tilde{3}}\right). \quad (19)$$

where $\Gamma := (S_{123}, S_{\tilde{1}5\tilde{6}}, S_{\tilde{4}\tilde{2}6}, S_{4\tilde{5}\tilde{3}})$. The quantities $S_{\alpha\beta\gamma}$ have the following symmetry properties [3]: they are invariant with respect to the even permutation of their low indices. In case of the odd permutation and conjugation their sign gets opposite: $S_{\alpha\beta\gamma} = - S_{\alpha\gamma\beta} = - S_{\beta\alpha\gamma} = - S_{\gamma\beta\alpha} = - S_{\tilde{\alpha}\tilde{\beta}\tilde{\gamma}}$. Since the parameters of each $S_{\alpha\beta\gamma}$ are in unambiguous correspondence with a definite set of three SU(3)-spins, the order of their disposition in a set Γ is not essential. Therefore we shall change this order sometimes if it proves convenient for some reasons. In particular this convention will be useful in formulating the 6Λ -symbol symmetry properties. Let now the main properties of the 6Λ -symbols be discussed.

1°. Nonvanishing condition. The vector coupling rules for SU(3)-spins impose definite limitations on the parameters of the 6Λ -symbol. Namely, the 6Λ -symbol $\left\{ \begin{matrix} \Lambda_1 & \Lambda_2 & \Lambda_3 \\ \Lambda_4 & \Lambda_5 & \Lambda_6 \end{matrix} \middle| \Gamma \right\}$ is nonvanishing, if the "triangle" rule (11) is satisfied for each set of the parameters $(\Lambda_1 \Lambda_2 \Lambda_3)$, $(\tilde{\Lambda}_1 \Lambda_5 \tilde{\Lambda}_6)$, $(\tilde{\Lambda}_4 \tilde{\Lambda}_2 \Lambda_6)$, $(\Lambda_4 \tilde{\Lambda}_5 \tilde{\Lambda}_3)$. The quantities S_{123} , $S_{\tilde{1}5\tilde{6}}$, $S_{\tilde{4}\tilde{2}6}$, $S_{4\tilde{5}\tilde{3}}$ must belong to the spectrum of the corresponding classification operators.

2°. Classical symmetry properties. These properties of the 6Λ -symbols can readily be obtained from the eq.(19) making allowance for the symmetry properties of the $3\Lambda\varkappa$ -symbols.

a) The 6Λ -symbol is invariant with respect to the even permutation of its columns and with respect to the odd permutation followed by conjugation of the first row. For example

$$\left\{ \begin{matrix} \Lambda_1 & \Lambda_2 & \Lambda_3 \\ \Lambda_4 & \Lambda_5 & \Lambda_6 \end{matrix} \middle| \Gamma \right\} = \left\{ \begin{matrix} \tilde{\Lambda}_2 & \tilde{\Lambda}_1 & \tilde{\Lambda}_3 \\ \Lambda_5 & \Lambda_4 & \Lambda_6 \end{matrix} \middle| \Gamma \right\} \tag{20}$$

b) The 6Λ-symbol is invariant with respect to the replacement of any two arbitrary parameters in the first row by two corresponding parameters from the second row and vice versa if this procedure is followed by the conjugation of parameters in two columns of the 6Λ-symbol, namely

$$\left\{\begin{matrix}\Lambda_1 & \Lambda_2 & \Lambda_3\\ \Lambda_4 & \Lambda_5 & \Lambda_6\end{matrix}\middle|\Gamma\right\} = \left\{\begin{matrix}\Lambda_4 & \tilde{\Lambda}_5 & \tilde{\Lambda}_3\\ \Lambda_1 & \tilde{\Lambda}_2 & \Lambda_6\end{matrix}\middle|\Gamma\right\} = \left\{\begin{matrix}\tilde{\Lambda}_4 & \tilde{\Lambda}_2 & \Lambda_6\\ \tilde{\Lambda}_1 & \tilde{\Lambda}_5 & \Lambda_3\end{matrix}\middle|\Gamma\right\} = \left\{\begin{matrix}\tilde{\Lambda}_1 & \Lambda_5 & \tilde{\Lambda}_6\\ \tilde{\Lambda}_4 & \Lambda_2 & \tilde{\Lambda}_3\end{matrix}\middle|\Gamma\right\}. \tag{21}$$

c) After the conjugation of all the parameters in the 6Λ-symbol we obtain

$$\left\{\begin{matrix}\Lambda_1 & \Lambda_2 & \Lambda_3\\ \Lambda_4 & \Lambda_5 & \Lambda_6\end{matrix}\middle|\Gamma\right\} = (-1)^{\Psi} \left\{\begin{matrix}\tilde{\Lambda}_1 & \tilde{\Lambda}_2 & \tilde{\Lambda}_3\\ \tilde{\Lambda}_4 & \tilde{\Lambda}_5 & \tilde{\Lambda}_6\end{matrix}\middle|\Gamma' = -\Gamma\right\} \tag{22}$$

where

$$\Psi = \Phi(\Lambda_1\Lambda_2\Lambda_3) + \Phi(\tilde{\Lambda}_1\Lambda_5\tilde{\Lambda}_6) + \Phi(\tilde{\Lambda}_4\tilde{\Lambda}_2\Lambda_6) + \Phi(\Lambda_4\tilde{\Lambda}_5\tilde{\Lambda}_3).$$

All the above mentioned symmetry properties are independent and show the relationships between $3! \times 4 \times 2 = 48$ different (in the general case) 6Λ-symbols.

3°. Unitarity relations. The unitarity of recoupling coefficients for three SU(3)-spins gives the following orthogonality condition for the 6Λ-symbols:

$$\sum_{\substack{\Lambda_3, S_{123},\\ S_{4\bar{5}\bar{3}}}} \dim\Lambda_3 \cdot \left\{\begin{matrix}\Lambda_1 & \Lambda_2 & \Lambda_3\\ \Lambda_4 & \Lambda_5 & \Lambda_6\end{matrix}\middle|\Gamma\right\} \cdot \left\{\begin{matrix}\Lambda_1 & \Lambda_2 & \Lambda_3\\ \Lambda_4 & \Lambda_5 & \Lambda'_6\end{matrix}\middle|\Gamma'\right\} = \frac{\delta_{\Lambda_6,\Lambda'_6}}{\dim\Lambda_6} \times$$
$$\times\ \delta_{S_{\bar{1}5\bar{6}}, S'_{\bar{1}5\bar{6}}} \cdot \delta_{S_{\bar{4}\bar{2}6}, S'_{\bar{4}\bar{2}6}} \cdot \{\tilde{\Lambda}_1\Lambda_5\tilde{\Lambda}_6 | S_{\bar{1}5\bar{6}}\} \cdot \{\tilde{\Lambda}_4\tilde{\Lambda}_2\Lambda_6 | S_{\bar{4}\bar{2}6}\} \tag{23}$$

where

$$\Gamma' = (S_{123}, S'_{\bar{1}5\bar{6}}, S_{\bar{4}\bar{2}6}, S_{4\bar{5}\bar{3}}).$$

Concluding this section, we shall write the explicit

expression for the 6Λ-symbol with one vanishing parameter

$$\begin{Bmatrix} 0 & \Lambda_2 & \Lambda_3 \\ \Lambda_4 & \Lambda_5 & \Lambda_6 \end{Bmatrix}\!\Big|\,\Gamma \Big\} = \frac{(-1)^{\lambda_2+\mu_5}}{\sqrt{\dim \Lambda_2 \cdot \dim \Lambda_5}}\, \delta_{\Lambda_2,\Lambda_3}\, \delta_{\Lambda_5,\Lambda_6}\, \delta_{S_{\tilde{4}\tilde{5}6},\, S_{45\tilde{3}}} . \quad (24)$$

4. <u>The 9Λ-symbols</u>. Four SU(3)-spins $\vec{\Lambda}_1, \vec{\Lambda}_2, \vec{\Lambda}_3, \vec{\Lambda}_4$ may be coupled into the total SU(3)-spin $\vec{\Lambda}$ with the projection $\varkappa$ at least in two different ways:
I) $\vec{\Lambda}_1 + \vec{\Lambda}_2 = \vec{\Lambda}_{12}$, $\vec{\Lambda}_3 + \vec{\Lambda}_4 = \vec{\Lambda}_{34}$, $\vec{\Lambda}_{12} + \vec{\Lambda}_{34} = \vec{\Lambda}$ and
II) $\vec{\Lambda}_1 + \vec{\Lambda}_3 = \vec{\Lambda}_{13}$, $\vec{\Lambda}_2 + \vec{\Lambda}_4 = \vec{\Lambda}_{24}$, $\vec{\Lambda}_{13} + \vec{\Lambda}_{24} = \vec{\Lambda}$.

The states corresponding to these coupling schemes may be written as follows:

$$|\Lambda_1 \Lambda_2 (S_{12} \Lambda_{12}),\ \Lambda_3 \Lambda_4\ (S_{34} \Lambda_{34}) : S_{12,34}\, \Lambda\, \varkappa \rangle =$$
$$= \sum (\Lambda_1 \varkappa_1 \Lambda_2 \varkappa_2 | S_{12} \Lambda_{12} \varkappa_{12})(\Lambda_3 \varkappa_3 \Lambda_4 \varkappa_4 | S_{34} \Lambda_{34} \varkappa_{34}) \times$$
$$\times (\Lambda_{12} \varkappa_{12} \Lambda_{34} \varkappa_{34} | S_{12,34} \Lambda \varkappa) |\Lambda_1 \varkappa_1\rangle |\Lambda_2 \varkappa_2\rangle\, |\Lambda_3 \varkappa_3\rangle\, |\Lambda_4 \varkappa_4\rangle, \quad (25)$$

$$|\Lambda_1 \Lambda_3 (S_{13} \Lambda_{13}), \Lambda_2 \Lambda_4 (S_{24} \Lambda_{24}) : S_{13,24}\, \Lambda\, \varkappa \rangle =$$
$$= \sum (\Lambda_1 \varkappa_1 \Lambda_3 \varkappa_3 | S_{13} \Lambda_{13} \varkappa_{13})(\Lambda_2 \varkappa_2 \Lambda_4 \varkappa_4 | S_{24} \Lambda_{24} \varkappa_{24}) \times$$
$$\times (\Lambda_{13} \varkappa_{13} \Lambda_{24} \varkappa_{24} | S_{13,24} \Lambda \varkappa) |\Lambda_1 \varkappa_1\rangle |\Lambda_2 \varkappa_2\rangle\, |\Lambda_3 \varkappa_3\rangle\, |\Lambda_4 \varkappa_4\rangle \quad (26)$$

where the summation over all "projections" $\varkappa_\alpha$ (except for $\varkappa$) is implied.

The 9Λ-symbol may be determined by the following relation

$$\langle \Lambda_1 \Lambda_2 (S_{12} \Lambda_{12}), \Lambda_3 \Lambda_4 (S_{34} \Lambda_{34}) : S_{12,34}\, \Lambda\, \varkappa \,|\, \Lambda_1 \Lambda_3 (S_{13} \Lambda_{13}),$$
$$, \Lambda_2 \Lambda_4 (S_{24} \Lambda_{24}) : S_{13,24}\, \Lambda\, \varkappa \rangle = \delta_{\varkappa,\varkappa'}\, \delta_{\Lambda,\Lambda'}\, (-1)^{\varphi} \times$$
$$\times \sqrt{\dim \Lambda_{12} \cdot \dim \Lambda_{34} \dim \Lambda_{13} \cdot \dim \Lambda_{24}} \left\{ \begin{array}{ccc|c} \Lambda_1 & \Lambda_2 & \tilde{\Lambda}_{12} & S_{12} \\ \Lambda_3 & \Lambda_4 & \tilde{\Lambda}_{34} & S_{34} \\ \tilde{\Lambda}_{13} & \tilde{\Lambda}_{24} & \Lambda & S_{13,24} \\ \hline -S_{13} & -S_{24} & -S_{12,34} & \end{array} \right\} \quad (27)$$

where

$$\varphi = \lambda_1 + \mu_1 + \lambda_4 + \mu_4 + \lambda + \mu + \lambda_{24} + \lambda_{34} + \mu_{12} + \mu_{13} + \\ + \Phi(\Lambda_1 \Lambda_3 \bar{\Lambda}_{13}) + \Phi(\Lambda_2 \Lambda_4 \bar{\Lambda}_{24}) + \Phi(\Lambda_{13} \Lambda_{24} \bar{\Lambda}).$$

The expression of the 9Λ-symbol in terms of $3\Lambda\varkappa$-symbols may readily found from Eq.(27):

$$\left\{\begin{array}{ccc|c} \Lambda_{11} & \Lambda_{12} & \Lambda_{13} & S^1 \\ \Lambda_{21} & \Lambda_{22} & \Lambda_{23} & S^2 \\ \Lambda_{31} & \Lambda_{32} & \Lambda_{33} & S^3 \\ \hline S_{\bar 1} & S_{\bar 2} & S_{\bar 3} & \end{array}\right\} = \sum_{\varkappa_{ik}} (-1)^{\theta(\varkappa)} \left(\begin{array}{ccc|c} \Lambda_{11} & \Lambda_{12} & \Lambda_{13} & \\ \varkappa_{11} & \varkappa_{12} & \varkappa_{13} & S^1 \end{array}\right) \left(\begin{array}{ccc|c} \Lambda_{21} & \Lambda_{22} & \Lambda_{23} & \\ \varkappa_{21} & \varkappa_{22} & \varkappa_{23} & S^2 \end{array}\right) \times \tag{28}$$

$$\left(\begin{array}{ccc|c} \Lambda_{31} & \Lambda_{32} & \Lambda_{33} & \\ \varkappa_{31} & \varkappa_{32} & \varkappa_{33} & S^3 \end{array}\right) \left(\begin{array}{ccc|c} \tilde{\Lambda}_{11} & \tilde{\Lambda}_{21} & \tilde{\Lambda}_{31} & \\ \tilde{\varkappa}_{11} & \tilde{\varkappa}_{21} & \tilde{\varkappa}_{31} & S_{\bar 1} \end{array}\right) \left(\begin{array}{ccc|c} \tilde{\Lambda}_{12} & \tilde{\Lambda}_{22} & \tilde{\Lambda}_{32} & \\ \tilde{\varkappa}_{12} & \tilde{\varkappa}_{22} & \tilde{\varkappa}_{32} & S_{\bar 2} \end{array}\right) \left(\begin{array}{ccc|c} \tilde{\Lambda}_{13} & \tilde{\Lambda}_{23} & \tilde{\Lambda}_{33} & \\ \tilde{\varkappa}_{13} & \tilde{\varkappa}_{23} & \tilde{\varkappa}_{33} & S_{\bar 3} \end{array}\right)$$

where $\theta(\varkappa) = \sum_{i,k=1}^{3} (j_{ik} - \frac{1}{2}\mu_{ik} + t_{ik2})$.

Let the main properties of 9Λ-symbols be evaluated.

1°. Nonvanishing conditions. The 9Λ-symbols (28) are nonvanishing only if all sets of the parameters:

$$(\Lambda_{11}\ \Lambda_{12}\ \Lambda_{13}),\ (\Lambda_{21}\ \Lambda_{22}\ \Lambda_{23}),\ (\Lambda_{31}\ \Lambda_{32}\ \Lambda_{33}),$$
$$(\tilde{\Lambda}_{11}\ \tilde{\Lambda}_{21}\ \tilde{\Lambda}_{31}),\ (\tilde{\Lambda}_{12}\ \tilde{\Lambda}_{22}\ \tilde{\Lambda}_{32}),\ (\tilde{\Lambda}_{13}\ \tilde{\Lambda}_{23}\ \tilde{\Lambda}_{33})$$

satisfy the triangle condition (11). Each of the eigenvalues $S^1, S^2, S^3, S_{\bar 1}, S_{\bar 2}, S_{\bar 3}$ must belong to the spectra of the respective classification operator.

2°. Classical symmetry properties may be obtained from Eq (28) taking account of the symmetry properties of the 9Λ-symbols.

a) Permutations of rows or columns.

$$\left\{\begin{array}{ccc|c} \Lambda_{11} & \Lambda_{12} & \Lambda_{13} & S^1 \\ \Lambda_{21} & \Lambda_{22} & \Lambda_{23} & S^2 \\ \Lambda_{31} & \Lambda_{32} & \Lambda_{33} & S^3 \\ \hline S_{\bar 1} & S_{\bar 2} & S_{\bar 3} & \end{array}\right\} = (-1)^{p_{i\ell k}\cdot\varepsilon} \left\{\begin{array}{ccc|c} \Lambda_{i1} & \Lambda_{i2} & \Lambda_{i3} & S^i \\ \Lambda_{\ell 1} & \Lambda_{\ell 2} & \Lambda_{\ell 3} & S^\ell \\ \Lambda_{k1} & \Lambda_{k2} & \Lambda_{k3} & S^k \\ \hline \gamma\cdot S_{\bar 1} & \gamma\cdot S_{\bar 2} & \gamma S_{\bar 3} & \end{array}\right\} \tag{29}$$

$$
= (-1)^{p_{ilk}\eta} \cdot \left\{ \begin{array}{ccc|c} \Lambda_{1i} & \Lambda_{1l} & \Lambda_{1k} & \gamma\cdot S^1 \\ \Lambda_{2i} & \Lambda_{2l} & \Lambda_{2k} & \gamma\cdot S^2 \\ \Lambda_{3i} & \Lambda_{3l} & \Lambda_{3k} & \gamma\cdot S^3 \\ \hline S_{\tilde{i}} & S_{\tilde{l}} & S_{\tilde{k}} & \end{array} \right\} \qquad (30)
$$

where

$$
\varepsilon = \sum_{i=1}^{3} \Phi(\Lambda_{1i}\Lambda_{2i}\Lambda_{3i}), \quad \eta = \sum_{i=1}^{3} \Phi(\Lambda_{i1}\Lambda_{i2}\Lambda_{i3}), \quad \gamma = (-1)^{p_{ilk}}.
$$

b) Transposition and conjugation

$$
\left\{ \begin{array}{ccc|c} \Lambda_{11} & \Lambda_{12} & \Lambda_{13} & S^1 \\ \Lambda_{21} & \Lambda_{22} & \Lambda_{23} & S^2 \\ \Lambda_{31} & \Lambda_{32} & \Lambda_{33} & S^3 \\ \hline S_{\tilde{1}} & S_{\tilde{2}} & S_{\tilde{3}} & \end{array} \right\} = (-1)^{\varepsilon+\eta} \left\{ \begin{array}{ccc|c} \Lambda_{11} & \Lambda_{21} & \Lambda_{31} & -S_{\tilde{1}} \\ \Lambda_{12} & \Lambda_{22} & \Lambda_{32} & -S_{\tilde{2}} \\ \Lambda_{13} & \Lambda_{23} & \Lambda_{33} & -S_{\tilde{3}} \\ \hline -S^1 & -S^2 & -S^3 & \end{array} \right\} \qquad (31)
$$

$$
= (-1)^{\varepsilon+\eta} \left\{ \begin{array}{ccc|c} \tilde{\Lambda}_{11} & \tilde{\Lambda}_{12} & \tilde{\Lambda}_{13} & -S^1 \\ \tilde{\Lambda}_{21} & \tilde{\Lambda}_{22} & \tilde{\Lambda}_{23} & -S^2 \\ \tilde{\Lambda}_{31} & \tilde{\Lambda}_{32} & \tilde{\Lambda}_{33} & -S^3 \\ \hline -S_{\tilde{1}} & -S_{\tilde{2}} & -S_{\tilde{3}} & \end{array} \right\} \qquad (32)
$$

All the above mentioned symmetry properties are independent and show the relationships between $3! \times 3! \times 2 \times 2 = 144$ different 9Λ-symbols.

3°. Unitarity relations. Because of the unitarity of the recoupling coefficients for four SU(3) spins we have the following orthogonality condition for 9Λ-symbols:

$$
\sum_{\substack{\Lambda_{13},\Lambda_{23} \\ S^1, S^2, S_{\tilde{3}}}} \dim\Lambda_{13} \dim\Lambda_{23} \left\{ \begin{array}{ccc|c} \Lambda_{11} & \Lambda_{12} & \Lambda_{13} & S^1 \\ \Lambda_{21} & \Lambda_{22} & \Lambda_{23} & S^2 \\ \Lambda_{31} & \Lambda_{32} & \Lambda_{33} & S^3 \\ \hline S_{\tilde{1}} & S_{\tilde{2}} & S_{\tilde{3}} & \end{array} \right\} \left\{ \begin{array}{ccc|c} \Lambda_{11} & \Lambda_{12} & \Lambda_{13} & S^1 \\ \Lambda_{21} & \Lambda_{22} & \Lambda_{23} & S^2 \\ \Lambda'_{31} & \Lambda'_{32} & \Lambda_{33} & S^3 \\ \hline S'_{\tilde{1}} & S'_{\tilde{2}} & S_{\tilde{3}} & \end{array} \right\} =
$$

$$
= \delta_{\Lambda_{31},\Lambda'_{31}} \cdot \delta_{\Lambda_{32},\Lambda'_{32}} \delta_{S_{\tilde{1}},S'_{\tilde{1}}} \delta_{S_{\tilde{2}},S'_{\tilde{2}}} \delta_{S^3,S'^3} (\dim\Lambda_{31})^{-1} \times
$$

$$\times (\dim \Lambda_{32})^{-1} \{\Lambda_{11}\Lambda_{21}\Lambda_{31}|S_I\}\{\Lambda_{12}\Lambda_{22}\Lambda_{32}|S_{II}\}\{\Lambda_{31}\Lambda_{32}\Lambda_{33}|S^3\}. \tag{33}$$

5. <u>The graphic technique in the SU(3)-spin theory</u>. The graphic technique of the angular momentum theory [1,2] may be transferred to the SU(3) spin theory with minimum modifications. Because the volume of the paper is limited, we shall give here only the graphic images of the quantities considered above.

1) The state vector $|\Lambda\varkappa\rangle$ may be pictured in the following two ways:

a) by a line emerging from the node and enumerated by numbers $\Lambda\varkappa$, b) by a line entering the node and enumerated by numbers $\tilde{\Lambda}\tilde{\varkappa}$:

a) ○—$\Lambda\varkappa$—→ , b) ○—$\tilde{\Lambda}\tilde{\varkappa}$—← (34)

2) $3\Lambda\varkappa$-symbol $\begin{pmatrix}\Lambda_1 & \Lambda_2 & \Lambda_3\\ \varkappa_1 & \varkappa_2 & \varkappa_3\end{pmatrix}S)$ is pictured by an orientated node connecting three lines that correspond to the states $|\Lambda_i\varkappa_i\rangle$, $i=1,2,3$. If the three lines are displayed in the counterclockwise order, the node must be marked by sign "+"; in the opposite case it is marked by sign "-". Besides, the eigenvalue S must be indicated near the node. Some of the feasible diagrams of the $3\Lambda\varkappa$-symbol are shown below

$$\begin{pmatrix}\Lambda_1 & \Lambda_2 & \Lambda_3\\ \varkappa_1 & \varkappa_2 & \varkappa_3\end{pmatrix}S\Big) = \ldots \tag{35}$$

3) If in some analytical expression there is a pair $\Lambda_i\varkappa_i$ and $\tilde{\Lambda}_i\tilde{\varkappa}_i$ and the summation over $\varkappa_i$ with the weight $(-1)^{j_i-\frac{1}{2}\mu_i+t_{iz}}$ is implied, then in the graphic language this summation is equivalent to unification of free tails of Λ_i and $\tilde{\Lambda}_i$ lines into one continuous line. Therefore the following diagrams correspond to Eqs. (19) and (28) for 6Λ- and 9Λ-symbols:

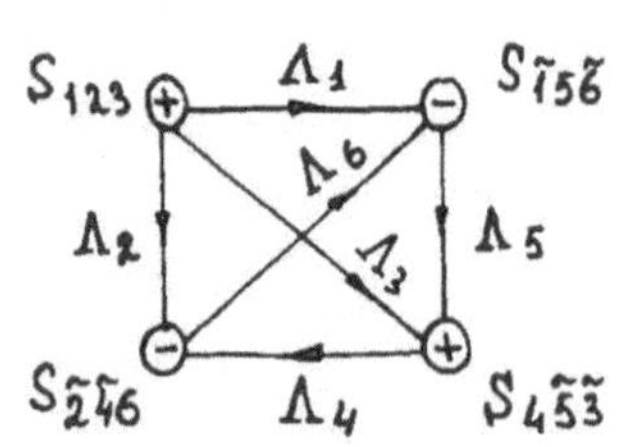

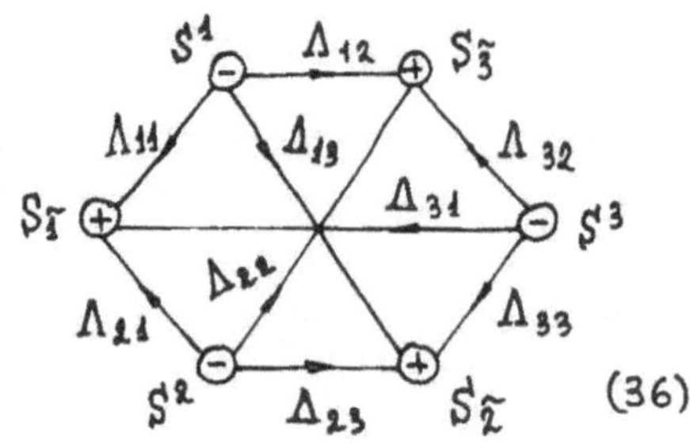

(36)

As in the angular momentum theory [1,2], three main properties are valid for such diagrams, namely, a) the distortion of a diagram does not change the $3n\Lambda$-symbol if the orientations of all nodes and lines remain the same; b) the change of the orientation of the node connecting the $\Lambda_1, \Lambda_2 \Lambda_3$ lines must be compensated by the phase factor $(-1)^{\phi(\Lambda_1\Lambda_2\Lambda_3}$; c) the value of the $3n\Lambda$ symbol pictured by a diagram does not change if the change of the orientation of some inner line Λ is followed by the substitution $\Lambda \to \tilde{\Lambda}$.

In conclusion we wish to underline that the relevant convenient selection of the classification operator makes the SU(3) spin theory very similar to well-known quantum theory of angular momenta.

References

1. Yucis, A.P., Levinson, I.B. Vanagas, V.V. (1960). Mathematical apparatus of the angular momentum theory, Vilnjus, Mintis. (Russian).
2. Varshalovich, D.A., Moskalev, A.N., Hersonskii, V.K. (1975). Quantum theory of angular momentum, Nauka, Leningrad (Russian).
3. Guseva, I.S., Smirnov, Yu.F., Tolstoy, V.N., Kharitonov, Yu.I. (1981). LIYaF preprints: 678, 42; (1983). 337, 36.
4. Pluhař, Z., Smirnov, Yu.F., Tolstoy, V.N. (1981). Charles University preprint, Praha, 69.
5. Tolstoy, V.N., Smirnov, Yu.F., Pluhař, Z. (1983). In: Group-theoretical methods in physics, Nauka, Moscow,

2, 523 (Russian).

6. Asherova, R.M., Smirnov, Yu.F., Tolstoy, V.N. (1979), Mat. USSR Zametki, 1979, 26, 15 (Russian).
7. Tolstoy, V.N. Preprint. (1985). P.N. Lebedev Physical Institute, Moscow, 27, 52.
8. Tolstoy, V.N. (1985). Yad. Fiz. 42, 706 (Russian).

MULTIPLICITIES OF ANGULAR MOMENTA FOR $N_1 \times N_2 \times N_3$-DIMENSIONAL OSCILLATOR AND THE REDUCTION $SU(N) \supset O(3)$

V.V.Mikhailov

Kazan Physical and Technical Institute of the USSR Academy of Sciences, Kazan 420029

1. Introduction. After Jordan fifty years ago [1] introduced his field quantization technique with the help of matrix representation by boson operator-functions and Schwinger [2] gave the detailed description of angular momentum (AM) in terms of two harmonic oscillators (HO), boson representations (BR) have been widely used in describing of different groups and algebras. For this N-dimensional HO (vector boson) [3] and 3N-dimensional [4] N x N-dimensional [5] HO (matrix boson) had been used. A complete review of BR of different groups is presented in [6].

One of the important problems solved with the help of BR, is the expansion of irreducible representations (IR) of SU(N) into IR's of O(3) group. It appears particularly in classification of states of n-particle system, when each particle has $2S+1=N$ energy levels, over quantum numbers of total AM. SU(3) O(3) reduction is discussed in more detail in [7,8]. The generally accepted technique of SU(N) IR expansion over O(3) IR is described in [9, 10, 11].

An alternative method of determination of AM multiplicities in the SU(N) symmetric. IR, to which the Yaung frame with one row corresponds, was described in [12], where generating functions (GF), recurrence relations and tables of AM multiplicities were obtained. Later the author got the same results for these multiplicities, using a vector boson for AM representation [13] allowing us to derive AM basis states and AM coherent states.

In [14] a representation of AM is used in terms of the matrix boson $a_{\mu_1\mu_2}$, where μ_1 and μ_2 can equal $N_1=2s_1+1$ and $N_2=2s_2+1$. Particularly, the representation $Q_n^{s_1s_2}$ based on all n-quantum states of $N_1\times N_2$ HO was studied. CF, recurrence relations and the tables in the simple cases 2x3, 2x4, 3x3, 3x4 HO were obtained for AM multiplicities $q_{nj_1j_2}^{s_1s_2}$ in Q_n; Multiplicities $q_{nj_1j_2}$ for 2x3 HO coincide with the multiplicities from [4]. Furthermore in [14] the comparison of $q_{nj_1j_2}$ with the tables of AM multiplicities in IR F_λ^s of SU(N) was made and as a consequence the relation

$$Q_n^{s_1s_2}=\sum_\lambda F_\lambda^{s_1}F_\lambda^{s_2} \qquad (1)$$

we obtained. In (1) the sum runs all possible Young frames $\lambda=[\lambda_1,\lambda_2,\ldots,\lambda_n]$.

In this paper a similar AM representation is studied in terms of volume matrix boson, i.e. $N_1\times N_2\times N_3$ HO. The discussion is based on the study of multiplicities $q_{nj_1j_2j_3}^{s_1s_2s_3}$ which generalize the AM multiplicities in the matrix BR. GF and reccurence relations for the numbers C are derived and the method of multiplicity determination from the numbers C is presented. The decomposition of F_λ^s into the O(3) IR D_j is the main result of the paper:

$$Q_n^{s_1s_2s_3}=\sum_\lambda A_{\lambda_1\lambda_2\lambda_3}F_{\lambda_1}^{s_1}F_{\lambda_2}^{s_2}F_{\lambda_3}^{s_3} \qquad (2)$$

generalizing (1). Then we describe the general properties of a volume matrix A and give a method of matrix construction for any in the case 2x2x2 HO.

<u>2. Representation of AM in terms of $N_1\times N_2\times N_3$ -dimensional HO.</u> Let us consider the $N_1\times N_2\times N_3$ pairs of the creation and annihilation boson operators $a_{\mu_1\mu_2\mu_3}$

and $\bar{a}_{\mu_1\mu_2\mu_3}$. Sometimes we will write a_μ instead of $a_{\mu_1\mu_2\mu_3}$. Operators a_μ satisfy the usual commutation relations

$$[\bar{a}_\mu, a_{\mu'}] = \delta_{\mu_1\mu_1'}\delta_{\mu_2\mu_2'}\delta_{\mu_3\mu_3'}, \quad [\bar{a}_\mu, \bar{a}_{\mu'}] = [a_\mu, a_{\mu'}] = 0$$
$$\mu_i = -s_i, -s_i+1, \dots s_i, \quad i = 1,2,3, \tag{3}$$
$$N_i = 2s_i + 1, \quad s_i = 0, 1/2, 1, 3/2, \dots$$

We define the particle number operator n

$$n = \sum_{\mu_1} n_{\mu_1} = \sum_{\mu_1\mu_2\mu_3} n_{\mu_1\mu_2\mu_3} \equiv \sum a_\mu \bar{a}_\mu \tag{4}$$

and the operators of AM

$$J_1^+ = \sum_\mu \sqrt{(s_1-\mu_1)(s_1+\mu_1+1)}\, a_{\mu_1+1,\mu_2\mu_3}\, \bar{a}_{\mu_1\mu_2\mu_3}$$
$$J_1^0 = \sum_{\mu_1} \mu_1 n_{\mu_1}, \quad J_1^- = (J_1^+)^+ \tag{5}$$

The operators $J_2^\pm$, J_2^0, $J_3^\pm$, J_3^0 are similarly determined. The operators J_i satisfy the AM commutation relation

$$[J_i^0, J_i^\pm] = \pm J_i^\pm, \quad [J_i^+, J_i^-] = 2J_i^0,$$
$$[J_i, n] = 0, \quad [J_i^\mu, J_k^\nu] = 0 \quad i \neq k \tag{6}$$

As a result we have three sets of operators $J_i^\pm$, J_i^0, $i = 1,2,3$, which generate three independent O(3) algebras. We can label now the states of $N_1 \times N_2 \times N_3$HO by the seven quantum numbers n, J_1, m_1 and J_2, m_2, J_3, m_3 where $J_i(J_i+1)$ and m_i are eigenvalues of the operators $J_i^0(J_i^0-1) + J_i^+J_i^-$ and J_i^0 . For any given indices J_i and m_i can take the following values

$$j_i = ns_i,\ ns_i - 1, \ldots 0 \text{ or } 1/2, \quad m_i = -j_i, -j_i + 1, \ldots j_i \tag{7}$$

Of course this number of indices is in sufficient for the full description of all the states of the system. For example, in the most simple case of the 2x2x2 HO multiplicity of the state even for small numbers $n = 6$, $j_1 = j_2 = j_3 = 1$ equals 2. But for the case of 2x5x5 HO multiplicity of the state for $n = 5$, $j_1 = 1/2$, $j_2 = j_3 = 3$ equals 210. Our aim is to study these multiplicities.

All states with given n are linear combinations of the monomials

$$a_{\mu^{(1)}} a_{\mu^{(2)}} \ldots a_{\mu^{(n)}} |0\rangle \tag{8}$$

Following [13] and [14] we can graphically interpret the number $C^{s_1 s_2 s_3}_{n m_1 m_2 m_3}$ of different nonequivalent states (8) as the number of different routes in the four-dimensional "tree" growing from the origin corresponding to $|0\rangle$. Each vertex is numbered by the indices n, m_1, m_2, m_3 and from it $N_1 \times N_2 \times N_3$ segments run upwards. Each segments corresponds to some boson creation operator.

Let us consider a representation $Q_n^{s_1 s_2 s_3}$ of $O(3) \times O(3) \times O(3)$. Its representation space is spanned by n-quantum states of $N_1 \times N_2 \times N_3$ HO. $Q_n^{s_1 s_2 s_3}$ can be expanded into the representations $D_{j_1 j_2 j_3}$ of $O(3) \times O(3) \times O(3)$

$$Q_n^{s_1 s_2 s_3} = \sum_j q n^{s_1 s_2 s_3}_{j_1 j_2 j_3} D_{j_1 j_2 j_3} \tag{9}$$

where $D_{j_1 j_2 j_3} = D_{j_1} \otimes D_{j_2} \otimes D_{j_3}$, $D_{j_1}, D_{j_2}, D_{j_3}$ are O(3) IR. The dimension $M_n^{s_1 s_2 s_3}$ of the $Q_n^{s_1 s_2 s_3}$ is

$$\mu_n^{s_1 s_2 s_3} = \binom{n + N_1 \cdot N_2 \cdot N_3 - 1}{n} \tag{10}$$

$$= \sum_m C^{s_1 s_2 s_3}_{n m_1 m_2 m_3} = \sum_j (2j_1+1)(2j_2+1)(2j_3+1) q^{s_1 s_2 s_3}_{n j_1 j_2 j_3} \tag{11}$$

We construct basis of $D_{j_1 j_2 j_3}$ beginning from highest states

$$J_1^+|\psi\rangle = J_2^+|\psi\rangle = J_3^+|\psi\rangle = 0, \quad |\psi\rangle = |n, j_1 j_1, j_2 j_2, j_3 j_3\rangle \tag{12}$$

$$|n j_1 m_1 j_2 m_2 j_3 m_3\rangle \propto (J_1^-)^{j_1 - m_1} (J_2^-)^{j_2 - m_2} (J_3^-)^{j_3 - m_3} |\psi\rangle \tag{13}$$

The multiplicities q and the numbers C are non-negative integers. For the case of $s_2 = s_3 = 0$, i.e. $N_2 = N_3 = 1$, they related via [13]

$$q^s_{nj} = C^s_{nm} - C^s_{n,m+1}\Big|_{m=j} \tag{14}$$

It is obtained 14 for a more general case of $s_3 = 0$, $s_1 \geq 1/2$, $s_2 \geq 1/2$ that multiplicities are given by

$$q_{n j_1 j_2} = C_{n j_1 j_2} - C_{n, j_1+1, j_2} - C_{n, j_1, j_2+1} + C_{n, j_1+1, j_2+1} \tag{15}$$

Here and in the similar relations below we omit the upper indices s_1, s_2, s_3. It can be shown that for $s_1 \geq 1/2$, $s_2 \geq 1/2$, $s_3 \geq 1/2$

$$q_{n j_1 j_2 j_3} = C_{n j_1 j_2 j_3} - C_{n, j_1+1, j_2, j_3} - C_{n, j_1, j_2+1, j_3} - C_{n, j_1, j_2, j_3+1} + C_{n, j_1+1, j_2+1, j_3} + C_{n, j_1+1, j_2, j_3+1} + C_{n, j_1, j_2+1, j_3+1} - C_{n, j_1+1, j_2+1, j_3+1} \tag{16}$$

From relations (14-16) it is not difficult to obtain further generalization for the case of the $N_1 \times N_2 \times N_3 \times N_4$ HO. In order to calculate multiplicities in the case of the $N_1 \times N_2 \times N_3$ HO we can use a more convenient method which is a consequence of (16). From the plane lattice of numbers $C_{nj_1j_2j_3}$ where n and j_3 are fixed, we subtract the lattice C_{nj_1,j_2,j_3+1} and obtain an auxiliary lattice

$$C'_{nj_1j_2j_3} = C_{nj_1j_2j_3} - C_{nj_1,j_2,j_3+1} \quad (17)$$

to which (15) can be applied.

So we can see that the problem of determining AM multiplicities $q_{nj_1j_2j_3}$ in the representation Q_n is reduced to calculation of the number of routes C_{nj_1,j_2,j_3}.

<u>3. The number of routes $C_{nj_1j_2j_3}$</u>. For $N_3 = 1$ it was shown [14] that GF for the number of routes C can be constructed with the help of two pairs of combinatorial variables x_1, y_1 and x_2, y_2. For the general case when $N_3 > 1$ we use three pairs of variables. Then GF takes the form

$$\prod_{M} \left(1 - x_1^{s_1+M_1} y_1^{s_1-M_1} x_2^{s_2+M_2} y_2^{s_2-M_2} x_3^{s_3+M_3} y_3^{s_3-M_3}\right)^{-1} =$$
$$= \sum_{n=0}^{\infty} \sum_{m_1=-ns_1}^{ns_1} \sum_{m_2} \sum_{m_3} C_{nm_1m_2m_3}^{s_1s_2s_3} x_1^{ns_1+m_1} y_1^{ns_1-m_1} \cdot \quad (18)$$
$$\cdot x_2^{ns_2+m_2} y_2^{ns_2-m_2} x_3^{ns_3+m_3} y_3^{ns_3-m_3}$$

From (18) it is possible ro derive many recurrences for determining numbers . But for our purpose it sufficies to obtain only a reccurrence allowing to construct all three-dimensional lattices and not unnecessarily complicated. For this we factor (18) into two multiplies: in the first one $M_3 = -s_3, -s_3+1, \ldots s_3-1$, in the second one $M_3 = s_3$. The first one is the GF for $N_1 \times N_2 \times (N_3 - 1)$ HO, the second one is GF for $N_1 \times N_2$

HO. Equating coefficients of the same powers of x_1 and y_1 we obtain the recurrence

$$C^{s_1 s_2 s_3}_{n m_1 m_2 m_3} = \sum C^{s_1 s_2 (s_3 - 1/2)}_{n' m_1' m_2' m_3'} \; C^{s_1 s_2}_{n'' m_1'' m_2''} \tag{19}$$

where

$$n = n' + n'', \; m_1 = m_1' + m_1'', \; m_2 = m_2' + m_2'', \; m_3 = m_3' + n' s_3 + n\, 1/2 \tag{20}$$

A method of constructing plane lattices, i.e. numbers for $N_3 = 1$ is given in [14]. Using these lattices we can obtain from (19) three-dimensional lattices for $N_1 \times N_2 \times 2$ HO, then for $N_1 \times N_2 \times 3$ HO, etc.

In the most general case the numbers are integers, each is placed at a fixed node of "cubical crystal lattice". The numbers in the lattice are symmetrical: $C_{n m_1 m_2 m_3} = C_{n, -m_1, m_2, m_3}$ etc. External planes of this volume lattice coincide with the plane lattices $C^{s_1 s_2}_{n m_1 m_2}$, $C^{s_1 s_3}_{n m_1 m_3}$, $C^{s_2 s_3}_{n m_2 m_3}$

<u>4. Comparison of multiplicities in the reduction SU(N)⊃ O(3) and in the decomposition $Q_n \supset D_{j_1 j_2 j_3}$</u>. Obtaining first the number of routes C and then multiplicities q (an example of these multiplicities for 3x3x3 HO and 2x5x5 HO is given in Appendix) we can compare the latter with the multiplicities $f^{s}_{\lambda j}$ which appear in the reduction SU(N)⊃ O'3):

$$F^{s}_{\lambda} = \sum_{j} f^{s}_{\lambda j} D_j \tag{21}$$

Here F^{s}_{λ} is the IR of SU(N), $\sum \lambda_i = n$. An example of such decomposition for $n = 4$, $s = 2$ is presented in the following table :

λ	N_λ	$j=0$	1	2	3	4	5	6	7	8	L^s_λ
4	1	1		2		2	1	1		1	70
3,1	3		2	2	3	2	2	1	1		105
2,2	2	2		2	1	2		1			50
$2,1^2$	3		2	1	2	1	1				45
1^4	1			1							5
	P^s_{nj} →	5	12	16	17	15	10	6	3	1	$\sum=625=5^4$

Here P^s_{nj} are the numbers of total AM j of a system of atoms with spin [15], N_λ the dimension IR of the symmetric group with Young tableau λ, L^s_λ the dimension of IR F^s_λ. Dimensions and multiplicities introduced here satisfy the relations

$$P^s_{nj} = \sum_\lambda f^s_{\lambda j} N_\lambda \tag{22}$$

$$\sum_j (2j+1) f^s_{\lambda j} = L^s_\lambda , \tag{23}$$

$$\sum_\lambda L^s_\lambda N_\lambda = \sum_j (2j+1) P^s_{nj} = (2s+1)^n \equiv N^n \tag{24}$$

Multiplicities have been studied in [12, 13] for $N_2=N_3=1$. From the results of these papers we derive

$$q^{s_1}_{nj_1} = f^{s_1}_{[n],j_1} \tag{25}$$

i.e. the N_1 -dimensional BR of AM which can be derived from (3-5) putting $s_2=s_3=0$ desribes only basis states of symmetric IR of the group SU(N).

For the case of $N_1 \times N_2$ HO ($N_3=1$ in (3-5)) the plane lattices of multiplicities $q_{nj_1j_2}$ are connected with $f^s_{\lambda j}$ in the following way [14]

$$q_{n j_1 j_2}^{s_1 s_2} = \sum_{\lambda} f_{\lambda j_1}^{s_1} f_{\lambda j_2}^{s_2} \tag{26}$$

which means that the representation Q_n of $N_1 \times N_2$ HO is a convolution of IR's $F_{\lambda}^{s_1}$ and $F_{\lambda}^{s_2}$ of SU(N_1) and SU(N_2)

$$Q_n^{s_1 s_2} = \sum_{\lambda} F_{\lambda}^{s_1} F_{\lambda}^{s_2} \tag{27}$$

In the case of $N_1 \times N_2 \times N_3$ HO we have the following generalization of (26)

$$q_{n j_1 j_2 j_3}^{s_1 s_2 s_3} = \sum_{\lambda_1 \lambda_2 \lambda_3} A_{\lambda_1 \lambda_2 \lambda_3} f_{\lambda_1 j_1}^{s_1} f_{\lambda_2 j_2}^{s_2} f_{\lambda_3 j_3}^{s_3} \tag{28}$$

Putting (28) in (9) and using (21) we can take the sum over j_1, j_2, j_3 and get a generalization of (27)

$$Q_n^{s_1 s_2 s_3} = \sum_{\lambda_1 \lambda_2 \lambda_3} A_{\lambda_1 \lambda_2 \lambda_3} F_{\lambda_1}^{s_1} F_{\lambda_2}^{s_2} F_{\lambda_3}^{s_3} \tag{29}$$

To illustrate explicitly matrices A in some cases.

a) Matrices A for 2x2x2 HO

n=2 $A_{[2]} = \begin{vmatrix} 1 & \\ & 1 \end{vmatrix}$ $A_{[1,1]} = \begin{vmatrix} & 1 \\ 1 & \end{vmatrix}$ n=3 $A_{[3]} = \begin{vmatrix} 1 & \\ & 1 \end{vmatrix}$ $A_{[2,1]} = \begin{vmatrix} & 1 \\ 1 & 1 \end{vmatrix}$

n=4 $A_{[4]} = \begin{vmatrix} 1 & & \\ & 1 & \\ & & 1 \end{vmatrix}$ $A_{[3,1]} = \begin{vmatrix} & 1 & \\ 1 & 1 & 1 \\ & 1 & \end{vmatrix}$ $A_{[2,2]} = \begin{vmatrix} & & 1 \\ & 1 & \\ 1 & & 1 \end{vmatrix}$

n=5 $A_{[5]} = \begin{vmatrix} 1 & & \\ & 1 & \\ & & 1 \end{vmatrix}$ $A_{[4,1]} = \begin{vmatrix} & 1 & \\ 1 & 1 & 1 \\ & 1 & 1 \end{vmatrix}$ $A_{[3,2]} = \begin{vmatrix} & & 1 \\ & 1 & 1 \\ 1 & 1 & 1 \end{vmatrix}$

n=6 $A_{[5,1]} = \begin{vmatrix} & 1 & & \\ 1 & 1 & 1 & \\ & 1 & 1 & 1 \\ & & 1 & \end{vmatrix}$ $A_{[4,2]} = \begin{vmatrix} & & 1 & \\ & 1 & 1 & \\ 1 & 1 & 2 & \\ & 1 & & 1 \end{vmatrix}$ $A_{[3,3]} = \begin{vmatrix} & & & 1 \\ & 1 & 1 & 1 \\ 1 & & 1 & \end{vmatrix}$

n=7

$$A_{[6,1]}=\begin{vmatrix} & 1 & & \\ 1 & 1 & 1 & \\ & 1 & 1 & 1 \\ & & 1 & 1 \end{vmatrix} \quad A_{[5,2]}=\begin{vmatrix} & & 1 & \\ & 1 & 1 & 1 \\ 1 & 1 & 2 & 1 \\ & 1 & 1 & 1 \end{vmatrix} \quad A_{[4,3]}=\begin{vmatrix} & & & 1 \\ & & 1 & 1 \\ & 1 & 1 & 1 \\ 1 & 1 & 1 & 1 \end{vmatrix}$$

n=8

$$A_{[7,1]}=\begin{vmatrix} & 1 & & & \\ 1 & 1 & 1 & & \\ & 1 & 1 & 1 & \\ & & 1 & 1 & 1 \\ & & & 1 & \end{vmatrix} \quad A_{[6,2]}=\begin{vmatrix} & & 1 & & \\ & 1 & 1 & 1 & \\ 1 & 1 & 2 & 1 & 1 \\ & 1 & 1 & 2 & \\ & & 1 & & 1 \end{vmatrix}$$

$$A_{[5,3]}=\begin{vmatrix} & & & 1 & \\ & & 1 & 1 & 1 \\ & 1 & 1 & 2 & \\ 1 & 1 & 2 & 1 & 1 \\ & 1 & & 1 & \end{vmatrix} \quad A_{[4,4]}=\begin{vmatrix} & & & & 1 \\ & & & 1 & \\ & & 1 & & 1 \\ & 1 & & 1 & \\ 1 & & 1 & & 1 \end{vmatrix}$$

b) Matrices A for 2x5x5 HO

n=2

$$A_{[2]}=\begin{vmatrix} 1 & \\ & 1 \end{vmatrix} \quad A_{[1,1]}=\begin{vmatrix} & 1 \\ 1 & \end{vmatrix}$$

n=3

$$A_{[3]}=\begin{vmatrix} 1 & & \\ & 1 & \\ & & 1 \end{vmatrix}$$

$$A_{[2,1]}=\begin{vmatrix} & 1 & \\ 1 & 1 & 1 \\ & 1 & \end{vmatrix}$$

n=4

$$A_{[4]}=\begin{vmatrix} 1 & & & & \\ & 1 & & & \\ & & 1 & & \\ & & & 1 & \\ & & & & 1 \end{vmatrix} \quad A_{[3,1]}=\begin{vmatrix} & 1 & & & \\ 1 & 1 & 1 & 1 & \\ & 1 & & 1 & \\ & 1 & 1 & 1 & 1 \\ & & & 1 & \end{vmatrix}$$

$$A_{[2,2]}=\begin{vmatrix} & & 1 & & \\ & 1 & & 1 & \\ 1 & & 1 & & 1 \\ & 1 & & 1 & \\ & & 1 & & \end{vmatrix}$$

n=5

$$A_{[4,1]}=\begin{vmatrix} & 1 & & & & & \\ 1 & 1 & 1 & 1 & & & \\ & 1 & 1 & 1 & 1 & & \\ & 1 & 1 & 1 & 1 & 1 & \\ & & 1 & 1 & 1 & 1 & \\ & & & 1 & 1 & 1 & 1 \\ & & & & & 1 & \end{vmatrix}$$

$$A_{[3,2]}=\begin{vmatrix} & & 1 & & & & \\ & 1 & 1 & 1 & 1 & & \\ 1 & 1 & 1 & 1 & 1 & 1 & \\ & 1 & 1 & 2 & 1 & 1 & \\ & 1 & 1 & 1 & 1 & 1 & 1 \\ & & 1 & 1 & 1 & 1 & \\ & & & & 1 & & \end{vmatrix}$$

n=6

$$A_{[5,1]} = \left|\begin{array}{cccccccccc}
 & 1 & & & & & & & & \\
1 & 1 & 1 & 1 & & & & & & \\
 & 1 & 1 & 1 & 1 & 1 & & & & \\
 & & 1 & & & 1 & & & & \\
 & 1 & 1 & & 1 & 1 & & 1 & & \\
 & & 1 & 1 & 1 & 2 & 1 & 1 & 1 & \\
 & & & & & 1 & & & 1 & \\
 & & & & 1 & 1 & & 1 & 1 & 1 \\
 & & & & & 1 & 1 & 1 & 1 & 1 \\
 & & & & & & & 1 & 1 & 1
\end{array}\right| \quad A_{[4,2]} = \left|\begin{array}{cccccccccc}
 & & 1 & & & & & & & \\
 & 1 & 1 & 1 & 1 & 1 & & & & \\
1 & 1 & 2 & & 1 & 2 & 1 & 1 & & \\
 & 1 & & 1 & 1 & 1 & & & 1 & \\
 & 1 & 1 & 1 & 2 & 2 & & 1 & 1 & \\
 & 1 & 2 & 1 & 2 & 3 & 1 & 2 & 2 & 1 \\
 & & 1 & & & 1 & 1 & 1 & & 1 \\
 & & 1 & & 1 & 2 & 1 & 2 & 1 & 1 \\
 & & & 1 & 1 & 2 & & 1 & 2 & 1 \\
 & & & & & 1 & 1 & 1 & 1 & 1
\end{array}\right|$$

$$A_{[3,3]} = \left|\begin{array}{cccccccccc}
 & & & 1 & & & & & & \\
 & & 1 & & & 1 & & & & \\
 & 1 & & 1 & 1 & 1 & & & 1 & \\
1 & & 1 & & & & 1 & 1 & & \\
 & & 1 & & 1 & 1 & 1 & 1 & & \\
 & 1 & 1 & & 1 & 2 & & 1 & 1 & 1 \\
 & & & 1 & 1 & & & & 1 & \\
 & & & 1 & 1 & 1 & & 1 & 1 & \\
 & & 1 & & & 1 & 1 & 1 & & 1 \\
 & & & & & 1 & & & 1 &
\end{array}\right|.$$

c) Matrices A for 3x3x3 HO

n=2 $A_{[2]} = \left|\begin{array}{cc} 1 & \\ & 1 \end{array}\right|$ $A_{[1,1]} = \left|\begin{array}{cc} & 1 \\ 1 & \end{array}\right|$ n=3 $A_{[2,1]} = \left|\begin{array}{ccc} & 1 & \\ 1 & 1 & 1 \\ & 1 & \end{array}\right|$ $A_{[1^3]} = \left|\begin{array}{ccc} & & 1 \\ & 1 & \\ 1 & & \end{array}\right|$

n=4

$$A_{[3,1]} = \left|\begin{array}{cccc} & 1 & & \\ 1 & 1 & 1 & 1 \\ & 1 & & 1 \\ & 1 & 1 & 1 \end{array}\right| \quad A_{[2,2]} = \left|\begin{array}{cccc} & & 1 & \\ & 1 & & 1 \\ 1 & & 1 & \\ & 1 & & 1 \end{array}\right| \quad A_{[2,1^2]} = \left|\begin{array}{cccc} & & & 1 \\ & 1 & 1 & 1 \\ & 1 & & 1 \\ 1 & 1 & 1 & 1 \end{array}\right|$$

Zero elements are omitted. Colimns and rows of these matrices are numbered from left to right and from top to bottom by the Young tabreaux λ [$\lambda_1, \lambda_2, \ldots \lambda_n$] which are ordered in the following way: at first one-row, then two-row, then three-row Young tableaux and so on. Tableaux with the same number of rows are ordered lexicographically. As an example we write ordering for 2x5x5 HO and n=6: 6; 5,1; 4,2; 3^2; $4,1^2$; 3,2,1; 2^3; $3,1^3$; $2^2 1^2$; $2,1^4$. This ordering is convenient to us, but differs from pure lexicographic ordering.

5. Construction of matrices A and their properties.

First, all matrices $A_{\lambda_1\lambda_2\lambda_3}$ for λ_1 or λ_2 or $\lambda_3 = n$ are unit matrices. This is a consequence of the fact that numbers C of internal planes of volume lattices coincide with the numbers of the plane lattices for $N_2 \times N_3$, $N_1 \times N_3$, $N_1 \times N_2$ HO.

Matrices A can be obtained from (28), where multiplicities $q^{S_1S_2S_3}$ and f^{nS} are known. This may be called an inverse problem. It can be uniquely solved at least for small n (≤ 6). Explicit examples of matrices A from previous section show that we can obtain matrices A for 2x2x2 HO from matrices A for 2x5x5 HO, both for the same number n , if in the latter we cancel all columns and rows numbered by Young tableaux with three and more rows. It is natural because of the well-known fact that only two-row Young tableaux are permitted for SU(2). In the general case the following lemma holds.

Lemma. For each n there exist a general cubic matrix $A_{\lambda_1\lambda_2\lambda_3}$ in dependent of N_1, N_2 or N_3 and in which indices λ_1 , λ_2 and λ_3 correspond to all Young tableaux (from $[n]$ to $[1^n]$). In concrete cases the numbers N_1, N_2, N_3 impose restrictions on the indices λ_1, λ_2, λ_3 and from a general matrix A we must retain only the part in which number of rows in the Young frame λ_i is not greater than N_i $(i=1,2,3)$.

There are some features that look like princople difficulties. For n=7, $S_1 = 1/2$, , $S_2 = S_3 = 2$ (i.e. 2x5x5 HO) we have several solutions for matrix A , if we consider eq. (28) separately. But it is clear from Lemma that the greater part of A is determined uniquely if we solve successively eq. (28) for the cases of 2x2x5, 2x3x5 and 2x4x5 HO. Then the problem for 2x5x5 HO, n=7 is solved simply and uniquely.

Determining matrices A is facilitated by the fact that for 2x2x2 HO there is the recurrence

$$A_{\lambda_1\lambda_2}[\kappa,\ell] = A_{\lambda_1\lambda_2}[\ell+1,\ell] \mathring{+} A_{\lambda_1\lambda_2}[\kappa-2,\ell] \quad (30)$$

which allows us to construct all the matrices . Here means the following matrix addition. Zero column is added to the left-hand side of the matrix $A[\kappa-2,\ell]$, zero row is added to the top of it. After that the matrix $A[\ell+1,\ell]$ is superimposed in the left upper corner of the expanded matrix $A[\kappa-2,\ell]$ and both matrices are added. Constructing the full set of matrices it is useful to have in mind that

$$A_{[\ell,\ell]} = \left|\begin{matrix} & & & & 1 \\ & & & 1 & 0 \\ & & 1 & 0 & 1 \\ \cdots & \cdots & \cdots & \cdots & \cdots \\ 1\ 0\ 1\ 0\ 1 & \cdots & & & \end{matrix}\right| \qquad A_{[\ell+1,\ell]} = \left|\begin{matrix} & & & & 1 \\ & & & 1 & 1 \\ & & 1 & 1 & 1 \\ \cdots & \cdots & \cdots & \cdots & \cdots \\ 1\ 1\ 1 & \cdots & & 1 & 1 \end{matrix}\right|$$

Summing up note that BR of the SU(N) carried out with the help of vector boson is unsufficient for the description of all states. It describes only fundamental IR's with the one-row Young tableaux.

BR of the SU(N_1) x SU(N_2) constructed with the help of matrix boson ($N_1 \times N_2$ HO) sufficies describe basis states of all representations $F_\lambda^{g_1} \otimes F_\lambda^{g_2}$ and is not redunant.

BR of the SU(N_1) x SU(N_2) x SU(N_3) constructed with the help of volume matrix boson ($N_1 \times N_2 \times N_3$ HO) is redunant. The multiplicities, completely described by the matrices A, appear here. We can state now that matrices A can be uniquely determined at least for $n \leq 10$ and for any N_1, N_2, N_3.

Appendix. Multiplicities for 2x5x5 HO.

n=1

$$j_1 = 2,\ 1,\ 0 \qquad \begin{vmatrix} & & 1 \\ & & \\ & & \end{vmatrix} \qquad j_3 = 1/2$$

$$j_2 = 0\ 1\ 2$$

n=2

$$j_1 = 4,\ 3,\ 2,\ 1,\ 0 \qquad \begin{vmatrix} 1 & & 1 & & 1 \\ & 1 & & 1 & \\ 1 & & 1 & & 1 \\ & 1 & & 1 & \\ 1 & & 1 & & 1 \end{vmatrix} \quad j_3 = 1 \qquad \begin{vmatrix} & 1 & & 1 & \\ 1 & & 1 & & 1 \\ & 1 & & 1 & \\ 1 & & 1 & & 1 \\ & 1 & & 1 & \end{vmatrix} \quad j_3 = 0$$

$$j_2 = 0\ 1\ 2\ 3\ 4 \qquad\qquad j_2 = 0\ 1\ 2\ 3\ 4$$

n=3

$$\begin{vmatrix} 1 & & 1 & 1 & 1 & & 1 \\ & 1 & 2 & 1 & 1 & 1 & \\ 1 & 1 & 3 & 2 & 2 & 1 & 1 \\ 1 & 2 & 3 & 3 & 2 & 1 & 1 \\ 1 & 2 & 5 & 3 & 3 & 2 & 1 \\ & 2 & 2 & 2 & 1 & 1 & \\ 1 & & 1 & 1 & 1 & & 1 \end{vmatrix} \quad j_3 = 3/2 \qquad \begin{vmatrix} & 1 & 2 & 1 & 1 & 1 & \\ 1 & 2 & 3 & 3 & 2 & 1 & 1 \\ 1 & 3 & 5 & 4 & 3 & 2 & 1 \\ 1 & 4 & 7 & 5 & 4 & 3 & 1 \\ 2 & 5 & 8 & 7 & 5 & 3 & 2 \\ 1 & 3 & 5 & 4 & 3 & 2 & 1 \\ & 1 & 2 & 1 & 1 & 1 & \end{vmatrix} \quad J_3 = 1/2$$

Multiplicities for 3x3x3 HO

n=2

$$\begin{vmatrix} 1 & & 1 \\ & 1 & \\ 1 & & 1 \end{vmatrix} \quad j_3 = 2 \qquad \begin{vmatrix} & 1 & \\ 1 & & 1 \\ & 1 & \end{vmatrix} \quad j_3 = 1 \qquad \begin{vmatrix} 1 & & 1 \\ & 1 & \\ 1 & & 1 \end{vmatrix} \quad j_3 = 0$$

n=3

$$\begin{vmatrix} & 1 & & 1 \\ & 1 & 1 & \\ & 2 & 1 & 1 \\ 1 & & & \end{vmatrix} \quad J_3 = 3 \qquad \begin{vmatrix} & 1 & 1 & \\ 1 & 2 & 1 & 1 \\ 1 & 3 & 2 & 1 \\ & 1 & 1 & \end{vmatrix} \quad J_3 = 2$$

$$\begin{vmatrix} & 2 & 1 & 1 \\ 1 & 3 & 2 & 1 \\ 1 & 5 & 3 & 2 \\ 1 & 1 & 1 & \end{vmatrix} \quad j_3 = 1 \qquad \begin{vmatrix} 1 & & & \\ & 1 & 1 & \\ 1 & 1 & 1 & \\ & 1 & & 1 \end{vmatrix} \quad J_3 = 0$$

n=4

$$\begin{vmatrix} 1 & & 1 & & 1 \\ & 1 & 1 & 1 & \\ 2 & 1 & 3 & 1 & 1 \\ & 2 & 1 & 1 & \\ 2 & & 2 & & \end{vmatrix} \quad J_3 = 4 \qquad \begin{vmatrix} & 1 & 1 & 1 & \\ 2 & 2 & 3 & 1 & 1 \\ 2 & 5 & 5 & 3 & 1 \\ 3 & 4 & 5 & 2 & 1 \\ & 3 & 2 & 2 & \end{vmatrix} \quad J_3 = 3$$

$$\begin{vmatrix} 2 & 1 & 3 & 1 & 1 \\ 2 & 5 & 5 & 3 & 1 \\ 7 & 8 & 12 & 5 & 3 \\ 3 & 10 & 8 & 5 & 1 \\ 5 & 3 & 7 & 2 & 2 \end{vmatrix} \quad \dot{J}_3 = 2 \qquad \begin{vmatrix} & 2 & 1 & 1 & \\ 3 & 4 & 5 & 2 & 1 \\ 3 & 10 & 8 & 5 & 1 \\ 6 & 8 & 10 & 4 & 2 \\ & 6 & 3 & 3 & \end{vmatrix} \quad \dot{J}_3 = 1 \qquad \begin{vmatrix} 2 & & 2 & & 1 \\ & 3 & 2 & 2 & \\ 5 & 3 & 7 & 2 & 2 \\ & 6 & 3 & 3 & \\ 5 & & 5 & & 2 \end{vmatrix} \quad \dot{J}_3 = 0$$

References

1. Jordan, P. (1935). Z.Physik, 94, 531.
2. Schwiger, J. (1965). In: Quantum Theory of Angular Momentum (eds. Biedenharn L.C. and van Dam H.) Academic P., N.Y., p.229.
3. Louck, J.D. (1965). J.Math. Phys., 6, 1786.
4. Bargmann, V., Moshinsky, M. (1960). Nucl. Phys., 18, 697.
5. Moshinsky, M. (1963). J. Math. Phys., 4, 1128.
6. Biedenharn, L.C., Louck J.D. (1984). Angular Momentum in Quantum Physics.
7. Judd, B.R., Miller, W.Jr., Patera,J., Winternitz, P. (1974). J.Math. Phys., 15, 1787.
8. Moshinsky, M., Patera, J., Sharp, R.T., Winternitz,P. (1975). Ann. Phys., 95, 139.
9. Hamermesh, M. (1962). Group Theory. Addison-Wesley, Reading, Mass.
10. Kaplan, I.C. Theory of many-Electron systems. Moscow: Nauka, 1969.
11. Elliot, J., Dober, P. Symmetry in Physics.
12. Büttner, H. (1967). Z.Phys., 198, 494.
13. Mikhailov, V.V. (1973). Theor. Math. Phys., 15, 367; (1974), 18, 342 (Russian).
14. Mikhailov, V.V. (1978). J.Phys.A:Math.Gen.. 11, 443.
15. Mikhailov, V.V. (1977). J.Phys.A:Math.Gen., 10, 147; (1981), 14, 1107.

CASIMIR OPERATORS OF THE GENERALISED POINCARÉ AND GALILEI GROUPS

L.Barannik[+], W.Fushchich
V.G.Korolenko Pedagogical Institute, Poltava [+]
Mathematical Institute of the Ukrainian SSR Academy of Sciences, Kiev

1. Introduction. The study of Casimir operators of a Lie group may be reduced to the description of its associated group invariants [1]. Such invariants are well known for the classical Lie groups [2]. Casimir operators of P(1,4) and their spectra were found in [3,4]. The same problem was solved for ISL (G,C) in [5] and for IU(n) with various subgroups of translations in [6]. The generalised Casimir operators (rational, transcendental) for the subgroups of P(1,3) are found in [7], and for the subgroups of the optical group Opt(1,3) in [8]. The articles [9,10] are devoted to the investigation of Casimir operators for nonhomogeneous groups of arbitrary dimension of associated groups for many nonhomogeneous classical groups is given in [11].

In this paper the maximal systems of algebraically independent polynomial Casimir operators (basic operators) of the generalised Poincaré group P(p,q), Galilei group G(n) and extended Galilei group $\tilde{G}(n)$ are found in an explicit form.

2. Casimir operators of the generalised Poincaré group. Poincaré algebra AP(p,q) is defined by the following commutation relations:

$$|J_{ab},J_{cd}| = g_{ad}J_{bc}+g_{bc}J_{ad}-g_{ac}J_{bd}-g_{bd}J_{ac},$$
$$|P_a,J_{bc}| = g_{ab}P_c-g_{ac}P_b,\ J_{ba}= -J_{ab},\ |P_a,P_b| = 0, \quad (1)$$

where $g = \mathrm{diag}\,\{\underbrace{1,\dots,1}_{p}, \underbrace{-1,\dots,-1}_{q}\}$, $(a,b,c,d = 1,2,\dots,n$, $n = p+q)$.

Let $\varepsilon_{a_1a_2\dots a_n}$ be the rank n completely antisymmetric tensor, $\varepsilon_{12\dots n} = 1$ and

$$g^{ab} = \delta^{ak}\,\delta^{bl}g_{kl}\ ,\quad J^{ab} = \delta^{ak}\,\delta^{bl}J_{kl}\ ,\quad P^{a} = \delta^{ak}P_k\ ,$$

$$W_{a_1\dots a_{n-1}} = \varepsilon_{a_1\dots a_n}P^{a_n}$$

$$W_{a_1,\dots a_r} = W_{a_1\dots a_r a_{r+1} a_{r+2}}J^{a_{r+1}a_{r+2}} \quad (r< n-3),$$

$$W = W_{a_1a_2}J^{a_1a_2} \quad \text{for } n \text{ odd.}$$

From 1 one deduces that the following relations are well-defined

$$|J_{ab},W_{d_1\dots d_t}| = g_{aa}g_{bb}(g_{bd_1}W_{ad_2\dots d_t}+\dots+ \\ + g_{bd_t}W_{d_1d_2\dots a}-g_{ad_1}W_{bd_2\dots d_t}-\dots-g_{ad_t}W_{d_1d_2\dots b}), \tag{2}$$

Theorem 1. Let $n = p+q$, $n >3$, $m =\left|\frac{n-1}{2}\right|$. The elements

$$T_o = g^{ab}P_aP_b$$

$$T_k = \frac{1}{(n-2k-1)!2^{2k}(k!)^2}\, g^{a_1b_1}\dots g^{a_{n-2k-1}b_{n-2k-1}}\,\cdot$$

$$\cdot\ W_{a_1\dots a_{n-2k-1}}W_{b_1\dots b_{n-2k-1}}$$

for $k <m$ or $k = m$ and n is even and

$$T_m = \frac{1}{2^m m!}\,W \quad \text{for } n \text{ is odd}$$

form a maximal algebraically independent system of Casimir operators for the algebra AP(p,q).

The invariance of the operators follows from (2). They are independent since so are the symmetrized Casimir operators of P(p,q), constructed in the same way.

Example: the basic Casimir operators for P(1,6):

$$P_1^2-P_2^2-P_3^2-P_4^2-P_5^2-P_6^2-P_7^2 \;;$$

$$\frac{1}{96}\, g^{a_1b_1}\cdots g^{a_4b_4}\,\varepsilon_{a_1\ldots a_7}\varepsilon_{b_1\ldots b_7} J^{a_5a_6}P^{a_7}J^{b_5b_6}P^{b_7} \;;$$

$$\frac{1}{128}\, g^{a_1b_1}g^{a_2b_2}\,\varepsilon_{a_1\ldots a_7}\varepsilon_{b_1\ldots b_7} J^{a_3a_4}J^{a_5a_6}P^{a_7}J^{b_3b_4}\cdot J^{b_5b_6}P^{b_7};$$

$$\frac{1}{48}\,\varepsilon_{a_1\ldots a_7} J^{a_1a_2}J^{a_3a_4}J^{a_5a_6}P^{a_7}$$

3. Casimir operators of the extended Galilei group. The Lie algebra $A\tilde{G}(n)$ of the extended Galilei froup $\tilde{G}(n)$ of n-dimensional euclidean space is defined as the real algebra with the following commutation relations:

$$\begin{aligned}
&|J_{ab},J_{cd}| = \delta_{ad}J_{bc}+ \delta_{bc}J_{ad}- \quad {}_{ac}J_{bc}- \quad {}_{bd}J_{ac} \;; \\
&|P_a,J_{bc}| = \delta_{ab}P_c- \delta_{ac}P_b \;; \qquad |P_a,P_b| = 0 \qquad\qquad (3)\\
&|G_a,J_{bc}| = \delta_{ab}G_c- \delta_{ac}G_b \;; \qquad |G_a,G_b| = 0 \;; \\
&|P_o,J_{bc}| = |P_o,P_a| = 0 \;; \qquad |G_a,P_o| = P_a \;; \\
&|G_a,P_b| = \delta_{ab}M \;; \quad |M,J_{ab}| = |M,P_o| = |M,P_a| = |M,G_a|=0 \\
&(a,b,c,d = 1,2,\ldots,n \;;\; n>3\,).
\end{aligned}$$

Let $\tilde{U}$ be its universal enveloping algebra, $Z\tilde{U}$ the centre of $\tilde{U}$.

Theorem 2. Let $m = |n/2|$, $C_1,C_2,\ldots,C_m$ a system of homogeneous algebraically independent invariant operators of U((n,R)), ${}_{ab} = MJ_{ab}-(P_aG_b-P_bG_a)$, C_j - the element of $\tilde{U}$ obtained from C_j replacing J_{ab} by Γ_{ab}. A maximal system of algebraically independent elements of $Z\tilde{U}$ is formed by the elements

$$M, \tilde{G}_o = 2MP_o - \delta^{ab}P_aP^b, \tilde{C}_1, \ldots, \tilde{C}_m .$$

Proof, From (3) we get

$$|P_o, \Gamma_{bc}| = |P_a, \Gamma_{bc}| = |G_a, \Gamma_{bc}| = 0, \Gamma_{ba} = -\Gamma_{ab},$$

$$|J_{ab}, \Gamma_{cd}| = \delta_{ad}\Gamma_{bc} + \delta_{bc}\Gamma_{ad} - \delta_{ac}\Gamma_{bd} - \delta_{bd}\Gamma_{ac} \quad (4)$$

$$|\Gamma_{ab}, \Gamma_{cd}| = M(\delta_{ad}\Gamma_{bc} + \delta_{bc}\Gamma_{ad} - \delta_{bd}\Gamma_{ac} - \delta_{ac}\Gamma_{bd} .$$

Let

$$\hat{P}_a = M^{-1}P_a, \hat{J}_{ab} = M^{-1}\Gamma_{ab} = J_{ab} - (\hat{P}_aG_b - \hat{P}_bG_a),$$

$\tilde{Q}$ the R-subalgebra of (AG(n)) generated by 1, P_o, $\hat{P}_a, G_b, M, \hat{J}_{ab}$. Since by (4) the elements $\hat{J}_{ab}(a,b=1,\ldots,n)$ generate (n, R) and

$$|G_a, \hat{P}_b| = \delta_{ab}, |\hat{P}_a, \hat{J}_{cd}| = 0, |G_a, \hat{J}_{cd}| = 0 ,$$

$$|G_a, P_o| = M\hat{P}_a ,$$

then

$$Q \tilde{=} Q_1 \times_R Q_2 \times_R Q_3 \times_R U\ (AO(n, R)) \quad (5)$$

where R - algebras Q_i are generated by 1 and M; 1 and $2P_o - M\hat{P}_a\hat{P}^a$; 1, $\hat{P}_a, G_b$ (a,b=1,...,n) for i=1,2,3. Since is the Weyl algebra, its centre coincides with R1. From (5) and what is known of ZU((n, R)) we conclude that tranc. deg (ZQ) = m+2 and $M, 2P_o - M\hat{P}_a\hat{P}^a$, C_j (j = 1,2,...,m) are algebraically independent.

Since $\tilde{U}$ Q and M Z(Q), then $Z(\tilde{U})$ Z(Q). Hence there are no more than m+2 algebraically independent elements in $Z\tilde{U}$.

At the same time $M, M(2P_o - M\hat{P}_a\hat{P}^a)$, $M^{t_j}C_j$ where t_j is the homogeneity degree of C_j $(j = 1,2,\ldots,m)$ belong to $\tilde{Z}U$ and are algebraically independent. The theorem is proved.

4. <u>Casimir operators of the Galilei group.</u> The commutation relations for the basis elements of the Galilei algebra $AG(n)$ we get from (3), putting $M=0$. We preserve the same notations. Let C_{ij}^k $(i,j,k = 1,\ldots s)$ be structure constants of a Lie algebra L,

$$B_{ij} = \sum_{k=1}^{s} C_{ij}^k z_k, \quad B = (B_{ij}), \quad r(L) = \sup_{(z_1,\ldots,z_s)} \operatorname{rank} B$$

From 1 and the results on the number of fundamental solutions of the 1st order partial differential equations we see that the trans. deg $Z(U(L)) < \dim L - r(L)$.

<u>Lemma</u>. $r\,(AG(n)) > \frac{1}{2}\,(n^2+3n) - |\,n/2\,|$.

Proof. Arrange the basis elements of $AG(n)$ as follows

$$P_0, P_1, \ldots, P_{n-1}\,;\; G_1, \ldots, G_{n-1};\; J_{12}, \ldots, J_{1,n-1};$$

$$J_{23}, \ldots, J_{2,n-1}, \ldots, J_{n-2,n-1};\; P_n, G_n, J_{1n}, J_{2n}, \ldots, J_{n-1,n}$$

and assign to them real variables

$$x_0, x_1, \ldots, x_{n-1};\; y_1, \ldots, y_{n-1};\; z_{12}, \ldots, z_{1,n-1};\; z_{23}, \ldots,$$

$$z_{2,n-1}, \ldots, z_{n-2,n-1};\; x_n, y_n, z_{1n}, z_{2n}, \ldots, z_{n-1,n}\ .$$

Put $x_1 = x_2 = \ldots = x_{n-1} = 0$, $x_n = 1$, $y_n = 0$, $z_{in} = 0$ $(i=1,\ldots,n-1)$. The matrix B is of the form

$$\left(\begin{array}{c|c|c} 0 & 0 & \begin{matrix} 0 & -1 & & 0 \\ & 1 & \ddots & \\ 0 & & \ddots & 1 \\ & & & 0 \end{matrix} \\ \hline 0 & B' & 0 \\ \hline \begin{matrix} 0 & & & \\ 1 & 0 & & 0 \\ & -1 & \ddots & \\ 0 & & -1 & 0 \end{matrix} & 0 & \begin{matrix} 0 & \cdots & 0 \\ \vdots & \mathcal{D} & \\ 0 & & \end{matrix} \end{array}\right)$$

(the first block row and the last block row each consist of n rows)

where B' corresponds to $AE(n-1)$, generated by $G_1,\dots,G_{n-1}, J_{12},\dots,J_{n-2,n-1}$. It is known that $r(AE(n-1))=n(n-1)/2$. Therefore $r(AG(n))$ $n(n-1)/2+2n = = n(n+3)/2$.
Lemma is proved.

Theorem 3. Let $m=|n/2|$, $K_{ab} = P_aG_b-P_bG_a$,

$$W_{a_1\dots a_{n-2}} = \varepsilon_{a_1\dots a_n} K^{a_{n-1}a_n}$$

$$W_{a_1\dots a_r} = W_{a_1\dots a_r a_{r+1} a_{r+2}} K^{a_{r+1}a_{r+2}} \quad (r< n-4),$$

$$W_{a_1} = W_{a_1a_2a_3}K^{a_2a_3} \quad \text{for } n \text{ odd},$$

$$W = W_{a_1a_2}K^{a_1a_2} \quad \text{for } n \text{ even}$$

The elements

$$T_o = P_aP^a,\ T_1 = W_{a_1\dots a_{n-2}}W^{a_1\dots a_{n-2}},\dots,\ T_m = = W_{a_1}W^{a_1} \ (\text{or } W)$$

form a maximal system of algebraically independent invariant operators of $AG(n)$.

This theorem is proved using Lemma and by the same arguments as Theorem 2.

Statement 1. Let $m=|n/2|$, $n>3$, $C_1,\dots,C_m$ - homogeneous generators of the centre $U(AO(n, R))$, $\tilde{C}_j$ the element of $U(A\tilde{G}(n))$ obtained from C_j replacing J_{ab} by Γ_{ab}. The elements

$$M, \tilde{C}_o = 2MP_o - \delta^{ab}P_aP_b,\ \tilde{C}_1,\dots,\tilde{C}_m$$

form a system of generators of $Z(U(A\tilde{G}(n)))$.

Proof. Denote $<M>$ the two-sided ideal generated by M. It is known that $U(A\tilde{G}(n))/<M> \cong U(AG(n))$. Identifying the corresponding elements we may assume that a

$\overline{P}_a = P_a + \langle M \rangle$, $\overline{G}_b = G_b + \langle M \rangle$, $\overline{P}_o = P_o + \langle M \rangle$, $\overline{J}_{ab} =$

$= J_{ab} + \langle M \rangle$ form a basis of $AG(n)$. Let $\overline{C}_j$ be the element of $U(AG(n))$ obtained from C_j replacing J_{ab} by $K_{ab} = \overline{P}_a \overline{G}_b - \overline{P}_b \overline{G}_a$. Then $\overline{C}_j = \tilde{C}_j + \langle M \rangle$ and

$$\delta^{ab}\overline{P}_a \overline{P}_b \text{ , } \overline{C}_1, \ldots, \overline{C}_m \quad U(AG(n))) \tag{7}$$

Since $T_1, T_2, \ldots, T_m$ constructed in Theorem 3 may be obtained from Casimir operators of $AO(n, R)$ replacing J_{ab} by K_{ab}, we have $T_j = f_j\ (\overline{C}_1, \ldots, \overline{C}_m)$ $(j+1, \ldots, m)$. Hence, since $T_o, T_1, \ldots, T_m$ are algebraically independent then so are (7).

The arguments in the proof of Theorem 2 imply that every element $Z(U(A\tilde{G}(n)))$ can be expressed in the form $M^{-k}f(M, \tilde{C}_o, \tilde{C}_1, \ldots, \tilde{C}_m)$. If elements (6) had not generated $Z(U(A\tilde{G}(n)))$ then for some nonzero polynomial with real coefficients $(x_o, x_1, \ldots, x_m)$ we would have had M^{-1} $(\tilde{C}_o, \tilde{C}_1, \ldots, \tilde{C}_m)$ $ZU(A\tilde{G}(n))$. Therefore $(\tilde{C}_o, \tilde{C}_1, \ldots, \tilde{C}_m)$ $\langle M \rangle$ and $(\delta^{ab}\overline{P}_a\overline{P}_b, \overline{C}_1, \ldots, \overline{C}_m) = 0$. The contradiction proves Statement.

Statement 2. Let $m = |n/2|$, $n > 3$, $\overline{C}$ an element of $U(AG(n))$ obtained from C $U(AO(n, R))$ replacing J_{ab} by $P_aG_b - P_bG_a$. For every system of algebraically independent homogeneous Casimir operators $C_1, \ldots, C_m$ of $O(n, R)$ the elements $\delta^{ab}P_aP_b, \overline{C}_1, \ldots, \overline{C}_m$ form a maximal system of algebraically independent Casimir operators of $G(n)$ for $n = 3,4,5,6$.

References

1. Gelfand,I.M. (1950). Mat. Sbornik, 26, 103.

2. Barut,A., Raczka,P. Representations theory of groups and its applications.

3. Fushchich,W.I., Krivsky,I.Yu. (1969). Nucl. Phys. B14, 321.

4. Fushchich,W.I. (1970). Theor. Math. Phys. 4, 360 (Russian).

5. Kadyshevsky,V.G., Todorov,I.T. (1966).Yad. fiz.(Soviet Nucl. Phys.), 3, 135 (Russian),

6. Mirman,R. (1968). J. Math. Phys. 9, 39.

7. Patera,J., Sharp,R.T., Winternitz,P., Zassenhaus,H. (1976). J. Math. Phys., 17, 977.

8. Burdel,G., Patera,J., Perrin,M., Winternitz,P. (1978). J. Math. Phys., 19, 1758.

9. Demichev,A.P., Nelipa, N.F. (1980). Vestnik MGU. Ser. phys. astron., 21, N 2, 3; N 2, 7; 21, N 4, 23 (Russian).

10. Demichev,A.P., Nelipa,N.F., Chaichian,M. (1980). Vestnik MGU, Ser. phys. astron., 21, N 4, 27; 21, N 5, 20 (Russian).

11. Perroud,M. (1983). J. Math. Phys., 24, 1381.

Part III
GAUGE THEORIES

CONFORMAL INVARIANCE IN QUANTUM NONABELIAN GAUGE FIELD THEORY

M.Ya.Palchik

Institute of Automation and Electrometry, Novosibirsk

1. Conformal symmetry in gauge theories. Fradkin and the author have shown in [1,2] that the generating functional for gauge theories (in a Euclidean space):

$$Z = \frac{1}{N} \int dA^a_\mu \det|\partial\nabla| \exp\{\int dx [\frac{1}{2\alpha}(\partial_\mu A^a_\mu)^2 - \frac{1}{4}(F^a_{\mu\nu})^2]\} \qquad (1.1)$$

is invariant under the nonlinear infinitesimal transformation (see also [3])

$$A_\mu(x) \rightarrow A'_\mu(x) = A_\mu(x) + \varepsilon_\rho K_\rho A_\mu(x) - $$

$$- 4\varepsilon_\rho \nabla_\mu \frac{1}{\partial\nabla} A_\rho(x) \qquad (1.2)$$

where ε_ρ is a small parameter and $\nabla_\mu \equiv \nabla^{ab}_\mu = \delta^{ab}\partial_\mu + t^{ab}_c A^c_\mu(x)$; This transformation is a combination of the special conformal transformation

$$A_\mu(x) \rightarrow A'_\mu(x) = A_\mu(x) + \varepsilon_\rho K_\rho A_\mu(x) \qquad (1.3)$$

where

$$K_\rho A_\mu(x) = (x^2\partial_\rho - 2x_\rho x_\tau \partial_\tau - 2d_A x_\rho) A_\mu(x) + 2x_\mu A_\rho(x) -$$

$$- 2\delta_{\mu\rho} x_\tau A_\tau(x) \qquad (1.4)$$

and the field-dependent gauge transformation $\delta A_\mu(x) = -\nabla_\mu \omega(x)$ where

$$\omega(x) = 4\varepsilon_\rho \frac{1}{\partial\nabla} A_\rho(x)$$

(The functional (1.1) is not invariant under the linear transformation (1.3) because of the factor $\det|\partial\nabla| \times \exp[\int dx \frac{1}{2\alpha}(\partial_\mu A_\mu)^2]$).

A symmetry of this type is present also in conformal gravity and in supersymmetric gauge theories. It is a common property of all the gauge theories and it is interesting to investigate it.

A conformal solution in quantum nonabelian gauge theories is possible either for fixed values of the charge, zeroes of $\beta(g)$, or for $\beta(g) \equiv 0$, as for example in N = 4 supersymmetric Yang-Mills theory [4] and in N = 4 conformal supergravity [5]. The Study of conformal solutions is of particular interest in connection with the confinement problem.

The following results are obtained in this paper:

1. It is shown that transformation (1.2) defines a nonlinear realization of conformal algebra (n°2). Only the canonical value $d_A = 1$ is admissible for the scale dimension, otherwise the algebra is not closed.

2. The conformal Ward identities are considered and conformal invariant propagators $\langle 0|A_\mu A_\nu|0\rangle$, $\langle 0|F_{\mu\nu}A_\tau|0\rangle$ and $\langle 0|F_{\mu\nu}F_{\sigma\tau}|0\rangle$ are obtained.

3. The transformations of the ghost fields are derived (n°4).

4. The conformal bootstrap method for calculating the dimension and coupling constants for $F_{\mu\nu}$ are discussed (see also 2 for more details), n 5.

The transformation law (1.2) in the abelian case becomes linear and agrees with the law obtained earlier in [6-8], where conformal QED was investigated from other considerations.

2. Nonlinear realization of conformal algebra. Let us show that (1.2) defines a nonlinear realization of special conformal transformations if in the latter we put

$$d_A = 1 \tag{2.1}$$

To prove this check that the nonlinear operators defined

as

$$\tilde{K}_\rho A_\mu(x) = K_\rho A_\mu(x) - 4\varepsilon_\rho \nabla_\mu \frac{1}{\partial\nabla} A_\rho(x) \tag{2.2}$$

satisfy the commutation relations of the algebra of the conformal group:

$$\begin{aligned} &[D, \tilde{K}_\rho] = \tilde{K}_\rho, \ [\tilde{K}_\rho, \tilde{K}_\tau] = 0, \\ &[\tilde{K}_\rho, M_{\tau\nu}] = \delta_{\rho\tau}\tilde{K}_\nu - \delta_{\rho\nu}\tilde{K}_\tau, \\ &[\tilde{K}_\rho, P_\tau] = 2(\delta_{\rho\tau} D + M_{\rho\tau}) \end{aligned} \tag{2.3}$$

Here $D = (d_A + x_\nu\partial_\nu)$, $M_{\tau\nu} = x_\tau\partial_\nu - x_\nu\partial_\tau + \Sigma_{\tau\nu}$ and $P_\tau = \partial_\tau$ are the usual generators of dilatations, Lorentz rotations and translations respectively. They remain unchanged in a nonlinear realization. It is slightly more difficult that in the usual case to calculate the commutators in (2.3) because of nonlinearity of the generators $\tilde{K}_\rho$.

Consider e.g. the first commutator in (2.3). It is conveniant to introduce the infinitesimal transformations $U(\lambda) = 1 + \lambda D$, where $\lambda \ll 1$, and $U(\varepsilon) = 1 + \varepsilon_\rho K_\rho$. Let us represent the $U(\lambda)$-action as follows:

$$U(\lambda)A_\mu(x) = A_\mu(x+\lambda x) + \lambda d_A A_\mu(x) \tag{2.4}$$

For small λ one has: $A_\mu(x+\lambda x) = (1 + \lambda x_\rho\partial_\rho)A_\mu(x)$. Similarly, define the $\tilde{U}(\varepsilon)$-action:

$$\tilde{U}(\varepsilon)A_\mu(x) = U(\varepsilon)A_\mu(x) - 4\varepsilon_\rho\nabla_\mu \frac{1}{\partial\nabla} A_\rho(x) \tag{2.5}$$

Here $U(\varepsilon) = 1 + \varepsilon_\rho K_\rho$ is the linear part of $\tilde{U}(\varepsilon)$

$$U(\varepsilon)A_\mu(x) = A_\mu(x^\varepsilon) - 2\varepsilon_\rho x_\rho A_\mu(x) + 2\varepsilon_\rho x_\mu A_\rho(x) - $$

$$- 2\varepsilon_\mu x_\rho A_\rho(x)$$

with

$$(x^\varepsilon)_\nu = \frac{x_\nu + \varepsilon_\nu x^2}{1+2\varepsilon x+\varepsilon^2 x^2} = x_\nu + \varepsilon_\nu x^2 - 2\varepsilon_\tau x_\tau x_\nu$$

The commutator $[D, \tilde{K}_\rho]$ is determined by the equality

$$\lambda\varepsilon_\rho [D, \tilde{K}_\rho] A_\mu(x) = U(\lambda)\tilde{U}(\varepsilon)A_\mu(x) - \\ - \tilde{U}(\varepsilon)U(\lambda)A_\mu(x). \qquad (2.6)$$

Therefore to check the first of the relations (2.3) it is necessary to prove that

$$U(\lambda)\tilde{U}(\varepsilon)A_\mu(x) - \tilde{U}(\varepsilon)U(\lambda)A_\mu(x) = \lambda\varepsilon_\rho\tilde{K}_\rho A_\mu(x). \qquad (2.7)$$

The linear operator $U(\varepsilon)$ satisfies this relation. In remains to verify that

$$U(\lambda)\nabla_\mu \frac{1}{\partial\nabla} A_\rho(x) - \nabla^B_\mu \frac{1}{\partial\nabla^B} B_\rho(x) = \lambda\nabla_\mu \frac{1}{\partial\nabla} A_\rho(x) \qquad (2.8)$$

where $B_\rho(x) = A_\rho(x) + \lambda(d_A + x_\nu\partial_\nu)A_\rho(x)$. From (3.4) we have

$$U(\lambda)\nabla_\mu \frac{1}{\partial\nabla} A_\rho(x) = \nabla'_\mu \frac{1}{\partial'\nabla'} A_\rho(x') + \lambda d_A\nabla_\mu \frac{1}{\partial\nabla} A_\rho(x)$$

where $x'_\nu = x_\nu + \lambda x_\nu$, $\partial'_\nu = \partial/\partial x'_\nu$, $\nabla'_\mu = \partial'_\mu + A_\mu(x') =$

$= (1+\lambda)\nabla_\mu - \lambda(1+x_\nu\partial_\nu) A_\mu(x)$. Keeping in equation (2.8) only the coefficients of λ we find that this equation holds if $_A = 1$. Thus (2.7), and consequently the first of the r ıtions (2.3), is proved. The proof of the remaining relations in (2.3) is similar.

The above calculations show that in the present nonlinear realization of conformal algebra the scale dimension of a gauge field can take on only the canonical value (2.1). This property is proved for classical fields. In the quantum case one might expect the appearance of anomalous dimensions d_A, as nonlinear terms in (2.2) contain products of quantum fields at coinsident points and are therefore ambiguous. Indeed, as shown in [3] a solution of the invariance conditions for the propagator

$\langle o|A_\mu\ A_\nu\ |o\rangle$ exists for any d_A. However, in the articles [1,2] arguments are presented that the value $d_A = 1$ is a consequence of the quantum Yang-Mills equations.

It is also shown there that the field equations admit anomalous dimension d_F of the tensor $F_{\mu\nu}$.(This is only true for nonabelian fields; in the abelian case $d_F = 2$.) The infinitesimal transformation for $F_{\mu\nu}$ in the nonlinear realization is of the form

$$\delta F_{\mu\nu}(x) = \varepsilon_\rho \tilde{K}_\rho F_{\mu\nu}(x) = \varepsilon_\rho K_\rho F_{\mu\nu}(x) + 4\varepsilon_\rho\left[\frac{1}{\partial\nabla} A_\rho(x), F_{\mu\nu}(x)\right] \tag{2.9}$$

where K_ρ is the generator of a linear transformation of $F_{\mu\nu}$ with anomalous dimension d_F. It can be shown that the nonlinear generator $\tilde{K}_\rho$ defined in (2.9), also satisfies the commutation relations (2.3). This does not impose any constraints on the dimension d_F, as $F_{\mu\nu}$ lineary enters (2.9).

3. Conformal Ward identities. Invariant propagators.

Introdusing the sources of the fields A_μ and $F_{\mu\nu}$ in the original functional (1.1) and differentiating the transformed functional with respect to ε_λ one gets (nonlinear) equations for the Green functions which express their invariance with respect to nonlinear transformations (1.2) and (2.9). Here we confine ourselves to the analysis of these equations for 2-point Green functions and find their expli cit expressions.

Consider the propagator

$$D_{\mu\nu}(x_{12}) = \langle o|A_\mu(x_1)\ A_\nu(x_2)|o\rangle$$

The invariance condition is

$$\langle o|\delta A_\mu(x_1)\ A_\nu(x_2)|o\rangle + \langle o|A_\mu(x_1)\ \delta A_\nu(x_2)|o\rangle = 0$$

where δA_μ is taken from (1.2). Substituting the explicit expression set

$$(K_\lambda^{x_1} + K_\lambda^{x_2})\, D_{\mu\nu}(x_{12}) - \tag{3.1}$$

$$-4\{\langle o|\nabla_\mu \frac{1}{\partial\nabla} A_\lambda(x_1) A_\nu(x_2)|o\rangle + \langle o|A_\mu(x_1)\nabla_\nu \frac{1}{\partial\nabla} A_\lambda(x_2)|o\rangle\} = 0,$$

where $K_\lambda^{x_1}$ acts on x_1 and the index μ, while $K_\lambda^{x_2}$ on x_2 and ν_2 by (1.4). Unlike the usual invariance condition this equation is nonlinear. The terms in bracket represent the sum of highest Green functions in which all arguments except two coincide. This sum, however, can be expressed via $D_{\mu\nu}$ using the scale invariance. Indeed, the most general scale invariant expression for $D_{\mu\nu}$ is

$$D_{\mu\nu}(x_{12}) = A(\delta_{\mu\nu} - \frac{\partial_\mu \partial_\nu}{\Box}) \frac{1}{x_{12}^2} + B \frac{\partial_\mu \partial_\nu}{\Box} \frac{1}{x_{12}^2} \tag{3.2}$$

where A is a certain constant, and B is fixed by the Ward identity: $B = \alpha$. Consider the nonlinear terms

$$G_{\mu\lambda\nu}(x_{12}) = \langle o|\nabla_\mu \frac{1}{\partial\nabla} A_\lambda(x_1)\; A_\nu(x_2)|o\rangle \tag{3.3}$$

From scale invariance we have

$$G_{\mu\lambda\nu}(x_{12}) = (\tilde{A}_1 \delta_{\nu\lambda} \partial_\mu^{x_1} + \tilde{A}_2 \delta_{\mu\nu} \partial_\lambda^{x_1} + \tilde{A}_3 \delta_{\mu\lambda} \partial_\nu^{x_1} +$$

$$+ \tilde{A}_4 \frac{\partial_\mu^{x_1} \partial_\lambda^{x_1} \partial_\nu^{x_1}}{\Box_{x_1}}) \frac{1}{\Box_{x_1}} (x_{12}^2)^{-1} \tag{3.4}$$

where $\tilde{A}_i$ are certain constants. By definition (3.3) we have

$$\partial_\mu^{x_1} G_{\mu\lambda\nu}(x_{12}) = D_{\lambda\nu}(x_{12}) \tag{3.5}$$

implying $\tilde{A}_1 = A$. Redefining the coefficients in (3.4) we obtain

$$G_{\mu\lambda\nu}(x_{12}) \frac{\partial_\mu^{x_1}}{\Box_{x_1}} D_{\lambda\nu}(x_{12}) + A_1 \frac{\partial_\lambda^{x_1}}{\Box_{x_1}} D_{\mu\nu}(x_{12}) +$$

$$+ A_2 \frac{\partial_\nu^{x_1}}{\Box_{x_1}} D_{\mu\lambda}(x_{12}) + A_3 \frac{\partial_\mu^{x_1}\partial_\lambda^{x_1}\partial_\nu^{x_1}}{\Box^2_{x_1}} \frac{1}{x_{12}^2} \tag{3.6}$$

First consider the transverse gauge. In this case there is an additional condition

$$\partial_\nu^{x_2} G^{tr}_{\mu\lambda\nu}(x_{12}) = \partial_\nu^{x_2} D^{tr}_{\mu\nu}(x_{12}) = 0 \tag{3.7}$$

From this one can find

$$G_{\mu\lambda\nu}(x_{12}) = \frac{\partial_\mu^{x_1}}{\Box_{x_1}} D^{tr}_{\lambda\nu}(x_{12}) + A_1 \frac{\partial_\lambda^{x_1}}{\Box_{x_1}} D^{tr}_{\mu\nu}(x_{12}),$$

$$A_2 = A_3 = 0 \tag{3.8}$$

Substitute this to (3.1). The second term ($\sim A_1$) does not contribute to the sum of the terms in brackets

$$G^{tr}_{\mu\lambda\nu}(x_{12}) + G^{tr}_{\nu\lambda\mu}(x_{21}) = \\ = \frac{\partial_\mu^{x_1}}{\Box_{x_1}} D^{tr}_{\lambda\nu}(x_{12}) + \frac{\partial_\nu^{x_2}}{\Box_{x_2}} D^{tr}_{\lambda\mu}(x_{12}). \tag{3.9}$$

We get the linear integro-differential equation

$$(K_\lambda^{x_1} + K_\lambda^{x_2}) D_{\mu\nu}(x_{12}) - 4\left[\frac{\partial_\mu^{x_1}}{\Box_{x_1}} D_{\lambda\nu}(x_{12}) + \right. \\ \left. + \frac{\partial_\nu^{x_2}}{\Box_{x_2}} D_{\lambda\mu}(x_{12}) \right] = 0 \tag{3.10}$$

It can be easily varified that the first term of (3.2) is the transverse solution of this equation;

Consider the generalized α gauge. Additional constraint (3.7) is absent now and the constants A_2, A_3 remain free. We can show, however, that the second term in (3.2) also satisfies (3.10) (it suffices to note that

$$\frac{\partial^{x_1}_{\mu}}{\Box_{x_1}} D^1_{\lambda\nu}(x_{12}) + \frac{\partial^{x_2}_{\nu}}{\Box_{x_2}} D^1_{\lambda\mu}(x_{12}) = 0,$$ where

$D^1_{\mu\nu}(x) \sim \frac{\partial_\mu \partial_\nu}{\Box} \frac{1}{x^2}$, and equation (3.10) reduces to the usual condition of the conformal invariance). It follows that equations (3.8) fits the α gauge as well, only now the condition $A_2 = 0$ appears to be a consequence of equation (3.1).

Analogously from the other Ward identities

$$\langle o|F_{\mu\nu}(x_1)\delta A_\tau(x_2)|o\rangle + \langle o|\delta F_{\mu\nu}(x_1)A_\tau(x_2)|o\rangle = 0$$

$$\langle o|F_{\mu\nu}(x_1)\delta F_{\sigma\tau}(x_2)|o\rangle + \langle o|\delta F_{\mu\nu}(x_{12})\, F_{\sigma\tau}(x_2)|o\rangle = 0$$

where δA_μ and $\delta F_{\mu\nu}$ are from (1.2) and (2.9), we get [2]:

$$\langle o|F_{\mu\nu}(x_1)A_\tau(x_2)|o\rangle = C(\delta_{\mu\tau}\partial_\nu - \delta_{\nu\tau}\partial_\mu)(x_{12}^2)^{-\frac{d_F}{2}} \quad (3.11)$$

$$\langle o|F_{\mu\nu}(x_1)F_{\sigma\tau}(x_2)|o\rangle = C_1\left[g_{\mu\sigma}(x_{12})g_{\nu\tau}(x_{12}) - \right.$$

$$\left. - g_{\mu\tau}(x_{12})g_{\nu\sigma}(x_{12})\right](x_{12}^2)^{-d_F} +$$

$$+ C_2(\delta_{\mu\sigma}\delta_{\nu\tau} - \delta_{\mu\tau}\delta_{\nu\sigma})(x_{12}^2)^{-d_F},$$

where $g_{\mu\nu}(x) = \delta_{\mu\nu} - 2\frac{x_\mu x_\nu}{x^2}$; C_1, C_1 and C_2 are certain constants.

4. Transformations of ghost fields. A transformation law of type (1.2) for ghost fields can be derived in the following way. Consider the BRST-transformations

$$\delta A_\mu(x) = - \nabla_\mu C(x)\varepsilon\,, \quad \delta C^a(x) = \frac{1}{2} t^{abd} C^b(x)\, C^d(c)\varepsilon\,, \quad (4.1)$$

$$\delta\bar{C}^a(x) = -\frac{1}{\alpha}\partial_\mu A^a_\mu(x)\mathcal{E},$$

Introduce combined transformations of ghost fields, consisting of the linear conformal transformations $\delta C(x) = \varepsilon_\rho K_\rho C(x)$, $\delta\bar{C}(x) = \varepsilon_\rho K_\rho \bar{C}(x)$ and BRST transformations (4.1). Choose the anticommuting parameter ε in (4.1) so that the BRST transformation of A_μ is equivalent to the nonlinear gauge transformation in (1.2). Then

$$4\varepsilon_\rho \frac{1}{\partial\nabla}A_\rho(x) = C(x)\varepsilon.$$

Substituting $\frac{1}{\partial\nabla}\ \delta(x-y) \to C(x)\bar{C}(y)$ we find:

$$\varepsilon = 4\varepsilon_\rho \int dy\ \bar{C}^a(y)\ A^a_\rho(y) \tag{4.2}$$

(Such BRST transformations with a field-dependent parameter $\mathcal{E}$ were considered in [9] and [10]. Performing these nonlinear transformations one should take into account changes of the integration measure in the functional integral.) The transformation law in (4.1) takes the form

$$A_\mu(x) \to A_\mu(x) + \varepsilon_\rho K_\rho A_\mu(x) - 4\varepsilon_\rho \nabla_\mu\ C(x)\int dy\bar{C}^b(y) \times A^b_\rho(y) \tag{4.3}$$

and for ghost fields we get

$$C^a(x) \to C^a(x) + \varepsilon_\rho\ K_\rho\ C^a(x) + 2\varepsilon_\rho t^{abd}C^b(x)\int dy\ \bar{C}^f(y)\times A^f_\rho(y) \tag{4.4}$$

$$\bar{C}^a(x) \to \bar{C}^a(x) + \varepsilon_\rho K_\rho \bar{C}^a(x) - \frac{4}{\alpha}\varepsilon_\rho \partial_\mu A^a_\mu(x)\int dy A^b_\rho(y)\ \bar{C}^b(y) \tag{4.5}$$

Under linear conformal transformations the fields C and $\bar{C}$ transform as conformal scalars with scale dimensions d_C and $d_{\bar{C}}$. Since C and $\bar{C}$ are not complex conjugates, the dimensions d_c and $d_{\bar{c}}$ may be different. The only restriction $d_c + d_{\bar{c}} = 2$ follows, as usual, from the scale invariance of the ghost action. Fradkin and the author in [1,2] have shown that the linear conformal transformations and the nonlinear BRST transformations compensate each other only if $d_c = 0$, $d_{\bar{c}} = 2$ and

$$K_\rho C(x) = (x^2 \partial_\rho - 2x_\rho x_\tau \partial_\tau)\, C(x)$$

$$K_\rho \bar{C}(x) = (x^2 \partial_\rho - 2x_\rho x_\tau \partial_\tau - 4x_\rho)\, \bar{C}(x) \tag{4.6}$$

The generating functional (1.1) rewritten in terms of ghost fields is invariant under (4.3) - (4.6).

5. The calculating field dimensions and "coupling" constants by the conformal bootstrap method. (A more detailed descussion of the conformal bootstrap for the nonabelian theory is given in [2].) Traditional formulation of the conformal bootstrap based on skeleton expansions requires a refinement in case of the Yang-Mills theory. The fact is that skeleton graphs are not invariant with respect to gauge or special conformal transformations taken separately. Whether they are invariant with respect to combined transformations (1.2) depends on the choice of vertex functions. In order to find them one should solve conformal Ward identities for 3-point Green functions. Because of the nonlinearity of transformation (1.2) these identities contain nonlinear terms which should be calculated in the same order in which the skeleton bootstrap equations are solved. A successive embodyment of this program is the subject of a separate work. Here we confine ourselves to discussing another, simple version of the bootstrap program.

Write down the skeleton equations in the first-order formalism in terms of the vertices Γ_{AAA}, Γ_{FAA} and the propagators D^{AA} and D^{FF} . This means that all insertions that make contribution to the Green function $<o|F_{\mu\nu} A_\tau|o>$ should be included in the corresponding vertices. Assume that

$$<o|F_{\mu\nu}(x_1)\ A_\tau(x_2)|o> = 0 \ , \text{ but } \Gamma_{AAA} \neq 0 \qquad (5.1)$$

(thus suppose that the constant C in (3.11), which in general should be calculated from the bootstrap equations, is O). Note, that due to the absence of bare vertices for Γ_{AAA} in the first-order formalizm, the existence of the latter vertex is a nonpertubative effect. It appears as a solution of homogeneous (without a bare term) equations. Assume that

$$<o|F^a_{\mu\nu}(x_1)A^b_\rho(x_2)A^c_\tau(x_3)|o> = t^{abc}P^{x_2}_{\lambda\rho}\ P^{x_3}_{\tau\sigma}\ \{g^F_1\ (x^2_{12}x^2_{13})^{-1} \times$$

$$\times\ [g_{\mu\rho}(x_{12})\ g_{\nu\tau}(x_{13}) - g_{\nu\rho}(x_{12}) \cdot \qquad (5.2)$$

$$\cdot\, g_{\mu\tau}(x_{13})] \ + g^F_2 \left[(x^2_{12})^{-1}\ g_{\mu\rho}(x_{12})\lambda^{x_1}_\nu\ (x_2x_3)\lambda^{x_3}_\tau(x_1x_2) \ + \right.$$

$$+\ (x^2_{13})^{-1}\ g_{\nu\rho}(x_{13})\lambda^{x_1}_\mu\ (x_2x_3)\ \lambda^{x_2}_\tau\ (x_3x_1) \ -$$

$$\left.(\mu \leftrightarrow \nu)\right]\}\ \ \left(\frac{x^2_{23}}{x^2_{12}x^2_{13}}\right)^{\frac{d_F-2}{2}} ,$$

$$<o|A^a_{\mu_1}\ (x_1)\ A^b_{\mu_2}(x_2)\ A^c_{\mu_3}\ (x_3)|o> = t^{abc}g^A\ P^{x_1}_{\mu_1\nu_1}\ \ P^{x_2}_{\mu_2\nu_2}$$

$$P^{x_3}_{\mu_3\nu_3}\ \lambda^{x_1}_{\nu_1}\ \ (x_2x_3)\lambda^{x_2}_{\nu_2}\ (x_3x_1)\ \ \lambda^{x_3}_{\nu_3}\ (x_1x_2) \qquad (5.3)$$

where $P^x_{\mu\nu} = \delta_{\mu\nu} - \frac{\partial_\mu\partial_\nu}{\Box}$, $\lambda^{x_1}_\mu\ (x_2x_3) = -\frac{1}{2}\partial^{x_1}_\mu \ln\frac{x^2_{12}}{x^2_{13}}$, t^{abc} are structure constants and $g^F_{1,2}$ and G^A are "coupling constants" calculated from the bootstrap equations.

The author is grateful to professor E.S.Fradkin for numerous discussions.

References

1. Fradkin, E.S., Palchik, M.Ya. (1984).Phys.Lett. B 147 86.
2. Palchik, M.Ya., (1985). Yad. Fiz. 42, 522 (Russian J. Nucl. Phys.).
3. Fradkin, E.S., Palchik, M.Ya. (1985). Doklady Akad. Nauk SSSR 280, 79 (Russian).
4. Avdeev, L.V., Tarasov, O.V., Vladimirov, A.A. (1980). Phys. Lett. 96B, 94; Novikov, V.A., Shifman, M.A., Vainstein, A.I., Zakharov, V.I. (1983). Nucl. Phys. B229, 381.
5. Fradkin, E.S., Tseytlin, A.A. (1983). Preprint N 185, Lebedev Phys. Inst., Moscow; (1984). Phys. Lett. 134B, 307.
6. Palchik, M.Ya. (1983). J. Phys. A16, 1523.
7. Fradkin, E.S., Kozhevnikov, A.A., Palchik, M.Ya., Pomeransky, A.A. (1983). Commun. Math. Phys. 91, 529.
8. Kozhevnikov, A.A., Palchik, M.Ya., Pomeransky, A.A. (1983). Yad. Fiz. 37, 481 (Russian J. Nucl. Phys.).
9. Batalin, I.A., Vilkovsky, G.A. (1977). Phys.Lett. B69, 303.
10. Fradkin, E.S., Fradkina, T.E. (1978). Phys. Lett. B72, 343.

ON TOPOLOGICAL PROPERTIES OF ORBIT SPACE IN GAUGE FIELD THEORIES

M.A.Soloviev

Lebedev Physical Institute, Moscow

During recent years analytic aspects of Yang-Mills fields have been intensively studied and their relation to the topologic and geometric structure of the gauge theory has been investigated. Note in particular, [1-3] treating the removable point singularity problem. In all that works Sobolev spaces of gauge potentials and Sobolev groups of gauge transformations G^p_q are used, with parameters p, q so as to provide a Lie group structure. This range of the parameters is called the Sobolev range since its existence is a consequence of Sobolev's embedding theorems. However in elaborating e.g. the most weak conditions existence is a consequence of Sobolev's embedding theorems. However in elaborating e.g. the most weak conditions under which gauge potentials have integer topological numbers one has to consider parameters belonging to the boundary of Sobolev range where there is no Lie group structure. A clear picture of different functional classes of gauge transformations also is needed in constructing the Feynman-Kac integral on the orbit space [4] and in the Gribov ambiguity problem [5-7]. We consider here the groups G^p_q with arbitrary p, q. This enables us to reveal general topological properties of the orbit space independent of differential geometric structure. The consideration is based on exploiting the Gagliardo-Nirenberg inequalities and Rellich's theorem. The gauge group is assumed to be any compact one but is realized as a subgroup of SU(N). A domain of gauge potentials is $\mathbb{R}^n$ but the results and the arguments are preserved after its compactification.

Let $M(N, C)$ be the space of $N \times N$ matrices with the scalar product $\mathrm{tr}\, a b^{+}$, $\|\cdot\|_{p,U}$ the Lebesgue norm on an open set U. If $U = R^n$ then the index U is omitted, e.g. for derivatives of a gauge transformation we write

$$\|g^{(k)}\|_p = \left\{\int |g^{(k)}(x)|^p dx\right\}^{\frac{1}{p}}, \text{ where } |g^{(k)}|^2 = \sum_{|\varkappa|=k} \mathrm{tr} D^{\varkappa} g D^{\varkappa} g^{+},$$

$\varkappa$ is a multiindex of the partial derivative, $|\varkappa| = \varkappa_1 + \ldots + \varkappa_n$. The Sobolev space L^p_q consists of the elements with weak p-integrable derivatives up to order q. Since the gauge transformations are non-integrable at infinity we use the spaces $\hat{L}^p_q$ consisting of the functions with finite norm

$$\|f\|_{p,|x|<1} + \sum_{1 \le k \le q} \|f^{(k)}\|_p . \tag{1}$$

Note that the convergence in $\hat{L}^p_q$ implies the convergence in the norm $\|\cdot\|_{p,|x|<R}$ for any R [8]. Let G^p_q be the set of functions of $\hat{L}^p_q$ mapping R^n into M with values in the gauge group. The topology of G^p_q is induced by that of $\hat{L}^p_q$.

Theorem 1. ([8]). G^p_q is a topological group for any p, q and a Lie group for $p < n < pq$.

The condition $n < pq$ is that of the Sobolev range [1,2]. The additional restriction $p < n$ is due to the fact that in our case the domain of gauge potentials is non-compact unlike [1, 2].

Let $\mathcal{L}^p_{q-1}$ be the set of gauge potentials of $L^{pq} \cap L^p_{q-1}$ considered as an affine space with the norm

$$\|A-B\| = \|A-B\|_{pq} + \|A-B\|_p + \sum_{1 \le k \le q-1} \|(A-B)^{(k)}\|_p . \tag{2}$$

By the Sobolev theorem, we have $L^p_{q-1} \subset L^{pq}$ for $n \leq pq$. Therefore $\mathcal{L}^p_{q-1}$ coincides with the usual Sobolev space L^p_{q-1} for $n \leq pq$ or $q=1$. The border case $n = pq$ draws particular attention [1,3]. For $1 < q < n/p$ the space L^p_{q-1} is not G^p_q-invariant and $\mathcal{L}^p_{q-1}$ is adequate to this group.

Theorem 2. The operators $g \in G^p_q$ transform any $\mathcal{L}^p_{q-1}$ into itself. If $A, \tilde{A} \in \mathcal{L}^p_{q-1}$ are gauge equivalent, then every gauge transformation carrying A to $\tilde{A}$ belongs to G^p_q. If A, $\tilde{A}$ run bounded sets in $\mathcal{L}^p_{q-1}$ then these transformations form a bounded set.

A proof of this theorem in [8] is based on the Gagliardo-Nirenberg inequalities

$$\|g^{(k)}\|_r \leq C\|g\|_s^{1-\frac{k}{q}} \|g^{(q)}\|_p^{\frac{k}{q}} \quad \left(\frac{q}{r} = \frac{q-k}{s} + \frac{k}{p}\right) \tag{3}$$

$$\|A^{(k)}\|_r \leq C\|A\|_s^{1-\frac{k}{q-1}} \|A^{(q-1)}\|_p^{\frac{k}{q-1}} \left(\frac{q-1}{r} = \frac{q-1-k}{s} + \frac{k}{p}\right). \tag{4}$$

Since the gauge group is compact, the norm $\|g\|_\infty$ is finite and (3) implies $g^{(k)} \in L^{pq/k}$. Furthermore (4) with $s = pq$ implies $A^{(k)} \in L^{pq/(k+1)}$. So the additional restriction $\|A\|_{pq} < \infty$ ensures the same integrability properties of the derivatives of $A_\mu \in L^p_{q-1}$ as those of $g^+ \partial_\mu g$, $g \in G^p_q$. Note, that (4) with $r = s = p$ implies also that the norm (2) is equivalent to the norm $\|\cdot\|_{pq} + \|\cdot\|_p + \|\cdot^{(q-1)}\|_p$.

Theorem 3. ([9]). The G^p_q-action on $\mathcal{L}^p_{q-1}$ is continuous and proper.

This theorem shows that every orbit is closed in $\mathcal{L}^p_{q-1}$ and the space of orbits is separated.

Theorem 4. The gauge orbit space $\mathcal{L}^p_{q-1}/G^p_q$ is a regular topological space.

Regularity is a stronger form of the separation property

means that for every closed set and a point outside it there exist neighbourhoods with empty intersection. It suffices to prove that the orbits of the gauge potentials contained in a closed ball $\|A_0 - A\| \leqslant \varepsilon$ form a closed set in $\mathcal{L}^p_{q-1}$. Let $B = \lim B_i$, $B_i = A^{g_i}$, $\|A_0 - A_i\| \leqslant \varepsilon$. Show that $B = A^g$, where $\|A_0 - A\| \leqslant \varepsilon$. Suppose that $p \neq 1$. Then $\mathcal{L}^p_{q-1}$ and $\hat{L}^p_q$ are reflexive and we can select a weakly convergent subsequences of the bounded sequences A_i, g_i. For brevity we use the same notation for them and let A, g be their limits. Then g is a gauge transformation. Indeed, by Rellich's theorem, $g_i \to g$ in the norm $\|\cdot\|_{p, |x| < R}$ for any R and we can select another subsequence tending to g almost everywhere. Hence $g(x)$ belongs to the gauge group almost everywhere since the group is closed. Furthermore, $\|A_0 - A\| \leqslant \varepsilon$ and we must only verify that

$$\partial_\mu g = g B_\mu - A_\mu g. \tag{5}$$

It suffices to show that (5) holds when integrated over any test function $\varphi \in \mathcal{D}$. Clearly, $\int g_i \partial_\mu \varphi \, dx \to \int g \partial_\mu \varphi \, dx$. Further,

$$\int (A^i_\mu g^i - A_\mu g) \varphi \, dx = \int A^i_\mu (g^i - g) \varphi \, dx + \int (A^i_\mu - A_\mu) g \varphi \, dx.$$

The second integral tends to zero due to the weak convergence of A_i and the first one due to the Rellich's theorem since $(g_i - g) \to 0$ for all the norms $\|\cdot\|_{p, |x| < R}$, $R < \infty$, implies convergence in the norms $\|\cdot\|_{z, |x| < R}$ with any $z < \infty$ as it is proved in [9] taking into account the uniform boundedness of $(g_i - g)$. Hence we can use Hölder's inequality determining z by $z^{-1} + p^{-1} = 1$. The same arguments are applicable to the sequence $g_i B_i$ which however tends to 0 even in the norm $\|\cdot\|_p$. Thus, for $p \neq 1$ the theorem is proved.

Corollary. The orbit space $\mathcal{L}^p_{q-1} / G^p_q$ $(p \neq 1)$ is metrizable.

Indeed, $\mathcal{L}^p_{q-1}$ is separable and so its regularity implies metrizability by the Urysohn-Tikhonoff theorem.

Now let $p=1$. First consider $\mathcal{L}^1_0 = L^1$. Define the distance between orbits $[A]$, $[B]$ by the formula

$$\rho([A],[B]) = \inf_g \|A - B^g\|_1 . \qquad (6)$$

Since the norm $\|\cdot\|_1$ is gauge invariant (6) is well defined. Indeed, (6) vanishes only for $[A]=[B]$ because the orbits are closed. It is essential here that this fact can be established without using reflexivity. The arguments are the same as in proving Theorem 4 but simpler since $A_i = A_0$. The Rellich theorem enables us to select a subsequence from g_i which converges in $\|\cdot\|_{p,|x|<R}$. Its limit belongs to G^p_q by Theorem 2.

A formula similar to (6) defines a metric for any L^p . For $p=2$, such a definition coincides with that in [10] but in [10] the strict positivity of ρ is proved differently. For $q \neq 1$ the situation is complicated by the fact that the norm (2) involving derivatives is not invariant under gauge transformations. This difficulty may be circumvent by assigning a proper norm to the tangent space at every point A. This norm is obtained from (2) by replacing $\partial_\mu \rightarrow \nabla_\mu(A)$. We omit also the intermediate derivatives in accordance with the above remark. We denote the new norm by $F(A,\cdot)$ emphasizing that the space of gauge potentials is thereby provided with a Finslerian structure. The length of a curve $A(t)$, $t \in [0,1]$ is defined as $\int_0^1 F(A,\dot{A})\,dt$, the distance d between A and B is the infimum of the lenghts over all the piecewise smooth curves joining these points and $\varrho([A], [B])$ is defined as $\inf d(A, B^g)$.

Theorem 5. The matric d determined by F is compatible with the original topology of $\mathcal{L}^p_{q-1}$ and ϱ is compatible with that of the orbit space.

Let us show that the new norms on tangent spaces are equivalent to the initial one. Denoting tangent vectors by τ we have

$$\|\nabla_{\nu_1} \ldots \nabla_{\nu_{q-1}} \tau_\mu\|_p \leqslant \|\partial_{\nu_1} \ldots \partial_{\nu_{q-1}} \tau_\mu\|_p + \|\Delta\|_p$$

where Δ is the sum of terms with derivatives of A and τ of orders 0 to $(q-2)$ as factors. By (4), the inequality $\|\partial_{\nu_1} \ldots \partial_{\nu_k} A_\nu\|_r \leqslant C\|A\|$ and an analogous inequality for τ hold for $r = pq/(k+1)$. If i labels the factors, then $\sum_i r_i^{-1} = p^{-1}$ for every summand in Δ and this enables us to apply the Hölder inequality to $\|\Delta\|_p$. We get

$$F(A, \tau) \leqslant C(1 + \|A\|)^{q-1}\|\tau\|. \tag{7}$$

This estimate shows not only finiteness but also continuity of $F(A, \tau)$ since the map $\mathcal{L}^p_{q-1} \times \ldots \times \mathcal{L}^p_{q-1} \to L^p$ assigning a particular summand in Δ to the collection $\{A_{\nu_1}, \ldots, A_{\nu_m}, \tau_\mu\}$ is polylinear and hence the map is bounded if and only if it is continuous. Thus the norm $F(A, \tau)$ is not stronger than the initial one. Let us show that it is not weaker, too. First, apply (4) to the factors A in Δ. This gives

$$\|\tau\| \leqslant CF(A, \tau) + C_0(1 + \|A\|)^{q-1}\|\tau\|_{pq} + \ldots + C_{q-2}(1 + \|A\|)\|\tau^{(q-2)}\|_{\frac{pq}{q-1}} \tag{8}$$

Now apply an additive form of (4) to $\tau^{(k)}$

$$\|\tau^{(k)}\|_{\frac{pq}{k+1}} \le C\left(h^{-\theta}\|\tau\|_{pq} + h^{1-\theta}\|\tau^{(q-1)}\|_p\right) \quad \left(\theta = \frac{k}{q-1}\right)$$

Here h is any positive number. Replacing $\|\tau\|_{pq}$ by $F(A,\tau)$, $\|\tau^{(q-1)}\|_p$ by $\|\tau\|$ and determining h_k by $C_k C(1+\|A\|)^{q-1-k}\, h_k^{1-\theta} = 1/2(q-2)$,

we obtain

$$\|\tau\| \le C(1+\|A\|)^{q-1} F(A,\tau) \tag{9}$$

From (7) it follows that the ball $\|A_0 - A\| < \delta$ is contained in the neighbourhood $d(A_0, A) < \varepsilon$ if δ is sufficiently small. Indeed, join A_0 and A by the line segment. Its length, $\int F(A,\tau)\,d\tau$ is majorized by $C(1+\|A_0\|+\delta)^{q-1}\delta$ and is less than ε under a suitable choice of δ. Conversely, show that $d(A_0, A) < \delta$ with sufficiently small δ implies $\|A_0 - A\| < \varepsilon$. Take B external to the ε-ball and join it with A_0 by a curve. By (9), the length of its part lying in the ball is greater than $\varepsilon / C(1+\|A_0\|+\varepsilon)^{q-1}$. Since the curve is arbitrary the distance $d(A_0, B)$ is estimated by the same number, and so δ can be taken to equal it. Theorem 5 is proved.

Besides topology, the metric d on $\mathcal{L}^p_{q-1}$ determines a uniform structure which need not be equivalent to the original one since the estimates (7), (9) are not uniform in A. Nevertheless each set bounded in the metric is also bounded in the original norm at least if its diameter is small enough and each d-fundamental sequence is $\|\cdot\|$-fundamental. (The converse is obvious.) To verify this, let us improve (9) using the inequality $d(0,A) \geqslant \|A\|_{pq}$. Substituting it in (4) gives

$$\|\partial_{\nu_1} \dots \partial_{\nu_k} A_\nu\|_{\frac{pq}{k+1}} \leqslant C d^{1-\frac{k}{q-1}} \|A\|^{\frac{k}{q-1}}$$

which leads to replacing $\|A\| \to d + \|A\|^{\frac{1}{q-1}}$ in (8) and results in

$$\|\tau\| \leqslant C'(1 + d^{q-1} + \|A\|) F(A,\tau) \tag{10}$$

instead of (9). Show that any set of diameter $\delta < 1/C'$ in d-metric is also bounded in the norm $\|\cdot\|$. Fix A_o in this set and suppose that for any R there is B lying outside the ball $\|A_o - A\| < R$ and such that $d(A_o, B) < \delta$. Then we immediately get contradiction because (10) gives

$$d(A_o,B) \geqslant \frac{R}{C'(1 + (d(0,A_o) + \delta)^{q-1} + \|A_o\| + R)} \tag{11}$$

where the right-hand side tends to $1/C'$ as $R \to \infty$. Let now A_i be fundamental in the metric d. We may assume that its diameter δ is $< 1/C'$ without loss of generality. Substituting $\|A_i - A_j\|$ for R in (11) enables us to estimate this norm by $d(A_i, A_j)$. Use the inequalities $d(0, A_i) \leqslant d(0, A_o) + \delta$, $\|A_i\| \leqslant \|A_o\| + R\delta$ and obtain a uniform estimate proving the fundamentality of A_i in the original norm.

Theorem 6. The gauge orbit space is complete in the metric determined by F.

Indeed, let $[A_i]$ be fundamental and i_k a sequence of indices such that $\rho([A_i],[A_j]) < 1/2^k$ for $i,j \geqslant i_k$. Select elements $\tilde{A}_k$ of the orbits

$[A_{i_k}]$ so that $d(\tilde{A}_k, \tilde{A}_{k+1}) < 1/2^k$. Then the sequence $\tilde{A}_k$ is fundamental, hence convergent since $\mathcal{L}^p_{q-1}$ is complete. Thus $[A_i]$ is convergent.

Remark. The most extensively employed Sobolev spaces are L^2_{q-1} with $2q > n$. They are Hilbert spaces and admit a gauge invariant Riemannian metric [4]. In the same manner as before one can investigate if the orbit spaces are complete in this metric. This is go if $n < 2$. However Theorem 6 shows that for any n there exists a metric which is not Riemannian but determines the same topology and in which L^2_{q-1} / G^2_q is complete.

The main result of this work is that a number of important topological properties are retained on the boundary of the Sobolev range in spite of the lack of Lie group structure. Moreover, they are retained outside this range as well, if the definition of Sobolev spaces is property modified.

The author is grateful to V.Ya.Fainberg for helpful discussions.

References

1. Uhlenbeck, K.K. (1982). Commun. Math. Phys., 83, 11.
2. Parker, T. (1982). Commun. Math. Phys., 85, 563.
3. Sibner, L.M., Sibner, R.J. (1984). Commun. Math. Phys., 93, 1.
4. Asorey, M., Mitter, P.K. (1981). Commun. Math. Phys., 80, 43.
5. Gribov, V.N. (1978). Nucl. Phys., B139, 1.
6. Singer, I.M. (1978). Commun. Math. Phys., 60, 1.
7. Soloviev, M.A. (1983). JETP Lett., 38, 504.
8. Soloviev, M.A. (1985). Lebedev Inst. Preprint N 60 (Russian).
9. Soloviev, M.A. (1985). Sov. Phys. - Lebedev Inst. Reports N 3.
10. Atiyah, M.F., Hitchin, N.J., Singer, I.M. (1978). Proc. R.Soc. London, A362, 425.

[illegible]

The author is grateful to [illegible] for helpful discussions.

References

1. [illegible] (1982), Commun. Math. Phys. [illegible]
2. [illegible] (1987), [illegible] Phys. [illegible]
3. [illegible] (1980), Commun. Math. Phys. [illegible] 1.
4. [illegible] (1981), Commun. Math. Phys. [illegible]
5. [illegible] (1981), [illegible] Phys. [illegible]
6. [illegible] (1976), Commun. Math. Phys. [illegible]
7. [illegible] (1983), JETP Lett., [illegible], 504.
8. [illegible] (1982), Lebedev Inst. Preprint N 60 (Russian).
9. [illegible]
10. [illegible] (1982), Proc. R. Soc. London, A382, [illegible].

STRING INTERACTION FROM THE PATH GROUP APPROACH

M.B.Mensky

State Standards Committee, Moscow

1. Introduction. In recent years the quantum string theory closely related to the contour approach in gauge theory, specifically in QCD, has become a topic of encreasing interest. In some works strings were supposed to be elementary extended entities, while in most cases they were introduced as a subsidiary instrument for treating ordinary matter, such as a gauge (specifically gluon) field. The interaction of quantum strings was usually understood as their topological rearrangement: cutting, fusion, closing etc. [1]. Sometimes special fields were supposed to transmit the string interaction, e.g. the second-rank antisymmetric tensor field [2].

The geometric method based on a connection in the space of loops was developed to investigate quantum strings and elaborate the contour approach in QCD [3-5]. The loops arise in this case as a way to describe gauge field, hence the curvature of this connection is zero.

In works of the author an algebraic method was developed for describing gauge field and gauge charged particles, making use of an induced representation of the so-called group of paths [6,7]. Later on the group of "paths in the space of paths" was proposed to describe quantum strings [7,8]. A path in the space of paths is called "2-path" to distinguish it from an ordinary path (say, in the Minkowski space). A 2-path is essentially an ordered surface in the original space (say, in the Minkowski space). The definition of the path group permits one to introduce a group operation (multiplication) for such surfaces. The resulting group of 2-paths enables one to develop the string theory completely by analogy with the theory of gauge charged particles. A nonlocal field presented by a functional of paths and therefore string-like, was predicted to transmit string interaction. This field was called a 2-gauge field [7,8]. The 2-gauge fields of topological nature were shown possible [8] leading to interference effects for strings of the Aharonov-Bohm type.

In the paper some detailes of this approach to the string theory are expounded. Specifically, the concepts of generalized gauge transformation (2-gauge transformation) and of generalized covariant derivative (2-covariant derivative) are introduced in the group-theoretical framework. The present group-theoretical approach based on the group of paths is closely related to the geometrical approach. Namely a 2-gauge field, defined as a representation of the group of 2-loops can be viewed as a connection in the space of paths. Unlike the case when a connection of this type was originated by a gauge field [3-5], a nonlocal 2-gauge field leads to a connection with non-zero curvature (the latter playing the role of strength for the 2-gauge field). Vanishing of the curvature (on the subspace of loops) for some concrete 2-gauge field means that this field reduces effectively to an ordinary gauge field reproducing the situation considered in [3-5].

The following plan is adopted in the paper. In n°2 an algebraic approach to gauge theory based on the group of paths is outlined. In n°3 the group of 2-paths is defined and the main principles of description of strings by presetations of this group are formulated. In n°4,5 there are given more detailed characteristics of the corresponding representations that characterize a 2-gauge field and its action on the string in detail. The generalized gauge transformation is analized in n$^{\circ}$6, and the special case of 2-gauge field presented by two local fields is considered in n$^{\circ}$7 (this special 2-gauge field should be identified with the generalized gauge field introduced by Nambu [2]). The short résumé is presented in n$^{\circ}$8.

2. <u>An alternative approach to gauge theory</u>. The group of paths may serve as a base for an alternative approach to gauge theory giving the group-theoretical interpretation of a covariant derivative. This approach contains gauge transformations as secondary and even not necessary element.

In order to come to gauge theory one neads the group of paths in the Minkowski space M. Denote this group by $P=P(M)$. Its element, a path $p=[x]\in P$ is a class of continuous curves in the Minkowski space, $\{x(\tau)\in M \mid 0\leqslant\tau\leqslant 1\}$, differing by parametrization, by general shifts, $x'(\tau)=x(\tau)+a$, and by inclusion of "appendices" (the latter being a curve passed in one direction and then backwards).

The key role in gauge theory is played by the subgroup of loops $L=L(M)\subset P(M)$ consisting of closed paths, $x(1)=x(0)$. It turns out [6,7] that each configuration of the (classical) gauge field is described by a representation $\ell\mapsto\alpha(\ell)$ of the subgroup L, while a quantum particle moving in this field is determined by the representation of P induced from this subgroup, i.e. by the representation $U_{\alpha}(P)=\alpha(L)\uparrow P$.

It is convenient to make use of ordered exponentials over curves

$$\hat{\alpha}(\{x\}) = P\exp\{i\int d\tau\, \dot{x}^{\mu}(\tau)\, A_{\mu}(x(\tau))\},$$

which form a representation of the groupoid of fixed paths [7]. (A fixed path $\hat{p}=p_{x}^{x'}=p_{x}$ is a class of continuous curves differing in parametrization and appendices. The source and the target of each fixed path are fixed, $x(0) = x$, $x(1) = x'$; Hence one can multiply two fixed paths only if the target of the first of them coincides with the source of the second one. In this sense fixed paths form a groupoid, not a group.) Denote this groupoid by $\hat{P}$. There exists a natural projection of $\hat{P}$ onto the group P. Conversly, a path $p \in P$ together with a point $x \in M$ determine a fixed path $p_x \in \hat{P}$. The ordered exponential (as can be shown) does not depend on the concrete curve, but rather on the class containing this curve. The product of the classes corresponds to the product of exponentials, $\hat{\alpha}(\hat{p}\hat{p}') = \hat{\alpha}(\hat{p})\,\hat{\alpha}(\hat{p}')$, so that the exponentials form a representation of $\hat{P}$.

If a fixed loop $\ell_0 \in \hat{P}$ with the source O is naturally associated with each loop $\ell \in L$, then the representation $\hat{\alpha}(L)$ discribing a gauge field may be defined by the formula $\alpha(\ell) = \hat{\alpha}(\ell_0)$. The induced representation $\alpha(L)\uparrow P$ acts in the space of functions $\psi(x)$ as follows:

$$(U_\alpha(p)\psi)(x) = \hat{\alpha}(p^x_{x'})\psi(x').$$

The operator $U_\alpha(p)$ can be expressed in terms of the covariant derivative $\nabla_\mu = \partial_\mu - iA_\mu$ in the form of an ordered exponential

$$U_\alpha(p) = P\exp\{-\int_p dx^\mu\, \nabla_\mu\},$$

thus the covariant derivative is a generator of the representation $U_\alpha(P)$. This is how the key elements of gauge theory arise in the path group representation theory.

One representation $\alpha(L)$, i.e. one gauge field, can be described by different potentials A_μ, A'_μ, i.e. by different representations $\hat{\alpha}(\hat{P})$, $\hat{\alpha}'(\hat{P})$ of the fixed path groupoid. A relation between different descriptions is just a gauge transformation [6,7]:

$$\psi' = V\psi \ , \quad A'_\mu = VA_\mu V^{-1} - i(\partial_\mu V)V^{-1}.$$

Hence, gauge transformations characterize ambiguity existing in local description of a gauge field or a gauge charged particle. The canonical realization of the induced representation $U_\alpha(P)$ however, (by left shifts of functions on the group) leads to an unambiguous description. It essentially coincides with path-dependent Mandelstam's formalizm, the latter thus acquiring a group-theoretical interpretation. A particle state is characterazed then by a function on the group of paths, $\Psi(p)$, satisfying a subsidiary, so-called structural condition,

$$\Psi(p\ell) = \alpha(\ell^{-1})\Psi(p), \text{ for any } p\in P, \ \ell\in L.$$

<u>3. The group of 2-paths and a quantum string</u>. A group of paths can be defined not only for paths in the Minkovski space (or any other linear space) but also for paths in any (topological) group [7]. A path $[g]$ in the group G is a class of continuous curves in this group, $\{g(\tau)\in G \mid 0\leq \tau \leq 1\}$, differing in a parametrization, in "appendices" and right shifts $g'(\tau) = g(\tau)g_0$. All such classes form the group $P(G)$ of paths in G. The subgroup $L(G)$ of loops, or closed paths, is singled out by condition $g(1) = g(0)$.

To construct the theory of strings in the framework of the present group-theoretical formalism take the group $G = P(M)$ for the original space M and then, in turn, define paths in G. We get $P(G) = P(P(M))$, The elements of $S = P(P(M))$ will be called the 2-paths in M, and the elements of $K = L(P(M))$ the 2-loops in M. Each 2-path is an oriented surface in M and 2-loops are closed oriented surfaces. The original space M may be the Minkowski space, or the 3-dimensional space, or a linear space of any other dimension, or even a Lie group space. The concrete choice of M determines the nature of a string under consideration.

In analogy with the above construction of gauge theory suppose that the induced representations $\chi(K)\uparrow S$ describe a quantum string, while the representations $\chi(K)$ of the subgroup of 2-loops describe a sort of field acting on the string, and, in a more general context, transmitting interaction between strings. This field generalizes a gauge field and hence will be called a 2-gauge field. The object described by a representation $\chi\uparrow S$ will be called a quantum string, because its states are presented by functions $\psi(p)$ in the space of paths. Probably it is more appropriate to call this object a 2-particle, the term "string" traditionally having somewhat more limited sense.

We saw above that a differential form $A = A_\mu(x)dx^\mu$ in M is needed to construct a representation of $P = P(M)$. This form determines the representation $\hat{\alpha}(\hat{P})$, then the representation $\alpha(L) = \hat{\alpha}(L_0)$ and the induced representation $\alpha(L)\uparrow P$, Let us apply the same scheme now for constructing the representation of $S = P(P(M))$. In this case one should evidently start from a 1-form in $P(M)$ and then take the following steps:

A 1-form in the space of paths $P(M)$

The representation of the groupoid of fixed 2-paths $\hat{S} = \hat{P}(P(M))$

The representation of the group of 2-loops $K = L(P(M))$

The induced representation of the group of 2-paths $S = P(P(M))$.

<u>4. Description of 2-gauge field</u>. A curve $\{x(\sigma) \in M | 0 \leq \sigma \leq 1\}$, $x(0) = 0$, determines the "coordinates" of a point p in $P(M)$. Denote these coordinates by $p^a = x^\nu(\sigma)$, where $a = (\nu, \sigma)$ is a collective index, Then 1-form in $P(M)$ may be expressed as follows:

$$h(p) = h_a(p)\delta p^a = \int_0^1 d\sigma h_\nu(p,\sigma)\delta x^\nu(\sigma).$$

The form h does not depend on a parametrization of a curve if $h_\nu(p,\sigma) = \dot{x}^\mu(\sigma)h_{\mu\nu}(p,x(\sigma))$. Integrating this 1-form over a 2-path $\hat{s} \in \hat{S}$ and taking the ordered exponential of such an integral we get the representation of the groupoid of fixed paths:

$$\hat{\chi}(\hat{s}) = P_\tau \exp\{i\int_0^1 d\tau \int_0^1 d\sigma h_{\mu\nu}(p(\tau),x)\frac{\partial x^\mu}{\partial\sigma}\cdot\frac{\partial x^\nu}{\partial\tau}\}. \tag{1}$$

Here the function $x = x(\sigma,\tau)$ determines the 2-path $\hat{s} = \{p(\tau) \in P(M) \,|\, 0 \leq \tau \leq 1\}$ presenting each of its constituent paths, $p(\tau) = \{x(\sigma,\tau) \in M \,|\, 0 \leq \sigma \leq 1\}$. The boundary condition $x(0,\tau) = 0$ is imposed on this presentation.

To determine now a representation of the group K, let us make use of the natural mapping $S \to \hat{S}$. Associate with each 2-path $s \in S$ a fixed 2-path $s_1 \in \hat{S}$ starting from a fixed point of the space of paths, namely from the unit $1 \in P(M)$ (it plays the role of the origin of this space). If a 2-path s is a class of curves in P(M) then s_1 is a subclass of curves starting from $1 \in P(M)$. Now put $\chi(k) = \hat{\chi}(k_1)$ for each 2-loop $k \in K \subset S$. Thus the representation $\chi(K)$ is determined. By definition, this representation describes a configuration of a 2-gauge field. Hence such a configuration is determined by the functional $h_{\mu\nu}(p,x)$ depending on a path and a point of this path. Dependence of this functional on a path means that a 2-gauge field is itself a string of a certain type.

5. Action of a 2-gauge field on a string. The induced representation $U_\chi(S) = \chi(K)\uparrow S$ acts in the space of fucntions of a path $\psi(p)$ by the formula

$$(U_\chi(s)\psi)(p) = \hat{\chi}(s^p_{p'})\psi(p'). \tag{2}$$

The operator $U_\chi(s)$ is expressed in the form of an ordered exponential:

$$U_\chi(s) = P_\tau \exp\{ - \int_0^1 d\tau \int_0^1 d\sigma \frac{\partial x^\nu(\sigma\ \tau)}{\partial \tau} \nabla_\nu \quad (\sigma) \} \qquad (3)$$

where the generalized covariant derivative or the 2-covariant derivative is

$$(\nabla_\nu(\sigma)\psi)(\{x(\sigma)\}) = (\frac{\delta}{\delta x^\nu(\sigma)} - i\dot{x}^\mu(\sigma) h_{\mu\nu}(p, x(\sigma)))$$

$$\psi(\{x(\sigma)\}). \qquad (4)$$

The 2-covariant derivative allows one to determine how a 2-gauge field interacts with a srting, provided the theory of a free string is known. The interaction term arises if the 2-covariant derivatives are substituted for variational derivatives in the free string action. Any other formulation of the free string dynamics can be used instead of an action. We are not interested here in the concrete form of this dynamics.

6. The generalized gauge transformation. Two different functionals, $h_{\mu\nu}(p,x)$ and $h'_{\mu\nu}(p,x)$, i.e. the two different representations of the groupoid of 2-paths, $\hat{\chi}(\hat{S})$ and $\hat{\chi}'(\hat{S})$, may correspond to the same 2-gauge field, i.e. the same representation $\chi(K)$. Then there arises a relation generalizing the gauge transformation:

$$\psi'(p) = V(p)\psi(p),$$

$$\dot{x}^\mu(\sigma) h'_{\mu\nu}(p, x(\sigma)) = V(p)\dot{x}^\mu h_{\mu\nu}(p, x(\sigma)) V^{-1}(p) -$$

$$- i \frac{\delta V(p)}{\delta x^\nu(\sigma)} V^{-1}(p). \qquad (5)$$

This formulas may be called a 2-gauge transformation.

The invariant formalism for a string containing no 2-gauge ambiguity arises if the induced representation $\chi(K)\uparrow S$ is realized canonically, by left shifts of functions on the group:

$$(U_\chi(s)\Psi)(s') = \psi(s^{-1}s').$$

In this case the states of the string are characterized by functions $\Psi(s)$ of a 2-path satisfying the structure condition:

$$\Psi(sk) = \chi(k^{-1})\Psi(s) \quad \text{for any } s\in S,\ k\in K.$$

These functions replace functions $\psi(p)$ of a path in ordinary formalism local in $P(M)$. The resulting 2-path-dependent formalism for the string theory generalizes Mandelstam's path-dependent one from the case of particles to the case of strings.

7. The local 2-gauge field. The representation $\chi(K)$ is equivalent to a connection in the space of paths (playing the role of the holonomy group for this connection). The curvature of the connection is generally nonzero, being in essence a 2-gauge field strength.

Let us consider a special case of a 2-gauge field, when it can be determined by local differential forms $A = A_\mu(x)dx^\mu$ and $H = H_{\mu\nu}(x)dx^\mu \wedge dx^\mu$. Let us construct the following path-dependent form via these local forms:

$$\underline{H}_{\mu\nu}(p) = \hat{\alpha}_A(p_\circ)^{-1} H_{\mu\nu}(pO)\hat{\alpha}_A(p_\circ),$$

where pO is the target of the path p (with the source O). Define then the functional $h_{\mu\nu}(p,x) = \underline{H}_{\mu\nu}(p_\circ(x))$ where $p_\circ(x)$ denotes an initial piece of the path p_O, from O to x. Thus the representation $\hat{\chi}_{H,A}(\hat{S})$, i.e; a 2-gauge field, is defined by the formula (1). Most likely, the 2-gauge field $\hat{\chi}_{H,A}(K)$ is identified with the generalized gauge field introduced by Nambu [2] for description of the string unteraction. The form $H_{\mu\nu}$ then plays the role of a tensor potential of this field.

In a more special case, when $H = DA = \nabla_\mu A_\nu dx^\mu \wedge dx^\nu$, we have

$$\chi_{DA,A}(k) = \alpha_A(\partial k), \tag{6}$$

where $\partial k = \{p(\tau)O|0\leq\tau\leq 1\}$ is a boundary of the 2-loop k, defined as the loop covered by the ends of the paths $p(\tau)$ consisting the 2-loop. eq.(6) is a compact form of the non-abelian Stokes theorem [6,9,7]. It can be shown with its aid that $\chi_{DA,A}(K)$ defines a connection with (over the subspace of loops) zero curvature. Indeed, if $k = \{\ell(\tau)\in L(M)|0\leq\tau\leq 1\}$ is a "loop in the space of loops" then $\partial k = 1$ as a consequence of $\Delta\ell(\tau) = 0$. Therefore, (6) implies $\chi_{DA,A}(k) = 1$ so that the curvature of the connection vanishes over the space of loops. This flat connection in the space of loops is just the one considered in [3-5] in the framework of QCD.

8. Concluding remarks. Thus, with the aid of representations of the group of 2-paths, the "two-dimensional" analogues of the main concepts and objects of gauge theory are obtained: a string (or a 2-particle) and a 2-gauge field. This analogy convinces that interaction of strings should be transmitted by a non-local 2-gauge field. Two-dimensional generalization of the covariant derivative determines the form of interaction of strings from the free string dynamics.

The description of a string or a 2-gauge field by functions in the space of paths, $\psi(p)$ and $h_{\mu\nu}(p,x)$, contains a 2-gauge ambiguity, disappearing after transition to functions of 2-paths, $\Psi(s)$, to characterize string states and the representation $\chi(K)$ to describe a 2-gauge field configuration. This gives a two-dimensional generalization of Mandelstam's formalizm.

The 2-gauge field of a special kind, $\chi_{H,A}(K)$, determined by two local forms, A and H, can be plausibly identified with the generalized gauge field with tensor potential $H_{\mu\nu}$. Such a field was introduced by Nambu [2] to describe the string interaction. In general, however, the interaction is transmitted by a non-local 2-gauge field.

References

1. Artru,X. (1983). Phys. Reports, 97, 147.
2. Nambu,Y. (1976). Phys. Reports, 23, 250.
3. Polyakov,A.M. (1980). Nucl. Phys., B164, 171;
4. Aref'eva,I.YA. (1979). On the integral formulation of gauge theories, preprint No. 480, Wrocław.
5. Aref'eva,I.YA. (1979). Lett. Math. Phys., 3, 242.
6. Mensky,M.B. (1983). The group of paths: measurements, fields, particles, Nauka, Moscow, (in Russian, the English translation to appear in Gordon and Breach).
7. Mensky,M.B. (1979). Lett. Math. Phys., 3, 513.
8. Mensky,M.B. (1983). Generalized gauge fields in the space with non-Euclidean topology. In: Contributed papers, 10th Intern; Conf; on Gen. Rel; and Grav., Padova, 1, Padova, (1983), p.583.
9. Aref'eva,I.YA. (1980). Teor. Mat. Fiz., 43, 111. (Russian).

LARGE N-LIMIT IN U(N) ONE-PLAQUETTE MODELS

S.I.Azakov
Institute of Physics, Baku

In spite of the fact that nowdays Monte Carlo simulations in lattice gauge theories allow one to make numerical estimations of the hadron spectrum and other physical magn tudes, it is still important to investigate this theory by analytical methods. One of the nonperturbative methods in the quantum theory of gauge fields with a gauge group U(N) is the so-called 1/N expansion [1], which gives not only the realistic picture in QCD [2], but with certain foundations pretends to the role of quantitative scheme of calculations [3].

The leading order of the 1/N expansion is described by the infinite sum of Feynman diagrams of the simplest topology, the so-called planar diagrams, the summation of which for 4-dimensional theories has not succeeded yet. So usually the 1/N expansion is investigated in the theories of lower dimensions and in the case of lattice gauge theories also for some models defined on the lattice of a finite size, where the problem is considerably simpler. For example, in the two-dimensional U(N) lattice gauge theories in the framework of Euclidean approach it is possible not only to find the leading order of 1/N [4] but also get the explicit expressions for the next orders [5].

The large N-limit in the two-dimensional U(N) gluodynamics with the Wilson action was considered for the first time in the paper [4], where it was discovered, that in this limit in the theory there is a third order phase transition. Then in the papers [6,7] it was shown, that in this theory with the Manton action [8], and with the nonabelian generalization of the Villain action [9] this phase transition disappears and there is an analytic dependence of physical quantities in the coupling constant.

Another working model is the so-called one-plaquette 2+1 model [10,11] considered in the framework of the Kogut-Susskind Hamiltonian approach [12] to the lattice gauge theories. According to Euclidean approach this model is three-dimensional but the lattice is finite, i.e. the gauge field is defined on links of a rectangular box, the cross-section of which is a space-like plaquette. Periodic boundary conditions are usually imposed on time direction, along which the larger edges of the box are directed.

The large N-limit in this model with the Wilson action was investigated in [10,11] and with the Manton action in [13-16]. In both cases the existence of the 3rd order phase transition demarcating the strong-coupling and weak-coupling phases was again demonstrated.

In the lattice gauge theories it is extremely important to ascertain which properties are independent of the choice of a lattice action or a gauge group, i.e. are in some sense universal. We have found [17] that in the one-plaquette 2+1 model with a gauge group U(N) there always exists a 3rd order phase transition in the large N limit if a one-plaquette action satisfies eq. (11) (see below) and is a finite class function.

In the first part of this report the large N-limit is considered for the 2+1 one-plaquette model in pure gluodynamics, and in the second part the topological expansion for the one-plaquette U(N) model with fermions is investigated.

<u>1. Large N-limit in the 2+1 one-plaquette model.</u> The one-plaquette Hamiltonian with a variant action in the theory with a gauge group G is given by

$$\hat{H} = \frac{g^2}{2a} \sum_{l=1}^{4} \hat{E}^{\alpha}(l)\hat{E}^{\alpha}(l) + \frac{1}{ag^2} S(U_p), \tag{1}$$

where g is a coupling constant, a the lattice spacing, U_p an element of the group G, which corresponds to the plaquette: $U_p = U(4)U(3)U(2)U(1), U(l)\in G$, $l = 1, \ldots, 4$, are the basic degrees of freedom, $\hat{E}^{\alpha}(l)$ the operators (in quantum mechanical sense) of electric field (the variables conjugate to the basic degrees of freedom U(l)), obeying

$$\left[\hat{E}^{\alpha}(l), \hat{E}^{\beta}(l')\right] = if^{\alpha\beta\gamma}\hat{E}^{\gamma}(l)\delta_{ll'}, \tag{2}$$

$\alpha,\beta,\gamma = 1, \ldots, d(G)$, $d(G) = \dim G$ and $f^{\alpha\beta\gamma}$ are its structure constants, i.e. the generators t^{α} of G obey the following commutation relations

$$\left[t^{\alpha}, t^{\beta}\right] = if^{\alpha\beta\gamma}t^{\gamma}, \tag{3}$$

and any element $U\in G$ can be expressed as $U = \exp(it^{\alpha}\theta_{\alpha})$. If U_R is the element of the representation R of G corres= ponding to $U\in G$ then

$$\left[\hat{E}^{\alpha}(l), (\hat{U}_R(l'))_{ab}\right] = \delta_{ll'}(T^{\alpha}_R)_{ac}(\hat{U}_R(l))_{cb} \tag{4}$$

a, b, c = 1, ..., d(R) is the dimension of the representation R and matrices T^{d}_{R} are the generators in this repre sentation.

The lattice action S(U) is a real class function on G $S(U) = S(VUV^{-1})$, $V\in G$.

The physical, gauge invariant states, must obey the Gauss law or their wave functions must depend on the variables U_p only, and be the class functions on the group. Therefore the physical state function always has the form

$\Psi(U_p)$, and under gauge transformations

$$\Psi(U_p) = \Psi(U'_p) \tag{5}$$

If we continue ourselves to the physical sector, the problem reduces to the solution of the Schrödinger equation for Ψ, which in the stationary case has the following form:

$$\hat{H}\Psi(U_p) = E\Psi(U_p). \tag{6}$$

It is not difficult to show, that only the plaquette variable enters the eg.(6), in spite of the fact, that in eg.(1) there appear the operators $\hat{E}^\alpha(1)$ defined on the separate links [17].

The physical state wave function $\Psi(U_p)$ being the function on the classes (or the function on the maximal torus) of G has an expansion in terms of the characters of the irreducible unitary representation of

$$\Psi(U_p) = \sum_{\{R\}} a_R \chi_R(U_p). \tag{7}$$

Substituting this expression in (6) we get

$$\sum_{\{R\}} a_R\left[\frac{g^2}{2a} 4C_2(R) + \frac{1}{ag^2} S(U_p)\right] \chi_R(U) =$$

$$= E \sum_{\{R\}} a_R\chi_R(U_p), \tag{8}$$

where $C_2(R)$ is an eigenvalue of the quadratic Casimir operator in the representation R.

From this we see that

$$\sum_{1\leq l\leq 4} \sum_\alpha \hat{E}^\alpha(1)\hat{E}^\alpha(1) = -4\Delta , \tag{9}$$

where Δ is the Beltrami-Laplace operator on G. On the other hand the physical state function depends only on the gauge invariant variables $\theta_1, \ldots, \theta_{r_G}$, where r_G is the rank of the group (dimension of the maxmal torus) and eq.(6) can be rewritten in the form

$$- \frac{2g^2}{a} \overset{\circ}{\Delta} \Psi(\theta_1, \ldots, \theta_{r_G}) + \frac{1}{ag^2} S(\theta_1, \ldots, \theta_{r_G}) \Psi(\theta_1, \ldots, \theta_{r_G}) = E\Psi(\theta_1, \ldots, \theta_{r_G}), \tag{10}$$

where $\overset{\circ}{\Delta}$ is a radial part of the Beltrami-Laplace operator, written in variables $\theta_1, \ldots, \theta_{r_G}$. The explicit expressions of the $\overset{\circ}{\Delta}$ for different groups are given in [17].

In the present paper we will consider only the forms of the action $S(\theta)$ that don't depend on the other parameters and satisfy the condition

$$S(\theta_1, \ldots, \theta_{r_G}) = \sum_{i=1}^{r_G} S(\theta_i). \tag{11}$$

The known forms of the action of Wilson, Manton and Jurkiewicz-Zalewski [18] are of this type.

The large N-limit of the eq.(10) can be investigated with the help of the quantum collective field method [19]. In particular for the ground state energy we have the following expression [17]

$$\left.\frac{E_0(\lambda)}{N^2}\right|_{N^2\to\infty} = \frac{\pi^2\lambda}{a}\mu(\lambda) - \frac{2\pi^2\lambda}{3a}\int_{-\pi}^{\pi} d\sigma \left(\mu(\lambda) - \frac{S(\sigma)}{\pi^2\lambda^2}\right)^{3/2} \theta\left(\mu(\lambda) - \frac{S(\sigma)}{\pi^2\lambda^2}\right) - \frac{\lambda}{12a}, \tag{12}$$

where $\mu(\lambda)$ satisfies

$$\int_{-\pi}^{\pi} d\sigma \left(\mu(\lambda) - \frac{S(\sigma)}{\pi^2\lambda^2}\right)^{1/2} \theta\left(\mu(\lambda) - \frac{S(\sigma)}{\pi^2\lambda^2}\right) = 1. \tag{13}$$

In [17] we have proved that if $S(\theta)$ is a finite function on $[-\pi, \pi]$ then $d^3E_0(\lambda)/d\lambda^3$ has a discontinuity second order at λ_c, i.e. in the theory there is a third order phase transition. But if $S(\theta)$ is not bounded on $[-\pi, \pi]$

the critical point is absent and there is no phase transition.

The presence of such phase transition does not correspond to the divergence of any correlation lenth, but, rather tells us about nonanalyticaly of physical quantities in the model as $N\to\infty$ on the coupling constant, i.e. about the imposibility to spread the results achieved in the strong coupling approximation to the weak coupling region.

2. Topological expansion in the U(N) one-plaquette model with fermions. An introductions of quarks into lattice gauge theories, i.e. considering the complete Lagrangian of OCD on a lattice is connected with additional difficulties for both Monte Carlo simulations and analytic calculations. Many simple models, in particular the two-dimensional U(N) chromodynamics, are not xactly solvable even in the $N\to\infty$ limit.

In the paper [20] we have investigated the simplest model, in which the gauge field is defined on the finite lattice consisting on a single plaquette and interacts with quarks situated in the sites of this plaquette.

As it is known in a $U(N_c)$ gauge theory with a global $U(N_f)$ flavour symmetry the contribution of fermions into the effective action is proportional to $\zeta = N_f/N_c$ and in the $N_c\to\infty$ limit it dissappears. It means, that if we could construct the systematic $1/N_c$ expansion fermions would give contribution only into the quantities of higher orders in $1/N_c$. But if one considers the so-called topological expansion [21], as $N_c\to\infty$ and $N_f\to\infty$ so that ζ and g^2N_c are fixed, then fermions contribute into the leading order of a such modified $1/N_c$ expansion.

The partition function of quantum chromodynamics on a D-dimensional hypercubic lattice has the following form

$$Z = \int \prod_{n,\mu} dU(n,\mu) d\bar{\Psi}(n) d\Psi(n) 1^{-S_G(g^2,U) \; -S_F(\bar{\Psi},\Psi,U)} , \quad (14)$$

where $S_G(g^2,U)$ is an action, containing only the gauge field $U(n,\mu) \in U(N)$, defined on each lattice link, $n = (n_o, \ldots, n_{D-1})$, $\mu \doteq (0,1, \ldots, D-1)$, g is a coupling constant, $S_F(\bar{\Psi},\Psi,U)$ is an action, discribing lattice fermions interacting with the gauge field $U(n,\mu)$.

We consider the so-called Wilson form of the action for the lattice fermions

$$S_F(\bar{\Psi},\Psi,U) = \sum_n \{\bar{\Psi}^i_{\alpha,a}(n) \Psi^i_{\alpha,a}(n) -$$

$$- K \sum_\mu [\bar{\Psi}^i_{\alpha,a}(n)(1-\gamma_\mu)_{\alpha\beta} U_{ab}(n,\mu) \Psi^i_{\beta,b}(n+\hat{\mu}) +$$

$$+ \bar{\Psi}^i_{\alpha,a}(n+\hat{\mu})(1+\gamma_\mu)_{\alpha\beta} \, U^+_{ab}(n,\mu) \Psi^i_{\beta,b}(n)] \}, \quad (15)$$

Here $\Psi^i_{\alpha,a}(n)$ ($\bar{\Psi}^j_{\beta,b}(n)$) is an (anti)quark field, defined on lattice sites, where the colour (a, b), Dirac (α,β) flavour (i, j) indices run 1 to N_f, $C = 2^{[D/2]}$ and N_f respectively, γ_μ are the Euclidean (Hermitian) γ-matrices ($\{\gamma_\mu,\gamma_\nu\} = 2\delta_{\mu\nu}$), K is the 'hopping' parameter, $\hat{\mu}$ is a vector of lenth of a lattice spacing along the μ-th direction.

Since (15) is bilinear in the fermionic fields $\bar{\Psi}$ and Ψ the fermion integration can be performed, and we obtain the partition function of the pure gluodynamics with an effective action in which the 'trace-log' term, where the trace operation involves colour, Dirac and flavour indices, is added to the action $S_G(g^2,U)$.

For the simplest lattice under consideration consisting only of a single plaquette the partition function contains a finite number of integrations:

$$Z = \int dU_{12} dU_{23} dU_{34} dU_{41} \prod_{1 \le i \le 4} d\bar{\Psi}(i)\, d\Psi(i)\, l^{-S_G(g^2,U) - S_F(\bar{\Psi},\Psi,U)} \quad (16)$$

Integrating over fermion fields one can represent it in the form of the integral over U(N) group

$$Z = \text{const} \int dU \cdot l^{-S_G(g^2,U) + N_f \,\text{tr}\, \ln[1/N_c + \frac{\xi}{2}(U+U^+)]} \quad (17)$$

where $\xi = 8K^4/(1+16K^8)$.

Since the exponent in the integrand in (17) is a gauge invariant (a function on cosets of the gauge group), we can express the partition function as an integral over angular variables θ_k, where $l^{i\theta_k}$ $(k = 1, \ldots, N_c)$ are eigenvalues of matrix U:

$$Z = \text{const} \int \prod_{i=1}^{N_c} d\theta_i\, l^{-N_c^2 S_{eff}(\theta_1, \ldots, \theta_{N_c})}$$

where $S_{eff} = -\frac{2}{N_c^2} \ln|\Delta(l^{i\theta_j})| + \frac{1}{N_c} S_G(\lambda;\theta_1,\ldots,\theta_{N_c}) - \frac{N_f}{N_c} \sum_{i=1}^{N_c} \frac{1}{N_c} \ln(1+\xi\cos\theta)$, and $\Delta(x_j) = \prod_{i>j}(x_i - x_j)$ is Vandermond's determinant, $\lambda = g^2 N_c$.

As an action $S_G(g^2,U)$ we consider the one-plaquette actions of Wilson $S_W(g^2,U) = \frac{1}{g^2} \text{tr}\,[1 - \frac{1}{2}(U+U^+)]$ and Manton $S_M(g^2,U) = \frac{1}{2g^2}\{\frac{N}{3}\pi^2 + 2\sum_{k=1}^{\infty} \frac{(-1)^k}{k^2} \text{tr}\,(U^k + U^{+k})\}$.

Both actions are expressed as follows; $S_G(g^2,\theta_1, \ldots, \theta_{N_c}) = (N_c/\lambda) \sum_{i=1}^{N_c} S(\theta_i)$, where $S_W(\theta) = 1 - \cos\theta$, $S_M(\theta) = \frac{1}{2}\theta^2$.

The action $S_{eff} \sim O(1)$ in N_c, provided that parameters $\zeta = N_f/N_c$ and $\lambda = g^2 N_c$ are fixed, and one can consider a WKB expansion in the small parameter $1/N_c^2$ for the evaluation of Z.

As $N_c \to \infty$ the free energy of the model per degree of freedom can be evaluated in the 0th approxmimation:

$$f(\lambda, \xi; \bar{\theta}_1, \ldots, \bar{\theta}_{N_c}) = - S_{eff}(\lambda, \xi; \bar{\theta}_1, \ldots, \bar{\theta}_{N_c})$$

where the arguments $\bar{\theta}_1, \ldots, \bar{\theta}_{N_c}$ are chosen so as to minimize S_{eff}.

Introducing a density distribution $\bar{\rho}(\theta)$ we can replace the summation in S_{eff} by integrations and

$$S_{eff}[\lambda; \xi; \bar{\rho}] = \frac{1}{\lambda} \int_0^{2\pi} d\theta \bar{\rho}(\theta) S(\theta) - \xi \int_0^{2\pi} d\theta \bar{\rho}(\theta) \ln(1+\xi\cos\theta) -$$

$$- \int_0^{2\pi} d\theta \int_0^{2\pi} d\theta' \rho(\theta) \bar{\rho}(\theta') \ln \left| 2 \sin\frac{\theta-\theta'}{2} \right|,$$

where $\bar{\rho}(\theta)$ is the solution of the equation

$$\left. \frac{\delta S_{eff}[\lambda, \xi; \rho]}{\delta\rho(\theta)} \right|_{\rho=\bar{\rho}} = 0, \tag{18}$$

and satisfies the conditions

$$\bar{\rho}(\theta) \geq 0, \quad \theta \in [0, 2\pi], \tag{19}$$

and

$$\int_0^{2\pi} \bar{\rho}(\theta)\, d\theta = 1 \tag{20}$$

Using the explisit form for S_{eff} the eq. (18) may be rewritten as

$$\frac{1}{\lambda} S(\theta) - \zeta \ln(1+\cos\theta) - 2 \int_0^{2\pi} d\theta' \rho(\theta') \ln \left| 2 \sin\frac{\theta-\theta'}{2} \right| = 0,$$

Differentiating it with respect to θ we obtain the singular integral equation

$$\frac{1}{\lambda} \frac{\partial S(\theta)}{\partial \theta} + \zeta\xi \frac{\sin\theta}{1+\xi\cos\theta} + \int_0^{2\pi} d\theta' \rho(\theta') \mathrm{ctg}\frac{\theta-\theta'}{2} = 0, \tag{21}$$

Thus for each pair (λ,ξ) one must find the function $\bar{\rho}(\Theta)$ whuch maximizes $f[\lambda,\xi;\rho]$ and satisfies the conditions (19), (20) and then calculate the corresponding maximum $\tilde{f}(\lambda,\xi) = f[\lambda,\xi;\rho]\,|_{\rho=\bar{\rho}}$, that depends on (λ,ξ) only. The system undergoes the k-th order phase transition at $(\lambda,\xi) = (\lambda_c,\xi_c)$ if at this point the first k-1 derivatives of $\tilde{f}(\lambda,\xi)$ are continuous, while the k-th derivative is not.

The singular equation (21) has the following solutions satisfying (19) and (20):

The Wilson action. When $\lambda > 1$, $0<\xi<\Phi(\lambda,\zeta)$:

$$\bar{\rho}_W(\Theta) = \frac{1}{2\pi}\left(1+\zeta+\frac{1}{\lambda}\cos\theta - \frac{\zeta(1-\xi^2)^{1/2}}{1+\xi\cos\theta}\right),\quad \theta\in[0,2\pi] \tag{22}$$

$$\Phi(\lambda,\zeta) \equiv \frac{(\lambda-1)(\lambda-1+2\lambda\zeta)}{\lambda^2\zeta^2 + (\lambda-1+\lambda\zeta)^2},$$

when $\lambda > 1$ and $\xi > \Phi(\lambda,\zeta)$ and when $0<\lambda<1$, and $\xi\in[0,1]$ is arbititirary:

$$\bar{\rho}_W(\Theta) = \frac{(\cos\Theta - \cos\alpha_c)^{1/2}}{\sqrt{2}\pi}\cos\frac{\theta}{2}\left[\frac{1}{\lambda} - \frac{\zeta(1-c_1)}{R_1(c_1)(\cos\theta+1/\xi)}\right], \tag{23}$$

where the angle α_c is determined from the equation

$$1 + \zeta = \frac{\zeta(1-C_1)}{R_1(C_1)} + \frac{1}{2\lambda}(1 - \cos\alpha_c),$$

where $C_1 = -\frac{1}{\xi} + \sqrt{\frac{1}{\xi^2} + 1}$, and $R_1(C_1) = -\sqrt{C_1^2 - 2C_1\cos\alpha_c + 1}$

On the curve $\xi = \Phi(\lambda,\zeta)$ for $\lambda > 1$, the solution (22) turns into (23) and vice versa.

The Manton action. When $\lambda > 0$, $0<\xi<1$:

$$\bar{\rho}_M(\theta) = \frac{1}{2\pi\lambda}\ln\frac{1-\cos\alpha_c+2\cos\theta+2\sqrt{2(\cos\theta-\cos\alpha_c)}\cos\frac{\theta}{2}}{1-\cos\alpha_c} -$$

$$- \frac{\zeta\cos\frac{\theta}{2}\sqrt{\cos\theta-\cos\alpha_c}\,(1-C_1)}{\sqrt{2}\pi R_1(C_1)\,(\cos\theta+1/\xi)}$$

and the angle α_c is determined from the equation

$$(1+\zeta) + \frac{\zeta(1-C_1)}{R_1(C_1)} + \frac{1}{\lambda}\ln\frac{(1+\cos\alpha_c)}{2} = 0.$$

Investigating the leading order of the topological expansion in a theory with fermions we take into account much larger set of diagrams, than we do, considering just the expansion in the inverse number of colours. One can expect that considering larger set of diagrams will change the phase structure of the theory. But the example of the considered one-plaquette model shows that this is not the case, and the phase picture remains unchanged. Namely, in the case of the Wilson action there are two phases, divided by the line of the 3rd order phase transition, and in the case of the Manton action the phase transition is is absent, as it takes in the pure gluodynamics.

References

1. 't Hooft, G. (1974).Nucl. Phys. B72, 461; (1975). B75, 461.

2. Aref'eva, I.Ya., Slavnov, A.A. Lectures on XIV School of Young Scientists on High Energy Physics, JINR, D2-81-158, Dubna, 1981.

3. Migdal, A.A. (1983). Phys.Rep. 102, 199.

   Makeenko, Yu.M. (1983). Large N, ITEP-preprint 16, Moscow.

   Aref'eva, I.Ya. (1983). In: XV International School of Young Scientists on High Energy Physics", JINR, D2, 4-83-179, Dubna.

4. Gross, D.J., Witten, E. (1980). Phys. Rev. D21, 446.

5. Goldschmidt, Y.Y. (1980). J.Math.Phys. 21, 1842.

6. Lang, C.B., Salomonson P., Skagerstam, B. (1981). Phys.

Lett. 100B, 23; (1981). Nucl. Phys. B190, FS3 , 337.

7. Onofri, E. (1981). Nuovo Cim. A66, 293.

Menotti, P., Onofri, E. (1981). Nucl. Phys. B190 FS3 228.

8. Manton, N.S. (1980). Phys. Lett. 96B, 328.

9. Villain, J. (1975). J.Phys.(Paris) 36, 581.

10. Wadia, S. (1980). Phys. Lett. 93B, 403.

11. Jevicki, A., Sakita, B. (1980). Phys. Rev., D22, 467.

12. Kogut, J., Susskind, L. (1975). Phys. Rev., D11, 395.

13. Azakov, S.I. (1982). The 1/N expansion in the one-plan-quette model with Manton's action. Preprint 46, Baku, IFAN.

14. Rodrigues, J.R. (1982). Phys. Rev;, D26, 2833.

15. Shimamune, Y. (1982). Phys.Lett., 108B, 407.

16. Trinhammer, O. (1983). Phys. Lett., 129B, 324.

17. Azakov, S.I. (1985). Theor. Math. Phys., 62, 222 (Russian).

18. Jurkiewicz, J., Zalewski, K. (1982). Phys. Lett., 115B, 143.

19. Jevicki, A., Sakita, B. (1980). Nucl. Phys. B165, 511.

20. Azakov, S.I., Aliev, E.S. (1985). The one-plaquette model with fermions in the framework of the topological expansion. Preprint 121, IFAN, Baku.

21. Veneziano, G. (1976). Nucl.Phys., B117, 519.

GAUGE MODEL GENERATED BY NON-P-EVEN FUNCTIONS

L.M.Slad

Institute of Nuclear Physics, Moscow State University, Moscow 119899

Is the standard gauge model [1-3] which has repeatedly been verified in experiment, the final version of electroweak interaction theory or is it a mere approximation to the future theory? This question is pertinent because the original non-equivalence of the left- and right-handed spinors (isospinors and isoscalars) in the standard model has no satisfactory explanation. This non-equivalence is an important problem of the Kaluza-Klein theories (see e.g. [4]). While settling some issues, the most popular schemes with the left-right symmetry [5-7] propose, in my opinion, some new question. In these schemes, the left- and right-handed spinors transform according to the equivalent representations of $SU(2)_L$ and $SU(2)_R$, respectively; and the functions determining transformations of each group are independent. The latter is not compatible with the geometrical approach to the gauge theories. Besides, the weak-interaction Lagrangian, for example,

$$\mathcal{L}_{int} = \frac{g_L}{\sqrt{2}} \bar{n} \gamma_\mu \frac{1+\gamma_5}{2} p W^-_{L\mu} + \frac{g_R}{\sqrt{2}} \bar{n} \gamma_\mu \frac{1-\gamma_5}{2} p W^-_{R\mu} + H.c. \quad (1)$$

while providing the P-invariance of the cross-sections of high-energy processes for equal constants $g_L = g_R$, is not itself P-invariant because the condition $PW^\pm_{L\mu} = W^\pm_{R\mu}$ is not satisfied (even before the spontaneous symmetry breaking). Based on the setting and transformation properties of Higgs bosons proposed in [5-7], our approach to constructing an electroweak interaction model with the left-right symmetry leads to a P-invariant electroweak interaction up to spontaneous symmetry breaking. In the general case our model is not reduced to the schemes [5-7] which is briefly discussed in what follows.

After the pioneering work by Yang and Mills [8], it has been generally accepted that to the gauge transformation

$$\Phi \rightarrow \exp(iL_a f^a(x))\Phi \tag{2}$$

there correspond vector particles and therefore the functions $f^a(x)$ are the Lorentz scalars (P-even). We refuse to impose this constraint on the functions f^a and assume that they have no definite parity. As a result, the functions f^a can be represented as a sum of the scalar f_1^a and the pseudo-scalar f_2^a components: $f^a = f_1^a + f_2^a$ and $Pf^aP^{-1} = f_1^a - f_2^a$. We also assume that if an arbitrary field Φ transforms according to (2), the field $P\Phi$ transforms as

$$P\Phi \rightarrow \exp(iL_a P f^a P^{-1})P\Phi. \tag{3}$$

We construct separately the electroweak interaction of leptons and quarks. The assumed masslessness of neutrinos keeps us to the beaten track so as to follow the main features of the standard schemes. The fact that all quarks have masses forces us to make additional assumptions.

For the left-handed lepton isodoublets we assume the following gauge transformation

$$\psi_L \rightarrow \exp\left(-\frac{1}{2}ig_1\vec{\tau}\vec{x}_1 - \frac{1}{2}ig_2\vec{\tau}\vec{x}_2 + \frac{1}{2}ig_1'y_1 + \frac{1}{2}ig_2'y_2\right)\psi_L = U_L\psi_L, \tag{4}$$

where $\vec{x}_1$ and y_1 are the Lorentz scalars and $\vec{x}_2$ and y_2 are pseudoscalars, $\vec{\tau}$ are the Pauli matrices.

Since $P\psi_L = \psi_R$, according to (3), the gauge transformation of right-handed lepton isodoublets is

$$\psi_R \to \exp\left(-\frac{1}{2}ig_1\vec{\tau}\vec{x}_1+\frac{1}{2}ig_2\vec{\tau}\vec{x}_2+\frac{1}{2}ig_1'y_1-\frac{1}{2}ig_2'y_2\right)\psi_R = U_R\psi_R. \qquad (5)$$

The relations (4) and (5) provide the equivalence of gauge transformations of the left and right-handed spinors and the functions defining these transformations are independent. The relations (4) and (5) reduce to the transformations of the left-right symmetrical schemes of [5-7]:

$$-\frac{1}{2}ig_1\vec{\tau}\vec{x}_1-\frac{1}{2}ig_2\vec{\tau}\vec{x}_2=-\frac{1}{2}ig_L\vec{\tau}\vec{x}_L,$$

$$-\frac{1}{2}ig_1\vec{\tau}\vec{x}_1+\frac{1}{2}ig_2\vec{\tau}\vec{x}_2=-\frac{1}{2}ig_R\vec{\tau}\vec{x}_R,\ g_L=g_R,$$

$$U_L\in SU(2)_L\times U(1)\times U(1),$$

$$U_R\in SU(2)_R\times U(1)\times U(1).$$

if and only if $g_1^2-g_2^2=0$ (the orthogonality condition of the transformations and the corresponding gauge fields).

So, in the general case our model of electroweak interaction with the left-right symmetry is wider than the models proposed in [5-7]; it shows clearly the relation between the gauge transformations of left- and right-handed spinors and argees nicely with the Kaluza-Klein theories.

In order to construct the mass terms of charged leptons and to provide the spontaneous left-right symmetry breakdown we introduce, as in [5-7], the Higgs fields of three types

$$\varphi=\begin{pmatrix}\varphi^0 & \varphi^+\\ \varphi^- & \varphi'^0\end{pmatrix},\ \varphi_1=\begin{pmatrix}\varphi_1^+\\ \varphi_1^0\end{pmatrix},\ \varphi_2=\begin{pmatrix}\varphi_2^+\\ \varphi_2^0\end{pmatrix},$$

which satisfy the conditions

$$P\varphi P^{-1}=\varphi^\dagger, \qquad (6a)$$

$$P\varphi_1 P^{-1} = \varphi_2. \tag{6b}$$

The gauge transformations of these fields are (see (4)-(5))

$$\varphi \to U_L \varphi U_R^{-1}, \tag{7}$$

$$\varphi_1 \to \exp\left(-\frac{1}{2}ig_1\vec{\tau}\vec{x}_1 - \frac{1}{2}ig_2\vec{\tau}\vec{x}_2 + \frac{1}{2}ib_1g_1'y_1 + \frac{1}{2}ib_2g_2'y_2\right)\varphi_1, \tag{8}$$

$$\varphi_2 \to \exp\left(-\frac{1}{2}ig_1\vec{\tau}\vec{x}_1 + \frac{1}{2}ig_2\vec{\tau}\vec{x}_2 + \frac{1}{2}ib_1g_1'y_1 - \frac{1}{2}ib_2g_2'y_2\right)\varphi_2. \tag{9}$$

The requirement that the axial current of charged leptons and the vector current of neutrinos do not interact with massless neutral gauge bosons, gives $b_1 = -1$, $b_2 = 1$.

The P-invariant potential describing the self-interaction of Higgs fields φ, φ_1, and φ_2 is

$$V = \lambda_1^2 (\varphi_1^+ \varphi_1 + \varphi_2^+ \varphi_2)^2 - \mu_1^2 (\varphi_1^+ \varphi_1 + \varphi_2^+ \varphi_2) + \\ + \lambda_2^2 \left[Sp(\varphi^+ \varphi)\right]^2 - \mu_2^2 \, Sp(\varphi^+ \varphi). \tag{10}$$

Then V has a degenerate minimum on the set

$$\varphi_1^+ \varphi_1 + \varphi_2^+ \varphi_2 = \frac{\mu_1^2}{2\lambda_1^2}. \tag{11}$$

Owing to (6b), $\varphi_{1,2}$ can be represented as $\varphi_{1,2} = \varphi' \pm \varphi''$, where φ' is the Lorentz scalar and φ'' is the pseudoscalar. As will be seen in what follows, if the vacuum expectation of the pseudoscalar φ'' is non-zero, the left-right symmetry in weak processes is broken: the left-charged current dominates if $|\langle \varphi_2^0 \rangle| > |\langle \varphi_1^0 \rangle|$ and the right-handed charged current dominates if $|\langle \varphi_2^0 \rangle| < |\langle \varphi_1^0 \rangle|$. The question if the requirement $\langle \varphi_2^0 \rangle \neq \langle \varphi_1^0 \rangle$ removes the degeneracy (11) and leads to the absolute minimum with respect to

$\varphi_1^+ \varphi_1$ and $\varphi_2^+ \varphi_2$, provided the higher-order contributions to the potential are taken into account will be discussed elsewhere.

Let

$$\langle \varphi^\circ \rangle = 0,\ \langle \varphi'^\circ \rangle = \frac{f}{\sqrt{2}},\ \langle \varphi_1^\circ \rangle = \frac{f_1}{\sqrt{2}},\ \langle \varphi_2^\circ \rangle = \frac{f_2}{\sqrt{2}}. \quad (12)$$

The contribution of the right-handed currents to the low energy charged processes is small as compared with the left-handed current contribution if, as is easy to verify, $(f_2^2 + f^2)(f_1^2 + f^2)^{-1} \gg 1$.

The interaction among Higgs bosons and leptons is described by the Lagrangian

$$\mathcal{L}_{\ell\varphi} = \varkappa_\ell \bar{\Psi}_L \varphi \Psi_R + \varkappa_\ell \bar{\Psi}_R \varphi^+ \Psi_L . \quad (13)$$

Owing to (12) and (13), the neutrino is massless. The Higgs fields φ_1 and φ_2 are not involved in the mass formation of charged leptons.

Using the standard expressions for the kinematic Lagrangian terms applying to the Higgs fields, we obtain, in a regular way, the following orthonormalized gauge fields:

$$W^\pm_{1\mu} = N_1(k_1 g_1 x^\pm_{1\mu} + k_2 g_2 x^\pm_{2\mu}),\ W^\pm_{2\mu} = N_2(k_3 g_1 x^\pm_{1\mu} + k_4 g_2 x^\pm_{2\mu}),$$

$$Z_{1\mu} = N_3(z_1 g_1 x^3_{1\mu} + z_2 g_2 x^3_{2\mu} - z_1 g_1' y_{1\mu} + z_2 g_2' y_{2\mu}), \quad (14)$$

$$Z_{2\mu} = N_4(z_3 g_1 x^3_{1\mu} + z_4 g_2 x^3_{2\mu} - z_3 g_1' y_{1\mu} + z_4 g_2' y_{2\mu}),$$

$$A_\mu = N_5(g_1' x^3_{1\mu} + g_1 y_{1\mu}),\ B_\mu = N_6(g_2' x^3_{2\mu} - g_2 y_{2\mu}),$$

where $N_1, \dots, N_6$ are the normalyzing factors, the vector fields $\vec{x}_{1\mu}$, $y_{1\mu}$ and the axial fields $\vec{x}_{2\mu}$, $y_{2\mu}$ correspond to the scalar $\vec{x}_1$, y_1 and pseudoscalar $\vec{x}_2$, y_2 functions from the gauge transformations (4), (5), (8) and (9); $x^3_{1\mu}$ and $x^3_{2\mu}$ are the third components of the isovectors $\vec{x}_{1\mu}$ and $\vec{x}_{2\mu}$; the quantities $k_1, \ldots, k_4$, $z_1, \ldots, z_4$ satisfy the system

$$
\begin{gathered}
k_1^2+k_3^2=k_2^2+k_4^2=1+c^2,\ k_1k_2+k_3k_4=1-c^2,\\
k_1k_3g_1^2+k_2k_4g_2^2=0,\ z_1^2+z_3^2=1+c_1^2,\ z_2^2+z_4^2=1+c_1^2+c_2^2, \qquad (15)\\
z_1z_2+z_3z_4=1-c_1^2,\ z_1z_3(g_1^2+g_1'^2)+z_2z_4(g_2^2+g_2'^2)=0,\\
c^2=\frac{f_2^2+f^2}{f_1^2+f^2},\ c_1^2=\frac{f_2^2}{f_1^2},\ c_2^2=\frac{4f^2}{f_1^2}.
\end{gathered}
$$

Note (see (14)) that unlike the electromagnetic vector A_μ and axial B_μ fields the W- and Z-bosons have no definite P-parity.

The masses of gauge bosons are

$$
\begin{gathered}
M_{W_1}^2 \simeq \frac{(f_1^2+f^2)g_1^2g_2^2}{g_1^2+g_2^2},\ M_{W_2}^2\simeq\frac{1}{4}c^2(f_1^2+f^2)(g_1^2+g_2^2),\ M_A=M_B=0\\
M_{Z_1}^2\simeq\frac{(f_1^2+f^2)(g_1^2+g_1'^2)(g_2^2+g_2'^2)}{g_1^2+g_1'^2+g_2^2+g_2'^2}, \qquad (16)\\
M_{Z_2}^2\simeq\frac{1}{4}f_2^2(g_1^2+g_2'^2+g_2^2+g_2'^2),
\end{gathered}
$$

From here on the sign $\simeq$ means that only the high degree in c^2 or f_2^2 is preserved.

The Lagrangian describing the interaction of leptons with gauge fields is

$$
\begin{aligned}
\mathcal{L}_{lept\ int} &= \Big[\frac{1}{\sqrt{2}}\bar{e}\gamma_\mu\frac{1+\gamma_5}{2}\nu_e(g_{L1}W_{1\mu}^-+g_{L2}W_{2\mu}^-)+\\
&+\frac{1}{\sqrt{2}}\bar{e}\gamma_\mu\frac{1-\gamma_5}{2}\nu_e(g_{R1}W_{1\mu}^-+g_{R2}W_{2\mu}^-)+\text{H.c.}\Big]+\\
&+g_{L1}\frac{M_{Z_1}}{M_{W_1}}j_{1\mu}^e Z_{1\mu}+g_{R2}\frac{M_{Z_2}}{M_{W_2}}j_{2\mu}^e Z_{2\mu}-e\bar{e}\gamma_\mu eA_\mu+ \qquad (17)\\
&+g\bar{\nu}_e\gamma_\mu\gamma_5\nu_e B_\mu+(e\to\mu)+(e\to\tau).
\end{aligned}
$$

Here, the interaction constants are

$$g_{L1} \simeq \frac{2g_1g_2}{\sqrt{g_1^2+g_2^2}}, \quad g_{R1} \simeq -\frac{2g_1g_2(g_1^2-g_2^2)}{c^2(g_1^2+g_2^2)^{3/2}},$$
$$g_{L2} \simeq \frac{g_1^2-g_2^2}{\sqrt{g_1^2+g_2^2}}, \quad g_{R2} \simeq \sqrt{g_1^2+g_2^2}, \tag{18}$$

and the electric e and axial g charges are

$$e = g_1 g_1' / \sqrt{g_1^2 + g_1'^2}, \quad g = g_2 g_2' / \sqrt{g_2^2 + g_2'^2}. \tag{19}$$

The neutral current is

$$j^e_{1\mu} \simeq \frac{1}{2}\Big(1 - \frac{g_2'^2}{g_2^2+g_2'^2}\Big)\bar{\nu}_e\gamma_\mu \frac{1+\gamma_5}{2}\nu_e + \frac{1}{2}\cdot\frac{g_2'^2}{g_2^2+g_2'^2}\bar{\nu}_e\gamma_\mu\frac{1-\gamma_5}{2}\nu_e -$$
$$-\frac{1}{2}\Big(1-2\frac{g_1'^2}{2(g_1^2+g_1'^2)}\Big)\bar{e}\gamma_\mu\frac{1+\gamma_5}{2}e + \frac{g_1'^2}{2(g_1^2+g_1'^2)}\bar{e}\gamma_\mu\frac{1-\gamma_5}{2}e. \tag{20}$$

An explicit form of the neutral lepton current interacting with the intermediate boson Z_2 is not discussed in this paper.

Comparing (16)-(20) with the experimental data on weak neutral processes (see the review [8]) and masses of $W_1^\pm$ [9-10] and also with the constraint on the coupling constant of the axial photon with neutrino $g^2/4\pi \lesssim 0.9\cdot10^{-4}$ [11], we see that the constants g_1, g_2, g_1' are close and $g_2'^2 \lesssim 10^{-2} g_2^2$. Thus, the quantities $g_2'^2(g_2^2 + g_2'^2)^{-1}$ in (20) can be neglected, i.e. our model repreduces with good accuracy the structure of weak neutral current of the standard model in which case the quantity $d^2 = g_1'^2/2(g_1^2+g_1'^2)$ is associated with $\sin^2\theta_W$. Note, that d^2 can vary 0 to 1/2 unlike $\sin^2\theta_W$ varying 0 to 1. It is easy to see that other relations typical for the standard model are also valid for $g_1 = g_2$: $e/g_{L1} = d$, $M_{W_1} = \sqrt{1-d^2}\,M_{Z_1}$. Recall that in this case our scheme reduces to the schemes [5-7]. Future experiments will establish whether the equality $g_1 = g_2$ is exact or approximate.

Let us now discuss the quark sector. If all quarks have masses, there are three variants consistent with the main assumption (3) of our approach.

1) We may assume $g_2' = 0$, i.e. forbidding the axial electromagnetic field by reducing the set of functions in (2). It is also necessary to introduce a new Higgs field possessing the same transformation properties (7) as those of φ .

Variants 2) and 3) are based on the introduction of "mirror" quarks. Let a quark q be described by the spinors ψ_{Lq} and ψ_{Rq} and its "mirror" quark q' , by ψ'_{Lq} and ψ'_{Rq} . We assume that the space-inversion P acts as

$$P\psi_{Lq} = \psi'_{Rq}, \quad P\psi_{Rq} = \psi'_{Lq}, \tag{21}$$

i.e. the quarks have no definite P-parity: under P the quark q turns into the "mirror" quark q'.

2) All leftorright-handed quark spinors form isodoublets. For the spinors ψ_{Lq} and ψ'_{Lq} we can write the gauge transformation of the type (4) but with different coefficients at y_1 and y_2 that are not the same for ψ_{Lq} and ψ'_{Lq} : in this case a nonzero constant g_2' and the interaction of quarks with the field B_μ are permissible. Again, it is necessary to introduce an additional Higgs field with the transformation properties (7).

3) The spinors ψ_{Lq} and ψ'_{Rq} form the isodoublets

$$\psi_{Lq} = \begin{pmatrix} \psi_L^u \\ \psi_L^d \end{pmatrix}, \quad \psi'_{Rq} = \begin{pmatrix} \psi'^u_R \\ \psi'^d_R \end{pmatrix}$$ and the spinors

ψ_{Rq} and ψ'_{Lq} form the inosinglets ψ^u_R , ψ^d_R and ψ'^u_L , ψ'^d_L . This variant will be treated in more detail. The Lagrangian containing quarks is

$$\begin{aligned}\mathcal{L}_q = \sum_q \{ & i\bar{\Psi}_{Lq}\gamma_\mu\left(\partial_\mu - \tfrac{1}{2}ig_1\vec{\tau}\vec{\mathcal{X}}_{1\mu} - \tfrac{1}{2}ig_2\vec{\tau}\vec{\mathcal{X}}_{2\mu} - \tfrac{1}{6}ig'_1 y_{1\mu}\right)\Psi_{Lq} + \\ & + i\bar{\Psi}^u_R\gamma_\mu\left(\partial_\mu - \tfrac{2}{3}ig'_1 y_{1\mu} + \tfrac{1}{2}ig'_2 y_{2\mu}\right)\Psi^u_R + \\ & + i\bar{\Psi}^d_R\gamma_\mu\left(\partial_\mu + \tfrac{1}{3}ig'_1 y_{1\mu} - \tfrac{1}{2}ig'_2 y_{2\mu}\right)\Psi^d_R + \\ & + i\bar{\Psi}'^u_L\gamma_\mu\left(\partial_\mu - \tfrac{2}{3}ig'_1 y_{1\mu} - \tfrac{1}{2}ig'_2 y_{2\mu}\right)\Psi'^u_L + \\ & + i\bar{\Psi}'^d_L\gamma_\mu\left(\partial_\mu + \tfrac{1}{3}ig'_1 y_{1\mu} + \tfrac{1}{2}ig'_2 y_{2\mu}\right)\Psi'^d_L + \\ & + i\bar{\Psi}'_{Rd}\gamma_\mu\left(\partial_\mu - \tfrac{1}{2}ig_1\vec{\tau}\vec{\mathcal{X}}_{1\mu} + \tfrac{1}{2}ig_2\vec{\tau}\vec{\mathcal{X}}_{2\mu} - \tfrac{1}{6}ig'_1 y_{1\mu}\right)\Psi'_{Rd} + \\ & + \left[\varkappa_1(\bar{\Psi}_{Lq}\varphi_1)\Psi^d_R + \varkappa_1(\bar{\Psi}'_{Rq}\varphi_2)\Psi'^d_L + \varkappa_2(\bar{\Psi}_{Lq}\tilde{\varphi}_1)\Psi^u_R + \right. \\ & \left. + \varkappa_2(\bar{\Psi}'_{Rq}\tilde{\varphi}_2)\Psi'^u_L + \text{H.c.}\right]\}, \end{aligned} \tag{22}$$

where φ_1 and φ_2 are Higgs-field isodoublets (see (6b), (8) and (9)) and the isodoublets $\tilde{\varphi}_1$ and $\tilde{\varphi}_2 = P\tilde{\varphi}_1 P^{-1}$ are representable as $\tilde{\varphi}_1 = \begin{pmatrix} \bar{\varphi}^0_1 \\ \varphi^-_1 \end{pmatrix}$, $\tilde{\varphi}_2 = \begin{pmatrix} \bar{\varphi}^0_2 \\ \varphi^-_2 \end{pmatrix}$.

Since $|f_2| \gg |f_1|$ (this has already been discussed above), the mass of quark q, as follows from (22), is much less than that of its mirror partner q'.

Using (14) we present an explicit form of the interaction of quarks with gauge bosons:

$$\begin{aligned}\mathcal{L}_{q\,int} = \sum_q \Big\{ & \Big[\tfrac{1}{\sqrt{2}}\bar{\Psi}^d_L\gamma_\mu\Psi^u_L(g_{L1}W^-_{1\mu} + g_{L2}W^-_{2\mu}) + \\ & + \tfrac{1}{\sqrt{2}}\bar{\Psi}'^d_R\gamma_\mu\Psi'^u_R(g_{R1}W^-_{1\mu} + g_{R2}W^-_{2\mu}) + \text{H.c.}\Big] + \\ & + g_{L1}\frac{M_{Z_1}}{M_{W_1}} j^{q+q'}_{1\mu} Z_{1\mu} + g_{R2}\frac{M_{Z_2}}{M_{W_2}} j^{q+q'}_{2\mu} Z_{2\mu} + \end{aligned} \tag{23}$$

$$+e\left(\frac{2}{3}\bar{\Psi}^{u}\gamma_{\mu}\Psi^{u}-\frac{1}{3}\bar{\Psi}^{d}\gamma_{\mu}\Psi^{d}+\frac{2}{3}\bar{\Psi}'^{u}\gamma_{\mu}\Psi'^{u}-\frac{1}{3}\bar{\Psi}'^{d}\gamma_{\mu}\Psi'^{d}\right)A_{\mu}+$$

$$+g\left(\frac{1}{2}\bar{\Psi}^{u}\gamma_{\mu}\Psi^{u}-\frac{1}{2}\bar{\Psi}^{d}\gamma_{\mu}\Psi^{d}-\frac{1}{2}\bar{\Psi}'^{u}\gamma_{\mu}\Psi'^{u}+\frac{1}{2}\bar{\Psi}'^{d}\gamma_{\mu}\Psi'^{d}\right)B_{\mu}\Big\},$$

where

$$j^{q+q'}_{1\mu}\simeq\left(\frac{1}{2}-\frac{2}{3}d^{2}\right)\bar{\Psi}^{u}_{L}\gamma_{\mu}\Psi^{u}_{L}-\frac{2}{3}d^{2}\bar{\Psi}^{u}_{R}\gamma_{\mu}\Psi^{u}_{R}+\left(-1/2+1/3\,d^{2}\right)\bar{\Psi}^{d}_{L}\gamma_{\mu}\Psi^{d}_{L}+$$

$$+\frac{1}{3}d^{2}\bar{\Psi}^{d}_{R}\gamma_{\mu}\Psi^{d}_{R}-\frac{2}{3}d^{2}\bar{\Psi}'^{u}\gamma_{\mu}\Psi'^{u}+\frac{1}{3}d^{2}\bar{\Psi}'^{d}\gamma_{\mu}\Psi'^{d}, \quad (24)$$

$$d^{2}=g'^{2}_{1}/2(g^{2}_{1}+g'^{2}_{1}).$$

In (24) the small quantities $g'^{2}_{2}(g^{2}_{2}+g'^{2}_{2})^{-1}$ are ommited. The neutral current $j^{q+q'}_{2\mu}$ is not recorded.

It follows from (23) that the axial electromagnetic field B_{μ} interacts with the vector quark current, the interaction constants with quarks q and q' have different signs and provide therefore P-invariance of the interaction. If q is related to W-bosons by its left-handed current, then q' is related to W-bosons by the right-handed current. The neutral current $j^{q+q'}_{1\mu}$ of q-quarks has the same structure as in the standard model, d^{2} is associated with $\sin^{2}\theta_{W}$. The neutral current $j^{q+q'}_{1\mu}$ of q' quarks is P-even.

The experimental consequences of the above model of electroweak interaction will be discussed in detail elsewhere. Here we merely note that possible experimental consequences of the existence of the axial electromagnetic field B_{μ} have been analysed in [11-14].

References

1. Glashow, S.L. (1961). Nucl.Phys. 22, 579.

2. Weinberg, S. (1967). Phys.Rev.Lett., 19, 1264.

3. Salam, A. (1968). In: Elementary Particle Physics, Stocholm, p,367.

4. Witten, E. (1981). Nucl.Phys., B 186, 412.

5. Pati, J.C., Salam, A. (1974). Phys.Rev., D10, 275.

6. Mohapatra, R.N., Pati, J.C. (1975). Phys.Rev., D11, 566.

7. Senjanovic, G., Mohapatra, R.N. (1975). Phys.Rev., D12, 1502.

8. Kim, J.E., Langacker, P., Levine, M., Williams, H.H. (1981). Rev.Mod.Phys., 53, 211.

9. UA2 Collab., Bagnia, P. et al; (1983). Phys.Lett., 129B, 130.

10. UAI Collab., Arnison, G, et al. (1983). Phys.Lett., 129B, 273.

11. Slad, L.M. (1982). Dokl.Akad.Nauk SSSR, 265, 615 (Russian)

12. Slad, L.M. (1983). Pisma ZhETPh., 37, 115 (Russian)

13. Slad, L.M. Dokl.Akad.Nauk SSSR, (983).269, 1345 (Russian)

14. Slad, L.M. (1984). Nucl.Phys., 40, 1517 (Russian)

HIGGS MODELS RESULTING FROM SYMMETRIC SPACES

YU.A.Kubyshin, I.P.Volobujev
Institute of Nuclear Physics, Moscow State University

1. Our paper is devoted to elaboration of a general method of calculating potentials for scalar fields in dimensionally reduced theories. Recent development of the idea of dimensional reduction was caused by the intensive studies of gauge (as well as gravitation) theories with additional space-time symmetries.

First of all we describe the main features of this approach (see e.g. [1,2]). Consider a "free" gauge theory with the gauge group K defined on a space-time manifold $M = M^4 \times G/H$ (here M^4 is the Minkovski space, G and H are compact Lie groups). The action is standard:

$$S = \frac{1}{\lambda^2}\int \langle F_{\hat{\mu}\hat{\nu}}, F^{\hat{\mu}\hat{\nu}} \rangle dv_M ,$$

$$F_{\hat{\mu}\hat{\nu}} = \partial_{\hat{\mu}} A_{\hat{\nu}} - \partial_{\hat{\nu}} A_{\hat{\mu}} + [A_{\hat{\mu}}, A_{\hat{\nu}}], \qquad (1)$$

$$\hat{\mu}, \hat{\nu} = 0, 1, 2, 3, \ldots, 3 + \dim G/H$$

where $\langle \, , \rangle$ denotes the canonical scalar product and $\underline{k}$ is Lie algebra of K. Suppose that $A_{\hat{\mu}}$ is G-symmetric; i.e. invariant under the G-action up to a gauge [3]. Then such a theory may be consistently interpreted as a gauge theory with scalar fields in the space of lesser dimension M^4. Every symmetric gauge field configuration uniquely defines a homomorphism $\mathcal{T} : H \to K$ of the stationary subgroup H of the G-action on M into K. Besides, this field is in one-to-one correspondence with the pair (A_μ, Φ).

Here $A_\mu(x)$ is a gauge field on M^4 with the gauge group $C = \{c \in K,\ cJ(h)c^{-1} = J(h)$ for any $h \in H\}$, the centralizer of $J(H)$ in K, $\Phi(x)$ is for any $x \in M^4$ a linear equivariant map pertinent to the scalar fields and satisfying [1, 2]:

$$\Phi(x) : \underline{h}^{\perp} \to \underline{k}\,; \tag{2a}$$

$$\mathrm{Ad}J(h) \circ \Phi(x) = \Phi(x) \circ \mathrm{Ad}h, \quad h \in H. \tag{2b}$$

In these formulas $\underline{h}$ is Lie algebra of H, the space $\underline{h}^{\perp}$ is defined from the reductive decomposition $\underline{g} = \underline{h} \oplus \underline{h}^{\perp}$ of Lie algebra of G, with respect to AdG-invariant scalar product in G.

Let the metric on M be of the form $\gamma_M = \gamma_{M^4} \oplus \gamma_{G/H}$, where $\gamma_{G/H}$ is a G-invariant metric on the internal space G/H. Then the action (1) can be represented in the form

$$S = \frac{1}{\lambda'^2} \int d^4x \left\{ \langle F_{\mu\nu}, F^{\mu\nu} \rangle + 2\sum_k \langle D_\mu \Phi(e_k), D^\mu \Phi(e_k) \rangle - V(\Phi) \right\}, \tag{3}$$

with $\{e_k\}$ an orthonormal basis in $\underline{h}^{\perp}$, $D_\mu \Phi(e_k)$ the covariant derivative of the field $\Phi(e_k) \equiv (\Phi(x))(e_k) \in \underline{k}$ and

$$V(\Phi) = -\sum_{k,l} \langle T_{kl}, T^{kl} \rangle,$$

$$T_{kl} = J([e_k, e_l]_{\underline{h}}) + \Phi([e_k, e_l]_{\underline{h}^{\perp}}) - [\Phi(e_k), \Phi(e_l)]. \tag{4}$$

2. In order to calculate $V(\Phi)$ it is necessary to solve the constraints (2b) and to find the scalar fields explicitly. The condition (2b) means that Φ is an operator intertwining the representations of the group H in linear spaces $\underline{h}^{\perp}$ and $\underline{k}$: Ad(H) $\underline{h}^{\perp}$ and Ad $(\mathcal{T}(H))\underline{k}$. To construct Φ it is necessary to decompose these representations into irreducible ones. The Schur lemma states that the intertwining operator between nonequivalent representations is zero and invertible for equivalent representations. In the latter case the intertwining number (the dimension of the intertwining operator space) can be 1, 2 or 4, depending on the type of the representation [4].

It is known that for complex representations the intertwining number is always 2 [4]. Thus, it is convenient to complexify the spaces $\underline{h}^{\perp}$ and $\underline{k}$ and to continue Φ by linearity to these spaces: $\Phi^c(b_1+ib_2) = \Phi(b_1) + i\Phi(b_2)$, $b_1, b_2 \in \underline{h}^{\perp}$. The operator Φ^c must satisfy the reality condition $\bar{\Phi}^c(b) = \Phi^c(\bar{b})$, $b \in \underline{h}^{\perp c}$, and the intertwining condition (2b) taken in the infinitesimal form and linearly continued to $\underline{h}^c$, $\underline{h}^{\perp c}$. It can be easily shown that the required operator Φ equals Φ^c $\underline{h}^{\perp}$. Henceforth we skip the index c and deal only with complex algebras.

3. Let the gauge group K and the symmetry group G be simple classical Lie groups, $\underline{h} \subset \underline{g}$ and $\mathcal{T}(\underline{h}) \subset \underline{k}$ regular subalgebras of their Lie algebras.

To construct the intertwining operator Φ explicitly we use the root technique for complex Lie algebras as well as the Dynkin schemes which are graphic representations of simple root systems [5,6]. However, it is convenient for us to represent graphically not only the simple roots, but all the positive roots. There are natural orderings of positive roots of the classical Lie algebras which are generalizations of the Dynkin diagrams. We call them root lattices.

The nodes of a root lattice are in one-to-one correspondence with positive roots. For the series A_m (corresponding to SU(m+1)) these lattices have triangle form (Fig.1). The nodes of the lattice denoted by $\alpha_1, \alpha_2, \ldots, \alpha_m$ correspond to the simple roots. The segments starting from the points α_i and α_j $(i<j)$ intersect at the node corresponding to the positive root $\alpha(i,j) = \alpha_i + \alpha_{i+1} + \ldots + \alpha_j$. The root lattice for the algebras C_n (Sp(n)) is shown on Fig.2. Here we have positive roots of two types: $\alpha(i,j)$ and $\beta(i,j)$ $(i \leqslant j,\ i, j = 1, 2, \ldots, n)$. The root $\alpha(i,j)$ corresponds to the node inside the triangle $(\alpha_1, \alpha_n, \delta)$ at the intersection of the segments coming from α_i and α_j; $\alpha(i,j) = \alpha_i + \alpha_{i+1} + \ldots + \alpha_j$. The root $\beta(i,j)$ corresponds to the node inside the triangle $(\delta, \alpha_n, \beta_1)$ at the intersection of the segments coming from α_i and β_j; $\beta(i,j) = \alpha_i + \alpha_{i+1} + \ldots + \alpha_{j-1} + 2(\alpha_j + \ldots \ldots \alpha_{n-1}) + \alpha_n$. At last, $\beta_i \equiv \beta(i,i) = 2(\alpha_i + \ldots + \alpha_{n-1}) + \alpha_n$. Similar schemes correspond to the algebras B_n D_n (SO(2n+1) and SO(2n)).

Root lattices enable us to determine not only the types and the multiplicities of irreducible representations in the decomposition of the restriction of the adjoint representation to a regular subalgebra but also to obtain their explicit realization on root vectors. For example, consider G = SU(m+1), H = SU(m)xU(1). Then $\underline{g} = A_m$ and $\underline{h} = A_{m-1} + \mathbb{C}\tilde{h}$ is its regular subalgebra. The non-abelian part of it is generated by the simple roots

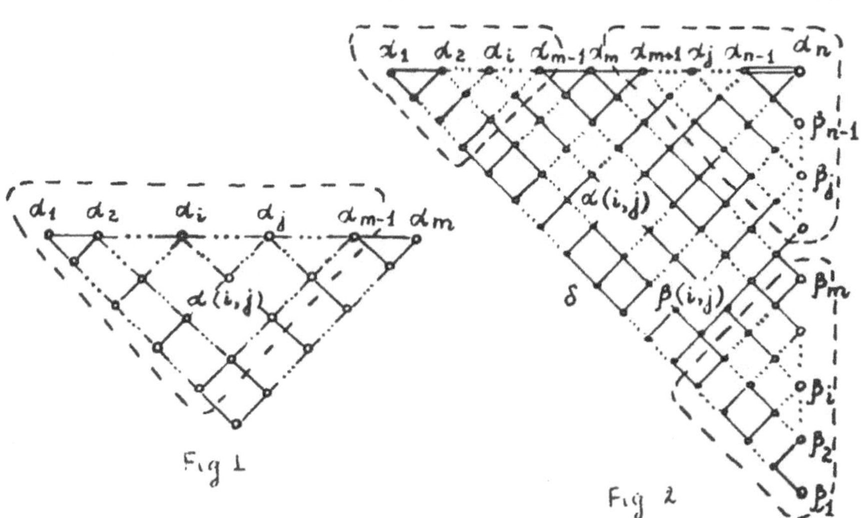

Fig 1

Fig 2

$\alpha_1, \alpha_2, \ldots, \alpha_{m-1}$ (Fig.1) and $\mathbb{C}\tilde{h}$ is one-dimensional subalgebra spanned by the element $\tilde{h} = \sum_{i=1}^{m} j h_{\alpha_j}$ of the Cartan subalgebra. It is convenient to characterize the irreducible representations of ad $\underline{h}$ by weight labels $(\nu_1, \ldots, \nu_{m-1};\ k)$, where $\nu_i = 2\langle \nu, \alpha_i\rangle / \langle \alpha_i, \alpha_i\rangle$, $k = \langle h_\nu, \tilde{h}\rangle / |\langle h_\nu, \tilde{h}\rangle|$ and ν is the maximal weight of the representation [7]. The nodes on the segment $(\alpha(1,m), \alpha_m)$ (Fig.1) correspond to the root vectors $e_{\alpha(\ell,m)}$ $(\ell = 1, \ldots, m)$ of the space $\underline{h}^{\perp}$ carrying the irreducible representation of weight $(1, 0, \ldots, 0;\ 1)$ and dimension m. The conjugate representation is realized on the root vectors $e_{-\alpha(\ell,m)}$ and its signature is $(0, \ldots, 0, 1; -1)$.

Let the gauge group of the initial theory be $K = Sp(n)$ and $\mathcal{T}(H) = SU(m) \times U(1)$ $(n > m)$. Its Lie algebra is $\underline{k} = C_n$, and the corresponding root lattice is represented on Fig.2. The Lie algebra $\mathcal{T}(\underline{h})$ is $A_{m-1} + \mathbb{C}\tilde{H}$, the root lattice of A_{m-1} is the triangle $(\alpha_1, \alpha_{m-1}, \alpha(1, m-1))$. The element $\tilde{H}$ of the Cartan subalgebra equals

$$\tilde{H}=\mathcal{J}(\tilde{h})=2(m+1)\Big(\sum_{j=1}^{m} jH_{\alpha_j}+m\sum_{j=m+1}^{n-1} H_{\alpha_j}+\frac{m}{2}H_{\alpha_n}\Big)$$

and $\langle \tilde{H}, H_{\alpha_j}\rangle = 0$ for all $j \neq m$. (We denote the root vectors and the elements of the Cartan subalgebra of $\underline{k}$ by the capital letters: E_α and H_{α_i}). The triangle $(\alpha_{m+1}, \alpha_n, \beta_{m+1})$ is the root lattice of the Lie algebra C_{n-m}. It does not contain simple roots connected with the simple roots of $\mathcal{J}(\underline{h})$ and therefore belongs to the trivial representation. Thus the centralizer of $\mathcal{J}(H)$ in K, i.e. the gauge group of the reduced theory, is $C = Sp(n-m) \times U(1)$. It can be easily verified that the segments $(\alpha(1,m+\ell-1), \alpha(m,m+\ell-1))$ and $(\beta(1,m+\ell), \beta(m,m+\ell))$, $\ell = 1,2,\dots, n-m$, correspond to $2(n-m)$ equivalent irreducible representations of signature $(1,0,\dots,0;1)$ and dimension m. The triangle $(\beta(1,m), \beta_m, \beta_1)$ describes the irreducible representation of dimension $(m+1)(m+2)/2$ and signature $(2,0,\dots,0;1)$. Intertwining the equivalent representations in the decompositions of ad $(\underline{h})\underline{h}^{\perp}$ and ad $(\mathcal{J}(\underline{h}))\,\underline{k}$ into irreducible representations and using the prescriptions of the section 2, we can find the intertwining operator Φ explicitly. For simplicity we set $e_j = e_{\alpha(j,m)}$, $E_j^s = E_{\alpha(j,m+s-1)}$ for $j = 1,2,\dots,m$, $s = 1,2,\dots,n-m$ and $E_j^s = E_{\beta(j,2m+s-n)}$ for $s = n-m+1,\dots,2(n-m)$.
Then in our case

$$\Phi(x)=\sum_{s=1}^{2(n-m)}\sum_{j=1}^{m}\Big(f_s(x)E_j^s\otimes e_j^* + \bar{f}_s(x)\bar{E}_j^s\otimes\bar{e}_j^*\Big),$$

where e^*_j and $\bar{e}^*_j$ are elements of $\underline{h}^{1*}$. Their realization depends on the choice of metric on the internal space. The functions f_s are the scalar fields of the reduced theory. We can show that the linear hull of the set $\{E^s_j, s=1,2,\ldots, 2(n-m)\}$ for a fixed j is an irreducible representation of AdC or, equivalently, the scalar fields $f_s(x)$ constitute an irreducible multiplet of C.

4. There is a class of internal spaces which are of particular interest to us. These are symmetric spaces [8, 9]. The following statement holds. Let G and K be simple classical groups, $H \subset G$ and $\mathcal{J}(H) \subset K$ their regular subgroups. If G/H is a symmetric space then the reduced theory contains only one irreducible multiplet of scalar fields.

All symmetric spaces G/H realized by the regular embeddings of classical Lie groups into simple classical Lie groups are listed in [8, chap.8, § 11]. This list provides us with all the possible symmetry groups G and their stationary subgroups H for which the reduced theory contains only one irreducible scalar multiplet. Combining various classical groups K and various embeddings $\mathcal{J}(H)$, we can find all the possible gauge groups C as the centralizers of $\mathcal{J}(H)$ in K. Using the root lattice technique, it is not difficult to verify that for all the symmetric spaces except $SU(m+1)/SU(m)\times U(1) = CP^m$ and $SO(m+2)/SO(m)\times SO(2) = G_{2,m+2}(R)$ (Grassmann manifold) the resulting scalar fields transform trivially under the non-abelian part of C. (Recall that C is the gauge group of the reduced theory).

5. Thus, to construct a gauge theory by the dimensional reduction method it is necessary to choose a group K (the gauge group of the initial theory) and an internal space G/H, then to construct explicitly the intertwining

operator Φ (according to the rules discussed in sections 2 and 3) and substitute it into the action (3).

We have got the general solution of the dimensional reduction problem for some class of "free" gauge theories. This class is characterized by the following features. The initial gauge group K and the symmetry group G are compact simple classical Lie groups, the internal space G/H is a symmetric one. Besides, H and τ(H) are regular subgroups of G and K respectively, which include only one direct factor U(1). Here we confine ourselves to the discussion of the case when the group C has also only one such factor.

Our calculations show that the dimensional reduction of theories of the above mentioned class can give us models with the gauge groups SU(n)xU(1), SO(n)xU(1), Sp(n)xU(1) only. If the gauge group of the reduced theory is C = SU(n)xU(1) or C = Sp(n)xU(1) the scalar field potential is of the form

$$V(\varphi(x)) = \frac{1}{4}\nu^2\Big(\sum_k |\varphi_k(x)|^2 - \mu^2\Big)^2 + R \qquad (5)$$

and leads to spontaneous symmetry breaking. The scalar fields $\varphi_k(x)$ form an irreducible multiplet with respect to the nonabelian part of C. Its dimension is n for C = SU(n)xU(1) and 2n for Sp(n)xU(1). If C = SO(n)x U(1) the scalar field transforms according to the vector representation of SO(n) and the potential $V(\varphi)$ is of the form (5) with the additional term $\varkappa\Big|\sum_{k=1}^{n}\varphi_k^2(x)\Big|^2$. Parameters ν, μ, R and $\varkappa$ are determined by the quantities λ, L, m, n, where λ is the coupling constant of the initial multidimensional theory (see (1)), L is the characteristic size of G/H, m the dimension of

G/H and n the rank of the gauge group of the reduced theory. The ratio of the gauge charges g_A and g, corresponding to the abelian and nonabelian components of the gauge field, are expressed through the discrete parameters m and n only.

For example, if K = Sp(n+m) and G/H = SU(m+1)/SU(m)x U(1) = CP^m, the gauge group of the reduced theory is C = Sp(n)xU(1). The scalar potential is

$$V(\varphi)=\frac{g^2(m+1)}{4m(n+1)}\left(\sum_{k=1}^{2n}|\varphi_k|^2-\frac{2}{L^2g^2}m(n+1)\right)^2+\frac{m(m+1)(n+1)}{g^2L^4}. \quad (6)$$

We have also the following relations between the gauge charges g and g_A and the coupling constant λ' (3):

$$g^2=\lambda'^2(n+1)/4; \quad (7a)$$

$$g_A/g=\sqrt{2/m(n+1)}\ . \quad (7b)$$

The potential in (6) is a Higgs potential and leads to the spontaneous symmetry breaking to Sp(n-1) x U(1).

Note that for the class of gauge theories described earlier we can also solve the inverse problem, i.e. the problem of reconstructing a multidimensional theory for a given Higgs model. We have obtained the following result. Consider a Higgs theory with the gauge group NxU(1), where N is a compact simple Lie group, and with a scalar field carrying the fundamental representation of N if N = SU(n) and the vector representation if N = = Sp(n) or N = SO(n). If the scalar potential is of the form $V(\varphi) = \nu^2(|\varphi|^2-\mu^2)^2/4$, there is a discrete set of ratios of the coupling constants, for which this theory can be obtained by the dimensional reduction from a multidimesional "free" gauge theory of the class described above. For example, if the gauge group C is Sp(n)xU(1) then the ratio of coupling constants is given by (7b) and the coupling constant λ of the multidimensional theory and size L of the internal space are determined by (6) and (7a).

6. Among the models which can be derived by the dimensional reduction from "free" multidimensional gauge theories, there is a Higgs model with the gauge group SU(2)x U(1) and the scalar field in the fundamental representation of SU(2), i.e. the boson sector of the Weinberg-Salam model. In the standard version of this model the boson sector has 4 parameters. We can choose them to be electric charge e and masses M_W, M_Z, M_H of the vector and the Higgs bosons. The model obtained by the dimensional reduction has only three independent parameters: λ' , L, m. Note that the Weinberg angle θ_W depends only on the integer m, the dimension of the internal space. This means that there should be a relation among M_W, M_H and M_Z enabling us to make physical predictions.

The most interesting results can be obtained in the case C = Sp(n)xU(1) for n=1 (Sp(1) is isomorphic to SU(2)). From (6) and (7), it is not difficult to get the following expressions for physical parameters

$$e = \lambda'/\sqrt{2(m+1)}; \qquad M_W = \sqrt{2m}/L;$$

$$M_Z = \sqrt{2(m+1)}/L; \quad M_H = \sqrt{2(m+1)}/L; \quad \sin^2\theta_W = 1/(m+1).$$

In this case we have $\sin^2\theta_W = 0.25$ for m=3 and $\sin^2\theta_W =$ 0.2 for m=4. The latter value is closer to the experimental one: $\sin^2 \theta_W^{exp} = 0.217 \pm 0,014$ [10]. The prediction $\sin^2 \theta_W = 0.2$ seems to be rather realistic since the experimental value is bound to decrease. For m=4 we can choose L to fit the experimental data for M_W and M_Z. A prediction of this model is the equality of the Higgs and Z-boson masses, $M_H = M_Z$. Here we have to note that some similar results were obtained earlier in [11].

The methods developed here can be used without any change for calculating scalar potentials in the cases when the reduced theory contains a number of scalar field multiplets carrying different representations of the gauge group. The root lattice technique can be generalized to the exceptional groups. To take into consideration nonregular embeddings of the stationary subgroup H into the symmetry group G and the initial gauge group K, one has to develop essentially new methods.

References

1. Forgacs, P., Manton, N.S. (1980). Commun.Math.Phys., 72, 15-34.
2. Volobujev, I.P., Rudolph, G. (1985). Theor.Math.Phys. 62, 388-399 (Russian).
3. Schwars, A.S. (1977). Commun. Math.Phys., 56, 79-86.
4. Kirillov, A.A. (1978). Elements of Representation Theory. 2nd ed. - Nauka, Moscow (Russian).
5. Goto, M., Grosshans, F.D. (1978). Semisimple Lie algebras. - Marcel Dekker, Inc: New York.
6. Dynkin, E.B. (1952). Mat.sb., 30, 349-462 (Russian).
7. Zhelobenko, D.P. (1970). Compact Lie groups and their representations. Nauka, Moscow (Russian).
8. Wolf, J.A. (1972). Spaces of constant curvature. Univ. of California, Berkley.
9. Kobayashi, S., Nomizu, K. (1963). Foundanations of differential geometry, 1. Interscience Publishers, N.Y.
10. Particle Data Group. Review of particle properties. (1984). Rev.Mod.Phys., 56, part II, S293-S294.

11.Manton, N.S. (1979). Nucl.Phys., B158, 141-153.

PROBLEM OF BOSE-CONDENSATION IN GAUGE THEORIES

E.Ferrer, V. de la Incera, A.E.Shabad [+]
IMACC, Havana
[+] Lebedev Physical Institute, Moscow

1. Introduction. It is known that the phenomenon of Bose-condensation reduces to the following. Consider the density matrix

$$\hat{\varrho} = \exp\{-\beta(\hat{H} - \mu\hat{Q})\} \quad (1.1)$$

where $\beta = 1/T$ is inverse temperature, $\hat{H}$ Hamiltonian, μ chemical potential associated to the conserved charge operator $\hat{Q}$.

Using $\hat{\varrho}$ one can calculate the thermodynamic potential $\Omega(\mu, \beta)$. From the equation

$$Q = \frac{\partial \Omega(\mu, \beta)}{\partial \mu} \quad (1.2)$$

the chemical potential can be determined as a function of the temperature and given expectation value $Q = T_r(\hat{Q}\hat{\varrho})$ of the operator $\hat{Q}$. The positivity condition of the Bose distribution function $n_k^{\pm} = 1/[\exp\beta(\varepsilon_k \mp \mu) - 1]$, where k is the total set of quantum numbers, ε_k the energy spectrum with the minimal value $(\varepsilon_k)_{min} = M$, demands that $|\mu| \leqslant m$. If, for a given Q, the temperature is lowered, or for a given T, the value of Q is raised, a solution of (1.2) can approach the boundary $\mu = \pm m$ of the region $|\mu| \leqslant m$ from inside. As a consequence the surplus of the charge starts following from the distribution into condensate, and the latter cannot described by the thermodynamic potential. In order to handle the situation it is necessary to perform the Bogolyubov transformation. This is equivalent to taking into account the condensate by considering the statistical expectation value $\bar{\phi} = \langle \phi \rangle$ of the field operator ϕ.

In the temperature Green functions method [1] the procedure is as follows. The interaction term $A_0(x)J_0d^3x$ with constant external current J_0 should be added to the Hamiltonian of (1.1), where A_0 is the 4-potential component, additive to μ in the Lagrangian. By solving the tree equations corresponding to T=0 an expression for μ and $\bar{\phi}$ can be found in terms of J_0. Then the charge of the condensate balances the given external charge J_0 . The thermic effect may be taken into account already within the one-loop approximation. For this perform a shift, integrate over the deviation and solve again the equation of motion for the modified Lagrangian with external source. The variation with respect to A_0 gives rise to an equation, equivalent to (1.2). One also obtains modified expressions for μ and $\bar{\phi}$ as functions of J_0, for the same value of the external charge $Q=VJ_0$, where V is the volume of the system. Now the total charge Q of the system is partially compensated by the charge of the condensate and, partially, by the overcondensate gas.

The above picture is adequate in the abelian case [2], but gives rise to some difficulties if we try to apply it to non-abelian gauge theories. The points is that the term of interaction with the external static source J_0A_0 breaks the gauge symmetry. This fact affects also the structure of constraints, and hence the quantization process should be improved [3] (the way to do this in Q.F.T. with non-abelian source is discussed in [4 , 5] and also in [6]). In performing the quantization there appears a new degree of freedom, pointed out at the clasical level by Kiskis [7] . Its possible interpretation as an excitation of the source contradicts the original idea of its independence and is hardly satisfactory.

It is therefore descrable to be able to introduce the interaction with the source in a gauge-invariant way. In other words, to stand with a neutral system without external charge, with two mutually balanced charged statistical quantum subsystems, then to take for one of them the limit of infinite mass in order to consider this part in the limit as a classical charge. However, this plan is again good only for abelian case (see Sect. 2), whereas for the non-abelian one it fails (at least as far as fermions are not involved), because in the non-abelian system the conserved total charge of the system cannot be divided into a sum of independently conserved charges.

2. Coexistence of two Bose-condensates in a neutral abelian system. Consider the Lagrangian of scalar electrodynamics with two complex scalar fields ϕ and χ

$$\begin{aligned}\mathcal{L}= & -\frac{1}{4}F_{\mu\nu}F^{\mu\nu}-(D_\mu\phi)^*(D^\mu\phi)-m^2\phi^*\phi-\lambda\,(\phi^*\phi)^2- \\ & -(D_\mu\chi)^*(D^\mu\chi)-m'^2\chi^*\chi-\lambda'(\chi^*\chi)^2\end{aligned} \tag{2.1}$$

$$F_{\mu\nu}=\partial_\mu A_\nu-\partial_\nu A_\mu,\quad D_\mu=\partial_\mu+ieA_\mu. \tag{2.2}$$

The invariance of lagrangian (2.1) under two independent phase transformations of the fields ϕ and χ implies the existance of two conserved charge densities. One of them is given by

$$j_0(\phi)=-i\left(\phi P_\phi-\phi^* P_{\phi^*}\right) \tag{2.3}$$

where P_ϕ, P_{ϕ^*} are canonical momenta conjugate to ϕ and ϕ^* respectively

$$P_\phi=\frac{\delta\mathcal{L}}{\delta\partial^0\phi}=(D_0\phi)^*,\quad P_{\phi^*}=\frac{\delta\mathcal{L}}{\delta\partial^0\phi^*}=D_0\phi. \tag{2.4}$$

The second charge density $J_o(x)$ has the same form, but with χ, χ^* instead of ϕ and ϕ^* respectively. Afrer subtracting

$$\mu\int j_o(\varphi)d^3x+\mu'\int j_o(\chi)\,d^3x$$

from the Hamiltonian of the system and performing the canonical integration in the expression for the grand partition function Z, we obtain Z as a functional integral over the fields. The effective Lagrangian in the integral differs from (2.1) by the replacement $D_v\phi \to \tilde{D}_v\phi$, $D'_v\chi \to \tilde{D}'_v\chi$, where

$$\tilde{D}_v=\partial_v+ie[A_v+(\mu/e)\delta_{v0}],\ \tilde{D}'_v=\partial_v+ie[A_v+(\mu'/e)\delta_{v0}]. \quad (2.5)$$

The equations corresponding to a system, neutral as a whole have the solution $A_v=0$, $\phi=\bar{\phi}$, $\chi=\bar{\chi}$, where $\bar{\phi}$ and $\bar{\chi}$ are real, and

$$\bar{\Phi}^2=\frac{\mu^2-m^2}{\partial\lambda},\ \bar{\chi}^2=\frac{\mu'^2-m'^2}{\partial\lambda'},\ \frac{\mu(\mu^2-m^2)}{\lambda}=\frac{-\mu'(\mu'^2-m'^2)}{\lambda'}. \quad (2.6)$$

It makes sense if $|\mu|\gg m$, $|\mu'|\gg m'$. The charges of the condensates are found using (2.3), (2.4) and substituting $D_v\to\tilde{D}_v$, $D'_v\to\tilde{D}'_v$ in (2,1): $j_o(\bar{\phi})=-\partial\mu\,\bar{\phi}^2$, $j_o(\bar{\chi})=-\partial\mu'\,\bar{\chi}^2$. The charge equilibrium condition $j_o(\bar{\phi})+j_o(\bar{\chi})=0$ is guaranteed by the last equation in (2,6). With the densities of the both condensates fixed, the chemical potentials can be obtained by solving a cubic equation.

3. Extra degree of freedom in the Yang-Mills-Higgs model in the presence of an external charge. Consider now an interaction of a SU(2) Yang-Mills field with a scalar triplet of Higgs scalars ϕ^a, and with an external static source $J^a=J_o\delta^{a3}\delta_{\mu o}$.

The lagrangian of this model is given by (2.1) with $\chi = 0$, $m^2 < 0$, and

$$F^a_{\mu\nu} = \partial_\mu A^a_\nu - \partial_\nu A^a_\mu - g\epsilon_{abc} A^b_\mu A^c_\nu, \quad D_\nu = \partial_\nu - g A^a_\nu \tau^a \tag{3.1}$$

minus the additional term $gA^3_o J_o$.

When we apply the Hamiltonian formalism to this system the consistency condition of the constraints $\pi^a_o \approx 0$ does not give rise to the usual first class constraints

$$\gamma^a = \vec{\partial}\cdot\vec{\pi}^a - g\epsilon_{abc}\vec{A}^b\cdot\vec{\pi}^c - g\rho\tau^a\phi \approx 0. \tag{3.2}$$

On the contrary, brackets of π^a_o with the Hamiltonian create the constraints $\tilde{\gamma}^a = \gamma^a - gJ_o\,\delta^{a3} \approx 0$. At the same time the consistency of $\tilde{\gamma}^a$ amounts to the two new constraints $\chi^b = \epsilon_{3bc}A^c_o J_o \approx 0$. Then, totally, we have the eight constraints π^a_o, $\tilde{\gamma}^a$, χ^b , from which only π^3_o and $\tilde{\gamma}^3$ are first class ones. This means that we must impose two gauge conditions, which we select in the form

$$A^3_o = 0, \quad F_\eta = \partial_\nu A^3_\nu + \frac{g}{\eta}\phi\tau^3\bar{\phi} - f(\vec{x},t) = 0 \tag{3.3}$$

where $\eta \to 1$, and $\bar{\phi}$ is the solution of the tree equations of the theory with external charge and chemical potential. Taking into account that we have now ten constraints, we easily see that 30 initial degrees of freedom in the phase space ((3x4 vectorial+3 scalar)x2), have been finally reduced to 20.

This means 10 degrees of freedom in the coordinate space, instead of the usual 9, when $J_o=0$.

Quantization in presence of first and second class constraints [8] is reduced to calculating the path integral for the grand partition function

$$Z = \int e^{iS}\sqrt{\det\{\Psi_n,\Psi_m\}}\;\delta(\Psi_n)\,dp\,dq \tag{3.4}$$

where ψ_n represent all the constraints (including the gauge conditions), dpdq measure over all the canonical variables. Substituting here the action in the Hamiltonian form with the source J_o and the term μj_o^3 included, where j_o^3 is the conserved charge density (it does not commute with the components $j_o^{1,2}$)

$$j_o^3 = \epsilon_{3bc} \vec{A}^b \cdot \vec{\pi}^c + p_\Phi \tau^3 \Phi \tag{3.5}$$

we obtain after integrating over all the momenta

$$Z = \int \exp\left[i \int_0^\beta dx_4 \int d^3x \mathcal{L}_{eff}\right] \sqrt{\det\{\chi^a, \chi^b\}}\, \delta(F_2) \det\{F_2, \tilde{\gamma}^3\} dA_\mu^a d\Phi \tag{3.6}$$

where χ^a are six first class constraints and L_{eff} is obtained by replacing $A_V^a \rightarrow A_V^a + \frac{\mu}{g} \delta^{a3} \delta_{V0}$ in (3.1) from the initial Hamiltonian. Consider the tree equation confining ourselves to the class of fields $A_\mu^a = 0$, $\bar{\phi}^a =$ =const. Equation $(\delta L_{eff}/\delta \phi) = 0$ are of the form

$$(-m^2 - 2\lambda \bar{\Phi}^2) \begin{pmatrix} \Phi_1 \\ \Phi_2 \\ \Phi_3 \end{pmatrix} + \mu^2 \begin{pmatrix} \Phi_1 \\ \Phi_2 \\ 0 \end{pmatrix} = 0 \tag{3.7}$$

while the equation $(\delta L_{eff}/\delta A_V^c) = 0$ is of the form

$$\Phi_1 \Phi_3 = 0, \quad \Phi_2 \Phi_3 = 0, \quad \mu (\Phi_1^2 + \Phi_2^2) = J_o . \tag{3.8}$$

Equations (3.7), (3.8) have the solution

$$\bar{\Phi}_3 = 0, \quad \xi^2 = \frac{-m^2 + \mu^2}{2\lambda}, \quad \mu = J_o / \xi^2 \tag{3.9}$$

where $\xi^2 = \phi_1^2 + \phi_2^2$, $m^2 < 0$.

In the same way as in [2] , the existence of the source implies that the Higgs mechanism produces a charged condensate, while the neutral component of the Higgs field, ϕ_3 , does not condense.

Note that in the non-abelian Higgs model without external source $(J_o=0)$, rgere exists a nonzero solution for the scalar field $\phi_1=\phi_2=0$, $\phi_3^2=m^2/\lambda$, but it is neutral with respect to the charge j_o^3 . Similarly, in the abelian model of the previous section, if we consider the Higgs case, $m^2<0$, there exists the solution $\bar{\phi}^2=|m^2|/2\lambda$, $\mu=0$ when $j_o(\ )=-j_o(\bar{\phi})=0$, but the charge of this state is zero because $\mu=0$.

Shifting the Higgs fields by the solution (3.4) and holding the quadratic contributions of the fields into the action, the thermodynamic potential can be obtained. It is given, as usual, by logarithms of Bose-distributions of the quasiparticles with the following spectrum of frequencies:

$$\omega_{1,2}^2=\frac{1}{2}\left[k^2+2\mu^2+g^2\xi^2\pm\sqrt{k^4+2g^2\xi^2k^2+g^2\xi^2(8\mu^2+g^2\xi^2)}\right]$$

$$\omega_{3,4}^2=\frac{1}{2}\left[\partial k^2+4\mu^2+(g^2+2\lambda)\xi^2\pm\sqrt{[4\mu^2+(2\lambda-g)\xi^2]^2+16\mu^2k^2}\right] \tag{3.10}$$

$$\omega_{5,6}^2=\frac{1}{2}\left[\partial k^2+\partial\mu^2+g^2\xi^2\pm\sqrt{g^4\xi^4+16\mu^2k^2}\right],\ \omega_7^2=k^2+g^2\xi^2.$$

Here the frequencies $\omega_{5,6,7}$ are twice degenerate, i.e. we have 10 modes altogether. In order to go to the no-source limit it suficies to put $\mu=0$, $\xi^2=m^2/2\lambda$. Then the frequency ω_2 tends to zero and new modes remain, as in the case when there was no source from the bagining.

4. <u>Nonseparation of the condensates in non-abelian theory.</u> Consider a non-abelian analogue to the model studied in Sect. 2 with the definition (3.1) in (2.1). The conserved current, corresponding to the global SU(2) symmetry, is formed now by the three parts

$$j_o^a=j_o^a(A)+j_o^a(\Phi)+j_o^a(\chi) \tag{4.1}$$

where

$$j_0^a(A)=\epsilon_{abc} A_i^b F_{0i}^c,\quad j_0(\Phi)=\frac{1}{2}\left[(D_0\Phi)^t\tau^a\Phi-\Phi^t\tau^a(D_0\Phi)\right]$$

and $j_0^a(\chi) = j_0^a(\phi \to \chi)$. They are not conserved separately, therefore we cannot introduce in this case two different chemical potentials associated twith two subsystems, each one with a nonzero charge, in so as the total charge were zero, Therefore, inlike the abelian case, in the non-abelian one, the difficulties with obtaining Bose condensation appear already at the dynamic level and do not reduce to forming the external charge by some limiting process: in the first place we cannot construct a dynamic state from which we could perform this limitiong transition. The abovesaid does not conceern non-abelian theories with fermions (cf. [9]).

The authors are grateful to prof. E.S.Fradkin for a thorough discussion of the problem and numerous suggessions. E.Ferrer and V.de la Incera are grateful to I.E.Tamm Department of Theoretical Physics FIAN for hospitality.

References

1. Fradkin,E.S. (1965). Proceedings of P.N.Lebedev Physical Institute 29, 7-138. Nauka, Moscow (Russian).
2. Kapusta,J. (1981). Phys. Rev. D, 24, 426-440.
3. Ferrer,E.F. , de la Incera,V. (1984). Rev. Cubana de Fisica, IV, 3.
4. Cabo,A., Shabad,A.E. (1980). Informe Cientifico Técnico N 167, ININTER, Ac. Ciencias de Cuba.
5. Cabo,A., Shabad,A.E. (1983). In: Group-Theoretical Methods in Physics, Proceedings of Zcenigorod 1982 Seminar, 1, 135-148, Nauka, Moscow (Russian).
6. Bialinki-Birula,I. (1986). In: Quantum Field Theory and Quantum Statistical Essays in Honor of the 60th Birthday of E.S.Fradkin, Adam Hitger Ltd. Bristol.
7. Kiskis,J. (1981). Phys. Rev.. D, 21, 1074-1091.
8. Fradkin,E.S. (1973). Acta Universitates Wratislaviensis N 207 Proceedings of the X-th Winter School of Theor. Phys.
9. Linde,A.D. (1979). Rep. Prog. Phys. 42, 389-437.; Phys. Lett. 86B, 39.

GAUGE INVARIANT GENERALIZATION OF THE WIGNER FUNCTION OF QUANTUM SYSTEM IN A NONUNIFORM MAGNETIC FIELD

V.V.Dodonov, V.I.Man'ko, D.L.Ossipov

P.N.Lebedev Physical Institute, Leninsky Prospect,53, Moscow 117924

As is well known, there are several formulations of quantum mechanics. One of them, closely connected with the Weyl representation of operators [1] and Wigner functions [2], use functions on the phase space to represent both observables and quantum states. For working in this representation it is necessary to replace operators by their Weyl symbols and density matrices by Wigner functions [3,4]. After fundamental works of Weyl, Wigner and Kirkwood [1,2,5] properties of the representation were studied and used to calculate the properties of various physical systems (see [6-23]), including systems with magnetic field [15-23].

The Wigner function is not gauge invariant neither are the density matrix and quantum state vector (see e.g. [22-25]). In calculations with a non-uniform magnetic field this leads to many gauge non-invariant terms with derivatives of vector potential [22,23]. That is why the desire to introduce a gauge-invariant generalization of the Wigner function is quite natural. For the first time this had been done by Stratonovich in 1956 [25]. Later this function was introduced by several authors in quantum mechanics [26-29] and quantum field theory [30], including non-abelian gauge fields [31]. Some properties of gauge invariant Wigner function, including a formula for the Weyl symbol of product of two operators for a special case (see (22), were obtained by Stratonovich [25] but we give a more detailed derivation of properties of modified Weyl transformation and illustrate them by some simple example. In our consideration we adhere to the accepted in [3] method of deriving usual Weyl representation properties.

One can write the following identity for an arbitrary operator $\hat{B}$ ($\bar{\mathcal{A}}(\bar{p},\bar{v})$ is defined by (4-6))

$$\hat{B} = (2\pi\hbar)^{-3N}\int d\bar{p}d\bar{q}\, d\bar{v}_1 d\bar{v}_2 \,|\, \bar{q}+\frac{\bar{v}_2}{2}\rangle\langle \bar{q}-\frac{\bar{v}_1}{2}|\hat{B}|\,\bar{q}+\frac{\bar{v}_1}{2}\rangle\langle\bar{q}-\frac{\bar{v}_2}{2}|\cdot\exp\left\{\frac{i}{\hbar}\left[\bar{v}_1(\bar{p}+\bar{\mathcal{A}}(\bar{q},\bar{v}_1)) + \bar{v}_2(\bar{p}+\bar{\mathcal{A}}(\bar{q},\bar{v}_2))\right]\right\} = (2\pi\hbar)^{-3N}\int d\bar{p}d\bar{q}\; b(\bar{p},\bar{q})\hat{D}(\bar{p},\bar{q}) \tag{1}$$

Hereafter any vector is 3N-dimensional. It is assumed that vector components can be broken into group of three components each, the i-th group describing some vector characteristics of the i-th particle:

$$\hat{D}(\bar{p},\bar{q}) = \int d\bar{v}\,\exp\left\{\frac{i}{\hbar}\bar{v}(\bar{p}+\bar{\mathcal{A}}(\bar{q},\bar{v}))\right\}|\bar{q}+\frac{\bar{v}}{2}\rangle\langle\bar{q}-\frac{\bar{v}}{2}| \tag{2}$$

$$b(\bar{p},\bar{q}) = \int d\bar{v}\,\exp\left\{\frac{i}{\hbar}\bar{v}(\bar{p}+\bar{\mathcal{A}}(\bar{q},\bar{v}))\right\}\langle\bar{q}-\frac{\bar{v}}{2}|\hat{B}|\bar{q}+\frac{\bar{v}}{2}\rangle \tag{3}$$

Call $b(\bar{p},\bar{q})$ [25] the modified Weyl symbol of an operator $\hat{B}$ (set $\rightleftarrows$ for the correspondence between an operator and its modified Weyl symbol).

Writw out some representations of $\bar{\mathcal{A}}(\bar{p},\bar{q})$ (see (1)) [25-27,29]:

$$\bar{\mathcal{A}}(\bar{q},\bar{v}) = \int_{-\frac{1}{2}}^{\frac{1}{2}} d\lambda \bar{A}(\bar{q}+\lambda\bar{v}) \tag{4}$$

$$= (\bar{v}\cdot\bar{\partial}(\bar{q}))^{-1}\left[\bar{A}(\bar{q}+\tfrac{1}{2}\bar{v})-\bar{A}(\bar{q}-\tfrac{1}{2}\bar{v})\right] \tag{5}$$

$$= \bar{A}(\bar{q}) + \frac{1}{24}\bar{A}_{ij}(\bar{q})v_i v_j + O(v^4) \tag{6}$$

$\bar{A}(\bar{q})$ is a 3N-dimensional vector potential formed as described above and multiplied by e/c, $\bar{\partial}(\bar{q})=(\partial_1(\bar{q}),\ldots,\ldots,\partial_{3N}(\bar{q}))=\frac{\partial}{\partial q}$ and (5) is understood in the sence of Fourier transform, i.e. one must take Fourier transform from (4) and (5) to obtain coinciding results. Let a) vector and tensor functions of co-ordinates carry superscripts $\bar{A}(\bar{q})=(A^1(\bar{q}),\ldots,A^{3N}(\bar{q}))$ and derivatives with respect to q subscripts $\partial_i(\bar{q})\bar{A}(\bar{q})=\bar{A}i(\bar{q})$; b) derivatives with respect to p carry superscripts $\partial_i(\bar{p})W(\bar{p},\bar{q})=W^i(\bar{p},\bar{q})$. One can use (5) directly to obtain the series (6), because of the absence of the 0-th term of the expansion of $\bar{A}(\bar{q}+\frac{1}{2}\bar{v})-A(\bar{q}-\frac{1}{2}\bar{v})$ in $\bar{v}$.

Note the properties of $\mathcal{A}(\bar{q},\bar{v})$ needed in sequel

1)

$$\bar{\mathcal{A}}(\bar{q},\bar{v})=\bar{\mathcal{A}}(\bar{q},-\bar{v}) \tag{7}$$

Due to this property a real function of $\bar{p}$ and $\bar{q}$ corresponds to a Hermitian operator.

2) Under the gauge transformation $\bar{A}(\bar{q})\longrightarrow\bar{A}(\bar{q})+\bar{\partial}(\bar{q})\chi(\bar{q})$ $\bar{v}\bar{\mathcal{A}}(\bar{q},\bar{v})$ turns into

$$\bar{v}\bar{\mathcal{A}}(\bar{q},\bar{v})+\chi(\bar{q}+\frac{\bar{v}}{2})-\chi(\bar{q}-\frac{\bar{v}}{2}) \tag{8}$$

To verify (1) integrate over $\bar{p}$ and then over one of the variables $\bar{v}$. The resulting operator is the operator $\hat{B}$ between two unity operators expanded over eigenbasis of position operator.

Using known identities [3] :

$$|\bar{q}+\frac{\bar{v}}{2}\rangle=e^{-\frac{i}{\hbar}\bar{v}\bar{p}}|\bar{q}-\frac{\bar{v}}{2}\rangle,\ |q\rangle\langle q|=$$

$$=(2\pi\hbar)^{-3N}\int d\bar{u}\ e^{i/\hbar(\bar{q}-\hat{\bar{q}})\bar{u}}$$

one can derive a more symmetric expansion for $\hat{D}(\bar{p},\bar{q})$:

$$\hat{D}(\bar{p},\bar{q})=(2\pi\hbar)^{-3N}\int d\bar{v}d\bar{u}\,\exp\left\{\frac{i}{\hbar}(\bar{v}\left[\bar{p}+\bar{\mathcal{A}}(\bar{q},\bar{v})-\hat{\bar{p}}\right]-\right.$$

$$\left.-\bar{u}\left[\bar{q}-\hat{\bar{q}}\right])\right\}.$$

(Do not confuse c-numbers $\bar{p}$ and $\bar{q}$ with operators $\hat{\bar{p}}$ and $\hat{\bar{q}}$). Substituting some simplest operators into (3), we obtain

$$\hat{I} \rightleftarrows 1 \tag{9}$$

$$U(\hat{\bar{q}}) \rightleftarrows U(\bar{q}) \tag{10}$$

$$\hat{\bar{p}} \rightleftarrows \bar{p} + \bar{A}(\bar{q}) \tag{11}$$

$$\hat{\bar{\pi}} \rightleftarrows \bar{p} \tag{12}$$

where $U(\hat{\bar{q}})$ is an arbitrary function of a position operator, $\hat{\bar{p}} = -i\hbar\bar{\partial}(\bar{q})$ is a canonical momentum operator, $\hat{\bar{\pi}} = \hat{\bar{p}} - \bar{A}(\hat{\bar{q}})$ a "physical" momentum operator.

With the aid of the identity

$$\langle\bar{p}'|\hat{B}|\bar{p}''\rangle = \mathrm{Tr}(\hat{B}|p''\rangle\langle p'|\ ) \tag{13}$$

and (2) we get a concise formula

$$b(\bar{p},\bar{q}) = \mathrm{Tr}(\hat{B}\,\hat{D}(\bar{p},\bar{q})) \tag{14}$$

Substituting $\hat{I}$ for $\hat{B}$ in (14) and using (9) we get:

$$\mathrm{Tr}(\hat{D}(\bar{p},\bar{q})) = 1 \tag{15}$$

From here we have for $\hat{B}$ using (1)

$$\mathrm{Tr}\,\hat{B} = (2\pi\hbar)^{-3N}\int d\bar{p}d\bar{q}\; b(\bar{p},\bar{q}) \tag{16}$$

Substituting $\hat{D}(\bar{p},\bar{q})$ for $\hat{B}$ in (14) we get

$$Tr(\hat{D}(\bar{p}',\bar{q}')\hat{D}(\bar{p},\bar{q})) = (2\pi\hbar)^{3N}\,\delta(\bar{p}-\bar{p}')\,\delta(\bar{q}-\bar{q}') \qquad (17)$$

$$\hat{D}(\bar{p}',\bar{q}') \rightleftarrows (2\pi\hbar)^{3N}\,\delta(\bar{p}-\bar{p}')\,\delta(\bar{q}-\bar{q}') \qquad (18)$$

The obtained formula allows us to express the trace of two arbitrary operators in representation (3)

$$Tr(\hat{B}\hat{C}) = (2\pi\hbar)^{-3N}\int d\bar{p}d\bar{q}\; b(\bar{p},\bar{q})\; c(\bar{p},\bar{q}) \qquad (19)$$

One can get a formula for the trace of three D-operators by introducing between them expansions of unity operators in eigenbasis of position operator:

$$Tr(\hat{D}(\bar{p},\bar{q})\hat{D}(\bar{p}',\bar{q}')\hat{D}(\bar{p}'',\bar{q}'')) =$$

$$= 2^{6N}\exp\Big\{\frac{2i}{\hbar}\Big[(\bar{q}''-\bar{q}')(\bar{p}+\bar{\mathcal{A}}(\bar{q},2(\bar{q}''-\bar{q}')) +$$

$$+ (\bar{q}-\bar{q}'')(\bar{p}'+\bar{\mathcal{A}}(\bar{q}',2(\bar{q}-\bar{q}'')))=$$

$$= (\bar{q}'-\bar{q})(\bar{p}''+\bar{\mathcal{A}}(\bar{q}'',2(\bar{q}'-\bar{q})))\Big]\Big\} \qquad (20)$$

or

$$= 2^{6N}\exp\Big\{\Big[\frac{2i}{\hbar}\Big[(\bar{q}'-\bar{q})(\bar{p}''-\bar{p}+\bar{\mathcal{A}}(\bar{q}'',2(\bar{q}'-\bar{q}))-$$

$$- (\bar{q},2(\bar{q}''-\bar{q}'))) - (\bar{q}''-\bar{q})(\bar{p}'-\bar{p}+\bar{\mathcal{A}}(\bar{q}',2(\bar{q}-\bar{q}'')) -$$

$$-\bar{\mathcal{A}}(\bar{q},2(\bar{q}''-\bar{q})))\Big]\Big\} \qquad (21)$$

A formula for a modified Weyl symbol of the product of two operators $G(\hat{\bar{\pi}})L$ was obtained by Stratonovich [25] for the case when the first operator depends only on $\hat{\bar{\pi}}$ in the following form

$$G(\hat{\bar{\pi}})\hat{L} \rightleftarrows G(\hat{\bar{\pi}})\,l(\bar{p},\bar{q}) \qquad (22)$$

where $G(\hat{\Pi})$ is the same function but of operator acting in the phase space $(\bar{p},\bar{q})$

$$\hat{\Pi}_m = p_m + \frac{\hbar}{2i}\partial_m(\bar{q}) + \frac{\hbar}{2i}\partial_e(\bar{p})\int_{-1}^{1} F^{em}\left(\bar{q} - \frac{\hbar\lambda}{2i}\bar{\partial}(p)\right)\frac{1+\lambda}{2}\,d\lambda \tag{23}$$

where $F^{ie}(\bar{q}) = A_i^e(\bar{q}) - A_e^i(\bar{q})$.

Note that formula (23) was also obtained by Bopp [32] some years later but without gauge field, i.e. for ordinary Weyl representatiom.

Here we give a general formula for the modified Weyl symbol of the product of two arbitrary operators without proof. One can get it using (1), (14), (20) (cf. the derivation of a similar formula for ordinary Weyl representation [3]):

$$\hat{B}\hat{C} \rightleftarrows b(\bar{p},\bar{q})c(\bar{p},\bar{q})\exp\{\mathcal{D}^{bc}\} \tag{24}$$

where

$$\begin{aligned}\mathcal{D}^{bc} = \frac{\hbar}{2i}\Big[&\bar{\partial}^b(\bar{p})\bar{\partial}^c(\bar{q}) - \bar{\partial}^b(\bar{q})\bar{\partial}^c(\bar{p}) + \\ &+\Big[\bar{\mathcal{A}}\left(\bar{q}, \tfrac{\hbar}{i}(\bar{\partial}^b(\bar{p})+\bar{\partial}^c(\bar{p}))\right) - \bar{\mathcal{A}}\left(\bar{q}+\tfrac{\hbar}{2i}\bar{\partial}^b(\bar{p}),\right. \\ &\left.\tfrac{\hbar}{i}\bar{\partial}^c(p)\right)\Big]\bar{\partial}^c(\bar{p}) + \Big[\bar{\mathcal{A}}\left(\bar{q}, \tfrac{h}{i}(\bar{\partial}^b(\bar{p})+\bar{\partial}^c(\bar{p}))\right) - \\ &- \bar{\mathcal{A}}\left(\bar{q} - \tfrac{\hbar}{2i}\bar{\partial}^c(\bar{p}), \tfrac{h}{i}\bar{\partial}^b(\bar{p})\right)\Big]\bar{\partial}^b(\bar{p})\end{aligned} \tag{25}$$

The differential operators act on the functions indicated by upper indices. For example $\bar{\partial}^c(\bar{p})$ acts only on $c(p,q)$. Note the analogy between the above convention and normally ordered operators in quantum field theory: after expanding an operator expression in $\hbar$ the operators act only on functions labeled by superscripts, it is not necessary to differentiate the functions of

$\bar{q}$, which one gets as coefficients of the expansion.

Now some properties of $\mathcal{D}^{bc}$:

a) with the aid of (8) one can show the gauge invariance of $\mathcal{D}^{bc}$;

b) from (25) one can see that $\mathcal{D}^{bc} - \mathcal{D}^{cb}$ contains only odd terms of expansion in $\hbar$ and $\mathcal{D}^{bc} + \mathcal{D}^{cb}$ only even terms. It follows that the transformations of the commutator and anticommutator of two operators are of the form:

$$[\hat{B},\hat{C}]_{-} \rightleftarrows b(\bar{p},\bar{q})\ c(\bar{p},\bar{q})\ od\{\mathcal{D}^{bc}\} \tag{26}$$

$$[\hat{B},\hat{C}]_{+} \rightleftarrows b(\bar{p},\bar{q})\ c(\bar{p},\bar{q})\ ev\{\mathcal{D}^{bc}\} \tag{27}$$

where $od\{\mathcal{D}\}^{bc}$ and $ev\{\mathcal{D}^{bc}\}$ denote operator series containing only odd or even terms of the $\hbar$-expansion of $\exp\{\mathcal{D}^{bc}\}$ respectively.

A generalized Wigner function is the modified Weyl symbol of a density operator $\hat{P}$. The average value $\bar{B}$ of $\hat{B}$ in quantum state described by $\hat{P}$ is:

$$\bar{B} = Tr(\hat{P}\hat{B}) = \int \frac{d\bar{p}d\bar{q}}{2\pi h}\ W(\bar{p},\bar{q})\ b(\bar{p},\bar{q}) \tag{28}$$

where $W(\bar{p},\bar{q})$ is the generalized Wigner function. Further we will need the expansion of $\mathcal{D}^{bc}$ up to $\hbar^2$:

$$\begin{aligned}\mathcal{D}^{bc} &= \frac{\hbar}{2i}\Big[\partial^{b}_{l}(\bar{p})\,\partial^{c}_{l}(\bar{q}) - \partial^{b}_{l}(\bar{q})\,\partial^{c}_{l}(\bar{p}) - \\ &- \partial^{b}_{l}(p)\ F^{li}\ \partial^{c}_{i}(p)\Big] + \frac{\hbar^2}{12}\Big[\partial^{c}_{i}(p)\,\partial^{b}_{m}(p)\ \cdot \\ &\cdot F^{il}_{m}\,\partial^{c}_{l}(p) + \partial^{b}_{l}(\bar{p})\ F^{il}_{m}\ \partial^{c}_{i}(\bar{p})\ \partial^{c}_{m}(\bar{p})\Big] + O(\hbar^4)\end{aligned} \tag{29}$$

Thus we have clearly illustrated the statement of gauge invariance of $\mathcal{D}^{bc}$.

With the help of eqs. (10), (12), (29) we get the transformation of Hamiltonian

$$\hat{H} = \frac{1}{2}M_{ik}\hat{\pi}_i\hat{\pi}_k + U(\hat{q}) \tag{30}$$

of N particles in magnetic field

$$\hat{H} \rightleftarrows h_{cl} = \frac{1}{2} M_{ik}p_i p_k + U(\bar{q}) \tag{31}$$

Note that the modified Weyl symbol of Hamiltonian (30) does not depend on magnetic field.

Obtain the Bloch equation for the equilibrium density matrix $\hat{P} = \exp\{-\beta\hat{H}\}$:

$$-\frac{\partial\hat{P}}{\partial\beta} = \hat{H}\hat{P} \tag{32}$$

(where $\beta = (kT)^{-1}$ is the inverse temperature) in representation (3)

$$-\frac{\partial W(\bar{p},\bar{q},\beta)}{\partial\beta} = h_{cl}(\bar{p},\bar{q})\, W(\bar{p},\bar{q},\beta)\ \mathrm{ev}\{\lambda^{hW}\} \tag{33}$$

(we have used the commutativity of $\hat{P}$ and $\hat{H}$). Note that eq. (33) for the uniform magnetic field takes the form (see (16)):

$$-\frac{\partial W}{\partial\beta} = h_{cl}\, W \cos\Big\{\frac{\hbar}{2}\Big[\partial^h_\ell(\bar{q})\partial^W_\ell(\bar{p}) - \partial^h_\ell(\bar{p})\partial^W_\ell(\bar{q}) + \\ + \partial^h_\ell(\bar{p})\, F^{li}\,\partial^W_i(\bar{p})\Big]\Big\} \tag{34}$$

This equation was obtained in [20,21,23] for ordinary Wigner function using the change of momentum variables. Hence it follows that the modified Weyl transformstion is necessary only in the presence of non-uniform field.

In the limit $\hbar = 0$ the solution of (33) with the initial value $W(\bar{p},\bar{q},\beta) = 1$ at $\beta = 0$ is the classical equilibrium distribution:

$$W_0(\bar{p},\bar{q},\beta) = e^{-\beta h_{cl}(\bar{p},\bar{q})} \tag{35}$$

Let us obtain quantum corrections to classical solution up to h^2 . It is natural to seek a solution in the form

$$W(p,q,\beta) = e^{-\beta h_{cl}} \; k(\bar{p},\bar{q},\beta) \tag{36}$$

where

$$k(\bar{p},\bar{q},\beta) = 1 + \hbar^2 k_2(\bar{p},\bar{q},\beta) + O(\hbar^4) \tag{37}$$

Expanding $\exp\{\mathcal{D}^{hW}\}$ up to $\hbar^2$ we get

$$\begin{aligned} -\frac{\partial W}{\partial \beta} = h_{cl}W - \frac{\hbar^2}{8}(M_{ie}U_{ie}+U_{ie}W^{ei} + \\ + 2M_{ie}F^{ek}W^k_i + M_{ie}F^{ek}F^{im}W^{km}) + \\ + \frac{\hbar^2}{12}(M_{im}F^{ie}_m W^l + M_{ek}p_k F^{ie}_m W^{im}) \end{aligned} \tag{38}$$

Substituting (36) in (38), separating different powers of $\hbar$, and integrating over β we get

$$\begin{aligned} k_2(\bar{p},\bar{q},\beta) = -\frac{\beta^2}{8}M_{ie}U_{ie} + \frac{\beta^3}{24}M_{ik}M_{em}U_{ie}p_k p_m + \\ + \frac{\beta^3}{24}M_{ie}U_i U_e - \frac{\beta^3}{12}M_{ei}M_{kq}F^{ek}U_i p_q - \\ - \frac{\beta^2}{16}M_{ei}M_{km}F^{ek}F^{im} + \\ + \frac{\beta^3}{24}M_{ei}M_{kp}M_{mq}F^{ek}F^{im}p_p p_q + \frac{\beta^2}{12}M_{im}M_{ek}F^{ie}_m p_k \end{aligned} \tag{39}$$

Although it has already been proved that it must be the case (see (28)), it is interesting that the verification yieldes the equality of O-th and first momenys of "physical" momentum which one can get with the aid of (39) and moments given by convolution with analogous expression for the ordinary Wigner function (see eq. (36) in [23]).

References

1. Weyl,H. (1927). Z. Physik. 46, 1.

2. Wigner,E.P. (1932). Phys. Rev., 40, 749.

3. de Groot,S.R., Suttorp,L.G. (1972). Foundations of Electrodynamics, North-Holland Publishing Company, Amsterdam, Chap. 6.

4. Tatarsky,V.I. (1983). Uspekhi Fiz. Nauk 139, 587 (Russi (Russian); Hillery,M., O'Connell,R.F., Scully,M.O., Wigner,E.P. (1984). Phys. Rep. 106, 121.

5. Kirkwood,J.G. (1933). Phys. Rev. 44, 31.

6. Groenwold,H.J. (1946). Physica 12, 405.

7. Moyal,J.E. (1949). Proc. Cambridge Phil. Soc. 45,99.

8. Imre,K., Ozizmir,E,, Rosembaum,M., Zweifel,P.F. (1967). J. Math. Phys. 8, 1097.

9. Leaj,B. (1968). J. Math. Phys. 9, 65.

10. Stratonovich,R.L. (1957). Zh. Eksp. Teor. Fiz. 4,891 (Russian).

11. Baker,G.A.(Jr.) (1958). Phys. Rev. 109, 2198.

12, Takabayasi,T. (1954). Progr. Theor. Phys. 11, 341.

13. Kubo,R. (1964). J: Phys. Soc. Japan 19, 2127.

14. Gibson, W.G. (1979). Physica, A98, 298.

15. Akhundova,E.A., Dodonov,V.V., Man'ko,V.I. (1982). Physica A115, 215.

16. Dodonov,V.V., Man'ko,V.I. (1983). Trudi FIAN 152, 145.

17. Feldezhof,B.U., Raval,S.P. (1976). Physica A82, 151.

18. Jennings,B.K., Bhaduri,R.K. (1976). Phys. Rev. B14, 1202.

19. Alastuey,A., Jancovici,B. (1979). Physica A97, 349.

20. Alastuey,A., Jancovici,B. (1980). Physica A102,327.

21. Jancovici,B. (1980). Physica A101, 324.

22. Dodonov,V.V., Man'ko,V.I., Ossipov,D.L. (1984). Prepr. FIAN N 91, Moscow, 30 p.

23. Dodonov,V.V., Man'ko,V.I., Ossipov,D.L. (1985). Physica A132, 269.

24. Landau,L.D., Lifshitz, E.M. (1974). Quantum Mechanics, Nauka, Moscow, p. 521 (Russian).

25. Stratonovich,R.L. (1956). Doklady Acad. Nauk USSR 109, 72 (Russian).

26. Fujita,S. (1966). Introduction to Non-Equilibrium Quantum Statistical Mechanics, Saunders Philadelphia,

Philadelphia, p. 75.

27. Bialynicki-Birula,I. (1977). Acta Phys. Austriaca Suppl. 18, 112.
28. Carruthers,P. (1983). Rev. Mod. Phys. 55, 245.
29. Badziag, P. (1985). Physica A130, 565.
30. Remler,E.A. (1977). Phys. Rev. D16, 3464.
31. Heinz,U. (1983). Phys. Rev. Lett. 51, 351.
32. Bopp,F. (1961). In: W.Heisenberg und die Physik Unserer Zeit, Vieveg, Braunschweig, p. 128.

Part IV
SPACE GROUPS AND SOLID STATE PHYSICS

(p,2)-SYMMETRY SPACE GROUPS

A.F.Palistrant
Kishinev State University

1. Introduction. The Shubnikov theory of antisymmetry [1] was a basis for profound investigations in geometric crystallography [2] as well as for other numerous new generalizations of classical symmetry theory [2-4] . Via two-colour understanding of antisymmetry N.V.Belov arrived at the notion of multicolour symmetry [5] called p-symmetry [4,6] .A.M.Zamorzaev introduced and developed another generalization of antisymmetry, i.e. the theory of multiple antisymmetry [2] , which extends antisymmetry assigning to the points of the transformed figure not one sign but ℓ qualitatively different signs, + and -.

These two generalizations resulted in the colour antisymmetry theory which is differently treated in [7] and [8] .

For the Pawley antisymmetry called (p')-symmetry in [4], colours and signs numbered by 1+,2+,...,p+ and 1-, 2-,...,p- are not independent qualities of various nature but can charge one into another according to the law of vertex mapping of rectangular-semiregular 2p-gon in the symmetry transformations of the latter [4] . Thus the Pawley colour symmetry, or (p')-symmetry, is set by the group of substitutions = $\{(1,2,\ldots,p)(\bar{p},\ldots,\bar{2},\bar{1}), (1,\bar{1}) \times (2,\bar{2})\ldots(p,\bar{p})\}$ determined by its two generating elements and the 2p properties generator, i.e. p "positive colours" i and p "negative colour" $\bar{i}$.

For the Belov-Neronova colour antisymmetry called (p,2)-symmetry [4] , the colours and signs numbered by similar numbers 1+,2+,...,p+ and 1-,2-,...,p- are independent qualities of various nature and can change into one another according to the law of vetrex mapping of a regular right prism with equally oriented p-angular bases in the symmetry transformation of the prism [4]. Hence the Neronova-Belov colour antisymmetry, or (p,2)-symmetry, is set by the P-group = $\{(1,2,\dots,p)(\bar{1},\bar{2},\dots$ $\dots,\bar{p}),\ (1,\bar{1})(2,\bar{2})\dots(p,\bar{p})\}$ which differs from the P-group determining the (p')-symmetry.

By late 70-ies all the two- and three-dimensional crystallographic (p')-symmetry groups had been studied [9]. This enabled the investigators not only to use the found (p')-symmetry groups in conjunction with classical crystallographic symmetry groups and their generalized antisymmetry and p-symmetry groups in the description of multi dimensional symmetry groups [10-12], but also advanced the problem of finding various magnetic symmetry space groups with crystallographic angles of spin vector rotation [13].

Crystallographic (p,2)-symmetry groups are also found both as physical (e.g. when analysing magnetoelectric structures [14]) and geometric supplements (e.g. when studying five- and six-dimensional groups with particular planes [12]). Only space (p,2)-symmetry $G_3^{1,P}$ groups are necessary to complete the scheme of the Neronova-Belov colour antisymmetry crystallographic groups. The main aim of this paper is to report the results of the complete determining these (p,2)-symmetry groups for p = 3,4,6 .

2. <u>Main concepts of the general (p,2)-symmetry</u>. We assign to each point of the figure, in addition to the index i ($i = 1,2,\dots,p;\ p \geqslant 3$), the sign + or - . By p- or (p,2)-symmetry transformation we mean an isometric transformation of the figure which carries each point with the index "i" to with index k+i (when $k+i \leqslant p$) or k+i-p (when $k+i > p$) and consequently

either preserves the signs of the points or changes them all. We can easily verify that the set of p- or (p,2)-symmetry transformations of the figure is a group which we call the (p,2)-symmetry group of the figure (cf. [4]).

Consequently any transformation of (p,2)-symmetry $g=s\varepsilon = \varepsilon s$, where s is the symmetry transformation, and ε is the index substitution from the $P= \{(1,2,\dots ,p)\} \times \{(+,-)\}$ group. A transformation group G of (p,2)-symmetry is a complete (p,2)-symmetry group if the group P_1 of index substitutions included in its transformations coincides with P. If P_1 is a non-trivial subgroup of P , G is a non-complete (p,2)-symmetry group [4] .

The theorem below indicates a method of deducing (p,2)-symmetry groups of a definite category from the derived p-symmetry group of the same category.

Any (p,2)-symmetry group which does not include p- or (p,2)-identical transformations (i.e. (p,2)-symmetry transformations that reserve the points of the figure and only [4] change the indices) may be derived from a certain p-symmetry group (p-generating group) by one of the following ways:

1) either add to it an anti-identical transformation this group is called the p-s nior group;

2) substitute all the elements of the coset modulo an index 2 subgroup for the corresponding (p,2)-symmetry transformations; this group is isomorphic to the p-generating group and is called the p-junior group.

The set of groups deduced from one p-generating group by these two ways is a p-family. Each p-family has one p-senior group, and all the other p-junior groups are isomorphic to the p-generating group. Various p-families do not include coinciding groups [4] .

3. <u>The review of the complete deduction of (p,2)-symmetry junior space groups</u>. To each point of the three-dimensional Euclidean we assign + or - in addition to

the index "i", which may take values. Consider the (p,2)-symmetry transformation group of this space with indices and signs at its points and call these groups (p,2)-homologous. We call such group a space (p,2)-symmetry if a) there is a sphere containing (p,2)-homologous images of a randomly chosen point in the space (space homogeneity); b) at least one point of the space is isolated in its homologous image class (local discreteness) (cf. chapter II in [4]).

Using general theory of the Shubnikov and Belov p-symmetry groups (when p=2,3,4,6) [2,4] we can show that generalizing any of the 817 junior space p-symmetry groups (completely described in App. P2 [4]) by methods 1 and 2 of the theprem we get the p-senior and a p-junior space (p,2)-symmetry groups $G_3^{1,p}$.

To get p-senior groups $G_3^{1,p}$ is trivial. As for the p-junior groups $G_3^{1,p}$, they can be conveniently derived from p-generating G_3^p groups by the Shubnikov method: the generator of each G_3^p group is replaced by the corresponding (p,2)-symmetry transformation. This may result in similar p-junior groups, and sometimes groups which contain (p,2)-identical transformations. Thus for example from a 3-generating $3^{(3)} : m$ point group by means of the Shubnikov method we can obtain a 3-junior $3^{(3)} : \underline{m}$ group and two similar 3-senior $\underline{3}^{(3)} : m = \underline{3}^{(3)} : \underline{m} = 3^{(3)} : m \times \underline{1}$ ones where $\underline{1}$ is the second order group generated by an anti-identical transformation

To avoid groups which include (p,2)-identical transformations when deriving junior $G_3^{1,p}$ groups by means of the Shubnikov method from the p-generating G_3^p groups, one should stick to the following rule: the generator of the p-generating group G_3^p may be replaces by the (p.2)-symmetry transformations only when the corresponding element in the initial classical group (the generating group for the given p-symmetry group G_3) was substituted for an antisymmetry transformation when deducing the Shubnikov junior G_3^1 groups [2] .

The validity of the rule is substantiated as follows. Since (p,2)-symmetry can be split into p-symmetry and antisymmetry [4] , the junior space (p,2)-symmetry $G_3^{1,p}$ group can be transformed into a junior space p-symmetry group when the $\pm$signs are neglected, and it can be transformed into the junior Shubnikov antisymmetry group when the indices assigned to points are neglected. Thus two junior space p-symmetry and antisymmetry groups may set the junior $G_3^{1,p}$ groups with a common classical generating Fedorov group for all the three groups (cf. with the method for finding polygonal symbols for junior point (p,2)-symmetry groups in [4]).

The number of junior $G_3^{1,p}$ groups in the family is considerably bounded by this rule. It would be rather cumbersome and impractical (to deduce likewise the junior Shubnikov and Belov groups from 230 G_3 [2,4]) to deduce all the p-junior groups from 817 groups G_3^p , and then find similar and remove the extra ones. So we split here the basic problem into simpler ones: 1) the derivation of junior and semijunior $G_3^{1,p}$ groups from classical point groups G_{30} ; 2) to solve the similar problem for 14 translational groups; 3) to complete the deduction of junior $G_3^{1,p}$ groups using the results of problems 1) and 2).

Problem 1) is completely solved in [2,4] . Crystal classes and the junior point antisymmetry groups which they generate are listed in [2] , and the junior point p- and (p,2)-symmetry groups (when p=3,4,6) are collected in [4] .

Problem 2) is also completely solved in [4,15] . All the 14 Bravais lattices and 22 their generalizing antitransition T groups are written in [2], 76 junior p-transitions $T^{(p)}$ groups (when p =3,4,6) are summarized in Table P1 of [4] , and 104 junior (p,2)-transition groups $\underline{T}^{(p)}$ (when p=3,4,6) are completely written in [15] . Note however, that every junior group $G_3^{1,p}$ has a trimetric (p,2)-transition sub-group; a junior $\underline{T}^{(p)}$, a semi-junior $T^{(p)}$ or $\underline{T}$, or a generating group T . In case T is a transition sub-group of a Frdorov group F , in general case not every junior $\underline{T}^{(p)}$ semi-junior group $T^{(p)}$ or T generated by T is a (p,2)- , p-transition or antitransition subgroup in at least one junior (p,2)-symmetry group generated by F.

In order for $\underline{T}^{(p)}$, $T^{(p)}$ or $\underline{T}$ generated by T fit their generating group T , the transition subgroup $T_o=T\cap\underline{T}^{(p)}$ and subgroups $T_1=T\cap T^{(p)}$ and $T_2=T\cap\underline{T}$ of T with intersection T_o should be connected with the same syngony. The junior groups $\underline{T}$, $T^{(p)}$ and $\underline{T}^{(p)}$ are deduced from the classical transition groups T to satisfy these criteria [2,4,15].

Now we may deduce junior groups $G_3^{1,p}$. Let F = T W be a symmorphous Fedorov group. Then any junior (p,2)-symmetry group generated by it may be of one of the following forms: $T\cdot\underline{W}^{(p)}$, $\underline{T}\cdot W^{(p)}$, $\underline{T}\cdot\underline{W}^{(p)}$, $T^{(p)}\cdot\underline{W}$, $\underline{T}^{(p)}\cdot W$, $\underline{T}^{(p)}\cdot\underline{W}$, $T^{(p_1)}\cdot\underline{W}^{(p_2)}$, $\underline{T}^{(p_1)}\cdot W^{(p_2)}$, $\underline{T}^{(p_1)}\cdot\underline{W}^{(p_2)}$, $T^{(p)}\cdot\underline{W}^{(p_2)}$, $\underline{T}^{(p)}\cdot W^{(p_2)}$, $\underline{T}^{(p)}\cdot\underline{W}^{(p_2)}$, where p is the smallest general multiple number of p_1 and p_2 , wheread $\underline{T}$, $T^{(p)}$, $\underline{T}^{(p)}$, $T^{(p_1)}$ and $\underline{T}^{(p_1)}$ are p-(p,2)-, p_1 and $(p_1,2)$-symmetry anti-transition groups which are generated by T and are consistent with the classes W,$\underline{W}$, $W^{(p_2)}$, $\underline{W}^{(p_2)}$, $W^{(p)}$ and $\underline{W}^{(p)}$. Various modifications of these products at a given p exhaust the junior (p,2)-symmetry groups of a given family; what remains to be done is to remove extra groups cf. [4,15] .

Here are several examples of deriving p-junior groups; p-generating groups G_3^p are set in symbols by A.M.Zamorzaev details see in [2,4]; the index in brackets in the right hand side above the element characterizes the order of p-symmetry "colour", the bar that it is substituted for the corresponding (p,2)-symmetry element.

Let us write down the p-generating symmorphous groups

$$\{a,b,c^{(3)}\}(2),\ \{a,b,c^{(4)}\}(2),\ \{a,b,c^{(4)}\}(2^{(2)}),$$

$$\{a^{(2)},b,c^{(4)}\}(2),\ \{a,b,c^{(3)}\}(2^{(2)}),\ \{a,b,c^{(6)}\}(2),$$

$$\{a^{(2)},b,c^{(3)}\}(2),\ \{a,b,c^{(6)}\}(2^{(2)}),\{a^{(2)},b,c^{(6)}\}(2)$$

of monocline syngony with the generating classical group P2 of Table P1 [4].

From the group $\{a,b,c^{(3)}\}(2)$ the following 3-junior groups are deduced: $\{a,b,c^{(3)}\}(\underline{2})$, $\{\underline{a},b,c^{(3)}\}(2) = \{\underline{a},b,c^{(3)}\}(\underline{2})$, $\{a,b,\underline{c}^{(3)}\}(2)$, $\{a,b,\underline{c}^{(3)}\}(\underline{2})$, $\{\underline{a},b,\underline{c}^{(3)}\}(2) = \{\underline{a},b,\underline{c}^{(3)}\}(\underline{2})$; 5 different groups altogether,

Similarly from $\{a,b,c^{(4)}\}(2)$ five different groups are deduced: $\{a,b,c^{(4)}\}(\underline{2})$, $\{\underline{a},b,c^{(4)}\}(2)$, $\{a,b,\underline{c}^{(4)}\}(2)$, $\{a,b,\underline{c}^{(4)}\}(\underline{2})$, $\{\underline{a},b,\underline{c}^{(4)}\}(2)$,

The following 4-junior groups: $\{a,b,c^{(4)}\}(\underline{2}^{(2)})$, $\{\underline{a},b,c^{(4)}\}(2^{(2)}) = \{\underline{a},b,c^{(4)}\}(\underline{2}^{(2)})$, $\{a,b,\underline{c}^{(4)}\}(2^{(2)})$, $\{a,b,\underline{c}^{(4)}\}(\underline{2}^{(2)})$, $\{\underline{a},b,\underline{c}^{(4)}\}(2^{(2)}) = \{\underline{a},b,\underline{c}^{(4)}\}(\underline{2}^{(2)})$ are deduced from $\{a,b,c^{(4)}\}(2^{(2)})$.

In turn $\{a^{(2)},b,c^{(4)}\}(2)$ generates seven 4-junior groups $\{a^{(2)},b,c^{(4)}\}(\underline{2}) = \{a^{(2)},b,c^{(4)}\}(\underline{2}^{(2)})$, $\{\underline{a}^{(2)},b,c^{(4)}\}(2) = \{\underline{a}^{(2)},b,c^{(4)}\}(\underline{2}^{(2)})$, $\{\underline{a}^{(2)},b,c^{(4)}\}(\underline{2}) = \{\underline{a}^{(2)},b,c^{(4)}\}(2^{(2)})$, $\{a^{(2)},b,\underline{c}^{(4)}\}(2) = \{a^{(2)},b,\underline{c}^{(4)}\}(2^{(2)})$, $\{a^{(2)},b,\underline{c}^{(4)}\}(\underline{2}) = \{a^{(2)},b,\underline{c}^{(4)}\}(\underline{2}^{(2)})$, $\{\underline{a}^{(2)},b,\underline{c}^{(4)}\}(2) = \{\underline{a}^{(2)},b,c^{(4)}\}(\underline{2}^{(2)})$, $\{\underline{a}^{(2)},b,\underline{c}^{(4)}\}(\underline{2}) = \{\underline{a}^{(2)},b,\underline{c}^{(4)}\}(2^{(2)})$, etc. The p-junior hemisymmorphous and asymmorphous $G_3^{1,p}$ groups are derived in a more complicated way but by the same principle.

All in all 3648 p-junior (p,2)-symmetry groups $G_3^{1,p}$ ($379\ G_3^{1,3} + 2049\ G_3^{1,4} + 1220\ G_3^{1,6}$) are derived from 817 p-generating group G_3^p via the Shubnikov method.

4. Conclusion. On the one hand the deduction of the space groups $G_3^{1,p}$ brought us closer to the completion of investigating magnetic symmetry space groups with crystallographic angles of vector spins [13] , so that (p,2)-symmetry (when p = 3,4,6) is analogous to crystallographic ($3\underline{1}$)-, ($4\underline{1}$)- and ($6\underline{1}$)-symmetries in the geometric classification (see [12,16]). On the other hand as follows from the classification of the space groups $G_3^{1,p}$ in terms of their generalizing P-symmetry groups, they ($G_3^{1,p}$) may be used in conjunction with the classical Fedorov groups to find all various 6-dimensional symmetry groups preserving a 5-dimensional plane with an embedded 3-dimensional one, i.e. to describe the categoty of groups G_{653} [12] .

References

1. Shubnikov,A.V. Symmetry and antisymmetry of finite figures. Moscow: Edited by the Academy of Sciences of the USSR. - 172 p. (Russian).

2. Shubnikov,A.V., Koptsik,V.A. (1974). Symmetry in Science, Art and Nature. Plenum-Presse, N.Y.

3. Zamorzaev,A.M. (1976). The theory of simple and multiple symmetry. (1976). Shtiintsa Kishinev (Russian).

4. Zamorzaev,A.M., Galyarskii,E.I., Palistrant,A.F. (1978). Colour symmetry and its generalization s and applications. Shtiintsa, Kishinev (Russian).

5. Belov,N.V., Tarkhova,T.N. (1956). Colour-symmetry groups. Cristallography, 1, 4-13 (Russian).

6. Zamorzaev,A.M. (1967). About quasi (P-symmetry) groups. Cristallography, 12, 819-825 (Russian).

7. Pawley,G.S. (1961). Mosaics for colour-antisymmetry groups. Cristallography, 6, 109-111 (Russian).

8. Neronova,N.N., Belov,N.V. (1961). Coloured antisymmetric mosaics. Cristallography, 6, 831-839 (Russian).

9. Palistrant,A.F. (1980). Space (p)-symmetry (Pawley) groups and their application in the deduction of fivemetric crystallographic symmetry groups. Doklady of the USSR Acad. Sci., 254, 1126-1130 (Russian).

10. Zamorzaev,A.M., Palistrant,A.F. (1980). Antisymmetry,

its generalizations and geometric applications. Z. Cristallographie, 151, 231-248.

11. Palistrant,A.F. (1980). Crystallographic P-symmetry groups and their applications in the deduction of polymeric symmetry groups. In: Group-theoretical methods in physics. Moscow: Nauka, 1, 97-105 (Russian).

12. Palistrant,A.F. (1981). Colour symmetry groups, their generalizations and applications. Doctor's thesis. Kishinev, 398 p. (Russian).

13. Zamorzaev,A.M., Palistrant,A.F., Karpova,Y.S. (1983). About space groups of crystal magnetic symmetry. In: Group theoretical methods in physics. Nauka, Moscow, 1, 346-352 (Russian).

14. Koptsik,V.A., Kotsev,I.N. (1974). Magnetic (spin) and magnetoelectric point P-symmetry groups. Communications of the Joint Institute for Nuclear Research, P4-8466, Dubna, 17 p.

15. Karpova,Y.S. (1983). About junior (p,2)- and (p2,2)-symmetry space groups. In: Research in modern algebra and geometry, Shtiintsa, Kishinev, p. 68-76 (Russian),

16. Zamorzaev,A.M., Palistrant,A.F. (1981). Geometric classification of P-symmetries. Doklady of the USSR Acad. Sci. 256, 856-859 (Russian).

AFFINE NORMALIZERS OF SPACE GROUPS AND THE GLOBAL SEARCH FOR OPTIMAL PACKINGS OF A MOLECULAR CRYSTAL

A.V. Dzyabchenko

Karpov Institute of Physical Chemistry,
ul. Obukha 10, Moscow 107120

1. Introduction. While the concept of the normalizer of a group is well established in mathematics [1] its significance in crystallography was not recognized until recently. To our knowledge, the normalizers of crystallographic groups were first applied by Zassenhaus in connection with his algorithm of calculating the space groups [2]. (This fact, as it seems did not draw due attention then since later a number of problems related to that of the normalizer was treated by crystallographers without an explicit indication to such a relation). Hirshfeld working on the problem of generating all possible arrangements of molecules in a crystal of known cell dimensions and given symmetry introduced the groups of derivative symmetry, the Cheshire groups [3], that are, in fact, the Euclidean normalizers of space groups. Koch and Fischer extended the idea of the Cheshire group and derived the automorphism groups of the space groups [4]. Burzlaff and Zimmermann in a paper on the choice of space-group origins [5] first determined the affine normalizers of space groups and showed their close relation to automorphism groups. At the same time, the role of normalizers of crystallographic point groups was independently established by Ascher in deriving the relativistic symmetry of a crystal [6] and by Galiulin in

the theory of simple forms of crystal polyhedra [7].

The normalizers of space groups form a mathematical base for some problems in descriptive crystallography. In particular, they are employed in calculating all equivalent descriptions of a crystal structure when a comparison of structures, classification or standardization of crystal-structure data is made [8].

Another application of normalizers is met when dealing with a multidimensional parameter space needed to describe all possible configurations of the infinite system of molecules. Physical properties of such a system are defined by potential-energy hypersurface which one could imagine to be in this space as a result of intermolecular interactions. If certain space symmetry of the system is postulated then the affine normalizer is one of the factors which relate the three-dimensional symmetry with the symmetry of the energy hypersurface in the parameter space. The latter just means the existence of equivalent, i.e. physically indistinguishable, descriptions of the system configurations.

The global search for optimal packings of a molecular crystal is a kind of problem that calls for consideration of the symmetry of the lattice-energy hypersurface in the space of crystal-structure parameters. Consequently, using normalizers here appears to be quite a reasonable and important thing. The method to be required has already been worked out, in the main, by Hirshfeld [3] and extended in our previous investigation [9] to a more general case of variable lattice constants. In the present work we shall consider a symmetry approach to a 'non-ideal' crystal system whose asymmetric part of the unit cell consists of two, or more, molecules. This case is most general. It is applicable to any system of molecules which possesses a three-dimensional periodicity, e.g., to those used in molecular dynamics and Monte-Carlo computer simulations of real crystals and liquids.

2. The affine normalizer. The affine normalizer, or more exactly, the normalizer $N_A(F)$ of a space group F in the affine group A of the three-dimensional space is the group of all elements $a \in A$ that map F onto itself:

$$aFa^{-1} = F.$$

Another formulation of the affine normalizer is a straightforward generalization of that given by Hirshfeld for the Cheshire group: $N_A(F)$ is the group of all affine transformations of the unit-cell vectors that preserve the coordinates of equivalent positions of F.

The two definitions are equivalent although in the former the object of transformations is the group F (active point of view) while in the latter a basis of F is transformed (passive point of view).

The Cheshire group is the group of all isometric transformations of a space group onto itself. It is deduced from the space group as the symmetry of the space-group geometric pattern. Namely, the space group is treated as a three-dimensional geometric structure whose elements are the space-group symmetry elements and whose unit cell is a characteristic (i.e. compatible with the crystal class of the space group) cell. The symmetry group of such a 'structure' is just the Cheshire group. One can now readily understand the origin of its name, for "it is the symmetry the structure would acquire if all its atoms were removed and only its symmetry elements left behind, like the grin of the vanishing Cheshire cat" [3].

For the space groups of symmetry higher than orthorhombic the affine normalizer and the Cheshire group are the same. For the orthorhombic space groups a method to derive the affine normalizers consists in deducing the derivative symmetry, as for the Cheshire group. However, when constructing the space-group pattern one should now use such a unit cell, instead of a characteristic one, that will result in the highest possible derivative symmetry. That is, if necessary, the tetragonal or cubic cell axes are to be applied. The resulting group of derivative symmetry, that we call the extended Cheshire group [9], contains all information on the affine normalizer and is isomorphic to it, although one should remember that the two groups are not strictly equal in geometric sense. (In fact, the affine normalizer of the orthorhombic space group is a group of "space homology" [10]).

In case of monoclinic and triclinic space groups there is no such isomorphism: there are elements of the normalizer that have no analogues in the space groups. A description of affine normalizers can be given in terms of restrictions on matrix elements and vector components of the operators of the normalizer. To derive these restrictions a "visual search" for all possible arrangements of the unit-cell vectors is performed on the space-group patterns.

Lists of the affine normalizers for the space groups of symmetry higher than monoclinic were first published by Burzlaff and Zimmermann [5], but very close results, the automorphism groups, had been reported earlier by Koch and Fischer [4]. The affine normalizers of monoclinic and triclinic space groups were listed explicitly in [11] by Billiet, Burzlaff and Zimmermann, they are also implicitly present in an earlier work [12]. A strictly mathematical algorithm to calculate the normalizers is proposed in [13].

An independent derivation of the affine normalizers has been also given in our previous paper [9]. The comparison of our results with those of others reveals that in Table 1 from [9] the space group Ccca erroneously assigned to the extended Cheshire group Pmmm should be

P4/mmm and the unit cell of the affine normalizer of the space group Cmm2 is written wrong: the correct cell is $a/2 \times b/2 \times \varepsilon c$. On the other hand, there is a misprint in the notation of the normalizer of the space groups P2/b and $P2_1/b$ in Table 1 from [11]: printed $\{(M_5, O_2)\}$, should be $\{(M_{5'}, O_8)\}$.

3. <u>Calculation of optimal crystal packings</u>. Prediction of a crystal structure by model energy calculation is a problem of great importance in solid-state physics and chemistry. Such a calculation consists in minimization of the crystal lattice energy with respect to the structural parameters. With molecular crystals the assumption the intramolecular structure to be unaffected by crystal forces is generally employed. If so, any crystal configuration of a given symmetry may be specified by the lattice constants $a, b, c, \alpha, \beta, \gamma$ together with sets of molecular parameters. The latter comprise the centre-of-mass coordinates u_i, v_i, w_i as well as the Eulerian angles $\varphi_i, \theta_i, \psi_i$, where $i = 1, \ldots, n$, n the number of symmetry-independent molecules.

The lattice energy is usually calculated as a sum of all the pair interactions between molecules. In case of polyatomic molecules an atom-atom approach that treats the energy of interaction between two molecules as a sum of pair atom-atom interactions is of common practice. Characteristics of the atom-atom potential energy functions are usually derived from experiment, in particular, by fitting the calculated lattice-energy minimum to the observed crystal-structure data and the heat of sublimation.

Even if a pretty simple form of intermolecular potential is employed a search for optimal structures is quite a laborious computational procedure because any calculation of the lattice energy requires treating a great number of atom-atom pairs. To predict crystal structure the global optimization of packings is necessary. That assumes a lot of local optimizations to be done with varied starting configurations and symmetries. The solution of the global optimization problem is a set of local energy minima, the lowest ones account for most probable predicted crystal polymorphs. Since much computational work is required, the global packing calculations were not really possible until recently. The study of optimal packings in solid benzene [14] is an example of such a calculation.

4. Symmetry in the space of crystal-structure parameters. When beginning systematic search for optimal packings one should set the grid of starting configurations that would guarantee obtaining all the lattice-energy minima. At the same time, the "hits" to the minima equivalent due to symmetry are to be avoided in order to prevent the loss of computer time. Consequently, the following question raises: what ranges must the structural parameters scan so as to generate all possible starting configurations with any of them occurring no more than once? In other words, an asymmetric unit in the space of structural parameters is to be determined. The problem defined is analogous to that considered in [3], but in this work a more general case, admitting the variable lattice constants and the presence of symmetry-independent molecules, is treated.

To derive an asymmetric unit one should consider all the independent factors which produce equivalent descriptions of a crystal structure and combine them to form a full list of equivalent sets of the structural parameters. These factors are

i) the transformations of the unit-cell axes by operations of the normalizer,

ii) the actions of the space-group symmetry operations on subsets of symmetry-independent molecules,

iii) the permutations of symmetry-independent molecules,

iv) the molecular symmetry.

As has been stated in [9] the use of the Niggli reduced-cell conditions confines the asymmetric-unit problem of the whole parameter space to that of the molecular parameters only. Further considerations are to be done mainly as in [3], with the extra equivalences due to ii) and iii) being taken into account. Let us take for example a crystal of space symmetry P$\bar{1}$ with two symmetry-independent particles in general position. Assuming for simplicity the spherical molecular symmetry one can define any crystal configuration by the structure parameters $a, b, c, \alpha, \beta, \gamma, u_1, v_1, w_1, u_2, v_2, w_2$, where indices 1 and 2 correspond to the first and second particles respectively.

As the first step to an asymmetric unit of the twelve-dimensional parameter space we apply the main conditions of the Niggli reduced cell [15]:

$$a \leq b \leq c; -b/2c \leq \cos\alpha \leq 0, -a/2c \leq \cos\beta \leq 0, -a/2b \leq \cos\gamma \leq a/2b;$$

if $\gamma > \pi/2$ then $c \leq (a^2+b^2+c^2+2ab\cos\gamma+2ac\cos\beta+2bc\cos\alpha)^{1/2}$.

This reduces the affine normalizer of the space group P$\bar{1}$ to the Cheshire group P$\bar{1}$, with its unit cell $a/2 \times b/2 \times c/2$. Considering this periodicity we can write preliminary ranges for the coordinates of the two particles:

$$0 \leq u_1, v_1, w_1 < 1/2, \quad -1/2 < u_2, v_2, w_2 \leq 1/2.$$

Our next step is to generate the equivalent sets of coordinates:

$$\tilde{u}_1, \tilde{v}_1, \tilde{w}_1, \tilde{u}_2, \tilde{v}_2, \tilde{w}_2; \quad (1)$$

$$-\tilde{u}_1, -\tilde{v}_1, -\tilde{w}_1, -\tilde{u}_2, -\tilde{v}_2, -\tilde{w}_2; \quad (2)$$

$$-\tilde{u}_1, -\tilde{v}_1, -\tilde{w}_1, \tilde{u}_2, \tilde{v}_2, \tilde{w}_2; \quad (3)$$

$$\tilde{u}_1, \tilde{v}_1, \tilde{w}_1, -\tilde{u}_2, -\tilde{v}_2, -\tilde{w}_2; \quad (4)$$

$$\tilde{u}_2, \tilde{v}_2, \tilde{w}_2, \tilde{u}_1, \tilde{v}_1, \tilde{w}_1; \quad (5)$$

$$-\tilde{u}_2, -\tilde{v}_2, -\tilde{w}_2, -\tilde{u}_1, -\tilde{v}_1, -\tilde{w}_1; \quad (6)$$

$$\tilde{u}_2, \tilde{v}_2, \tilde{w}_2, -\tilde{u}_1, -\tilde{v}_1, -\tilde{w}_1; \quad (7)$$

$$-\tilde{u}_2, -\tilde{v}_2, -\tilde{w}_2, \tilde{u}_1, \tilde{v}_1, \tilde{w}_1. \quad (8)$$

In this list the equivalent coordinate sets (2)-(8) are given in terms of the original set (1). It is assumed here that regardless of the indices supplied, the first triplet of coordinates within each equivalent set accounts for the first particle just as the second triplet for the second particle. The equivalent set (2) corresponds to a transformation of the unit-cell axes by $\bar{1}$ of the Cheshire group, (3) shows that the position of the particle 1 is replaced by a symmetry related one as a result of the space-group operation $\bar{1}$ local action, (4) corresponds to the same transformation for the particle 2.

In addition to this list one should keep in mind the equivalent sets arising as combinations of (2)-(8) with the Cheshire-group translations.

By selecting the new range for u_2, $0 \leqslant u_2 \leqslant 1/2$, we discard all the equivalent sets of the full list except (1), (5) and for the two sets produced by adding the Cheshire-group translation $1/2, 0, 0$ to (2) and (6):

$$1/2-\tilde{u}_1, -\tilde{v}_1, -\tilde{w}_1, 1/2-\tilde{u}_2, -\tilde{v}_2, -\tilde{w}_2; \quad (2')$$

$$1/2-\tilde{u}_2, -\tilde{v}_2, -\tilde{w}_2, 1/2-\tilde{u}_1, -\tilde{v}_1, -\tilde{w}_1. \quad (6')$$

The subsequent constraints $0 \leqslant u_1 \leqslant u_2 \leqslant 1/2$ exclude from the remaining lists the sets (5) and (2') and, at last, $0 \leqslant u_1 \leqslant u_2 \leqslant 1/2 - u_1$ retains just (1). Finally, we get:

$$0 \leqslant u_1 \leqslant 1/4, \; u_1 \leqslant u_2 \leqslant 1/2 - u_1, \; -1/2 < v_2, w_2 \leqslant 1/2.$$

These inequalities, together with the constraints on the lattice constants given above, define the asymmetric unit in the twelve-dimensional space of structural parameters. (Of course, it is not the only unit we could have selected). The conditions imply certain "hierarchy" for the structure parameters: thus, the range for u_2 has no numerical values until u_1 takes a value of (0,1/4).

5. Conclusion. The affine normalizer is the group of all affine transformations of a space group into itself. It plays an important role in deriving the symmetry of the multidimensional potential-energy hypersurface in the space of structure parameters. Other factors that define the symmetry in the parameter space are the permutations of symmetry-independent molecules, the actions of the space-group symmetry elements on subsets of symmetry-independent molecules and the molecular symmetry. There exists a systematic procedure to derive the asymmetric unit of the parameter space. This unit is a region to be scanned when calculating the global lattice-energy minimum.

References

1. Kurosh, A.G. (1960). Theory of groups. Chelsea, N.Y.
2. Zassenhaus, H. (1948). Comment. math. helv., 21, 117.
3. Hirshfeld, F.L. (1968). Acta crystallogr., A24, 301.
4. Koch, E., Fischer, W. (1975). Acta crystallogr., A31, 88.
5. Burzlaff, H., Zimmermann, H. (1980). Z. Kristallogr., 153, 151.
6. Ascher, E. (1974). Phys. status solidi B, 65, 677.
7. Galiulin, R.V. (1980). Proc. of the Steklov Math. Inst. 4, 81.
8. Parthe, E., Gelato, L.M. (1984). Acta crystallogr., A40, 169.
9. Dzyabchenko, A.V. (1983). Acta crystallogr., A39, 941.

10. Mikheyev, V.I. (1961). Homology of Crystals. Gostoptekhizdat, Leningrad (Russian).

11. Billiet, Y., Burzlaff, H., Zimmermann, H. (1982). Z. Krystallogr., 160, 155.

12. Sayari, A., Billiet, Y. (1977). Acta crystallogr., A33, 985.

13. Gubler, M. (1982). Z. Kristallogr., 158, 1.

14. Dzyabchenko, A.V. (1984).Zh. Strukt. Chem., 25(3), 85.(Russian).

15. Santoro, A., Mighell, A.D. (1970). Acta crystallogr., A26, 124.

APPLICATION OF THE PERMUTATION-INVERSION SYMMETRY GROUP TO THE ANALYSIS OF ORIENTATIONAL DISORDER IN CRYSTALS

A.Koroliev,V.Smirnov

Institute of Precise Mechanics and Optics, Leningrad

1. Introduction. Using of permutation-inversion symmetry for the construction of symmetry groups of non-rigid crystals was first suggested in [1] . The Hamiltonian of a system is invariant under the operations of permutations P of identical nuclei and inversion E^* of the coordinates of all particles (nuclei and electrons). The permutation-inversion group of a non-rigid crystal is determined as the group of permutation-inversion operations P, $P^* = PE^* = E^*P$, corresponding to feasible (experimentally observable) motions of the particles in the system. An example of a non-rigid crystal is a high temperature phase of a molecular crystal in which the centres of mass of molecules or molecular ions still form a regular crystal lattice while their orientations display some degree of disorder. Another example is a crystal with non-rigid impurity molecular centre. Permutation-inversion group can be also applied for analisis of the latter type of non-rigid crystals [2,3] .

An element P (or P^*) corresponding to the non-rigid motions connects two isoenergetic configurations of nuclei usually separated by a barrier. In case of a molecular crystal with rotating molecules such isoenergetic configurations correspond to the different orientations of molecules. If the barrier separating isoenergetic configurations is overcomed, the element P (or P^*) has to be included in the symmetry group of the non-rigid crystal.

Generation of the symmetry groups of the crystal by permutation-inversion elements meets some difficulties since it is practically impossible to realize the labelling of nuclei by cycles usually used in the theory of molecules [4,5]. The labelling by position vectors of nuclei is more suitable in this case. The permutations of nuclei in crystals can be described by means of point and space group elements [1,6]. These methods of labelling and the description of permutation-inversion symmetry elements allow one to obtain in general form the laws of coordinate transformations under the elements of permutation-inversion symmetry group [6].

$$\vec{X}'_{\vec{r}} = \hat{P}_x(f,\tilde{m})\,\vec{X}_{\vec{r}} = [f]\,\vec{X}_{f^{-1}\vec{r}} \quad \text{in CSA} \tag{1}$$

$$\Omega'_{\vec{r}} = \hat{P}\,\Omega(f,\tilde{m})\,\Omega_{\vec{r}} = [f]\quad \Omega_{f^{-1}\vec{r}}\,\Omega^{(o)}_{f^{-1}\vec{r}}\,\tilde{m}^{-1}\,\Omega^{(o)}_{\vec{r}}{}^{-1} \text{ in CSA} \tag{2}$$

$$\vec{\xi}'_{\vec{R}} = \hat{P}_\xi(f,\tilde{m})\,\vec{\xi}_{\vec{R}} = \tilde{m}\,\vec{\xi}_{\vec{R}'} \quad \text{in MSA} \tag{3}$$

$$(\vec{R}' = f^{-1}\vec{r} + \Omega^{(o)}_{f^{-1}\vec{r}}\,\tilde{m}\,\vec{d}\,)$$

$$\vec{q}'_j = \hat{P}_q(f,\tilde{m})\vec{q}_i = f\vec{q}_i \quad \text{in CSA} \tag{4}$$

Here the vector $\vec{\xi}_{\vec{R}}$ is the displacement of the nucleus $\vec{R}$ from its equilibrium position in the molecular system axes ($MSA_{\vec{r}}$); the matrix $\Omega_{\vec{r}}$ describes the rotation of the molecule $\vec{r}$ ($MSA_{\vec{r}}$) relative to the crystal system axes (CSA); the vector $\vec{X}_{\vec{r}}$ gives the displacement of the molecular centre of mass (of the $MSA_{\vec{r}}$) with respect to the CSA; $\vec{q}_j$ are electronic coordinates.The position vectors of the nuclei ($\vec{R}$) and the molecules

($\vec{r}$) in the initial crystal configurations are used to label these configuration.

2. <u>Development of the probability density function.</u> Let us consider the group formed by operators $\hat{P}_\Omega(f,\tilde{m})$. Now consider only elements $f \equiv s' \in S'$ which preserve a site $\vec{r}_1$ ($s'r_1 = r_1$). The group S' of site symmetry is statistically [7] realized. The group Q of operators $\hat{P}_\Omega(s',\tilde{m})$ is a subgroup of $\hat{P}_\Omega(f,\tilde{m})$.

By definition (2) the group $\hat{P}_\Omega(s',\tilde{m})$ is isomorphic to one of subgroups of the direct product $S' \times M$. For plastic phase of the crystals formed by molecules of spherical top type (for example CH_4) all motions are feasible. For this case the group $\hat{P}_\Omega(s',\tilde{m})$ is isomorphic to the considered in [8]. For low symmetry molecules (of symmetrical top type, for instance) the barriers hindering the rotation about symmetrically nonequivalent axes can be different. Therefore only rotation about one of these axes can appear to be feasible. The order of $\hat{P}_\Omega(s',\tilde{m})$ is in this case less than that of the group from [8]. Hence the development of the probability density function $P_o(\Omega)$ in the symmetry adopted basis of Wigner functions differs from that in [8]. $P_o(\Omega)$ is the probability density for a molecule to have orientation Ω (Ω is a matrix determined by three Euler angles describing the rotation which translates CSA to MSA). The development of $P_o(\Omega)$ is [8]:

$$P_o(\Omega) = \sum_{l\sigma\mu} A_l^{\sigma\mu} \Delta_l^{\sigma\mu}(\Omega) \qquad (5)$$

where $\Delta_l^{\sigma\mu}$ is symmetry adopted functions transforming according to the unity representation of $\hat{P}_\Omega(s',\tilde{m})$. The labels σ and μ are the names of the irreducible representations of the groups S' and M, different subspaces corresponding to the same representation and different basic functions for multidimensional representations, l the label of the Wigner function, $A_l^{\sigma\mu}$ a coefficient of development. The nonzero coefficients $A_l^{\sigma\mu}$ correspond to the irreducible representations of the group S'x M induced from the identity representation of the group $Q \subset S' \times M$.

As an example consider the molecular crystal $CCl(CH_3)_3$. In high temperature phase the site group is $S' = O_h$. The molecular group is $M = C_{3v}$. If only the rotation about a threefold axis is possible, the development (5) contains the following pairs (σ,μ) : (A_{1g},A_1), (A_{1u},A_2), (A_{2u},A_1), (A_{2g},A_2), (F_{2g},A_1), (F_{1u},A_1), (F_{1g},A_2), (F_{2u},A_2). The last six pairs are not included in the development (5) in [8]. The numbers of independent non-zero coefficients in the development (5) for this case are given in the following table :

	$A_{1g}A_1$	$A_{1u}A_2$	$A_{2u}A_1$	$A_{2g}A_2$	$F_{1u}A_1$	$F_{1g}A_2$	$F_{2u}A_2$	$F_{2g}A_1$
l=0	1	0	0	0	0	0	0	0
1	0	0	0	0	1	0	0	0
2	0	0	0	0	0	0	0	1
3	0	0	2	0	2	0	1	0
4	2	0	0	0	0	1	0	2
5	0	0	0	0	4	0	1	0
6	3	0	0	2	0	2	0	6
7	0	0	3	0	6	0	6	0
8	3	0	0	0	0	6	0	6
9	0	3	4	0	12	0	6	0

3. Transformation properties of coordinates for non-rigid impurity centres in crystal. Transformation laws (1)-(4) also are valid for crystals containing non-rigid impurity centres. For example, consider alkali halide crystal KCl containing NO_2^- impurity centre in an anion site. The axis of the uniaxial rotation of the ion passes

through its centre of mass and is parallel to the base of the NO_2^- triangle. The axis of the uniaxial rotation of NO_2^- is assumed to be one of the symmetry axes of the crystal lattice. Accordingly, there are three models of impurity centre C_2, C_3 and C_4. These models were throughly investigated in [2,3]. The transformation properties of coordinates were also derived in [2] and [3]. One can obtain the latters directly from (1)-(4). Let us consider C_3-model, as the most adequate to the experimental data [3]. The axis of the uniaxial rotation (OZ axis of MSA) NO_2^- is one of the threefold axes. The CSA are supposed to coincide with MSA (attached to the NO_2^-) in the initial configuration. The permutation-inversion operation $P(C_{2x},\tilde{C}_{2x})$ corresponds to a motion of NO_2^-. This operation and the identity one form a group isomorphic to C_2. The following operations have to be also included in the symmetry group

$$P(E,\tilde{E}),\ P(C_{3z},\tilde{E}),\ P(C_{3z}^2,\tilde{E}),\ P^*(S_6,\tilde{\sigma}),\ P^*(I,\sigma),$$
$$P^*(S_6^5,\tilde{\sigma}),\quad \text{where}\quad \tilde{\sigma} = I\tilde{C}_{2z} \quad . \tag{6}$$

These permutations connect isoenergetic configurations separated by the barrier surpassed by uniaxial rotation $P(C_{2x},\tilde{C}_{2x})$, elements (6) and their products form the symmetry group Q for the C_3-model of impurity ion NO_2^- in alkali halide crystal KCl. This group is isomorphic to the point group D_{3d}. The relations (1) and (3) show that the Wigner operation is induced on the vibrational displacements. The operators $\hat{P}_x(f,\tilde{m})$ form the group isomorphic to D_{3d} and the operators $\hat{P}_\xi(f,\tilde{m})$ form the one isomorphic to C_{2v}. Formula (2) determines the transformation law for the angle of uniaxial rotation. For example $\gamma_z' = \hat{P}_\Omega(C_{3z},\tilde{E})\ \gamma_z = C_{3z}\ \gamma_z = \gamma_z + \frac{2\pi}{3}$.

References:

[1] Zilich,A.G., Kiselev,A.A., Smirnov,V.P. (1981).Fizika

Tverdogo Tela (Solid body physics) 23, 808. (Russian)

2. Kiselev,A.A., Lüders,K. (1979). Vestnik Leningradsko-go Univ. N 16, 31 (Russian).

3. Kiselev,A.A., Lüders,K. (1981). In: Woprocy kwanto-woj teorii atomov i molekul. (Quantum theory of atoms and molecules). Leningrad, 2, 56 (Russian).

4. Longuet-Higgins,H.C. (1963). Mol. Phys. 6, 445.

5. Bunker,P.R. (1979). Molecular Symmetry and Spectroscopy, Acad. Press, N.Y.

6. Korolev,A.A., Smirnov,V.P. (1985). Phys. stat. sol. (b) 129, 41.

7. Pule,H., Mathieu, J.-P. (1970). Spectres de vibration et symmetric des cristaux, Gordon and Breach, Paris.

8. Yvinec,M., Pick,R.M. (1980). J. Physique 41, 1045.

TENSOR SYMMETRY OF THE MACROSCOPIC PHYSICAL PROPERTIES OF TWO-SUBLATTICE MAGNETICS

S.S.Girgel

Gomel State University

1. The properties of the majority of magnetics are known to be well described by a two-sublattice model. Magnetization vector $\vec{m}$ and antiferromagnetic vector $\vec{l}$, that are introduced in this model, transform independently according to some representations of a space symmetry groups of the paramagnetic phase of a crystal [2].

2. Physical macroscopic properties of anisotropic media have, as a rule, anisotropic nature and are described by tenosrs of different ranks. Let us consider a phenomenological description of tensorial characteristics of physical properties and phenomena in two-sublattice magnetics with coinciding chemical and magnetic cells.

Group-theoretical analysis suggests that physical phenomena in a crystal should be connected with vectors $\vec{m}$ and $\vec{l}$ of its magnetic subsystem. The magnetic symmetry and hence the value and orientation of vectors $\vec{m}$ and $\vec{l}$ can be easily changed under the influence of temperature, magnetic field, pressure, etc. In symmetry analysis it is convenient to separate the exchange interactions resulting in the appearance of a magnetic structure from the considerably weaker relativistic interactions which are responsible, in particular, for the orientation of spin configuration with respect to the crystallographic axes [3]. It is known that in the first approximation the exchange symmetry of the ionic ferro- or antiferromagnetic structure does not often change when the total symmetry of a crystal is changed. Therefore, tensor components of the physical property can be conveniently expanded in powers of components l_k , m_n if one regards them as external influence on a crystal with unchangeable symmetry [2,4,5].

In order to study the influence of magnetic ordering on the optical properties of the crystallic media we write e.g. the components ε^{-1}_{ij} of electric permittivity tensor in the following form [6]

$$\varepsilon^{-1} = \overset{\circ}{\varepsilon}{}^{-1} + \alpha^{m} \cdot \vec{m} + \alpha^{l} \cdot \vec{l} + \alpha^{mm} : \vec{m}\vec{m} + \alpha^{ll} : \vec{l}\vec{l} + \alpha^{lm} : \vec{l}\vec{m} + \ldots \quad (1)$$

Here $\overset{\circ}{\varepsilon}{}^{-1}$ characterizes optical properties of a crystal in the paramagnetic phase; α^{m} and α^{l} describe the Faraday effect; α^{mm} , α^{ll} and α^{lm} define for linear magnetic birefrigence.

In the first approximation we regard vectors $\vec{m}$, $\vec{l}$ as depending on external influence. The tensorial coefficients of expansion α^{m} , α^{l} , α^{mm}, α^{ll}, α^{lm} do not depend on external influence (e.g. magnetic field and are specified by the exchange symmetry of a crystal.

We study magnetics with collinear exchange magnetic structures. Magnetic vectors originating in these structures for exchange approximation transform according to one-dimensional real irreducible representation of the paramagnetic phase of a crystal [2, 3]. Vectors $\vec{m}$ and $\vec{l}$ in ferromagnetics with crystallographically non-equivalent sublattices possess the same transformational properties. But we can not say this about crystals with equivalent sublattices.

According to Turov's notations [2] symmetry operations permuting (non permuting) sublattices of a magnetic structure are called odd (even) operations with respect to this structure and are labelled by (-) and (+) respectively. It is very important to take into account that $\vec{l}$ transforms like $\vec{m}$ only under even symmetry operations. But under odd symmetry operations it changes the sign. This means, that along with the ordinary symmetry operations the exchange sublattice opeartion $1^{(-)}$ is implicitly introduced and actually used for magnetics with crystallographically equivalent magnetic sublattices. That is why the classical Shubnikov magnetic groups do not characterize fully even two-sublattice magnetics, and do not give full information on the symmetry of $\vec{m}$ and $\vec{\ell}$.

Here a question arises: how can one describe the symmetry of vectors $\vec{m}$ and $\vec{\ell}$ and the corresponding structures whose exchange symmetry groups have odd operations? In such cases one can introduce more general groups of colour symmetry [7, 3], interpreting the signs (+), (-) as symbols of additional load on the elements of magnetic point groups of a crystal. In this approach exchange symmetry of two-sublattice magnetics is characterized by P-groups of colour symmetry [3, 7].

For non-magnetic media it suffices to reparate tensors into even and odd types with respect to the space inversion $\bar{1}$. In the classical physics of magnetic crystals one takes into account an additional symmetry operation $\underline{1}$ (antiidentity), which inverts the signs of the currents and magnetic spins. It is often called time-reversal. In this case the crystals are specified by more general Shubnikov groups. Then, one distinguishes 4 types of tensors: even, electrical, magnetic and magneto-electric. They all transform viy irreducible representations of the group $\bar{1} \times \underline{1}$. The symmetry and transformational properties of these tensor values are known [1].

3. Generalizing the approach [1] and explicitly taking into account the new possible sublattice exchange operation, we see that any opeartion of magnetic point of a two-sublattice magnetics defined by the orthogonal group $\infty\infty\bar{1} \times \underline{1} \times 1^{(-)}$ can be expressed as

$$G(m,n,p,R)=(\bar{1})^{m}(\underline{1})^{n}(1^{(-)})^{p}R(\vec{k},\varphi) \tag{2}$$

where $\bar{1}$ and $\underline{1}$ are space inversion and time-reversal respectively, $\bar{1}^{(-)}$ the sublattice exchange operator; R the rotation by an angle φ, $\vec{k}$ wave vector. m, n, p = 0,1. Consequently, we can say that all tensors of the obtained P-groups of magnetic symmetry of two-sublattice magnetics transform via an irreducible representations of $\bar{1} \times \underline{1} \times 1^{(-)}$. That is why the number of tensors of different types defined by $\infty\infty\bar{1} \times \underline{1} \times 1^{(-)}$ is doubled (8 instead of 4) (table 1).

Using the isomorphism between magnetic groups and well investigated crystallographical groups it is possible to show that the components of n-rank tensors $\hat{A}$ of all eight types transform according to

$$A_{i_1 \ldots i_n}=(-1)^{\gamma} R_{i_1 k_1} \ldots R_{i_n k_n} A_{k_1 \ldots k_n} \tag{3}$$

where γ depends on the type of the tensor $A_{k_1 \ldots k_n}$ and on loads (m, n, p) of symmetry operation $G(m, z, p, R)$ is defined in Table 1, $R_{i_n k_n}$ is the matrix of the proper space rotations.

Developing terminology of [1], we call $\mathring{\mathcal{E}}^{-1}$, α^{mm}, α^{ll} tensors of even (+) type, α^{lm} those of even (-) type and tensors α^{m}, $\vec{m}$ and α^{l}, $\vec{l}$ tensors of magnetic (+) and (-) types respectively.

Table 1. Basis tensors and irreducible representations of the group $\bar{1} \times \underline{1} \times 1^{(-)}$

	Type of basis tensors	Symmetry operation 1	$\bar{1}$	$\underline{1}$	$1^{(-)}$	$\bar{1}^{(-)}$	$\underline{1}^{(-)}$	$\underline{\bar{1}}$	$\underline{\bar{1}}^{(-)}$	γ
1	Even (+)	1	1	1	1	1	1	1	1	0
2	Electric (+)	1	-1	1	1	-1	1	-1	-1	m

3	Magnetic (+)	1	1	-1	1	1	-1	-1	-1	n
4	Even (-)	1	1	1	-1	-1	-1	1	-1	p
5	Electric (-)	1	-1	1	-1	1	-1	-1	1	m+p
6	Magnetic (-)	1	1	-1	-1	-1	1	-1	1	n+p
7	Magneto-electric (+)	1	-1	-1	1	-1	-1	1	1	m+n
8	Magneto-electric (-)	1	-1	-1	-1	1	1	1	-1	m+n+p

Using Table 1 we can reduce the formal problems of determining non-zero tensor components of the six odd types (3-8) of two-sublattice magnetics properties to the already solved calculation problem of the form of non-magnetic media electric type tensors. According to (3) the symmetry operation G(m, n, p, R) transforms tensor $\hat{A}$ of arbitrary rank in the same way as $G(\gamma, 0, 0, R)$ acts on the corresponding tensor of electric type, the latter having the same intrinsic symmetry as $\hat{A}$. For non-magnetic crystals all the non-zero components of electric type tensors up to the 4th are given, e.g. in [8]; tensors of higher rank in [9].

The suggested procedure enables us to obtain explicit phenomenological expressions for invariants and thermodynamic potentials (in the phase transition theory), magneto-optical, piezomagnetic and other properties of the two-sublattice magnetics using group symmetry. It enables us to avoid tiresome calculations.

For example, the experimentally observed [10, 5] birefringence of antiferromagnetic crystals CoF_2 and Dy_3FeO_3 bilinear in vectors and is described by tensors α^{lm} of even (-) type. These crystals belong to the classes $4_z^{(-)} 2_d^{(+)} \bar{1}^{(+)}$ and $2_z^{(-)} 2_x^{(+)} \bar{1}^{(+)}$ respectively. Consequently α^{lm} is equivalent to the 4th rank electric tensors (pseudo-tensors) in the crystals of non-magnetic classes $\bar{4}_z 2_x m_d \underline{1}$ and $m_z m_y 2_x \underline{1}$ respectively. Therefore it is not difficult to extract all non-zero components of d^{lm} for all antiferromagnetic structures.

4. The suggested approach based on the isomorphism of magnetic and non-magnetic groups enables us to reduce many problems of the two-sublattice magnetics in phenomenological tensorial crystallophysics to the corresponding problems of non-magnetic media crystallophysics.

This approach can be applied to more complex magnetic structures.

The author is grateful to Dr. V.A.Koptsik for useful comments.

References

1. Sirotin, Yu.I. (1962). Crystallography, 7, 89 (Russian)

2. Turov, E.A. (1963). Physical properties of magnetically ordered crystals. Nauka, Moscow (Russian).

3. Baryakhtar, V.G., Jablonsky, D.A. (1980). Low Temp. Phys., 6, 345 (Russian).

4. Mitsek, A.I., Shavrov, V.G. (1964). Solid State Phys., 64, 210 (Russian).

5. Kharchenko, N.F., Gnatchenko, S.L. (1981). Low Temp. Phys., 7, 475 (Russian).

6. Pisarev, R.V. (1970). ZhETPh 58, 1421 (Russian).

7. Koptsik, V.A., Kotzev, J.N. (1974). Commun; JINR,P4-8466, Dubna.

8. Birss, R.R. (1965). Symmetry and Magnetism. Amsterdam.

9. Smith, G.F. (1970). Annals of the New York Acad. of Sci., 175, 57.

10. Kharchenko, N.F., Eremenko, V.V., Bely, L.I. (1982). ZhETh, 82, 827 (Russian).

THE SYMMETRY OF GROUP REPRESENTATIONS AND THE REDUCTION OF THE MULTIPLICITY PROBLEM

R. Dirl[+], P. Kasperkovitz[+], J.N. Kotzev, M.I. Aroyo, M.N. Angelova

[+]Institut fur Theoretische Physik, TU Wien
Faculty of Physics, Sofia University

1.Introduction. Consider the transformation group $Q=\{q|q: R_G \rightarrow R_G\}$ defined on the set R_G of all unitary matrix representations (reps or UMR) D of a given group G. The group Q, its subgroups and reps, as well as some of the operator groups generated by Q, contain useful information about some symmetry relations that appear as "accidental" or even escape one's notice in the application of group-theoretical methods in physics.

In this paper we show how the above-mentioned symmetry relations can help to solve one of the standard problems arising in the development of the method of irreducible tensorial sets for a given finite or compact continuous group G - the problem of the ambiguity in the determining the basic elements of method, such as Clebsch-Gordan coefficients, generalized and symmetrized Clebsch-Gordan coefficients (nj-symbols), etc. [1,2]. This problem can be put on more general terms - the ambiguity of the reducing matrix S; i.e. the matrix that reduces a reducible UMR R(g), $g \in G$, to the direct sum of irreducible components $D^k(g)$:

$$S^{-1} R(g) S = \bigoplus_{K} (e_K \otimes D^K(g)) \qquad (1)$$

where e_k is the unit matrix with $\dim e^K = m_K = (R|D^K)$, the multiplicity of D^k in R, and $D^{Km}(g) = D^K(g)$, for all $m = 1,\ldots, m_k$. It is more convenient to split the square matrix S into rectangular submatrices S^{km}, consisting of $n_\kappa (= \dim D^\kappa)$ columns and n_R (=dim R) rows, which gives for (1)

$$R(g)\, S^{\kappa m} = S^{\kappa m} D^\kappa(g), \quad D^\kappa \in R, \quad m = 1,2,\ldots m_\kappa$$

From the Schur lemma [3] it follows that (i) the blocks S^k for a given D^k are fixed up to an arbitrary m_k-dimensional unitary transformation, which belongs to the commutator algebra of the resulting subrepresentation $e_k \otimes D^k$ of R and (ii) there exist exactly m_k nonvanishing linearly independent blocks $S^{\kappa m}$, $m = 1, \ldots, m_\kappa$ for each D^k. They form the unitary space of dimension m_k with the scalar product

$$\langle S^{\kappa,m}, S^{\kappa,m'} \rangle = n_\kappa \delta_{mm'}. \tag{3}$$

Most of the existing approaches to the multiplicity problem are based on the transfer of the transformation properties of the involved reps (generated by operations such as tensoring by a one-dimensional rep, hereafter denoted as "association" [6-8], automorphism [7], complex conjugation [1,9,10], permutations [9,11], etc.) on the reducing matrices. Each of these operations gives some useful results, but their separate and independent application usually does not completely solve the problem. The crucial point of the present procedure is that the three operations: (a) associations, (b) automorphisms and (c) complex conjugation, are closed into a group, Q, and their correlated application is much more effective than an individual one. Moreover, we have succeeded to define operator groups associated with subgroups of Q, which allow one to reduce the multiplicity problem to group theoretical grounds only and to define consistent generating relations for the subsets of reducing matrices (Sect. 2). In the last section an example illustrating the general scheme is worked out.

2. <u>General scheme</u>. 1) <u>Group Q</u>. Introduce the auxiliary group Q consisting of bijective mapping of the set R_G of all unitary matrix reps of G.

The first of these mappings, called <u>associations</u> a_i, is realized by tensoring by one-dimensional reps D^i of G:

$$(a_i D)(g) = D^i(g) \otimes D(g) \qquad (4)$$

The associations a_i from an abelian group ASS.

The next class of mappings to each automorphism B_j of G assigns a mapping b_j of R_G via

$$(b_j D)(g) = D(B_j^{-1}(g)) \qquad (5)$$

The set of these mappings forms a group AUT. The inner automorphisms lead to equivalent reps, so we are mainly interested in the outer automorphisms.

The last mapping is the complex conjugation,

$$(cD)(g) = D(g)^* \qquad (6)$$

The operators c and $c^2 = e$ form the group CON $\cong Z_2$

From the above definitions it follows that

$$Q = ASS \ltimes (AUT \otimes CON) = \{q \mid q = (xyz)\} \qquad (7)$$

where $\ltimes$ denotes the semi-direct product and $\otimes$ the direct one.

The operators $q = (x, y, z)$ with z = c are antilinear and the corresponding groups are antiunitary (of the well-known Shubnikov group-type [4-5]), i.e. the corepresentation theory should be used.

The group Q induces an equivalence relation (Q equivalence) on classes of equivalent irreducible representations (irreps) and they can be united into disjoint sets. In order to define the standard irreps of G we have to choose one representative of each Q -class, say D^k, as a standard irrep, while the remaining inequivalent standard irreps D^l of the same Q -class are generated from D^k by applying suitable transformations of Q i.e. $D^l = q_l^{(k)} D^k$, $q_l^{(k)} \in Q$.

2) Groups Q^R and Q^K. The problem we want to solve here is to find a unitary matrix S(1). For this purpose it suffices to use the subgroups Q^R and Q^K of Q, defined as follows:

$$Q^R = \{q \mid qR \sim R,\ q \in Q\} \subseteq Q,\quad Q^K = \{q \mid qD^K \sim D^K,\ q \in Q^R\} \subseteq Q^R. \quad (8)$$

The corresponding coset decomposition of Q^R modulo Q^K for fixed representatives $q_i^{(K)}$:

$$Q^R = q_1^{(K)} Q^K \cup q_2^{(K)} Q^K \cup \ldots \quad (9)$$

allows us to define Q^R-equivalence classes [k] of irreps $D^K \in R$,

$$[K] = \{qK \mid q \in Q^R\} = \{q_1^{(K)}K,\ q_2^{(K)}K, \ldots\} \quad (10)$$

In this case the set of standard irreps is constructed as $D' = q_1^{(K)} D^K$, $q_K^{(K)} = q_0$ = identical transformation.

3) Matrix groups $\bar{Q}^R$ and $\bar{Q}^K$. For each $q \in Q^R$ there exists a unitary matrix U(q) such that

$$(qR)(g) = U(q)^+ R(g') U(q),\quad g \in G. \quad (11)$$

The repeated application of q_i's onto R(g) show that the matrix

$$U(q_1) * U(q_2) * \ldots * U(q_n) = U(q_1)(q_1 U)(q_2) \ldots (q_1 q_2 \cdots q_{n-1} U)(q_n) \quad (12)$$

transforms R into $(q_1 q_2 \cdots q_n R)$ if the matrices $U(q_1)$ satisfy equations of the type (11). In (12)

$$(qM) = \begin{cases} M & \text{for a linear } q; \text{ for } M = U(q) \\ M^* & \text{for an antilinear} \end{cases} \quad (13)$$

The product (12) is well-known in the corepresentation theory and is the comultiplication. The inverse of $U(q)$ with respect to comultiplication is

$$U^*_{(q)} = ({}_q U)^+(q) \quad (14)$$

and this matrix can be used to transform R into $q^{-1}R$ i.e.

$$U(q^{-1}) = U^*(q). \quad (15)$$

Now let $\langle q_x, q_y, \cdots \rangle$ be a set of generators of Q^R and

$U(q_x), U(q_y), \ldots, U(q_x^{-1}), U(q_y^{-1}) \ldots$ a set of matrices satisfying (11). It is possible to define a matrix cogroup Q^{-R} forming all coproducts of matrices $U(q)$, $q \in \{q_x, q_y, \ldots q_x^{-1}, q_y^{-1} \ldots\}$. Each element of Q^{-R} is denoted by a word $q = q_1 q_2 \ldots q_n$ composed of letters $q_i \in \{q_x, q_y, \ldots\}$, different words need not necessarily denote different elements of Q^{R}.

Note that there is a great freedom in definition of Q^{-R}, especially if some of the multiplicities m_k in (1) is greater than one. Also the structure of Q^{-R} strongly depends on the selection of the generating matrices $U(q_x)$, $U(q_y)$, etc. A more detailed discussion of these problems and the relation between Q^{R} and Q^{-R} is given elsewhere [12].

The same remarks apply also for the matrix cogroups Q^{-K} generated by the matrices $U^K(q)$, $q \in \{q_r \ldots, q_r^{-1}, \ldots\}$, where $\langle q_r, q_s, \ldots \rangle$ is a set of generators of Q^K. We will need one such group for each Q^R-class [k].

4) <u>Operator groups $\hat{Q}^K$</u>. Now consider the action of $q \in Q^K$ on both sides of (2). Using the convention (13) for $M = S^{km}$ we obtain

$$R(q)[T(q) S^{K,m}] = [T(q) S^{K,m}] D^K(q), \text{ for each } D^K \in R, \; g \in G, g \in Q^K \quad (16)$$

where

$$T(q) S^{Km} \equiv U(q)(q S^{km}) U^{K+}(q) \quad (17)$$

The righthand side of (16) defines a transformation of S^{km} for a general element $q \in Q^K$. In the unitary space spanned by the set of m_k solutions $S^{k,m}$ of (2), the transformation $S^{Km} \rightarrow T(q) S^{Km}$ defines a norm-preserving operator $T(q)$ which is linear for $q = (a, b, e) \in Q^K$ and antilinear for $q = (a, b, c) \in Q^K$.

It is worth mentioning that each operator $T(q)$ is defined by a pair of matrices $U(q) \in Q^{-R}$ and $U^K_{(q)} \in Q^{-K}$, where in both cases q has to be identified with a word composed of letters $q_x, q_y, \ldots$, and $q_r, q_s, \ldots$ respectively. To define $T(q)$ explicitly one can start with expressing the generators $q_r, q_s, \cdots$ of Q^K as words composed of the generators $q_x, q_y, \cdots$ of Q^ℓ and compute the corresponding matrices $U(q_r), U(q_S), \cdots$ of $\bar{Q}^\ell$. These matrices combined with the generators $U^K_{(q_K)}, U^K_{(q_s)}, \cdots$ of Q^{-K} then define the operators $T(q_r), T(q_s), \cdots$ for all the generators $q_r, q_s, \cdots$ of Q^K. This set of operators can be closed into a group $\tilde{Q}^K$ if the product $T(q_r), T(q_s), \cdots$ is defined by means of the co-products of the matrices $U(q_i)$ and $U^K_{(q_i)}$.

The space spanned by the blocks $S^{K,m}$, $m=1,\cdots,m_K$, is invariant under the operations of $\tilde{Q}^K$ but in general not irreducible, i.e. m_k-dimensional corepresentation T_K of $\tilde{Q}^K$ can be reduced into irreducible corepresentations T_K^i. A decomposition of this space into irreducible subspaces reduces the multiplicity problem since blocks transforming according to different irreducible corepresentations T_K^i of $\tilde{Q}^K$ and blocks transforming according to different rows of the same corepresentations are orthogonal. Therefore if the corepresentation of $\tilde{Q}^K$ carried by the m_k-dimensional vector space contains each irreducible corepresentation T_K^i no more than once, the multiplicity problem is resolved. But if a corepresentation T_K^i occurs in T_K, more than once, then the multiplicity problem is only reduced and one has to apply the Schmidt procedure.

Once the blocks S^k are known, the remaining blocks S^l, belonging to the same Q^ℓ -equivalence class [k] are easily generated: applying $q_l^{(K)}$ to (2) one gets

$$S^l = U(q_l^{(K)})(q_l^{(K)} S^K).$$

In order to construct the whole S-matrix we have to repeat the above-discussed procedure for all the Q^ℓ -classes and combine the blocks into the matrix S.

3. <u>Example</u>. Consider a 12-dimensional reducible rep R of the double point group ${}^dD_3$. As generators of ${}^dD_3$ we

will use the two elements C_{3z} and C_{2x}. The generating relations, the decomposition into classes of conjugate elements and the character table for the irreps $D^K (D^K = \Gamma_K)$ of this group can be found in [5].

If we take D^6 to define the generating element a_6 of $ASS \cong C_4$: $(a_6 D)(g) = D^6(g) \otimes D(g)$ and consider the involutive automorphism β: $\beta(C_{3z}) = C_{3z}$, $\beta(C_{2x}) = \bar{C}_{2x}$ which determines $AUT \cong C_2$, it follows that $Q \cong C_4 \ltimes (C_2 \otimes C_2)$. How the transformations of this group relate the irrep D^K of ${}^dD_3$ is seen from the table:

$D^K \equiv K$	1	2	3	4	5	6
$a_6 D^K$	6	5	4	3	1	2
$b\ D^K$	1	2	3	4	6	5
$c\ D^K$	1	2	3	4	6	5

The reducible rep R is the tensor product of two reps; $R(q) = D^P(q) \otimes D^4(q)$ where D^P is a six-dimensional permutation representation defined by

$$D^P(C_{3z}) = e_2 \otimes \begin{vmatrix} 0 & 0 & 1 \\ 1 & 0 & 0 \\ 0 & 1 & 0 \end{vmatrix}, \quad D^P(C_{2x}) = \begin{vmatrix} 0 & 1 \\ 1 & 0 \end{vmatrix} \otimes \begin{vmatrix} 1 & 0 & 0 \\ 0 & 0 & 1 \\ 0 & 1 & 0 \end{vmatrix}$$

where e_2 is 2x2 matrix unit and D^4 is one of the standard irreps:

$$D^P(C_{3z}) = \begin{vmatrix} \delta & 0 \\ 0 & \delta^* \end{vmatrix}, \quad \delta = \exp(i\pi/3), \quad D^4(C_{2x}) = \begin{vmatrix} 0 & 1 \\ -1 & 0 \end{vmatrix}$$

Using the character table we see that $R \sim 2D^5 \oplus 2D^6 \oplus 4D^4$.

This and the transformation properties of D^k (see the above table) imply that Q^R is a proper subgroup of Q and $Q^R \cong C_2 \otimes C_2 \otimes C_2 \cong D_2 \otimes \theta$

The matrices U(a), U(b) and U(c) corresponding to a,b,c $\in Q^R$ and transforming R into aR, bR and cR, may be chosen in the form

$$U(a) = U(b) = \begin{vmatrix} 1 & \\ & -1 \end{vmatrix} \otimes e_6 \quad U(c) = e_6 \otimes \begin{vmatrix} & 1 \\ -1 & \end{vmatrix}$$

Accordingly, $Q^{-R} \cong C_2 \otimes C_4 \cong C_{4h} (C_{2h})$.

The irreps contained in R belong to two Q^R -classes

with the corresponding groups: Q^K:

$[D^5]=\{D^5, D^6\}$ $Q^5 \cong \{e, ab, ac, bc\} \cong D_2(C_2)$, $q_5^{(5)}=e$, $q_4^{(5)}=a$

$[D^4]=\{D^4\}$ $Q^4 \cong Q^R \cong D_2 \otimes \Theta$

The blocks $S^{5,m}$, $m=1,2$ are column matrices for which $RS^5=S^5D^5$. Considering this equation for the generators of ${}^dD_3$ one finds that S^5 is uniquely determined by a column matrix $B=\begin{vmatrix} B_1 \\ B_2 \end{vmatrix}$:

$$(S^5)^t=\left| B, D^4(C_{3z}^{-1})B, D^4(\bar{C}_{3z})B, -iD^4(\bar{C}_{2x})B, -iD^4(\bar{C}_{2y})B, -iD^4(\bar{C}_{2xy})B \right|$$

Since $\dim D^5=1$, we may set $U^5(q)=1$ for all $q \in Q^5$. To define operators $T(q) \in \tilde{Q}^5$ we also need the matrices $U(q)$ for the generators of Q^5, i.e. $q_r=ab$, $q_s=ac$. We choose $U(ab)=U(a)U(b)$ and $U(ac)=U(a)U(c)$ which fixes the action of $T(ab)$ and $T(ac)$: $T(ab)S^5=U(ab)S^5=S^5$, $T(ac)S^5=U(ac)S^{5*}$, $(T(ac))^2S^5=(U(ac))^2S^5=-S^5$

It follows that $\tilde{Q}^5$ is generated by the only antiunitary operator T(ac) of order 4, i.e. $\tilde{Q}^5 \cong C_4(C_2)$. Taking into account the corepresentation character table of this group [13]

	E	C_2	$C_4\theta$	$C_4^3\theta$
T^1	1	1	1	1
T^2	2	-2	0	0

we see that the space of the blocks $S^{5,m}$, $m=1,2$, carries the two-dimensional corepresentation T^2 (Note that all the linear representations of C_4 are one-dimensional!). If we choose the basis:

$S^{5,1}: B_1=1, B_2=0$, $S^{5,2}: B_1=0, B_2=1$

then $S^{5,2}=T(ac)S^{5,1}=U(ac)S^{5,1*}$

We recover $S^{6,m}$ from the columns $S^{5,m}$ via

$$S^{6,m}=U(a)S^{5,m}.$$

In a similar way we can find the blocks $S^{4,m}$, $m=1,\dots,4$. In this case the auxiliary group $\tilde{Q}^4$ reduces the multiplicity problem but does not solve it completely, so the Schmidt's procedure is needed.

A more detailed version of the theory and two additional examples are given in [12].

References

1. Wigner, E.P. (1959). Group theory. Academic Press, N.Y.
2. Klimyk, A.U. (1979). Matrix elements and Clebsch-Gordan coefficients for group representations. Naukova Dumka, Kiev (Russian).
3. Zhelobenko,D.P. (1970). The compact Lie groups and their representations. Nauka, Moscow. (Russian).
4. Koptsik, V.A. (1966). Shubnikov Groups. MGU, Moscow. (Russian).
5. Bradley, C.J., Cracknell,A.P. (1972). The mathematical theory of symmetry in solids. OUP, Oxford.
6. Griffith, J.S. (1961). The theory of transition metal ions. Cambridge University Press.
7. Dirl, R. (1983). J. Math. Phys. 24, 1935.
8. Kotzev, J.N., Aroyo, M.I. (1984). J. Phys. A17, 727.
9. Butler, P.H. (1975). Phil. Trans. R. Soc. (London) A277, 454.
10. Derome, J.R., Sharp, W.T. (1965). J. Math. Phys. 6, 1584.
11. Chatterjee, R., Lulek, T. (1982). J. Math. Phys. 23, 922.
12. Dirl, R., Kasperkovitz, P., Aroyo, M.I., Kotzev, J.N., Angelova, M.N. (1985). J. Math. Phys. (submitted).
13. Kotzev, J.N., Aroyo, M.I. (1982). J. Phys. A. 15,711.

SYMMETRY CLASSIFICATION OF MULTICOMPONENT ORDER PARAMETERS WITH REGARD TO INCOMMENSURATE PHASES

Kraizman I.L., Shakhnenko V.P., Chechin G.M.
Rostov State University

Order parameters (OP) with different transformational properties may generate identical thermodynamic potentials (the phenomenological Hamiltonians), written as functions of OP components and external thermodynamic fields. This clearly enables one to solve the problem of thermodynamic analysis of phase transitions by grouping parameters c in types characterized by the same thermodynamic potentials $\Phi(c, p, T, \ldots)$. According to Landau's theory [1], the minima of Φ in components c represent low-symmetry phases appearing as a result of spontaneous symmetry breaking. Thus OP of the same type induce the same number of low-symmetry phases and the same sets of phase p,T-diagrams [2]. This lead to classification of two- and three-component OP satisfying Lifshitz' condition and an analysis of the corresponding phase diagrams [2]. Lifshitz' condition in such cases justifies stability of symmetric phases with respect to macroscopic inhomogeneous distortions.

However, the above classification proves inadequate for OP transforming according to multicomponent irreducible (IR) representations of dimension $l \geq 3$. Generally, for IR conforming to the Lifshitz condition there may exist invariants linear in the first spatial derivatives of OP, but of higher degree in OP: $c_1^r c_2^s \ldots c_f^t \frac{\partial c_q}{\partial z}$ $(r+s+\ldots+t+1=n)$. At condensation of certain components c in low-symmetry phases the invariants may assume the usual Lifshitz form, thus leading to instability of the phases and the occurrence of complicated incommensurate phases. For l=4 these phases may adjoin a symmetric phase at a multicritical point near which phase-transition sequences with alternating homogeneous and inhomogeneous phases are possible [4]. For l=3 incommensurate phases may border with a symmetric phase on second-order phase-transition lines.

Precise classification of OP should therefore take into account not only homogeneous in c terms in Φ , but all the possible gradient invariants, which may lead to the Lifshitz instability of low-symmetry phases. Such classification is presented below for all the three-component OP, describing structural phase transitions in crystals. Transformational properties of OP, important for solving the above mentioned problems are given in Table 1 by sets of generators of reducible representations, composed each of a critical IR and a vector representation (or any of its reducible components) of the corresponding space groups. Here E is the unit matrix; a generators of three-component IR of space groups; b,d, e generators of reducible representations of the hexagonal and rhombohedral space groups; g those for the cubic space groups : $\varepsilon = \exp(i\, 2\pi/3)$.

The resulting classification of the three-component OP for all space groups is given in Tables 2 and 3. In the first column of the Tables an OP symbol is given, where the first figure corresponds to a given set of homogeneous invariants included in the integrity basis of invariants, whereas the second defines a set of gradient invariants of the lowest degree, generating low-symmetry incommensurate phases. The second column contains generators of reducible representations for each type of OP (see Table 1), while the third, the space group symbol, followed by the corresponding numbers of

Table 1. Generators of the reducible representations for the three-component OP

	1	2	3	4	5	6	7
a	-1 1 -1	1 -1 -1	1 1 1	1 1 1	-1 -1 -1		
b	a_1 1 1	a_2 1 1	a_3 ε ε^2	a_4 -1 -1	a_5 -1 -1	E -1 -1	
d	a_1 1	a_2 1	a_3 1	a_4 -1	a_5 -1	E -1	
e	a_1 1 1 1	a_2 1 1 1	a_3 ε ε^2 1	a_4 -1 -1 -1	a_5 -1 -1 -1	a_5 -1 -1 -1	E -1 -1 1
g	a_1 a_1	a_2 a_2	a_3 a_3	a_3 $a_3 a_5$	a_5 a_5	a_5 E	

wave vectors and IR, respectively, is taken from [5].

For all types of three-component OP, there exist only five types of integrity bases of homogeneous invariants [2]:

1) I_1, I_2, I_3; 2) I_1, I_2, I_3, I_4'; 3) I_1, I_2, I_3, I_4; 4) I_1, I_2, I_5; 5) I_1, I_2, I_4, I_5 , where $I_1 = c_1^2 + c_2^2 + c_3^2$, $I_2 = c_1^4 + c_2^4 + c_3^4$, $I_3 = c_1^2 c_2^2 c_3^2$, $I_4' = I_4 \cdot I_5$, $I_4 = c_1^2(c_2^4 - c_3^4) + c_2^2(c_3^4 - c_1^4) + c_3^2(c_1^4 - c_2^4)$, $I_5 = c_1 c_2 c_3$.

The sets of gradient invariants are:

1) I_1^g; 2) I_2^g; 3) I_3^g; 4) I_4^g; 5) I_1^g, I_3^g; 6) I_1^g, I_4^g; 7) I_2^g, I_3^g; 8) I_2^g, I_4^g; 9) I_1^g, I_2^g; 10) I_3^g, I_4^g; 11) I_1^g, I_2^g, I_3^g; 12) $I_1^g, I_2^g, I_3^g, I_4^g$; 13) I_5^g; 14) I_6^g, I_7^g; 15) I_5^g, I_6^g, I_7^g,

where

$$I_1^g = \sqrt{3}\left(c_1 c_2 \frac{\partial c_3}{\partial x} - c_1 c_3 \frac{\partial c_2}{\partial x}\right) + c_1 c_2 \frac{\partial c_3}{\partial y} + c_1 c_3 \frac{\partial c_2}{\partial y} - 2 c_2 c_3 \frac{\partial c_1}{\partial y}$$

$$I_2^g = \sqrt{3}\left(c_1 c_2 \frac{\partial c_3}{\partial y} - c_1 c_3 \frac{\partial c_2}{\partial y}\right) - c_1 c_2 \frac{\partial c_3}{\partial x} - c_1 c_3 \frac{\partial c_2}{\partial x} + 2 c_2 c_3 \frac{\partial c_1}{\partial x}$$

$$I_3^g = c_1 c_2\left(c_1 \frac{\partial c_2}{\partial z} - c_2 \frac{\partial c_1}{\partial z}\right) + c_2 c_3\left(c_2 \frac{\partial c_3}{\partial z} - c_3 \frac{\partial c_2}{\partial z}\right) + c_3 c_1\left(c_3 \frac{\partial c_1}{\partial z} - c_1 \frac{\partial c_3}{\partial z}\right)$$

$$I_4^g = c_1 c_2(c_1^2 - c_2^2)\frac{\partial c_3}{\partial z} + c_2 c_3(c_2^2 - c_3^2)\frac{\partial c_1}{\partial z} + c_3 c_1(c_3^2 - c_1^2)\frac{\partial c_2}{\partial z}$$

for hexagonal and rhombohedral space groups and

$$I_5^g = \left(c_2 \frac{\partial c_3}{\partial x} - c_3 \frac{\partial c_2}{\partial x}\right) + \left(c_3 \frac{\partial c_1}{\partial y} - c_1 \frac{\partial c_3}{\partial y}\right) + \left(c_1 \frac{\partial c_2}{\partial z} - c_2 \frac{\partial c_1}{\partial z}\right)$$

$$I_6^g = (c_2^2 - c_3^2)\frac{\partial c_1}{\partial x} + (c_3^2 - c_1^2)\frac{\partial c_2}{\partial y} + (c_1^2 - c_2^2)\frac{\partial c_3}{\partial z}$$

$$I_7^g = c_1\left(c_2 \frac{\partial c_2}{\partial x} - c_3 \frac{\partial c_3}{\partial x}\right) + c_2\left(c_3 \frac{\partial c_3}{\partial y} - c_1 \frac{\partial c_1}{\partial y}\right) + c_3\left(c_1 \frac{\partial c_1}{\partial z} - c_2 \frac{\partial c_2}{\partial z}\right)$$

for cubic space groups.

Table 2. Classification of three-component order parameters of hexagonal and rhombohedral space groups

Type	Generators	IR	
1	$a_1 - a_5$	D_{6h}^{1-4}	12/4, 5, 7
		D_{6h}^{1}	14/1-8
		D_{3h}^{1-4}	12/4
		$D_{3h}^{3,4}$	14/1-4
		$D_{3d}^{1,3}$	14/1-4
		$C_{3v}^{1,2}$	14/1, 2
		C_{3v}^{5}	4/1, 2
		C_{6v}^{1}	14/1-4
1-1	$b_1 - b_5$	D_{6h}^{1-4}	12/8
		$D_{3d}^{3,4}$	12/4
		C_{6v}^{1-4}	12/3
1-2	$b_1 - b_3$, $b_4 \cdot b_6$, b_5	D_{6h}^{1-4}	12/6
		$D_{3d}^{1,2}$	12/4
		$D_{3d}^{5,6}$	5/4
		C_{6v}^{1-4}	12/4
1-3	$d_1 - d_4$, $d_5 \cdot d_6$	D_{6}^{1}	14/1-4
		$D_{3}^{1,2}$	14/1, 2
		D_{3}^{7}	4/1, 2
1-4	$d_1 - d_5$	D_{6h}^{1-4}	12/2

Table 2 (continued)

Type	Generators	IR	
1-4	d_1-d_5	D^{1-4}_{3h}	12/2
1-5	e_1-e_4, e_6	D^{1-6}_{6}	12/4
1-6	e_1-e_5	$D^{1,2}_{3d}$	12/2
		$D^{5,6}_{3d}$	5/2
1-7	$e_1-e_3, e_4 \cdot e_7, e_6$	D^{1-6}_{6}	12/3
1-8	$e_1-e_3, e_4 \cdot e_7, e_5$	$D^{3,4}_{3d}$	12/2
2	$a_1-a_3, a_4 \cdot a_5$	D^{1-4}_{6h}	12/3
		D^{1-4}_{3d}	12/3
		$D^{5,6}_{3d}$	5/3
2-1	$b_1-b_3, b_4 \cdot b_5$	$D^{3,4}_{3h}$	12/3
2-2	$b_1-b_3, b_4 \cdot b_5 \cdot b_6$	$D^{1,2}_{3h}$	12/3
2-3	$d_1-d_3, d_4 \cdot d_5 \cdot d_6$	D^{1-6}_{6}	12/2
2-4	$d_1-d_3, d_4 \cdot d_5$	C^{1-4}_{6v}	12/2
2-5	$e_1-e_3, e_4 \cdot e_6$	$D^{2,4,6}_{3}$	12/2
2-6	$e_1-e_3, e_4 \cdot e_5$	$C^{2,4}_{3v}$	12/2
2-7	$e_1-e_3, e_4 \cdot e_6 \cdot e_7$	$D^{1,3,5}_{3}$	12/2
2-8	$e_1-e_3, e_4 \cdot e_5 \cdot e_7$	$C^{1,3}_{3v}$	12/2

Table 2 (continued)

Type	Generators	IR	
3	$a_1 - a_3, a_5$	$C_{6h}^{1,2}$	12/3
		C_{6h}^{1}	14/1-4
		C_{3h}^{1}	14/1,2
		S_6^{1}	14/1,2
		S_6^{2}	4/1,2
3-3	$d_1 - d_3, d_5 \cdot d_6$	C_6^{1}	14/1,2
		C_3^{1}	14/1
		C_3^{4}	4/1
3-4	$d_1 - d_3, d_5$	$C_{6h}^{1,2}$	12/2
		$C_{3h}^{1,2}$	12/2
3-9	$b_1 - b_3, b_5$	$C_{6h}^{1,2}$	12/4
3-11	$e_1 - e_3, e_5$	S_6^{1}	12/2
		S_6^{2}	12/2
		C_6^{1}	12/2
4	$a_1 - a_4$	D_{6h}^{1-4}	12/1
		C_{6v}^{1-4}	12/1
		D_{3d}^{1-4}	12/1
		$D_{3d}^{5,6}$	5/1

Table 2 (continued)

Type	Generators	IR	
4-1	$b_1 - b_4$	$D^{1,2}_{3h}$	12/1
		$C^{1,3}_{3v}$	12/1
4-2	$b_1 - b_3, b_4 \cdot b_6$	$D^{3,4}_{3h}$	12/1
		$C^{2,4}_{3v}$	12/1
4-6	$e_1 - e_4$	$D^{1,3,5}_{3}$	12/1
		D^{7}_{3}	5/1
4-8	$e_1 - e_3, e_4 \cdot e_7$	$D^{2,4,6}_{3}$	12/1
4-10	$d_1 - d_4$	D^{1-6}_{6}	12/1
5	$a_1 - a_3$	$C^{1,2}_{6h}$	12/1
		S^{1}_{6}	12/1
		S^{2}_{6}	5/1
5-9	$b_1 - b_3$	C^{1}_{3h}	12/1
5-10	$d_1 - d_3$	C^{1-6}_{6}	12/1
5-12	$e_1 - e_3$	C^{1-3}_{3}	12/1
		C^{4}_{3}	5/1

Table 3. Classification of three-component order parameters of cubic space groups

Type	Generators	IR	
1	$a_1 - a_5$	O_h^{1-4}	12/10
		O_h^{5-10}	11/10
		O_h^{1-4}	13/7-10
		O_h^{9}	12/7-10
		O_h^{1}	10/1-8
		$O_h^{1,3}$	11/2,4,6,8
		O_h^{5}	10/2,4,6,8
		O_h^{6}	10/1,3,5,7
		$T_d^{1}{}^{1}$	10/1-4
		T_d^{1}	13/4,5
		T_d^{3}	12/4,5
		$O^{1,2}$	10/1-4
1-13	$g_1 - g_4$, g_6	$O^{1,2}$	13/4,5
		$O^{5,8}$	12/4,5
1-14	$g_1 - g_5$	O_h^{1-4}	12/8
		O_h^{5-10}	11/8
2	$a_1 - a_3$, $a_4 \cdot a_5$	O_h^{1-4}	12/9
		O_h^{5-10}	11/9
		O_h^{1}	11/3,5
		O_h^{3}	11/1,7

Table 3 (continued)

Type	Generators	IR	
2	a_1-a_3, $a_4 \cdot a_5$	O_h^5	10/3,5
		O_h^6	10/4,6
		T_d^1	11/2,4
		T_d^4	11/1,3
		O^1	11/2,3
		O^2	11/1,4
		$O^{3,4}$	10/2,3
		T_d^2	10/2,4
		T_d^5	10/1,3
2-13	g_1-g_3, $g_4 \cdot g_6$	$O^{1,2,6,7}$	12/5
		$O^{3,4,5,8}$	11/5
2-14	g_1-g_3, $g_4 \cdot g_5$	$T_d^{1,4}$	12/5
		$T_d^{2,3,5,6}$	11/5
3	a_1-a_3, a_5	T^1	10/1-4
		T_h^1	10/1-8
		$T_h^1 1$	11/2,4,6
		T_h^3	10/2,4,6,8
		$T_h^{1,2}$	13/7,8
		$T_h^{5,7}$	12/7,8

Table 3 (continued)

Type	Generators	IR	
	$g_1 - g_3$, g_5	$T_h^{1,2,6}$	12/8
		$T_h^{3,4,5,7}$	11/8
3-14	$g_1 - g_3$, g_6	T^1	13/4
		$T^{3,5}$	12/4
4	$a_1 - a_4$	O_h^{1-4}	12/7
		O_h^{5-10}	11/7
		O_h^1	11/1,7
		O_h^3	11/3,5
		O_h^5	10/1,7
		O_h^6	10/2,8
		$T_d^{1,4}$	12/4
		$T_d^{2,3,5,6}$	11/4
		O^1	11/1,4
		O^2	11/2,3

Table 3 (continued)

Type	Generators	IR	
4	$a_1 - a_4$	$O^{3,4}$	10/1,4
		T_d^2	10/1,3
		T_d^5	10/2,4
		T_d^1	11/1,3
		T_d^4	11/2,4
4-15	$g_1 - g_4$	$O^{1,2,6,7}$	12/4
		$O^{3,4,5,8}$	11/4
5	$a_1 - a_3$	T^1	11/1-4
		T_h^1	11/1,3,5,7
		T_h^3	10/1,3,5,7
		$T_h^{1,2,6}$	12/7
		$T_h^{3,4,5,7}$	11/7
		T^2	10/1-4
5-15	$g_1 - g_3$	$T^{1,4}$	12/4
		$T^{2,3,5}$	11/4

References

1. Landau, L.D., Lifshitz, E.M. (1964). Statistical Physics, Nauka, Moscow. (Russian).

2. Gufan, Yu.M., Sakhnenko, V.P. (1975)., 63, 1908, (1973). Sov. Phys. JETP, 36, 1009; (1975). ibid, 42, 728.

3. Lifshitz, E.M. (1941). ZhETPh, 11, 255 (Russian).

4. Kraizman, I.L., Sakhnenko, V.P., (1984). Sov. Phys. JETP Lett. 40, 931.

5. Kovalev, O.V. (1961). Irreducible Representations of Space Groups. UkSSR, Akad. Sci, Kiev (Russian).

ON THE WIGNER-ECKART THEOREM AND COUPLED TENSOR OPERATORS FOR COREPRESENTATIONS OF SHUBNIKOV GROUPS

M.N. Angelova, M.I. Aroyo and J.N. Kotzev
Faculty of Physics, Sofia University, Sofia-1126

1. Introduction. Matrix elements of irreducible tensor operators (ITO) used for quantum-mechanical calculations of atomic, nuclear, solid state physics etc., can be factorized due to the well-known Wigner-Eckart theorem [1]. However, in the case of magnetic symmetry, the wave functions and the components of ITO do not transform by the linear representations but by the Wigner corepresentations (coreps) of the corresponding antiunitary (AU) Shubnikov magnetic groups [1-3]. This leads to serious difficulties in the computing matrix elements. They result mainly from the generalized Schur lemma for coreps [4] and from the particular form of the orthogonality relations for coreps [5].

In this paper we solve the above-mentioned problems using the symmetrized coefficients for coreps (nD-symbols-analogues of the Wigner nj-symbols). The main results are the following: (i) the Wigner-Eckart theorem for coreps is generalized in terms of 3D-symbols; (ii) the relations among reduced matrix elements (RME) for two systems connected by group-subgroup relations are investigated using the symmetrized isoscalar factor (IF) for coreps; (iii) the RME of coupled tensor operators (CTO) for coreps are expressed by RME of their components with the help of 6D- and 9D-symbols.

2. 3D-symbol version of the Wigner-Eckart theorem. The generalization of the Wigner-Eckart theorem for the cause of coreps of finite AU groups in terms of Clebsch-Gordan

coefficients (CGC) for coreps (we shall refer to it as "CGC version") was given first by Kotzev [6,7] and Aviran and Zak [8]. There are some problems which cannot be solved in principle within the framework of the CGC version of the theorem and which limit its (theorem's) application in the general case: (i) the explicit form of that version is different for the different Wigner type of coreps; (ii) the RME are not completely independent of the basis functions (as it is in the classical case of linear representations). Using 3D-symbols [4,9] we obtain a new form of the Wigner-Eckart theorem for coreps (referred to as a 3D-symbol version)

$$\begin{cases} \langle \alpha_1 a_1 | T_q^{\varkappa} | \alpha_2 a_2 \rangle = \sum\limits_{\rho} \langle \alpha_1 \| T^{\varkappa} \| \alpha_2 \rangle_\rho \sum (K_{a_1 a_1^*}^{\alpha_1^*})^* \begin{pmatrix} \alpha_1^* & \varkappa & \alpha_2 \\ a_1^* & q & a_2 \end{pmatrix}_\rho^* , \\ \langle \alpha_1 \| T^{\varkappa} \| \alpha_2 \rangle_\rho = \langle \alpha_1 \| T^{\varkappa} \| \alpha_2 \rangle_\rho^* , \quad \rho = 1, \ldots, (\alpha_1^* \varkappa \alpha_2 | \alpha_0), \end{cases}$$

Here the unitary matrix $K^{\alpha_1^*} = \| K_{a_1 a_1^*}^{\alpha_1^*}$ transforms the corep $(D^{\alpha_1})^* = \{ D^{\alpha_1}(g)^*, g \in G_A \}$ to the equivalent corep $D^{\alpha_1^*} = \{ D^{\alpha_1^*}(g), g \in G_A \}$, belonging to the standard set

$$K^{\alpha_1^*}(g), g \in G_A \qquad = D^{\alpha_1^*}(g), \quad g \in G_A, \tag{3}$$

where the complex conjugation is applied if and only if $g \in G_A$ is an AU operator. By analogy with the Wigner 1j-symbols [10-12] the matrix elements of $K^{\alpha_1^*}$ are called 1D-symbols.

We define the 3D-symbols [4,9] as matrix elements of the matrix

$$V^{\alpha_1^* \varkappa \alpha_2} = \| V_{a_1^* q\, a_2, \rho}^{\alpha_1^* \varkappa \alpha_2} \| = \| \begin{pmatrix} \alpha_1^* & \varkappa & \alpha_2 \\ a_1^* & q & a_2 \end{pmatrix}_\rho \| \tag{4}$$

which reduce the direct product of coreps $D^{\alpha_1^*} \times D^{\varkappa} \times D^{\alpha_2}$ to the identity corep D^{α_0} with multiplicity $(\alpha_1^* \varkappa \alpha_2 | \alpha_0)$.

In the 3D-symbol version of the theorem a new definition of RME is introduced:

$$\langle \alpha_1 \| T^{\varkappa} \| \alpha_2 \rangle_g = \sum_{a_1 a_1^* q, a_2} \langle \alpha_1 a_1 | T^{\varkappa}_q | \alpha_2 a_2 \rangle K^{\alpha_1^*}_{a_1 a_1^*} \begin{pmatrix} \alpha_1^* & \varkappa & \alpha_2 \\ a_1^* & q & a_2 \end{pmatrix}_g \quad (5)$$

Now the RME do not depend on indices of the corep basis functions as they do in the CGC version [6-8]. The dependence on the Wigner type of D^{α_1} is included only in the multiplicity index $g = g(x_{\alpha_1}, x_0)$, $x_{\alpha_1} = 1, \dots, (\varkappa \alpha_2 | \alpha_1)$, $x_0 = 1, \dots, (\alpha_1 \alpha_1^* | \alpha_0)$, where $(\alpha_1 \alpha_1^* | \alpha_0) = 1, 4, 2$ for a, b, c types of the corep D^{α_1} respectively. It is important that as a result of the consideration of both unitary and AU operators of G_A and the new definition of RME (5), we obtain real RME (2) for all types of coreps D^{α_1} (not only for type a corep as we do in the CGC version).

The new form of the Wigner-Eckart theorem for coreps, expressed by (1) and (2), is valid for all kinds of magnetic groups (grey and black-and-white). Similar results but only for the special case of grey groups are presented in [12]. The 3D-symbol version of the theorem has the following advantages in comparison with CGC version: (i) the RME are real for all the cases; (ii) the matrix elements are completely factorized; (iii) the form of the theorem is considerably simplified and it does not depend on the type of coreps.

3. Relations among RME for group-subgroup chains. The relations among RME for an arbitrary group-subgroup chain $G_A \supset G_B$ are of great interest. If we express the matrix elements of ITO consecutively in the basis of coreps of the group G_A and its subgroup G_B using the Racah lemma for 3D-symbols [4], we get:

$$\langle \alpha_1 \beta_1 S_{\beta_1} \| T^{\varkappa \tau S_\tau} \| \alpha_2 \beta_2 S_{\beta_2} \rangle_{g_\beta} = \sum_{g_\alpha} \langle \alpha_1 \| T^{\varkappa} \| \alpha_2 \rangle_{g_\alpha} \sum_{\beta_1^* S_{\beta_1^*}} \chi^{\alpha_1 \alpha_1^*}_{\beta_1 S_{\beta_1}, \beta_1^* S_{\beta_1^*}} \begin{pmatrix} \alpha_1^* & \varkappa & \alpha_2 \\ \beta_1^* & \tau & \beta_2 \\ S_{\beta_1^*} & S_\tau & S_{\beta_2} \end{pmatrix}^{g_\alpha}_{g_\beta}$$

$$g_\beta = 1, \dots, (\beta_1^* \tau \beta_2 | \beta_0) \quad (6)$$

Here the subduction multiplicity indices S_{β_i} and S_τ run 1 to the multiplicity of D^{β_i} and D^{τ} in the subductions $(D^{\alpha_i}\downarrow G_B)$ and $(D^{\varkappa}\downarrow G_B)$ respectively. The quantities in the parenthesis are symmetrized outer isoscalar factors (OIF) for coreps. They appear as a result of the application of the generalized Schur lemma for reducible coreps [4] and they express the multiplicity arbitrariness. These coefficients do not depend on the partners of the basis and they realize the relation between the coupling schemes in G_A and its subgroup G_B. In the corep case OIF are real numbers while in the rep case they can be complex. The quantities χ have a similar meaning and they appear in the subduction of $K^{\alpha_i^*}$ to G_B

The relations (6) give the possibility for calculating RME for a subgroup if the corresponding RME for the group are known (e.g. the calculation of RME for magnetic point groups from the known RME in the full rotational group O(3) x Θ).

4. Factorization of RME of coupled tensor operators for AU magnetic groups. Using the advantages of the 3D-symbol version of Wigner-Eckart theorem we factorize the RME of coupled tensor operators (CTO) for coreps in terms of 6D- and 9D-symbols. Such relations are well-known for linear representations [11, 13] but essential peculiarities appear for the coreps.

In the general case when the CTO $T^{\varkappa r_\varkappa} = \{P^{\varkappa_1} Q^{\varkappa_2}\}^{\varkappa r_\varkappa}$, $r_\varkappa = 1, \ldots, (\varkappa_1 \varkappa_2 | \varkappa)$ as well as its components $P^{\varkappa_1}$ and $Q^{\varkappa_2}$ act in the same space, we get the following relation among RME:

$$\langle \alpha_1 \| \{P^{\varkappa_1} Q^{\varkappa_2}\}^{\varkappa r_\varkappa} \| \alpha_2 \rangle_{\rho_\alpha} = [D^{\varkappa}]^{1/2} \sum_{\alpha \rho_1 \rho_2} \langle \alpha_1 \| P^{\varkappa_1} \| \alpha \rangle_{\rho_1} \cdot \tag{7}$$

$$\langle \alpha \| Q^{\varkappa_2} \| \alpha_2 \rangle_{\rho_2} \sum_{\rho_2' \rho_1' \rho_\varkappa} \begin{Bmatrix} \varkappa_2^* & \varkappa & \varkappa_1^* \\ \alpha_1^* & \alpha^* & \alpha_2^* \end{Bmatrix}_{\rho_2' \rho_\alpha \rho_1' \rho_\varkappa} A_{\rho_2' \rho_1' \rho_\varkappa, \rho_2 \rho_1 r_\varkappa r_0}$$

In the derivation of the relations (7) sums of product of four 3D-symbols and the corresponding 1D-symbols appear which can be transformed to the form:

$$\sum_{a_i a_i^* q_i q_i^*} K^{\varkappa_2}_{q_2^* q_2} K^{\varkappa^*}_{q q^*} K^{\varkappa_1}_{q_1^* q_1} K^{\alpha_1}_{a_1^* a_1} K^{\alpha}_{\alpha^* a} K^{\alpha_2}_{a_2^* a_2} \begin{pmatrix} \varkappa_2^* & \alpha & \alpha_2^* \\ q_2^* & a & a_2^* \end{pmatrix}_{\rho_2'} \begin{pmatrix} \alpha_1^* & \varkappa & \alpha_2 \\ a_1^* & q & a_2 \end{pmatrix}_{\rho_\alpha} \begin{pmatrix} \alpha_1 & \alpha^* & \varkappa_1^* \\ a_1 & a^* & q_1^* \end{pmatrix}_{\rho_1'} \begin{pmatrix} \varkappa_2 & \varkappa^* & \varkappa_1 \\ q_2 & q^* & q_1 \end{pmatrix}_{\rho_\varkappa} =$$

$$\begin{Bmatrix} \varkappa_2^* & \varkappa & \varkappa_1^* \\ \alpha_1^* & \alpha^* & \alpha_2^* \end{Bmatrix}_{\rho_2' \rho_2 \rho_1' \rho_\varkappa}, \quad \rho_2 = 1,\dots,(\varkappa_2^* \alpha \alpha_2^* | \alpha_0);\ \rho_1 = 1,\dots,(\alpha_1^* \varkappa \alpha_2 | \alpha_0); \quad (8)$$
$$\rho_1 = 1,\dots,(\alpha_1 \alpha^* \varkappa_1^* | \alpha_0);\ \rho_\varkappa = 1,\dots,(\varkappa_1 \varkappa_2 \varkappa^* | \alpha_0)$$

The quantity (8) has symmetry properties (with respect to permutations of rows and columns and under complex conjugation) similar to the corresponding ones of the 6j-symbols. Because of the specific peculiarities for the case of coreps [14], we call it a 6D-symbol.

The last factor in (7) is a product of the inner isoscalar factors (IIF) [4]. Their origin and properties will be discussed elsewhere.

If the CTO space is the direct product of two subspaces and each of ITO $P^{\varkappa_1}$ or $Q^{\varkappa_2}$ acts only in one of the component spaces, then eqs. (7) are transformed as follows:

$$\langle (\beta_1 \gamma_1)\alpha_1 r_{\alpha_1} \| \{P^{\varkappa_1} Q^{\varkappa_2}\}^{\varkappa r_\varkappa} \| (\beta_2 \gamma_2)\alpha_2 r_{\alpha_2} \rangle =$$
$$= [D^{\alpha_1} \times D^{\varkappa} \times D^{\alpha_2}_{*}]^{1/2} \sum_{\rho_\beta \rho_\gamma} \langle \beta_1 \| P^{\varkappa_1} \| \beta_2 \rangle_{\rho_\beta} \langle \gamma \| Q^{\varkappa_2} \| \gamma_2 \rangle \rho_\gamma^* \quad (9)$$

$$\times \sum_{\rho_{\alpha_1} \rho_{\alpha_2} \rho_\varkappa \rho_\beta' \rho_\gamma'} \begin{Bmatrix} \beta_1^* & \gamma_1^* & \alpha_1 \\ \varkappa_1 & \varkappa_2 & \varkappa^*_* \\ \beta_2 & \gamma_2 & \alpha_2^* \\ \rho_\beta' & \rho_\gamma' & \rho_\alpha \end{Bmatrix}_{\rho_{\alpha_2}}^{\rho_{\alpha_1}\ \rho_\varkappa} B_{\rho_{\alpha_1} \rho_\varkappa \rho_{\alpha_2} \rho_\beta' \rho_\gamma', r_{\alpha_1} r_{0_1} r_\varkappa r_0 r_{\alpha_2} r_{0_2} \rho_\beta \rho_\gamma}$$

The 9D-symbol is

$$\begin{Bmatrix} \beta_1^* & \gamma_1^* & \alpha_1 \\ \varkappa_1 & \varkappa_2 & \varkappa^* \\ \beta_2 & \gamma_2 & \alpha_2^* \\ \rho_\beta' & \rho_\gamma' & \rho_\alpha \end{Bmatrix}_{\rho_{\alpha_2}}^{\rho_{\alpha_1}\ \rho_\varkappa} = \sum_{\substack{a_i a_i^* b_i b_i' \\ c_i c_i^* q_i q_i^*}} K^{\beta_1}_{b_1^* b_1} K^{\gamma}_{c_1^* c_1} K^{\alpha^*}_{a_1 a_1^*} K^{\varkappa_1^*}_{q_1 q_1^*} K^{\varkappa_2^*}_{q_2 q_2^*} K^{\varkappa}_{q^* q} K^{\beta_2^*}_{b_2 b_2^*} K^{\gamma_2^*}_{c_2 c_2^*} \quad (10)$$

$$\times \begin{pmatrix} \beta_1^* & \gamma_1^* & \alpha_1 \\ b_1^* & c_1^* & a_1 \end{pmatrix}_{\rho_{\alpha_1}} \begin{pmatrix} \varkappa_1 & \varkappa_2 & \varkappa^* \\ q_1 & q_2 & q^* \end{pmatrix}_{\rho_\varkappa} \begin{pmatrix} \beta_2 & \gamma_2 & \alpha_2^* \\ b_2 & c_2 & a_2^* \end{pmatrix}_{\rho_{\alpha_2}} \begin{pmatrix} \beta_1 & \varkappa_1^* & \beta_2^* \\ b_1 & q_1 & b_2^* \end{pmatrix}_{\rho_\beta'} \begin{pmatrix} \gamma_1 & \varkappa_2^* & \gamma_2^* \\ c_1 & q_2^* & c_2^* \end{pmatrix}_{\rho_\gamma'} \begin{pmatrix} \alpha_1^* & \varkappa & \alpha_2 \\ a_1^* & q & a_2 \end{pmatrix}_{\rho_\alpha}$$

where the multiplicity indices $\rho_{\alpha_1}, \rho_\varkappa, \rho_\beta'$ and ρ_γ' run 1 to $(\beta_1^* \gamma_1^* \alpha_1 | \alpha_0)$, $(\varkappa_1 \varkappa_2 \varkappa^* | \alpha_0)$, $(\beta_1 \varkappa_1^* \beta_2^* | \alpha_0)$ and $(\gamma_1 \varkappa_2^* \gamma_2^* | \alpha_0)$ correspondingly. The last factor in (9) is a product of IIF.

We note that for the case of grey groups relations

similar to (9) are obtained in [14] but the authors used lower symmetrized "quasi 9j-symbols".

The relations (7) and (9) are valid for all types of AU magnetic groups (grey and black-and-white). Selecting reasonably IIF matrices (7) and (9) can be brought into a form most appropriate to the corresponding relations in the linear representation case [11]. The character features of the coreps, however, lead to real RME defined by (5), real coefficient A and B and some differences of the properties of 6D- (8) and 9D-symbols (10).

The results obtained in this paper can be applied in the treatment of magnetic and non-magnetic interactions (crystal field, spin-orbit interactions, exchange interactions, etc.) for systems with magnetic symmetry.

References

1. Wigner, E.P. (1959). Group Theory, Academic Press, NY.
2. Koptsik, V.A. (1966). Shubnikov Groups, MGU, Moscow, (Russian).
3. Bradley, C.J., Cracknell, A.P. (1972). The Mathematical Theory of Symmetry in Solids, OUP, Oxford.
4. Kotzev, J.N., Aroyo, M.I. (1983). In: Group-Theoretical Methods in Physics (Zvenigorod 82), 1 (ed. Markov, M.A.), p. 362, Nauka, Moscow. (Russian).
5. Kotzev, J.N. (1974). Sov. Phys. Crystallography, 19, 286.
6. Kotzev, J.N. (1967). On the Wigner-Eckart theorem for corepresentations, Kharkov Univ. Kharkov. (Russian).
7. Kotzev, J.N. (1972). On the theory of corepresentations of magnetic groups, IRE AN USSR, Kharkov. (Russian).
8. Aviran, A., Zak, J. (1968). J. Math. Phys., 9, 2138.
9. Kotzev, J.N., Aroyo, M.I., Angelova, M.N. (1984). J. Mol. Str., 115, 123.
10. Derome, J.R., Sharp, W.T. (1965). J. Math. Phys., 6, 1584.
11. Butler, P.H. (1975). Phil. Trans. Roy, Soc. London. A277, 545.

12. Newmarch, J.D., Golding, R.M. (1981). J. Math. Phys., 22, 233.

13. Judd, B.R. (1963). Operator Techniques in Atomic Spectroscopy, McGraw Hill, NY.

14. Newmarch, J.D., Golding, R.M. (1983). J. Math. Phys., 24, 441.

LARGE UNIT CELL - SMALL BRILLOUIN ZONE MODEL IN THE OSCILLATION PROBLEMS OF CRYSTALLOPHYSICS AND CONTINUUM MECHANICS APPROXIMATION

V.A. Koptsik, S.A. Ryabtchikov
Moscow State University

1. Introduction. Methods of continuum mechanics with microstructure in solid state physic have been recently applied and extensively utilized in the spectroscopy of non-rigid molecules and crystals, in the theory of electronic structure of solids [1], in the phase transition theory, in the continuum theory of dislocations, etc.

The crystal lattice theory is in close relation to the non-local elasticity theory [2]. The latter allows us to take into account the translation and rotation invariance of energy. The discrete space symmetry group of a crystal Φ is actually a subgroup of the Euclidean group E(3) which is the semidirect product of the (normal) group of translations T(3) and O(3). This allows us to construct a microstructure model of a continuum with generalized symmetry group of a crystal and set the following problems:

1) to construct a model of a compound continuum corresponding to the ideal structure of a complicated (molecular) crystal;

2) to investigate symmetry of a continuous model and its compounds (subcontinua);

3) to study symmetries of dynamic substances (tensors of interaction) connecting these objects into a continuum model;

4) to simplify the Lagrangian using the symmetry laws and solving the equations of motion.

The dynamic theory of ionic crystals was first constructed by M. Born, who gave an explanation to the elasticity, piezoelectricity, dielectric properties, specific heat and optical properties using a uniform approach [3]. He took into account inner degrees of freedom (relative displacements of sublattices) in the dynamic equations of motion.

Then the development of the theory of crystal lattice bifurkated. At first there were investigations of real crystals and non-linear effects, and the trials to coordinate the results of the macroscopic and microscopic theories [4] (note the model with noncentral interaction [5]). Then, the linear Born theory was utilized for constructing dynamical models of non-rigid molecular crystals to embrace additional degrees of freedom, particularly the inner rotations.

The method of incorporating inner degrees of freedom by constructing large unit cells (LUC), which was earlier used in the theory of electronic structure of crystals [6], is widely used, in our case, tc investigate the oscillation spectrum of crystals [7,8].

2. The continuum mechanics. First we examined the model of the continuum with inner displacements of subcontinua, where according to Born's method the crystal lattice was considered as a composition of sublattices each one composed of a nonisotropic continuous medium. Thus, microstructure continuum is a system of penetrating subcontinua. This aggregate is examined with inner displacements $\vec{u}^{\alpha}$ of subcontinuum α as a rigid body (sublattice analogy).

Consider short wave oscillations with nonzero wave vector. In this case each translationally equivalent aggregate of atoms splits into its aggregates oscillating with the same phase. The latter form sublattices belonging to the same crystal syngory. The structure as a whole is described by LUC and small Brillouin zone (SBZ). In the three dimensional model the crystal symmetry is ordered according to oscillations (with $\vec{k} \neq 0$) of the star of vectors $\vec{k}$. In [9] the same method was applied to the oscillation problems with $\vec{k} = 0$. The same LUC-SBZ method (with $\vec{k} \neq 0$) can be applied to continuum with inner displacements and rotations describing the oscillations in molecular crystals. A number of translationary equivalent molecules forms a molecular sublattice of crystal. The structure of molecular crystal, as a whole, describes an aggregate of these sublattices.

The field of displacements of a complex continuum can be divided in a sum of the fields of displacements of the continuum (as a rigid body), subcontinua and local media. The latter can be expanded in the series in a neighbourhood of z' of a local medium:

$$v^{\alpha''}_{iq}(z,z') = u_i(z) + u^{\alpha}_i(z) + u^{\alpha'}_{iq}(z,z'),$$

where $u^{\alpha'}_{iq}(z,z') = \varphi^{\alpha'}_k(z)\delta_{ik_j}\delta z'_j + u^{\alpha'}_{(i/j)}(z)\delta z'_j,$

$$\overset{\circ}{\varphi}{}^{\alpha'}(z) = \frac{1}{2}\operatorname{rot}_{(z')}\vec{u}^{\alpha'}(z,z'), u^{\alpha'}_{(i/j)}(z) = \varepsilon^{\alpha'}_{ij}(z).$$

The first term results from the translational movement of the points of the local medium, and the second one is the distorsion tensor in the z'-space. Its antisymmetric part is dual to the axial vector $\overset{\circ}{\varphi}{}^{\alpha'}(z)$, and the symmetric one is a homogeneous (in z'-space) deformation of the local medium $\varepsilon^{\alpha'}_{ij}(z)$.

3. <u>The symmetry of compound continuum</u>. To use these models in complicated cases it is necessary to know the symmetry of compounds and the tensors of their interaction. The base of examination of the oscillation structure of crystals is the generalized Shubnikov-Cury principle.

In the group theory approach the connection of the microscopic lattice theory and the longwave theory of continuum with microstructure is devined by a homomorphism $\Phi \to G \leftrightarrow \Phi/T$. In this sense LUC model, constructed with the help of the group $\Phi_{k'} = T_{k'} \cdot G_{k'} \subset TG = \Phi = T_k G_k$, is placed between microscopic and macroscopic approximations: $\Phi \to F \leftrightarrow \Phi/T_k$ so far as with oversymmetry group $F = G_k \cdot G^c \supset G, T_k \subset T$.

The group theoretical investigation of the model allows us to find out the symmetry groups of its composites and outer and inner symmetries of tensors of interaction of the generalized molecular model of the composed continuum. With the help of groups $H^{\alpha\beta} \subset F$ this permits to elucidate not only the classical data of the microscopic components of these tensors (such as (non)vanishing of the relations among the components of each tensor) but also the similarity of the tensors.

In the limit the constructed models enable us to calculate and classify the frequencies of infrared (IRS) and Raman (RS) spectra.

It is shown that the equations of motion are divisible or not depending on the symmetry of the model e.g. in models with central symmetry in nondeformable subcontinua the system of equations of motions divides automatically.

In models without central symmetry continuum division into elastic and optic vibrations depends on the symmetry of the medium and its extension. For instance, the 3rd rank pseudotensor ℓ^{α}_{ijk}, symmetric up to two indices divides into two pseudovectors, one deviator and one pseudosector. Everywhere besides higher cubic singony there is either deviator or pseudovector. Subsequently, $\ell^{\alpha}_{ijk} = 0$, if the point symmetry group of the medium is 432, $\bar{4}3m$, $m\bar{3}m$.

If the point group of interaction $H^{\alpha} \cap H^{\beta}$ of the symmetry groups of two subcontinua belongs to the class 432 of the highest cubic singony, then the translation and orientation optical oscillations are completely divided.

4. The application of the model. The detailed description of eigenvalues of the general Christoffel tensor was carried out for the diamond structure as an example [10]. The choice of the diamond structure is explained for more complete comparison of the results obtained in the utilized model and other experiences and theories [11].

The wave equations are compiled, quasiacoustic and quasioptic frequences are calculated, and eigenvectors of the waves propagating along [001], [110], [111] are found out. For the wave vector 001 the dispersion curves of quasioptic and quasiacoustic waves are found with the wave vector $\vec{k} = \frac{1}{2}\vec{b}_3$.

In the models of piezoelectric crystals in the deformable local media approximation it is necessary to take into account the quadruple moment of subcontinuum. In the nonpiezoelectric crystals besides the tensor characterizing elasticity of continuum there are terms of interaction among the translated subcontinua in which the rotations of ionic components take place. In media with central symmetry the part of the tensor of piezoelectric coefficients depending only on the rotation of the local medium does not vanish.

The classification of the shortwave oscillations is used in the projective representations of Fedorov's groups [11, 12].

To the compound continuum with a crystal microstructure we assign the color space group isomorphic to the Fedorov group of the crystal. Isomorphic groups have the same set of irreducible representations that is also used to solve the classification problems.

The investigation of the interaction tensors is given in [13]. The agreement of the theoretical study of the diamond and cubic ZnS spectrums with the experiments [14] is rather good.

The detailed investigation of the symmetry of dynamic structure is carried out in the well-known spectrum of antracen at $\vec{k} = \frac{1}{2}(\vec{b}_1 + \vec{b}_2)$ of Brillouin zone, the basic model for calculation and classification of the frequencies of naphthalene [8]. In the approximation of non-interacting longwave oscillations and those on the boundary of Brillouin zone and taking into account the symmetry of the interaction tensors the high-frequency part of the spectrum of naphthalen was found.

The anisotropic continuum model with inner (translational and rotational) degrees of freedom was used for classifying the neutronographic oscillation spectrum of H_2O molecules in the $LiClO_4 \cdot 3H_2O$ structure [15]. In the normal coordinates found with the projection operators method the equations of motion were solved and the detailed group theoretic analysis of outer, translational and rotational oscillations of H_2O molecules was performed.

5. Conclusions. 1) A model of composed continuum with microtranslations and microrotations can be used to describe and calculate acoustic oscillations and high-frequency part of spectrum in the atomic and molecular structures of crystals. The tension and deformation parameters and tensors of interaction of structural parts of the model were investigated.

2) These models allow us to investigate the high-frequency part of the optical spectrum in the long-wave Born's approximation. It is shown that with the help of LUC-SBZ method one can calculate eigenvectors characterized by the star of wave vectors $\{\vec{k}\}$.

3) In a composed continuum model its generalized symmetry was obviously taken into account. It is shown how to find the symmetry group Γ of the dynamic model, which is the space group modulo $T^* \triangleleft T$.

4) The obtained subgroups of the symmetry group of the continuum characterize the symmetry of interaction of the continuum's components. The connection of the tensors of interaction and the material tensors of

crystallophysics is shown.

5) Taking into account the inner symmetry of dynamic continuum one can simplify the characteristic equation matrix. The continuum model was used to describe the piezoelectric effect in ionic crystals with radicals.

6) The constructed models were used to calculate RS of diamond and URS of cubic ZnS for the directions of wave propagation [001], [110], [111]. From the classification of RS of antracen the high-frequency spectrum of naphthalene was obtained. There were found the way of calculating H_2O oscillation frequencies in the structure of $LiClO_4 \cdot 3H_2O$ and the numerical data of interaction tensors of the following bonds: H_2O-ClO_4; H_2O-Li; H_2O-H_2O; Li-ClO_4; Li-Li. The spectrum of H_2O oscillations was identified.

References

1. Evarestov, R.A. (1982). Methods of quantum mechanics in solid state theory. Leningrad Univ. Leningrad (Russian).
2. Kunin, I.A. (1975). The theory of elastic continuum with microstructure. Nauka, Moscow. (Russian).
3. Born, M., Huang, K. (1954). Dynamic theory of crystal lattices. Oxford Univ., Oxford.
4. Leibfried, G. (1955). Handbuch der physik. Bd. VII/1, Springer, Berlin.
5. Kuvshinsky, E.V., Aero, E.L. (1963). Solid state physics. 5, 9, 2591-2598.
6. Smirnov, V.P., Evarestov, R.A. (1980). Vestnik Leningrad Univ., 4, 28-33. (Russian).
7. Koptsik, V.A., Evarestov, R.A. (1980). Crystallography, 25, 1, 5-13. (Russian).
8. Koptsik, V.A., Ryabtchikov, S.A., Sirotin, Y.I. (1978). Crystallography. 22, 2, 229-241. (Russian).
9. Newman, D.J. - J. Phys., C16, p.1179, 1983.
10. Ryabtchikov, S.A., Koptsik, V.A. (1984). Dep. VINITI, No.4939-84Dep. (Russian).
11. Birman, J.L. (1974). Theory of crystal space groups and infrared and Raman lattice processes of insulating

crystals. - Springer, Berlin et al.

12. Gorelik, V.S. (1982). Transactions of Lebedev Phys. Inst., 132, p.15-140. (Russian).
13. Ryabtchikov, S.A., Koptsik, V.A. (1984). Dep. VINITI, No.4938-84Dep. (Russian).
14. Tolpigo, K.B. (1960). Solid state physics, 2, 10, 2655-2665.
15. Govorova, E.Z., Ryabtchikov, S.A., Sirotin, Y.I. (1974). J. of Struct. Chem. 15, 2, 193-199.

SUPERSPACE GROUP APPROACH TO COMPLICATED CRYSTAL LATTICE DYNAMICS

I.I.Nebola, N.R.Kharkhalis
Uzhgorod State University

1. Introduction. The results of the theoretical research of the last years show the spread of methods of studying physical properties of solids based on the generalized (coloured) group theory [1,2]. In particular for investigating crystal vibrational structure a number of methods taking into consideration both external (positional) and internal crystal symmetry was suggested. The expanded unit cell method (EUC) is based on the composite continum model, where model's symmetry is realized by symmetry and antisymmetry groups [2-4] . The lattice dynamic study of incommensurate crystal phases becomes possible in a model that suppose nonincreasing symmetry of the basic structure introducing superspace group compensating automorphysms [5-7] .

The study of complicated crystal phonon spectra is connected with the well known difficultes,which may be partially eliminated taking into account the symmetry of internal regularity of crystal structure. As is shown in [8-9] describing complicated object is possible determining a crystal by a function:

$$\rho(\vec{r}) = \rho_o(\vec{r}) + \delta\rho(\vec{r}), \qquad (1)$$

where $\rho_o(\vec{r})$ describes matter distribution in the basic structure with the G_o symmetry, $\delta\rho(\vec{r})$ is deviation from the basic structure.

For example consider $AgGaS_2$, which has a chalcopyrite

lattice and is a superstructure as compared with zinc blend structure. Two subsystems of equivalent cation positions can be united into the single system corresponding to the basic structure $\varrho_o(\vec{r})$: $G_o = I\bar{4}m2(D^9_{2d})$. The change of this symmetry is taken into consideration by the character of the real atom occupation of the cation position subsystems:

$$\delta\varrho(\vec{\tau}) = \sum_{\nu=1}^{2} \hat{p}^{\nu}_{j}(\vec{q})\varrho^{\nu}(\vec{\tau})\exp(i\vec{q}\vec{\tau}_j) \tag{2}$$

where $\hat{p}_j(q)$ are Fourier components of some space-depended probability function of the ν-th subsystem, $\vec{q}$ modulation vector. The symmetry of such a crystal is a superspace group symmetry and the superspace group G is $(G_E, G_I) \ni (g_E, g_I) = (R_E, \vec{V}_E), (R_I, \vec{V}_I)$, where G_E and G_I are groups of automorphysms in external and internal space accordingly. There is quite a definite connection between points in the internal and the external space: $\vec{q} = \Delta^{*}\vec{b}^{*}$, where $\vec{b}^{*}$ is basic reciprocal internal space vector and Δ^{*} induces a homomorphic mapping of the internal space into the external one. $AgGaS_2$ crystal in the superstructure sence has the symmetry determined by the 4-dimensional [10-12] space group:

$$G(4) = W^{I\bar{4}m2}_{\bar{1}1\bar{1}} - I_{c'/2}\bar{4}^{(11)}m^{(11')}2^{(11')} = G^{(w)}(3). \tag{3}$$

2. <u>Equations of motion</u>. Potentional energy Φ of the supercrystal basic region is a function of 3sN displacements $U^{\vec{n}\vec{\tau}}_{\alpha j}$, where $\alpha = x, y, z$; j is the position of an atom in the n-th unit cell of basic structure, $\vec{\tau}$ the vector in the internal space representing the phase of modulation wave. Equations of motion have the form [6]

$$m_j \omega^2 U_{\alpha j}^{\vec{n}\vec{\tau}} = \sum_{\vec{n}',j',\beta} \Phi_{\alpha\beta}^{(2)}(\vec{n}\vec{n}',jj',\vec{\tau}) U_{\beta j'}^{\vec{n}'\vec{\tau}} . \tag{4}$$

Superspace periodicity allows one to express 3sN different atom displacement by means of the wave vector $\vec{k}$ in BZ of the basic structure and the parameter $\vec{b}^*$. In a special case crystal described by occupation wave can be interpreted as a composite geometric object with potential energy independent of $\vec{\tau}$. Then solutions of (4) can be expressed in the form

$$U_{\alpha j}^{\vec{n}\vec{\tau}} = m^{-1/2} A_{\alpha j}^{kb} \exp(-i\vec{k}\vec{n} + i\vec{b}^*\vec{\tau}). \tag{5}$$

Introduce the Fourier-transformed dynamical matrix $D_{\alpha\beta}(jj'|\vec{k}\,\vec{b}^*)$ and define it as follows

$$D_{\alpha\beta}(jj'|\vec{k}\vec{b}^*) = (m_j m_{j'})^{-1/2} \sum_{\vec{n}} \Phi_{\alpha\beta}^{(2)}(\vec{n},jj',\vec{\tau}) \exp(-i\vec{k}\vec{n} + i\vec{b}^*\vec{\tau}), \tag{6}$$

where in exp n stands for $(\vec{n}' - \vec{n})$.

After substituting (5), (6) into (4) we have

$$\omega^2 A_{\alpha j}^{\vec{k}\vec{b}^*} = \sum_{\beta,j'} D_{\alpha\beta}(jj'|\vec{k}\vec{b}^*) A_{\beta j'}^{\vec{k}\vec{b}^*} . \tag{7}$$

Enumerate solutions of (7) ascribing to ω index $\nu=1, 2,\ldots,3s$ formed by the triplet of numbers $\sigma a \lambda$, where σ is the number of irreducible representation (IR), a multiplicity, λ runs 1 to f_σ , the dimension of IR.

The frequencies $\{\omega^2(k,b,\nu)\}$ are the eigenvalues of $D(\vec{k},\vec{b}^*)$ and $\{A_{\alpha j}^{\vec{k}\vec{b}^*}\}$ the eigenvectors. To make explicit that the particular eigenfunctions $A_{\alpha j}^{kb^*}$ associate with the wave vector $\vec{k}(\pm \Delta^*\vec{b}^*)$ and have corresponding eigenvalues $\omega^2(\vec{k},\vec{b}^*,\nu)$ we rewrite the former as $e_\alpha(\vec{k},\vec{b}^*,\nu)$ which differ from the complex amplitudes in normalization only.

The wave vector group G_k is generated by elements of G which preserve $\vec{k}$: $R_E\vec{k} = \vec{k}$. It is shown in [13] that representations of G_k are formed by matrices:

$$T_{\alpha\beta}(jj'|\vec{k},\vec{b}^*,R) = R_{\alpha\beta}\delta(jj')\exp[i(\vec{k}-\Delta^*\vec{b}^*),(R_E\vec{\tau}_j-\vec{\tau}_{j'})]. \quad (8)$$

Consider atom vibrations of $AgGaS_2$- structure at k = 0(Γ); in [14] it is denoted as k_9. The group G_k is the full superspace group $W^{\bar{1}4m2}_{\bar{1}1\bar{1}}$. Since the unit cell of the basic structure of $AgGaS_2$ (2 atoms) is 4 times less than that of the real structure (8 atoms) the vector $\vec{b}^*$ admits 4 values $\vec{b}^* = \pm\ell b_1^*$ at Γ. Displacements can be divided into two classes: $\vec{b}^* = 0$ and $\vec{b}^* = \pm\ell b_1^*$. If $\vec{b}^* = 0$, we have

$$\Gamma(\vec{b}^*=0) = 2\Gamma_3 + 2\Gamma_5, \quad (9)$$

that corresponds to lattice vibrations of the basic structure i.e. nonmodulated crystal.

Further we consider the vibrations that apply to the star $\vec{b}^* = \pm\ell b_1^*$. Suppose l = 1. In this case the star consists of two vectors $\vec{b} = b_1^*$ and $\vec{b}^* = -b_1$ describing equivalent states:

$$\Gamma(\pm b_1^*) = 1\Gamma_1 + 2\Gamma_2 + 3\Gamma_4 + 3\Gamma_5. \quad (10)$$

In case $\ell = 2$ the vectors $\vec{b}^* = 2b_1^*$ and $\vec{b}^* = -2b_1^*$ are equivalent (mod $\vec{B}^* = \Delta\vec{K}$). The star consists of a single vector. Therefore

$$\Gamma(2b_1^*) = 2\Gamma_3 + 2\Gamma_5. \quad (11)$$

In the long run the full vibrational representation in the centre of BZ is the sum of (9), (10) and (11):

$$\Gamma = 1\Gamma_1 + 2\Gamma_2 + 4\Gamma_3 + 3\Gamma_4 + 7\Gamma_5. \quad (12)$$

The (12) concides with the classical result [15] .

It follows from this analysis that Γ_3 and Γ_5 vibrations, arised in the zink blend structure after transition of the chalcopyrite lattice, mark the beginning of Γ_1, Γ_2, Γ_4, Γ_5 modes. Normal modes $\vec{k} = 0$ of the real crystal can be obtained by a combination of $\{\vec{b}^* = 0\}$ and $\{\vec{b}^* \neq 0\}$ states belonging to the same IR. For a group-theoretical analysis of modulated structure vibrational spectrum it suffices to IR(G(3)) that allows to avoid a labour-consuming procedure of $G(4) \rightarrow G(3)$ construction and to use that $IR(G(3)) \leftrightarrow IR(G(3)^{(w)})$.

3. <u>Approximate solutions of equations of motion.</u> The form of eigenvectors can be obtained with projection operator technique [16,17] . An analogue of the usual projection operator in (3+d)s-dimensional space is 3s x 3s matrix $P^{(G)}_{\lambda\lambda'}(\vec{k},\vec{b}^*)$ determined as follows:

$$P^{(G)}_{\lambda\lambda'}(\vec{k},\vec{b}^*) = \frac{f_G}{h} \sum_R \tau^{(G)}_{\lambda\lambda'}(k,R)^* T(\vec{k},\vec{b}^*,R) . \tag{13}$$

If we denote by $\vec{\Psi}$ an arbitrary 3s-component vector the elements of which are $\Psi_\alpha(j)$, we can write for symmetry adopted vectors

$$\vec{E}(\vec{k},\vec{b}^*,G\lambda) = P^{(G)}_{\lambda\lambda'}(\vec{k},\vec{b}^*)\vec{\Psi} \tag{14}$$

and

$$D(\vec{k},\vec{b}^*)\vec{E}(\vec{k},\vec{b}^*,G\lambda) = \omega^2(\vec{k},\vec{b}^*,Ga)\vec{E}(k,\vec{b}^*,G\lambda). \tag{15}$$

Since the basic structure has 2 atoms per unit cell, the dynamic matrix is of the form

$$D(jj'|\vec{k},\vec{b}^*) = \begin{pmatrix} D(11|\vec{k},\vec{b}^*) & D(12|\vec{k},\vec{b}^*) \\ D(12|\vec{k},\vec{b}^*) & D(22|\vec{k},\vec{b}^*) \end{pmatrix} . \tag{16}$$

The symmetry of the interactional directions is described by $\bar{4}m2$ group from which it follows that we have to deal with six nonzero matrix elements and dynamic matrix is of the form

$$D(k_g,0)=\begin{pmatrix} A & & & C & & \\ & A & & & C & \\ & & B & & & D \\ C & & & M & & \\ & C & & & M & \\ & & D & & & L \end{pmatrix},\quad D(k_g,b_1^*)=\begin{pmatrix} A & & & & & K \\ & A & & K & & \\ & & B & & & \\ & K & & M & & \\ & & & & M & \\ K & & & & & L \end{pmatrix}. \tag{17}$$

The matrices (17) represent dynamic symmetry of the vibrational states in $\vec{k}-\Delta^*\vec{b}^*$ points according to the values of vector $\vec{b}^*=\pm 1b_1^*$.

Using the method of [14,16], it is easy to obtain the forms of the eigenvectors and eigenfrequency values corresponding to (17). The solutions (5) of equations of motion in the special case of modulated crystal correspond to longperiod crystal which has a potential energy of interaction independent of cation.

The account of regulation modulating perturbation in occupation wave form causes the possibility to expand the dynamic matrix as in (1):

$$D(jj'|\vec{k},\vec{b}^*) = D^{(0)}(jj'|\vec{k},\vec{b}^*) + \delta D(jj'|\vec{k},\vec{b}^*) \tag{18}$$

where $D^{(o)}(jj'|\vec{k},\vec{b}^*)$ reflects the interaction of atoms in the basic structure and $\delta D(jj'|\vec{k},\vec{b}^*)$ the additional perturbation connected with $\delta\varrho(\vec{r})$. The expression (1) using (2) leads to

$$m_j = m_j^{(0)} + \sum_{\nu=1}^{2} m_j^{\nu}\hat{\rho}_j^{\nu}(\vec{q})\exp(i\Delta^*\vec{b}^*\vec{n}), \tag{19}$$

where $m_j^{(o)}$ is an average cation mass in the basic structure. It follows from the analysis of the modulation function, that the value $\vec{b}^* = b_1^*$ corresponds to the vibrational state of neighbouring cells including different cations and $\vec{b}^* = 2b_1^*$ the same cations. Representing $\Phi^F_{\alpha\beta}(jj'|\vec{k},\vec{b}^*)$ as a constant and using (18) and (19) it is easy to obtain the following expressions for $\delta D(\vec{k},\vec{b}^*)$ of chalcopyrite type structures:

$$\delta D(jj'|\vec{k},\vec{b}^*) = \frac{m_j^o m_{j'}^o - m_j m_{j'}}{m_j m_{j'} m_j^o m_{j'}^o} \Phi^F(jj'|\vec{k},\vec{b}^*) . \qquad (20)$$

As an example of obtaining origin parameters for the calculated model let us use GaP. Since atom masses P and S do not differ much GaP vibrational spectrum can be compared with vibrations in the basic structure of $AgGaS_2$. According to [18] GaP is characterized at by the threefold degenerate vibration Γ_{15} with the frequency $\approx 390\ cm^{-1}$. Since the basic structure of $AgGaS_2$ is tetragonal, Γ_{15} splits to $\Gamma_5 \times \Gamma_3$. Using elastic modules of CdS and $AgGaS_2$ [18] and that $m_S/m^o =$ $= A/M = B/L = 0,46$, we have

$$A = 4.12\cdot10^4\ cm^{-2},\ M = 8.95\cdot10^4\ cm^{-2},\ B = 3.65\cdot10^4 cm^{-2}$$

$$L = 7.91\cdot10^4\ cm^{-2},\ C = 6.07\cdot10^4\ cm^{-2},\ D = 5.37\cdot10^4 cm^{-2}$$

$$K = 4.01\cdot10^4\ cm^{-2} \ .$$

We get corrections $\delta D(jj'|\vec{k},\vec{b}^*)$ from (20) using $m_1^o = m_{Ga}$, $m_1^1 = m_{Ag}$ and $m_2 = m_S$:

$$\delta D(11|k_9, b_1^*) = -0.2D^{(o)}(11\ k_9, b_1),$$
$$\delta D(11|k_9, 2b_1^*) = -0.3D^{(o)}(11\ k_9, 2b_1), \qquad (21)$$
$$\delta D(12|k_9, \vec{b}^*) = -0.2D^{(o)}(12\ k_9, \vec{b}^*).$$

For the splitting of equal frequencies of vibrations of Γ_4-type it is necessary to take into account perturbation arising in consequence of interaction between the vibrations corresponding to vectors $+\vec{b}^*$ and $-\vec{b}^*$ respectively. This interaction is described by $\delta D(11|k_9+b_1^*, 2b_1^*)$ which can be compared with $\delta D(11|k_9, 2b_1^*)$. Using (21) it is easy to calculate vibrational frequencies in for structure of $AgGaS_2$. These results and some known experimental and theoretical data are given in Table 1.

Table 1. Calculated and experimental vibrational frequencies of atoms of $AgGaS_2$

Mode	Theor. cm^{-1}	Exper. [19], TO cm^{-1}	Exper. [20], TO cm^{-1}	Theor. 20, TO cm^{-1}
Γ_5	362	365	368	372
	341	323	325	310
	332	225	226	223
	281	158	157	164
	109	95	90	115
	44	34	63	98
Γ_3	340	366	334	313
	321	215	93	100
	42	64	54	93
Γ_4	299	334	367	368
	171	179	212	231
	142	118	190	167
Γ_2	299			326
	281			212
Γ_1	299	295	295	263

4. <u>Conclusion</u>. The lattice vibrations analysis of chalcopyrite type structure in the framework of superspace group-theoretical method allows us to determine the character and to calculate the normal mode of frequencies. Some discrepancy as compared with the experimental data is perhaps due to the fact that the constructed model does not take into account the force constant changes by the basis-real structure transition. This approach, unlike the classical one, allows us to calculate the fundamental properties of the complicated crystal with the help of the basic structure possessing usually not many atoms per unit cell and proceeding from the basic structure properties to predict the properties of derivative compounds.

References

1. Koptsik,V.A. (1975). Kristal und Technik, 10, 231.
2. Koptsik,V.A. (1983). J. Phys. C, 16, 1.
3. Koptsik,V.A., Rjabchikov,S.A., Sirotin,J.I. (1977). Cristallography, 22, 229 (Russian).
4. Koptsik,V.A., Evarestov,R.A. (1980). Cristallography, 25, 5 (Russian).
5. Janner,A., Janssen,T. (1977). Phys. Rev. B, 15, 643.
6. Janssen,T.)1979). J. Phys. C, 12, 5381.
7. Shabanov,V.F., Vtjurin,A.I., Vetrov,S.J. (1979). Prepr. IFSO-103F, Krasnojarsk (Russian),
8. Koptsik,V.A., (1983). In: Group-theoretical methods in Physics 1, Nauka, Moscow (Russian).
9. Koptsik,V.A. (1983). J. Phys. C, 16, 23.
10. Nebola,I.I., Kharkhalis,N.R., Bercha,D.M. (1985). Crystallography, 30, 340 (Russian).
11. Nebola,I.I., Kharkhalis,N.R., Bercha,D.M. (1985). UkrNIINTI 141Uk-85Dep. (Russian).
12. Wolff,P.M., Janssen,T, Janner,A. (1981). Acta Cryst. A37, 625.
13. Kharkhalis,N.R., Nebola,I.I., Bercha,D.M. (1985). UkrNIINTI 142Uk-85Dep. (Russian).
14. Kovalev,O.V. (1961). Irreducible representations of the space groups. The USSR Acad.Sci.,Kiev (Russian).

15. Kaminov,I.P. (1970). Phys. Rev. B,2,960.

16. Maradudin,A.A., Vosko,S.H. (1968). Rev. Mod. Phys., 40, 1.

17. Warren,J.L. (1968). Rev. Mod. Phys., 40, 38.

18. Miller,A., MacKinnon,A., Weaire,D. (1981). Sol. St. Phys.: Adv. Phys. and Appl., 36, 119.

19. Lauwers,H.A., Herman,M.A. (1977). J. Phys. Chem. Sol. 38, 983.

20. Tjuterev,V.G., Skachkov,S.I., Brysneva,L.A. (1982). Solid State Phys., 24, 2236.

ON THE PHENOMENOLOGIC DESCRIPTION OF ELECTRONIC PHASE TRANSITIONS TO INCOMMENSURATE STATE

A.A.Gorbatsevich, Yu.V.Kopaev

P.N.Lebedev Physical Institute, Moscow

The Landau functional method proves to be a usual tool for investigating various types of phase transitions [1]. The method is based upon expansion of the free energy density $\mathcal{F}(\underline{z})$ in powers of the order parameter $\Delta(\underline{z})$:

$$\mathcal{F}_o(\underline{z}) = \alpha \left| \Delta(\underline{z}) \right|^2 + \beta \left| \Delta(\underline{z}) \right|^4 + \gamma \left| \nabla \Delta(\underline{z}) \right|^2. \qquad (1)$$

The order parameter $\Delta = \hat{\Delta}$ is in general a tensor quantity that by an irriducible representation called active and which characterizes the symmetry of the ordered phase. Components of the order parameter enter the r.h.s. of (1) in combinations, which are invariant with respect to the symmetry group of high-temperature nonordered (symmetric) phase. In symmetric phase all the coefficients in (1) are positive. An ordinary second order phase transition occures when α changes its sign due to variation of some internal parameter of the system (e.g. temperature θ). Then $\alpha = a(\theta - \theta_c)$, $a > 0$, where θ_c is the phase transition temperature. The free energy $F\{\Delta\}$ is minimal if

$$F\{\Delta\} = \int \mathcal{F}(\underline{z})\, d\underline{z}, \quad \delta F\{\Delta\}/\delta\Delta = 0 \qquad (2)$$

i.e. Δ (which is a scalar here) is

$$\Delta_o(\underline{z}) = \Delta = const\ , \ |\Delta_o|^2 = -\frac{\alpha}{2\beta}\ . \qquad (3)$$

The transition to inhomogeneous state $(\Delta(r) \neq const)$ is described via the Landau functional approach by the change of γ-coefficient sign. The period of $\Delta(r)$ is in general incommensurate with the period of symmetric phase. Stability of the system requires that the r.h.s. of (1) be supplemented with higher derivative term $\eta|\nabla^2\Delta|^2$ with $\eta > 0$. Lines of transitions to homogeneous and incommensurate phases fork at the point where

$$\alpha(\theta,\tau) = 0 \qquad \text{and} \qquad \gamma(\theta,\tau) = 0, \tag{4}$$

here τ is an internal parameter different from the temperature (e.g. pressure). A point satisfying (4) is called a Lifshits point [2].

The expansion (1) describes large variety of phase transitions and, in particular, a transition to a superconducting state. Expansion (1) in the superconductivity theory is called the Ginzburg-Landau (G-L) expansion. Minimization of (1) leads to the G-L equation for the order parameter:

$$\alpha\Delta + 2\beta\Delta^3 - \gamma\underline{\nabla}^2\Delta = 0. \tag{5}$$

Gorkov has proposed a procedure which makes it possible to express values of coefficients in (1) through the microscopic parameters of the Hamiltonian of the model [3]. The Hamiltonian of the BCS model of superconductivity is of the form:

$$\hat{H} = \sum_{k} \varepsilon_a(\underline{k}) a^+_{\underline{k}} a_{\underline{k}} + \varepsilon_b(\underline{k}) b^+_{\underline{k}} b_{\underline{k}} + g\Delta a^+_{\underline{k}} b^+_{\underline{k}} + h.c. \tag{6}$$

Here g is an effective constant of electron interactions (superconducting transition occurs for $g < 0$), a^+_k the creation operator for electron with spin up and momentum $\underline{k}$, b^+_k the creation operator for the electron with spin down and momentum $-\underline{k}$, Δ the order parameter,

$\varepsilon_a = \varepsilon_b = \varepsilon$ the proper energy of the Hamiltonian (5) with $\Delta = 0$ (the energy spectrum of the system above the transition). Let φ_a and φ_b be eigenfunctions of the Hamiltonian (5) in the absence of SU perconducting pairing. Then Hamiltonian with $\Delta = 0$ can be digonalized by a canonical transformation:

$$\psi(\underline{r}) = u_{\underline{k}\downarrow}(\underline{r})\varphi_{a\underline{k}}(\underline{r}) + v_{\underline{k}\uparrow}(\underline{r})\varphi_{b\underline{k}}(\underline{r}),$$

where the arrow denote spin direction's. The functions $u_{\underline{k}\downarrow}(\underline{r})$ and $v_{\underline{k}\uparrow}(\underline{r})$ play the role of quasiparticle wave functions in ordered state. These functions satisfy the following system:

$$\begin{cases} \varepsilon(\nabla)u_{\underline{k}}(\underline{r}) + \Delta(\underline{r})v_{\underline{k}}(\underline{r}) = E(\underline{k})u_{\underline{k}}(\underline{r}) \\ -\varepsilon(\nabla)v_{\underline{k}}(\underline{r}) + \Delta^*(\underline{r})u_{\underline{k}}(\underline{r}) = E(\underline{k})v_{\underline{k}}(\underline{r}) \end{cases} \qquad (7)$$

where $E(\underline{k})$ is the energy spectrum of the superconducting state.

The order parameter is determined by the self-consistency equation:

$$\Delta(\underline{r}) = g\sum_{\underline{k}} th\frac{E(\underline{k})}{2\theta} u_{\underline{k}}(\underline{r})v^*_{\underline{k}}(\underline{r}). \qquad (8)$$

Perturbation theory in powers of $\Delta(\underline{r})$ and its space derivatives applied to the r.h.s. of (8) immediatly yields (5) with the coefficients determined by the microscopic parameters in the r.h.s. of (8). If the energy spectrum in Gorkov's procedure is taken to be perfectly congruent ($\varepsilon_a = \varepsilon_b$) then $\gamma > 0$.

The coefficient γ can vanish however if the Fermi-surfaces for spins with opposite directions do not coincide (this occurs in an internal magnetic field)

$$\varepsilon_{a,b} = \varepsilon \pm h \qquad (9)$$

and the noncongruence parameter h reaches some critical value h_c. If $h > h_c$ superconductivity exists in inhomogeneous state with $\Delta(\underline{r}) \neq \mathrm{const}$ [4,5] .

The excitonic insulator (EI) [6] is another model which permits direct derivation of the coefficients in the free energy expansion (1). EI Hamiltonian is formally similar to BCS Hamiltonian, $b^+_{\underline{k}}$ being now an operator creating a hole with spin $\underline{\sigma}'$ in band 1 and $a^+_{\underline{k}}$ be an operator creating in electron with spin $\underline{\sigma}$ in band 2. Electron and hole spins $\underline{\sigma}$ and $\underline{\sigma}'$ can posses arbitrary values unlike superconductivity, where spins of electrons in Cooper pair are opposite because they are fermious. An order parameter has, in general, a complex structure [7] and can be written as a spin matrix with real and imaginary components

$$\Delta = \Delta^s_{Re} + i\,\Delta^s_{Im} + \vec{\underline{\sigma}}\left(\underline{\Delta}^t_{Re} + i\,\Delta^t_{Im}\right). \tag{10}$$

Real components of the order parameter describe the well-known types of symmetry breaking - charge density wave state (Δ^s_{Re}) and spin density wave state ($\underline{\Delta}^t_{Re}$). Much attention has been paid in recent years to be ordering connected with the real imaginary order parameter Δ^s_{Im}. If the momentum interband matrix element $\underline{P}_{12} = \langle 1 \,|\, \underline{\nabla} \,|\, 2 \rangle$ is nonzero, the state with $\Delta^s_{Im}(\underline{r}) \neq \mathrm{const}$ is characterized by the macroscopic current density $\underline{j}$ [8] :

$$\underline{j}(\underline{r}) = e\,\mathrm{rot}\,\mathrm{rot}\,\gamma_T\,\underline{P}_{12}\,\Delta^s_{Im}(\underline{r}), \tag{11}$$

here γ_T is a coefficient determined by microscopic parameters of the model, e electron's charge. Equation (1) makes it possible to introduce the toroidal dipole moment density [9] $\underline{T} = \frac{e}{c}\gamma_T \underline{P}_{12}\ \Delta^s_{Im}$. Toroidal moments [10] form the third independent family of electromagnetic multipoles and so the state with $\underline{P}_{12}\,\Delta^{s}_{Im}$ is called toroidal current state (TCS). (Here one has to distinguish the toroidal moments formed of orbital currents from those of spins; the latter are rather trivial objects [11]). Noncongruence of energy spectrum analogious to (9) in high temperature (symmetric) phase results in inhomogeneity of the ground state of EI. Physical reasons for this noncongruence can be caused by doping [12,13] or spectrum anisotropy [14] . An interesting property of inhomogeneous phase of TCS is its diamagnetic responce [15] . The susceptibility of TCS can reach values close to the ideal one - the value of the susceptibility of a superconductor. Description of TCS diamagnetism is fairly complicated and cannot be performed by minimization of the free energy functional obtained by Gorkov's procedure [15].

One of the authors has shown recently [16] that it is the rigidity of wave functions appearing as a result of bound state formation in inhomogeneous TCS that is responsible for TCS diamagnetism. Gorkov's procedure fails in the presence of bound states of perturbation expansion in powers of order parameter $\Delta(r)$, which plays the role of a potential in equations for wave functions (7). The description of all the properties of an inhomogeneous system in the presence of bound states in elementary excitations spectrum must be based upon the exact solutions [17] . The exact solutions of (7) satisfying (8) can be written for one-dimensional inhomogeneity with the help of elliptic functions. The order parameter has the form of kink-lattice:

$$\Delta(x) = \Delta_1\, sn\left(x\frac{\Delta_1}{\varkappa v_F}, \varkappa\right). \tag{12}$$

The energy spectrum E contains two gaps: $(-E_+,-E_-)$ and (E_-,E_+). The parameters Δ_1 and $\varkappa$ in (12) are

$\Delta_1 = E_+ - E_-$, $\varkappa = \frac{E_+ - E_-}{E_+ + E_-}$; V_F is Fermi velocity. We have

$$\Delta(x) \to \Delta_0 \, th \, \frac{\Delta_0}{V_F} x \quad \text{as} \quad \varkappa \to 1 \quad \text{(simple kink)} \tag{13}$$

and the central band $(-E_-, E_-)$ turns into the bound state level $E_0 = 0$ in the middle of a gap in the elementary excitation spectrum.

The perturbation theory for magnetic field which we need to calculate the marnetic responce should be built upon the basis of exact solutions of (7) with the potential (12), but not upon the basis of wave functions of high temperature phase as it is done in Gorkov's procedure. Strictly speaking, Gorkov's procedure cannot be applied whenever there are bound states in the inhomogeneous ordered phase that always occur for two- or one-dimensional inhomogeneity. To avoid misunderstandings, note that Gorkov's equation for normal and abnormal Green functions are always valid when the mean field approach works. The appearence of bound states breaks only the solution of these equations by perturbative expansion in powers of order parameter. The descrepancy obtained by applying perturbative procedure depends on the nature of physical problem. If we investigated a volume property not strictly connected with properties of bound states (e.g. free energy, kinetic coefficients etc.), the descrepance will be determined by specific volume occupied by bound states. But perturbative approach gives a quali tatively wrong answer for TCS diamagnetism which is due to the existence of bound states.

Even to describe inhomogeneous ground state in the absence of external field is a problem because of the non-perturbative nature of solution (12). But one might expect that the limits of applicability of the phenomenological expansion (1) are wider then the limits of perturbative Gorkov's approach and so they are. There exists a direct connection between phenomenological expansion (1) and exact solutions of the model with an inhomogeneous potential even in the presence of bound states. In [15] they noted that the system of Dirac-type equations (7) which give the exact description of the ground state of the microscopic model is a standard system adopted in the inverse scattering method of constructing nonreflective potentials [19,20]. The sum of the selfenergies E(k) which determines the exact value of the free energy is an integral of motion of some type of nonlinear differential equations [20]. In our case it is a modified Karteveg de Vries equation (MKdV):

$$V_F^2\nabla^3\Delta(x) - 6\Delta^2(x)\nabla\Delta(x) + \Delta_1(1+x^{-2})\nabla\Delta(x) = 0 \quad (14)$$

and its higher analogue [19]. Equation (14) is written so as to have a solution in the form (12). The remarkable fact is that the functional (2) is also an integral of motion of equation (14) or more precisely, is the sum of its two integrals of motion:

$$F_0 = \alpha I_{-1} + \beta I_0, I_{-1} = \int \Delta^2(x)\,dx, \; I_0 = \int (\Delta^4(x) + (\nabla\Delta(x))^2)\,dx. \quad (15)$$

The desired coincidence of coefficients in (15) and (2) can be obtained by the scaling of space variable in I_0 (15). So far as I_{-1} and I_0 are integrals of motion of (14), all the solutions of (14) satisfy the minimum condition for the functional (15). This fact becomes especially obvious if one notes that (14) can be obtained differentiating with respect to space variable the Euler equation (5) for the functional (2) and (15). One can unambigiously express phenomenological coefficients in (2) and in (15) in terms of microscopic parameters of the model taking into account the coincidence of the form of the Euler equation (5) for the functional (2) and of (14) which satisfies minimum condition for the exact functional. The first integral of the equation (5) has the form:

$$\alpha\Delta^2+\beta\Delta^4-\gamma(\nabla\Delta)^2=-\lambda \tag{16}$$

where λ is the integration constant a phenomenological parameter. The parameter λ is to be determined in terms of microparameters as well as other phenomenological coefficients: α, β, γ . Taking the sign of the r.h.s. of (16) to be negative ($\lambda>0$), we obtain a solution of (16):

$$\Delta(x)=\Delta\, sn\left(x\,\frac{\Delta}{qV_{eff}}\,,\,q\right) \tag{17}$$

where the parameters Δ , q and V_{eff} connected with phenomenological coefficients in (16) are the following ones

$$\Delta^2=\Delta_o^2-\sqrt{\Delta_o^4-\frac{\lambda}{\beta}}\,,\;q^2=\beta\,\frac{\Delta_o^4}{\lambda}\,,\;V_{eff}=\left(\frac{2\gamma}{\beta}\right)^{1/2}, \tag{18}$$

where Δ_o is determined in (3). Comparing the solutions (12) and (17) makes it possible to express unambigiously Δ , q and V_{eff} in terms of microscopic parameters Δ_1, $\varkappa$, V_F . Three equations in (18) determine four parameters α, β, γ and λ up to an arbitrary factor. The fourth equation can be obtained by substituting (17) in the functional (2), (15) and equating it to the exact value of the free energy known from the exact solution [17,16] . Thus it is possible to formulate the problem of determining ground state of inhomogeneous system in the presence of bound states as the minimum problem for the free energy functional (2). A transition to an inhomogeneous state in a commonly used scheme, where Gorkov's procedure is used to evaluate the coefficients in the functional and bound states formally results from the change of the sign of γ . The parameters of the inhomogeneous phase are obtained from the minimum condition for the functional (2) supplemented with a higher derivative term $(\nabla^2\Delta)^2$. The free energy density in the above approach is given by the unique expression for both homogeneous and inhomogeneous phases. The coefficients in (2) are evaluated by Gorkov's procedure in homogeneous phase and on the basis of equations (18) supplemented with the condition for the exact value of the free energy to equal its phenomenological expression in inhomogeneous phase. In homogeneous phase we have $\lambda = \lambda_o = \beta\Delta^4_o = \alpha^2/4\beta$ everywhere. The values of λ smaller then λ_o correspond to inhomogeneous phase. It is interesting to note that (5) can be turned into the form which permits one to describe excited states of microscopic model. It can be performed by equating the r.h.s. of (5) to a nonzero constant. In this case $\Delta(x)$ is obtained in the form of a double domain wall (polaron type solution) [21], which is always an excited state solution [22].

The energy of homogeneous phase equals to that of the phase with kink-like solution in an infinite system. So they are equivalent from the point of view of free energy functional and differ only in boundary values but not in the values of the functional coefficients. Therefore in the case of an isolated kink we have $\lambda = \lambda_o = \beta\Delta_o^4$ and the parameters are determined by Gorkov's procedure. Note that the criterion of a homogeneous phase instability is $\gamma = 0$.

Functional containing higher then (15) powers of the order parameter can also be reduced under proper conditions to integrals of motion of macroscopic nonlinear equations.

A polinomial integral of motion of the order next to (15) is of the form:

$$I_1 = \int(\Delta^6 + \Delta^2(\nabla\Delta)^2 + (\nabla^2\Delta)^2)\,dx \tag{18}$$

and result in the first order MKdV whose solutions are hyperelliptic functions. The functional F_o can be decomposed into the sum of integrals of motion I_{-1} and I_o with arbitrary relations among α, β and γ, which can be changed by normalizing Δ and space variable rescaling. Such representation is applicable only under proper restrictions on coefficients of the functional. Thus Landau functional written up to the 6 order describes an exact solution of the nonlinear macroscopic problem only if the coefficients of Δ^6, $\Delta^2(\nabla\Delta)^2$ and $(\nabla^2\Delta)^2$ are equal . In this case:

$$F_6 = F_o + \gamma I_1 . \qquad (19)$$

The necessity to supplement the functional with terms with higher spacial derivatives can be caused by an increase of the number of microparameters (but not by variation of γ). For example the first order MKdV equation satisfying the minimum condition for the functional (19) describes the ground state of the system in magnetic field with separated Fermi-surfaces of electrons with opposite spin directions. Separation of Fermi-surfaces results in the formation of inhomogeneous structure with two incommensurate periods [22] .

References

1. Landau,L.D., Lifshitz,E.M. (1976). Statistical physics, Nauka, Moscow (Russian; English translation by Pergamon).

2. Hornreich,R.M., Luban,M.,Shtrikman,S. (1975). Phys. Rev. Lett. 35, 1678.

3. Abricosov,A.A., Gorkov,L.P., Dzyaloshinsky.I.E.(1962). Methods of quantum field theory in statistical physics, Fizmatgiz, Moscow (Russian).

4. Fulde,P., Ferrel,R. (1964). Phys. Rev. 135, 550.

5. Larkin,A.I., Ovchinnikov,Yu.N. (1965). Sov. Phys. JETP 20, 762.

6. Keldysh,L.V., Kopaev,Yu.V. (1965). Sov. Phys. Solid State 6, 2219.

7. Halperin,B.I., Rice,T.M. (1968). Solid State Phys.

21, 115.

8. Volkov,B.A., Gorbatsevich,A.A., Kopaev,Yu.V., Tugushev,V.V. (1981). Sov. Phys. JETP 54, 391.

9. Ginzburg,V.L., Gorbatsevich,A.A., Kopaev,Yu.V., Volkov,B.A. (1984). Solid State Communs 50, 339.

10. Dubovik,V.M., Tosunyan,L.A. (1983). Sov. J. Part. Nuclei 14, 504.

11. Artamonov,Yu.A., Gorbatsevich,A.A. (1985). ZhETPh 89, 1092.

12. Kopaev,Yu.V. (1975). Proc. Lebedev Phys. 86, 3 (Russian).

13. Rice,T.M. (1970). Phys. Rev. 2B, 3619.

14. Gorkov,L.P., Mnatzakanov,T.T. (1982). ZhETPh 63,684 (Russian).

15. Volkov,B.A., Gorbatsevich,A.A., Kopaev,Yu.V. (1984). Sov. Phys. JETP 59, 1087

16. Gorbatsevich,A.A. (1985). ZhETPh Pisma 42, 399 (Russian).

17. Bragovsky,S.A., Gordyunin,S.A., Kirova,N.N. (1980). JETP Lett. 31, 456.

18. Fateev,V.A., Schwarz,A.S., Tyupkin,Yu.S. (1976).Preprint Lebedev Phys. Inst. N 155 (Russian).

19. Zakharov,V.E., Manakov,S.V., Novikov,S.P., Pitaevskii,L.P. (1980). Soliton Theory, Nauka, Moscow (Russian).

20. Ablowitz,M.J., Kaup,D.J., Newell,A.C., Segur,H. (1974). Stud. Appl. Math. 53, 249.

21. Brazovsky,S.A., Kirova,N.N. (1981). ZhETPh Pisma 33, 6 (Russian).

22. Brazovsky,S.A., Dzyaloshinsky,I.E., Kirova,N.N. (1981). ZhETPh 81, 2279 (Russian).

APPLICATIONS OF GROUP THEORY TO RELATIVISTIC COMPUTING IMPURITY IONS IN CRYSTALS

J.Brants

Latvian State University, Riga

In physics the most different problems dealing with symmetries of physical systems are solved by the group theory methods. To ensure successful application of the group theory it is often necessary to decompose a reducible representation of a symmetry group of the system into irreducible representation.

We encounter this problem studying zone structure of energetic electron-and-hole spectrum; interpreting optical absorption and luminescence spectra and paramagnetic resonance spectra; tracing symmetric coordinates of molecules and complexes and molecular orbitals.

In most cases we need wave functions realizing an irreducible representation of a crystal group. At the same time the role of reducible representations is played by irreducible representations of the rotation group. The necessity to have such functions appears also computing impurity ions in crystals.

If a relativistic discussion of a physical problem takes place, then we emplcy spin functions tbansforming via the representation $D^{(1/2)}$ of the orthogonal group O(3) or more general functions transforming via $D^{(j)}$ with half-integer j. In this case the representations are double-valued: basis vector changes the sign under the rotation by 2π ground any axis [1].

To obtain wave functions with a definite transformation symmetry the method of projections can be used [2]. According to this method the basis function of an irreducible representation Γ of a crystal group G transforming the μ-th row of the irreducible representation is built as a linear combination of basis functions $|jm\rangle$ of the representation $D^{(j)}$ of an orthogonal group in the form

$$|j\Gamma\mu a\rangle = \sum_{m=-j}^{j} \langle jm|j\Gamma\mu a\rangle |jm\rangle \tag{1}$$

where $\langle jm|j\Gamma\mu a\rangle$ are the coefficients of the linear combination at the basis function $|jm\rangle$, the index a numbers multiple representations in the decomposition.

The orthogonality relations follow from the unitarity of the transformation (1)

$$\sum_{m} \langle j\Gamma\mu a|jm\rangle\langle jm|j\Gamma'\mu' a'\rangle = \delta(\Gamma,\Gamma')\delta(\mu,\mu')\delta(a,a') \tag{2}$$

$$\sum_{\Gamma\mu a} \langle jm|j\Gamma\mu a\rangle\langle j'\Gamma\mu a|j'm'\rangle = \delta(j,j')\delta(m,m'), \tag{3}$$

where $\langle j\Gamma\mu a|jm\rangle = \langle jm|j\Gamma\mu a\rangle^*$

The method of projections uses the operator

$$\hat{P}^{(\Gamma)}_{\mu\mu'} = \frac{|\Gamma|}{|G|} \sum_{g\in G} \langle \Gamma\mu|\hat{R}|\Gamma\mu'\rangle^* \hat{R}, \tag{4}$$

which preserves the basis functions of Γ and maps all the other basis functions to 0. In equation (4) $\langle \Gamma\mu|\hat{R}|\Gamma\mu'\rangle$ is matrix element of Γ, $\hat{R}$ the operator of a 3-dimensional space turn, $|\Gamma|$ and $|G|$ the dimension of the representations Γ and G respectively.

Applying (4) to $|jm\rangle$ we can obtain an equation

for determining the basis functions of a given irreducible representation:

$$|j\Gamma\mu a\rangle = \frac{|\Gamma|}{|G|}\sum_{m'}\sum_{G}\langle\Gamma\mu|\hat{R}|\Gamma\mu'\rangle^{*} D^{(j)}_{mm'}(\hat{R})\,|jm\rangle, \tag{5}$$

where $D^{(j)}_{mm'}(\hat{R})$ are matrix elements of the transformation of spherical function determined e.g. in [3].

In the theory of coupled systems and at the output of selection rules an important role is played by the direct product of two representations.

Let us assume that we have n_μ basis functions $f_j^{(\mu)}$ of $D^{(\mu)}$ and n_ν basis functions $f_\ell^{(\nu)}$ of $D^{(\nu)}$. We are to find n_λ functions $f_s^{(\lambda)}$ which are linear combinations of the products $f_j^{(\mu)} f_\ell^{(\nu)}$ and form a basis of $D^{(\lambda)}$. The functions $f_s^{(\lambda)}$ represent a partner system belonging to $D^{(\lambda)}$. Such a system exists only if the representation occurs in the product $D^{(\mu)} \times D^{(\nu)}$. Hewever, it is possible to form several independent partner systems $f_s^{(\lambda)}$ for multiple representations in the decomposition of $D^{(\mu)} \times D^{(\nu)}$. To separate these partner systems we write $f_s^{(\lambda\tau_\lambda)}$ where τ_λ numbers multiple representarions. We have

$$f_s^{(\lambda\tau_\lambda)} = \sum_{j,\ell} f_j^{(\mu)} f_\ell^{(\nu)} \langle\mu j,\nu\ell\rangle|\lambda\tau_\lambda s\rangle \tag{6}$$

where $\langle\mu j,\nu\ell|\lambda\tau_\lambda\ s\rangle$ Clebsch-Gordan coefficients

Eq. (6) establishes dependence between bases $f_s^{(\lambda\tau_\lambda)}$ and $f_j^{(\mu)} f_\ell^{(\nu)}$ in the representation space of the direct product.

If we use the $\{j,\Gamma\}$ scheme (1), where the coefficients $\langle jm|j\Gamma\mu a\rangle$ are elements of a unitary matrix, we obtain an equation for the basis functions

$$|j\Gamma\mu a\rangle = \sum_{\substack{\Gamma_1\mu_1 a_1 \\ \Gamma_2\mu_2 a_2}} |j_1\Gamma_1\mu_1 a_1\rangle j_2\Gamma_2\mu_2 a_2\rangle \langle j_1\Gamma_1\mu_1 a_1, j_2\Gamma_2\mu_2 a_2|j\Gamma\mu a\rangle \tag{7}$$

where the coupling coefficients are defined by [4,5]

$$\langle j_1\Gamma_1\mu_1 a_1, j_2\Gamma_2\mu_2 a_2|j\Gamma\mu a\rangle = \sum_{m_1 m_2 m} \langle j_1 m_1, j_2 m_2|j m\rangle \times$$

$$\times \langle j_1\Gamma_1\mu_1 a_1|j_1 m_1\rangle \langle j_2\Gamma_2\mu_2 a_2|j_2 m_2\rangle \langle j\Gamma\mu a|j m\rangle^{*}. \tag{8}$$

By the Racah lemma the coefficients (8) are expressed via the Clebsch-Gordan coefficients of the corresponding group G

$$\langle j_1\Gamma_1\mu_1 a_1, j_2\Gamma_2\mu_2 a_2|j\Gamma\mu a\rangle = \sum_{b} \langle \Gamma_1\mu_1\Gamma_2\mu_2|\Gamma\mu b\rangle \times$$

$$\times (j_1\Gamma_1 a_1, j_2\Gamma_2 a_2|j\Gamma a b) \tag{9}$$

and after simplification we have

$$\langle \Gamma_1\nu_1, \Gamma_2\nu_2|\Gamma\nu b\rangle = \langle j_1\Gamma_1\nu_1 a_1, j_2\Gamma_2\nu_2 a_2|j\Gamma\nu a\rangle /$$

$$/(j_1\Gamma_1 a_1, j_2\Gamma_2 a_2|j\Gamma a b), \tag{10}$$

which may be used for calculating Clebsch-Gordan coefficients from the coefficients of the linear combination of basis functions $|j_1 m_1\rangle$, $|j_2 m_2\rangle$ and $|j m\rangle$.

A complex of programs has been worked out on the basis of the above arguments containing operations with irreducible representations of the point groups, calculates the coefficients $\langle jm | j\Gamma \mu a \rangle$ and Clebsch-Gordan coefficients for the point group.

Using the above formulas it is possible to get equations for matrix elements of the interaction with the crystalline field energy operator and compute the energetic spectrum and different properties of the impurity ion in a crystal.

References

1. Bethe, H. (1929). Ann. der Phys., 3, 133-208.

2. Petraschen, M.I., Trifonov, E.D. (1967). Application of the Group Theory in Quantum Mechanics. 307 (Russian).

3. Jucys, A.P., Bantaitis, A.A. (1965). Theory of Angular Momentum in Quantum Mechanics. Mintis, Vilnius, 463 (Russian).

4. Batarunas, I.V., Levinson, S.B. (1960). Trans. Lithuan. SSR, Acad.Sci., 2, 22, 15-32 (Russian).

5. König, E., Kremer, S. (1973). Teoret.Chim.Acta, 32, 27-40.

6. Racah, G. (1949). Phys. Rev., 76, 1352-1365.

NEW TOPOLOGICAL INVARIANTS OF LINKED LINEAR DEFECTS IN CONDENSED MATTER

M.I.Monastyrsky, V.S.Retakh
Institute of Theoretical and Experimental Physics,Moscow

This paper contains a general study of several linked singularities. Relevant theory from the topological point of view is the classical theory of links, but having in mind physical applications we should take into account new structures related to thermodynamical problems requiring introducing order parameter.

Our main results are the following:

1. We introduce a hierarchy of conservation laws in terms of differential forms corresponding to a sequence of linking invariants so that we can distinguish nontrivial links possessing zero Gauss linking coefficient.

2. We obtain a set of topological obstruction rules for links in nematics, cholesterics and superfluids ^3He and ^4He. In particular we would like to point out the case of ^3He [9] . Recall that the order parameter of superfluid ^3He is a complex 3 x 3 matrix subject to constraints required by the choice of superfluid phase and the minimization of the magnetic energy which can be written in the form $F_H = g_H A_{pi} A_{qi} H_p H_q$.

Next, if the matrix of the order parameter has three different eigenvalues, then under the above conditions the spin part of the order parameter is degenerate and consists of three interlinked curves corresponding to the eigenvalues of the matrix A_{pi} . Another example of a link in superfluid ^3He is provided by vortices corresponding to the orbital part of the order parameter. Here the topology of vortices is the same as that of singular lines in nematics.

The mathematical apparatus used in this article is rather nonconvential. However, we expect the approach worked out in this paper will be useful for physical problems in which the sophisticated structures of links, singular lines, surfaces and n-dimensional structures are of primary importance. The proof of the results see [14].

1. Links and defects. A set l of closed non self-intersecting curves $l_1,\dots,l_n$ in S^3 or R^3 is called a link. Let us fix a point $*$ outside l. The i-th meridian of l is a non self-intersecting loop m_i beginning at $*$ and going in the complement of the set formed by curves of the link so that l_i crosses the surface bounded by m_i at exactly one point. Evidently, the loop m_i can be choosen arbitraryly narrow.

Consider a manifold M, the physical space of the system, and a manifold V, the order parameter space. Let Φ be a continuous map of the complement of a subset $\Sigma \subset M$ into V. The set Σ is called the defect of Φ. The defect Σ is essential if there is no continuous extension of Φ in a neighbourhood containing points of Σ.

Suppose that $M = S^3$, and $\Sigma = l = (l_1,\dots,l_n)$ is a link.

The link l is called an m-essential defect if $\Phi(m_i)$ is not contractible in V for all $i \leqslant n$.

Note that an m-essential defect is essential. In many physically important cases an essential defect is m-essential. A deformation of Φ is a map $\Phi' : M \setminus \Sigma \to V$ such that Φ is homotopic to Φ' and coincides with it outside a sufficiently small neighbourhood of one of meridians.

An essential defect Σ of Φ is stable if Σ is an essential defect for any deformation Φ' of Φ.

Theorem 1.1. Let $M = S^3$ and an essential defect related to the link $l = \Sigma$ be stable for $\Phi : S^3 \setminus \Sigma \to V$. Then Σ is m-essential.

Recall that a loop $\bar{l}_i$ is the n-th parallel of a link $l = (l_1, \ldots, l_n)$ if it is obtained as follows. Consider a curve in the complement to l that starts at the point $*$ along a path p_i to a sufficiently narrow tubular neighbourhood of l_i then goes down into this neighbourhood along l_i and returns to $*$ along p_i. Suppose that the parallel $\bar{l}_i$ is contracted to a point of $S^3 \setminus l_i$. (Hereafter "contraction" means contraction into a point.) The defect $\Sigma = l$ is called strongly essential of $\Phi : S^3 \setminus l \to V$ if Σ remains essential for any map $\Phi : S^3 \setminus l \to V$ homotopic to Φ.

<u>Theorem 1.2</u>. If $\Sigma = l$ is a strongly essential defect of Φ and the curve l_i of a link l is not separated from other curves by a homeomorphic image of two-dimensional sphere then $\Phi(\bar{l}_i)$ is not contractible in V.

Let Σ be a defect of $\Phi : S^3 \setminus l \to V$ formed by a link l. Consider 4 model examples.

Example 1.1 (Fig. 1). In this case parallels of the link are contracted in $S^3 \setminus l$ hence, for any essential

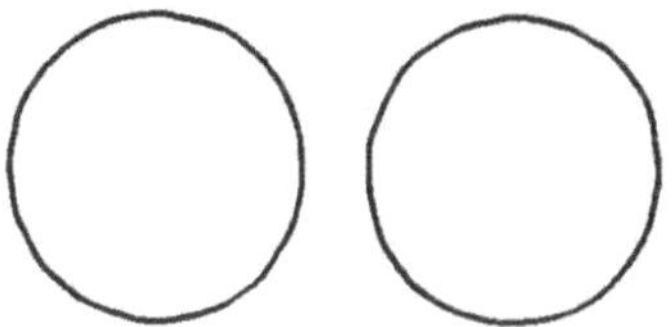

Fig. 1

defect their images are contracted.

<u>Example 1.2</u> (fig. 2). The link is constructed by means

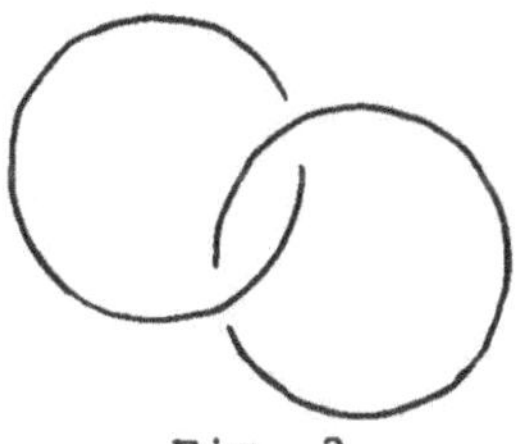

Fig. 2

of the simplest linking of two closed curves with Gauss coefficient 1. Ckearly a parallel to one of the curves is homotopic to a meridian of the other one. Hence in this case the m-essential defect results in the non-contracting of the parallel images in V.

Example 1.3 (Fig. 3). The link 1 is well-known Bor-

Fig. 3

romean rings. It consists of three closed curves that can be split by cutting one of them. Here the parallel of one of the curves e.g. $\bar{l}_3$, is homotopic to the commutator of the meridians of two others: $\bar{l}_3 \sim [m_1, m_2] = m_1 m_2 m_1^{-1} m_2^{-1}$. Clearly when the fundamental group $\pi_1(v)$ of the order parameter space is commutative the parallel's images are contracted in V. If $\pi_1(V)$ is non-commutative, the situation is more complicated.

For example, the fundamental group of biaxial nematic or cholesteric consists of 8 elements $\pm e_1$, $\pm e_2$, $\pm e_3$, J, 1 with relations $e_1^2 = e_2^2 = e_3^2 = J$, $Je_i = e_i J = -e_i$ $(i=1,2,3)$, $e_1 e_2 e_3 = J$. The centre of this group consists of the two elements, J and 1. Therefore, if a defect is Borromean rings and the images of all parallels and meridians are non-contractible in V, then the images of meridians m_1, m_2, m_3 are homotopic up to renumbering to $\pm e_1$, $\pm e_2$, $\pm e_3$ and $\Phi(\bar{l}_i)$ is homotopic to J for all i.

It is important that, although Gauss linking coefficient for any two curves of Borromean rings is zero, the properties of l drastically differ from the three unlinked closed curves.

Example 1.4 (Fig. 4). Consider the well-known Whitehead link l of two curves l_1 and l_2. In this case the Gauss coefficient of the link l is also zero, but the topological properties of this link are considerably different from the trivial link in example 1.8. The parallel l_2 in the Whitehead link is homotopic to the product of cummutators of meridians $[m_1^{-1}, m_2]\cdot[m_1,m_2]$. Consequently, if $\pi_1(V)$ is commutative $\Phi(\bar{l}_2)$ is contractible for any defect. However, $\Phi(\bar{l}_2)$ is also contractible for biaxial nematic and cholesteric.

Thus, the topological properties of the map depend essentially on the structure of the defect. In the following sections the so-called higher linking coefficients will be determined which allow one to tell Borromean, Whitehead and trivial link from one another.

2. General theory of links. Two links $l=(l_1,\dots,l_n)$, $l'=(l'_1,\dots,l'_n)$ are homotopic if there is a 1-parameter family of maps h_t from the space $C(n)$ consisting of n disjoint oriented circles into the three-dimensional sphere S^3 such that for each t disjoint circles have disjoint images and $h_0(C(n)) = l$, $h_1(C(n)) = l'$. These links l and l' are called isotopic if, additionally, $h_t(C(n))$ is a link for any t.

A curve l_i from a link $l = (l_1,\dots,l_n)$ is homotopic unlinkable to other components of l if l is homotopic to a link l' with l'_i contractible to a point. A link l is homotopic trivial if it is homotopic to a link consisting of points.

A curve l_i is isotopic unlinkable to other components of l if it can be separated by a homeomorphic image of the S^2 from other components. A link l is isotopic trivial if its every component is isotopic unlinkable to the other ones.

From the physical point of view "unlinkability" of

links means its isotopical "unlinkability". Homotopic unlinking allows self-intersection of curve l_i which "costs an energy". Homotopic "unlinkability" is rougher than the isotopic one.

Fig. 4

Example 2.1. The Borromean rings are not homotopic trivial though each pair of its components is isotopic trivial.

Example 2.2. The Whitehead link is not isotopic trivial but it is homotopic trivial.

The simplest linking invariant for two oriented closed curves l_1, l_2 in S^3 is the Gauss linking coefficient $k(l_1, l_2)$ [10] defined by the Gauss integral. For the link on Fig. 2 $k(l_1, l_2) = 1$ and for the link on Fig. 4 $k(l_1, l_2) = 0$. Note, that $k(l_1, l_2) = 0$ if and only if l_1 and l_2 are homotopic unlinked. However, the Whitehead link shows that for isotopic linked curves this coefficient may be zero. Below we describe some higher order linking coefficients which allow us to solve the problem of unlinking. The corresponding definitions are based on a generalization of the Gauss coefficient for properties of $S^3 \setminus l$.

Let $l = (l_1, \ldots, l_n)$ be a link in S^3. Choose a point $*$ in the complement to l. We assume every curve to be orientable. A parallel inherits the orientation of the curve and the meridian m_i is oriented so as

$k(l_i, m_i) = 1$.

For an arbitrary group consider a lower central series of subgroups $G = G_1 \supset G_2 \supset G_3 \ldots$. For a commutative group G we have $G_2 = \{1\}$. If $G =$ (V) then for a biaxial nematic $G_3 = \{1\}$.

Proposition [11]. For any q there exists a map α of a free group F in noncommuting variables $x_1, \ldots, x_n$ onto $G = \pi_1(S^3 \setminus l)$ such that:

i) $\alpha(x_i)$ is the homotopy class of m_i;

ii) α induces isomorphism $F/F_q \simeq G/G_q$;

iii) if $\alpha(y_i)$ is the homotopy class of a parallel $\bar{l}_i$ then $[x_i, y_i] = 1$.

Let $l = (l_1, \ldots, l_n)$ be a link in S^3 and $q=n$. Let $\mu(i,j)$ be the sum of powers of the generators x_i of the group F in the decomposition of the elements y_j for $i \neq j$. For example, for the Borromean rings we have $\mu(1,3) = \mu(2,3) = 1-1 = 0$. The numbers $\mu(i,j)$ are called 1 order Milnor coefficients of l. They do not depend on the choice of parallels and meridians and are equal to $k(i,j)$. A link of two curves l_1 and l_2 is homotopic unlinking if and only if $\mu(1,2) = 0$. The Whitehead link shows that for isotopic unlinking this condition is not sufficient.

The 1 order Milnor coefficients are, however, very rough characteristics of links. Indeed, for two-component links $\mu(1,2) = 0$ for both trivial and Whitehead link, which is isotopic unlinked. For the Borromean rings $\mu(i,j)=0$ for $1 \leqslant i, j \leqslant 3$ but this link is not trivial even in homotopic sense. In order to distinguish such links Milnor [12] introduced higher order linking coefficients.

In terms of Proposition, construct a homomorphism θ of the free group F with generators $x_1, \ldots, x_n$ to the multiplicative group of formal power series with integer coefficients in noncommuting variables $x_1, \ldots, x_n$ with constant term 1. Put $\theta(x_i) = 1+x_i$. Clearly $\theta(x_i^{-1}) = 1-x_i+x_i^2-\ldots$ for $1 \leqslant i \leqslant n$. Milnor coefficient of the

order $p-1$ for $p \geq 2$. Consider the set of indices $1 \leq i_1,\dots,i_p \leq n$. Let y_{i_p} be an element transformed into the homotopy class of the parallel l_{i_p}. We denote $\mu(i_1,\dots,i_p)$ the coefficient of the monomial $x_{i_1}\dots x_{i_p}$ in the formal series $\theta(y_{i_p})$. This is the $(p-1)$-th Milnor coefficient. Clearly the two ways of determining $\mu(1,2)$ coincide.

For the Borromean rings $\mu(1,2,3)=1$, $\mu(2,1,3)=-1$. For the Whitehead link $\mu(1,2,2,1)=\mu(1,1,2,2)=1$, $\mu(1,2,1,2) = 2$.

Milnor [12] has shown that numbers $\mu(i_1,\dots,i_p)$ are uniquely determined modulo the numbers $\mu(j_1,\dots,j_q)$, where $(j_1,\dots,j_q)\subset(i_1,\dots,i_p)$ is an arbitrary subset for $q<p$. In particular, if $\mu(j_1,\dots,j_q)=0$ for all subsets, then $\mu(i_1,\dots,i_p)$ is defined uniquely. In the general case we consider the image $\mu(i_1,\dots,i_p)$ in the group $\mathbb{Z}/\mu\mathbb{Z}$. Here μ is the maximal common divisor of $\mu(j_1,\dots,j_q)$'s for $q \leq p-1$. This image is denoted by $\bar{\mu}(i_1,\dots,i_p)$ and is called the Milnor coefficient of order $p-1$ for l.

The Milnor coefficients $\mu(i_1,\dots,i_p)$ are homotopy invariants of a link in case there are no equal numbers in the set $(i_1,\dots,i_p)$. If there are ones these coefficients are only isotopy invariants. Moreover, the link $l=(l_1,\dots,l_n)$ is homotopic unlinked if and only if $\mu(i_1,\dots,i_p)=0$ for any $p \leq n$ when there are no equal numbers in the set $(i_1,\dots,i_p)$. Therefore, for the homotopic unlinking it suffices to examine vanishing of only finite set of numerical invariants.

A different situation arises in the isotopic case. Clearly for isotopic trivial link $\mu(i_1,\dots,i_p)=0$ for any set $(i_1,\dots,i_p)$. The converse is not true in general [12]. For the relations between Milnor coefficients see [12]. The simplest one is: $\mu(i_1,\dots,i_p)=(-1)^p \mu(i_p,\dots,i_1)$.

Note that for a link of two curves vanishing of the 1st

order Milnor coefficient implies vanishing of the 2nd order one.

3. <u>The homotopy properties of the maps into the parameter order spaces</u>. In this section we consider a map $\Phi : S^3 \setminus \Sigma \to V$ with defect $\Sigma = l = (l_1, \dots, l_n)$. Using Milnor coefficients we study the homotopy classes of the images $\Phi(\bar{l}_i)$, where $\bar{l}_i$ is a parallel of l, and the properties of $\Phi(m_i)$, where m_i is a meridian of l_i. Let $(*) \in V$ be a fixed point.

<u>Theorem 3.1</u>. Let $\pi_1(V)$ have a finite central series. Then the homotopy classes of $\Phi(m_i)$ and $\Phi(\bar{l}_i)$ commute for every $1 \leqslant i \leqslant n$. If every $\Phi(m_i)$ can be contracted in V then so can every $\Phi(\bar{l}_i)$. If every Milnor coefficient for l vanishes then $\Phi(\bar{l}_i)$ can be contracted in V for $1 \leqslant i \leqslant n$.

Consider the following examples of parameter order spaces.

1. $\pi_1(V)$ is commutative. If $\pi_1(V) = \mathbb{Z}$ then the physical system is superfluid ^4He. Here the links consist of Abrikosov vortices. If $\pi_1(V) = \mathbb{Z} \oplus \mathbb{Z}$ then the corresponding physical system is A-phase of ^3He with regard to the spin-orbital interaction F_d and magnetic field. Here $V = S^1 \times S^1$ [13] .

2. $\pi_1(V) = \mathbb{Z}_2$. The corresponding physical system is nematic, $V = RP^2$, i.e. the real projective plane. The other example is A-phase in ^3He with regard to F_d. The order parameter space $V = SO(3)$.

3. $\pi_1(V) = \mathbb{Z}_4$. The corresponding system is A-phase for the ^3He.

4. $\pi_1(V)$ equals to the group Q of unit quaternions. The corresponding physical systems are biaxial nematic and cholesteric with $V = SO(3)/D_2$, where D_2 is the diedral group.

The identification of the order parameter space in these examples was obtained by many authors. The details can be found in the reviews [16,17] .

Theorem 3.2. Fix $1 \leqslant i \leqslant n$. If $\mu(i,j)=k(l_i,l_j)=0$ for $1 \leqslant j \leqslant n$, $i \neq j$ and $\pi_1(V)$ is commutative then the image of the parallel $\bar{l}_i$ can be contracted in V.

If $\pi_1(V) = Z_2$ and the defect Σ is m-essential then the loop $\varphi(\bar{l}_i)$ can be contracted in V if and only if

$$\sum_{j=1}^{i-1} \mu(i,j) + \sum_{j=i}^{n} \mu(i,j) \equiv 0 \pmod 2.$$

Let us illustrate the theorem for $\Sigma = l = (l_1, l_2)$. If $\varphi(\bar{l}_1)$ can be contracted in V then the linking coefficient is even. If $k(l_1,l_2) \leqslant 1$ under some energy considerations (only such links appeared in Bouligand experiments [4]) then $k(l_1,l_2)=0$ so l_1 and l_2 are homotopic unlinked.

Theorem 3.3. Let $\pi_1(V) = Z_4$. Let t be a loop in V whose homotopy class is a generator in $\pi_1(V)$. Let $\varphi(m_j)$ for every j is homotopic to the loop $t^{S(j)}$, where $0 \leqslant S(j) \leqslant 3$. Then $\varphi(\bar{l}_i)$ can be contracted in V if and only if

$$\sum_{j=1}^{n} S(j) \mu(i,j) \equiv 0 \pmod 4.$$

Theorem 3.4. Let $\pi_1(V) = Q$ (V is the order parameter space for biaxial nematic and cholesteric). Let $l=(l_1,l_2)$ and the defect $\Sigma = l$ be m-essential. Denote by h the homotopy class of $\varphi(m_1)$. Then $\varphi(\bar{l}_2)$ can be contracted in V if and only if

$\mu(1,2) \equiv 0 \bmod 4$ for $h = \pm e_i$ and $i=1,2,3$

$\mu(1,2) \equiv 0 \bmod 2$ for $h = J$.

Theorem 3.5. Let $\Sigma = (l_1,l_2,l_3) = l$ and $\varphi(m_1)$ and $\varphi(m_2)$ do not homotopically commute in V. Then $\varphi(\bar{l}_3)$ can be contracted in V if and only if $\mu(1,3)+$

$+ \mu(2,3) \equiv 0 \mod 4$.

4. The differential forms and the higher order linking coefficients. In this section we define, using the differential forms, many cohomological obstructions to unlinking of a link $l=(l_1,\dots,l_n)$ in S^3. V.Poenaru [21] noted that Sullivan's model based on studying differential forms on $S^3 \setminus l$ can be used for measuring the nontriviality of a link for $n=2$. This construction includes a special graded differential skew-commutative algebra. Some interesting applications of such construction to the problem of multi-valued functionals and selecting new important properties of rigidness were considered by S.P.Novikov [15] .

Let u_i be the differential 1-form the Alexander dual of the curve l_i, defined on $S^3 \setminus l_i$. Let v_i be the differential 2-form with compact support on S^3, the Poincaré dual to this curve, and B_i be the boundary of some sufficiently small tubular neighbourhood of l_i. For $n=2$

$$\int_{B_1} u_1 u_2 = -\int_{B_2} u_1 u_2 = \int_{S^3} u_1 v_2 = -\int_{S^3} v_1 u_2 - k(l_1,l_2) = \mu(1,2).$$

(We denote by $u\, v$ the usual skew product of differential forms u and v).

For $n=3$ suppose that $k(l_1,l_2)=k(l_2,l_3)=0$. Then there are 1-forms $u_{i\,i+1}$ on S^3 and 2-forms $v_{i,i+1}$, $v'_{i\,i+1}$ $(i=1,2)$ on S^3 such that $du_{12}=u_1u_2$, $dv_{12}= -v_1u_2$, $dv'_{12} = u_1v_2$. It is easy to verify that $\tilde{u}_{123}= u_{12}u_3 + u_1u_{23}$, $\tilde{v}_{123} = -v_{12}u_3 + v_1u_{23}$, $\tilde{v}'_{123} = u_{12}v_3 + u_1v_{23}$ are closed. We can select v_{12} and v_{23} so that v_{123} and v'_{123} are defined on the whole S^3.

One can show that regardless of the choice of $u_{i\,i+1}$, v_{ij}

$$\int_{B_1}\tilde{u}_{123} = -\int_{B_3}\tilde{u}_{123} = \int_{S^3}\tilde{v}_{123} = \int_{S^3}\tilde{v}'_{123} = \mu(1,2,3).$$

The cohomology classes of $\tilde{u}_{123}, \tilde{v}_{123}, \tilde{v}'_{123}$ are called Massey products $\langle u_1,u_2,u_3\rangle$, $\langle \breve{v}_1,u_2,u_3\rangle$, $\langle u_1,u_2,\breve{v}_3\rangle$ respectively.

This fact is generalized by

Theorem 4.1. Let one of the three Massey products $\langle u_{i_1},\dots,u_{i_p}\rangle$, $\langle v_{i_1},\dots,u_{i_p}\rangle$, $\langle u_{i_1},\dots,v_{i_p}\rangle$ be defined. Then so are two others and the integer

$$\int_{B_{i_1}}\langle u_{i_1},\dots,u_{i_p}\rangle = (-1)^p\int_{B_{i_p}}\langle u_{i_1},\dots,u_{i_p}\rangle =$$

$$= \int_{S^3}\langle v_1,\dots,u_{i_p}\rangle = \int_{S^3}\langle u_{i_1},\dots,v_{i_p}\rangle$$

is well-defined.

This integer is the (p-1)-th order linking coefficient of 1. The q-th order linking coefficient is defined if and only if every r-th order linking coefficient vanoshes for $r \leqslant q$. The Stallings conjecture proved in [11] and [18] implies that (p-1)-th order linking coefficient defined by Theorem 4.1 equals $\mu(i_1,\dots,i_p)$.

Conclusion. In this paper a topological approach to the recognition of non trivial links has been developed. The physical applications are mainly illustrative in nature and consist in descibing the restrictions on types of defect links in condensed mather. We hope however that our method will be useful in other physical problems e.g. in the statistics of polymer chains or in the turbulence in superfluids. The influence of linked vortical threads on the turbulence origin in a homogeneous superfluid ^{4}He was considered in [20] . Another example is magnetohydrodynamics where the formulas for linking coefficients from 4 play the role of topological conservation laws for magnet power threads frozen in a field.

References

1. Gennes de, P.G. (1974). The Physics of Liquid Crystals, Oxford, Clarendon Press.
2. Chandrasekhar,S. (1977). Liquid Crystals. Cambridge Univ. Press, Cambridge.
3. Friedel,G. (1922). Ann. Phys. 18, 273.
4. Bouligand,Y. (1974). J. Phys. 35, 959.
5. Toulouse,G., Kleman,M. (1976). J. Phys. 37, L149.
6. Poenaru,V., Toulouse,G. (1977). J. Phys. 38, 887.
7. Poenaru,V., Toulouse,G. (1979). J.Math. Phys. 20,13.
8. Janich,K., Trebin,H.R. (1981). Les Houches, Session XXXV, 1980, 421 in Physics of Defects, North Holland.
9. Golo,V.L., Monastyrsky,M.I. (1978). Lett. Math. Phys. 2, 373.
10. Rolfsen,D. (1976). Knots and Links. Publish or Perish, Willington.
11. Porter,R. (1980). Trans, Amer. Math. Soc. 257, 39.
12, Milnor,J. (1954). Ann. Math. 59, 177.
13, Mineyev,V.P., Volovik,G.E. (1978). Phys. Rev. B18, 3197.
14. Monastyrsky,M.I., Retakh,V.S. (1984). Preprint ITEP-184.
15. Novikov,S.P. (1984). Russian Math, Serveys 39, 5, 97 (Russian).
16. Leggett,A.J. (1975). Rev. Mod. Phys. 47, 331.
17. Mermin,N.D. (1979). Rev. Mod. Phys. 51, 591.
18. Turaev,V.G. (1976). Zapiski Sem. Lomi 66, 1989 (Russian).
19. Milnor,J.W., Stasheff,J.D. (1974). Characteristic Classes, Princeton Univ. Press, Princeton.
20. Schwarz,K. (1982). Phys. Rev. Lett. 49, 283.
21. Poenaru,V. (1979). Superalgebras and confinement in condensed matter physics. Preprint.

TENSOR FIELD REPRESENTATION IN THE NUCLEAR SOLID STATE PHYSICS

E.N.Ovchinnikova, R.N.Kuz'min

Moscow State University, Department of Physics

Nuclear γ- resonance spectroscopy and diffraction enables us to investigate the structures formed by the internal magnetic fields and the electrical field gradients (EFG) at the nuclei of Mössbauer isotopes in crystals or hyperfine field structures (HFS). It was shown [1, 2] that the color groups $G^{(p)}$ and $G^{(w)}$ should be used to describe the symmetry of HFS's. The difference between the symmetry of HFS and the spatial symmetry of crystals results in extra reflections in the Mössbauer diffraction pattern (MDP) as compared with the set of reflections, responding to the potential scattering of γ and X-rays. These reflections in accordance with the type of the HFS in crystal were called quadrupole, magnetic and combined. All these reflections were predicted theoretically [3-5]. Two of them were observed experimentally. The group-theoretical approach proposed in this paper not only gives a description of HFS's symmetry but in some cases allows one to determine the directions of the hyperfine fields on nuclei and the set of the MDP reflections. In this approach the tensor field representation (TFR) of the space groups introduced in 6, 7 is used. The relations of TFR with irreducible representations of space groups is also considered. In [8, 9] the permutational, mechanic, magnetic and tensor representations are constructed. All this was used in [8, 9] for the study of crystal structures possessing local physical properties described by 0-th and 1st rank tensors. In this paper we consider the structure formed by 2nd rank tensors placed into the points of the space group orbit, i.e. the structure of the EFG tensors on nuclei. The symmetry properties of the permittivity tensors (PT) of crystals with hyperfine fields for γ - rays frequencies are also considered. This approach is based on the isomorphism between the color and the classical groups, which allows us to use the set of irreducible representations of the space groups investigating the HFS symmetry.

It was shown in [10] that the propogation, scattering and diffraction of the Moessbauer radiation in crystals can be described by the PT:

$$\varepsilon_{ij}(\bar{k},\bar{k}',\omega)=\sum_{\bar{H}}\varepsilon_{ij}^{(\bar{H})}(\bar{k},\omega)\delta(\bar{k}-\bar{k}'-\bar{H}),$$

where $\varepsilon_{ij}^{(\bar{H})}$ are the diffraction components of the PT. The diffraction part of the polarization tensor $\chi_{ij} = (\varepsilon_{ij}-\delta_{ij})/4\pi$ is related to the structure amplitude $F(\bar{H})$ by the formula:

$$F(\bar{k}',\bar{e}',\bar{k},\bar{e})=e'^{*}_{i}\chi_{ij}^{(\bar{H})}e_{j},$$

where e_i and e_j are the polarization vectors of the incident and scattered radiation. If there are s Moessbauer nuclei in the crystal unit cell and the hyperfine fields are oriented differently with respect to the crystallographic axes then

$$\chi_{ij}^{(\bar{H})}=\sum_{s}\chi_{ij}^{s}e^{i\bar{H}\bar{z}_s}.$$

The tensors χ_{ij}^{s} can be represented as

$$\chi^{s}_{ij} = \sum_{\lambda} c_{\lambda} \chi_{ij}\left({k\nu \atop \lambda} \middle| s \right), \quad (1)$$

where $\chi_{ij}\left({k\nu \atop \lambda} \middle| s \right)$ are the basic functions of the ν representaion of the k star, c_{λ} are the unknown mixinq coefficients.

It was shown in [9], that the basic functions TBF) can be represented in the following form:

$$\chi_{ij}\left({k\nu \atop \lambda} \middle| s \right) = \sum_{g \in G^{o}_{k}} d^{*k\nu}_{\lambda[\mu]}(g) \exp[-i\bar{k}\bar{a}_{sp}(g)] \delta_{s,g[p]} [R^{h}_{i[\ell]} \times R^{h}_{j[m]}],$$

where $d^{*k\nu}_{\lambda[\mu]}(g)$ are the coefficients of the matricies, corresponding to the irreducible representation $k\nu$ of G^{o}_{k}, $\bar{k}$ the wave vector of the HFS, $\bar{a}_{sp}$ the set of the returning translations, $\delta_{s,g[p]} = 1$ for $\hat{g}\bar{z}_{p} = \bar{z}_{s} + \bar{a}_{sp}$ and 0 for $\hat{g}\bar{z}_{p} \neq \bar{z}_{s} + \bar{a}_{sp}$, $[R^{h}_{i[\ell]} \times R^{h}_{j[m]}]$ is the direct product of the matricies which describe the transformations of the coordinates under the rotational part h of $g \in G^{o}_{k}$.

Determining the reflections with $F(\bar{H}) \neq 0$ is essentially important for the study of the diffraction pattern. The systematic investigation of possible extinctions in the Moessbauer diffraction pattern has not been carried out.

It follows from (1), that the extinction occurs if $\hat{\chi}^{(\bar{H})} \equiv 0$, or $\hat{\chi}^{(\bar{H})}(k\nu, \lambda) = 0$, where $\hat{\chi}^{(\bar{H})}(k\nu, \lambda) = \sum_{s} \hat{\chi}\left({k\nu \atop \lambda} \middle| s \right) e^{i\bar{H}\bar{z}_{s}}$.

The BF's can be used for considering the PT in the presence of the HFS and for studying structures formed by the EFG tensors on nuclei. The EFG tensor $\varphi_{\alpha\beta}$ is the 2nd derivative of the crystal potential $\varphi(\bar{z})$ at points of the nuclei's location. The EFG is the deviator with the external symmetry ∞ / mmm if the asymmetry parameter η is or $2/mmm$ if $\eta \neq 0$. The existence of the FFG on nuclei may lead to the splitting of the Moessbauer line and occurence of the additional quadrupole reflections in the MDP.

The translation under the lattice vector does not change the directions of the main axes of the EFG (it is known, that they are determined by the crystal environment of the atom), so the wave vector $\bar{k}=0$ corresponds to this structure. Moreover the basic functions of the representation τ_1 must be choosed for describing the considered structure, because the crystal field is transformed by the identity representation of the G^{o}_{k} group.

We have calculated TBF's for the crystals with the shpinel structure (the space group O^7_h). The atoms of the Moessbauer isotope ^{57}Fe occupy the special position 16(d). Ten irreducible representations correspond to the $\bar{k}=0$ star. Four of them are one-dimensional. The representation τ_1 enters twice into the decomposition of the tensor representation. There are two linearly independent sets of BF's. The first of them does not correspond to the deviator, so we must use the second set of BF for the description of the EFG structure. The BF's for the nuclei 1-4 are:

$$\psi(01|1)=\begin{pmatrix}0&1&1\\1&0&1\\1&1&0\end{pmatrix},\ \psi(01|2)=\begin{pmatrix}0&-1&-1\\-1&0&1\\-1&1&0\end{pmatrix},\psi(01|3)=\begin{pmatrix}0&-1&1\\-1&0&-1\\1&-1&0\end{pmatrix},\psi(01|4)=\begin{pmatrix}0&1&-1\\1&0&-1\\-1&-1&0\end{pmatrix}.$$

The BF's for the remaining 12 nuclei coincide with the four listed above. The eigenvalues $\lambda_{1,2}=\pm 1$ and $\lambda_3 = \mp 2$ correspond to these tensors. The eigenvectors form the set of orthogonal vectors which determine directions of the main axes of EFG's. The eigenvectors corresponding to the multiple root are linearly dependent therefore any linear combination of them is also an eigenvector and the EFG tensor is axially symmetric. The eigenvectors corresponding to λ_3 determine an directions of EFG axes. For nuclei 1-4 they are $\pm[111]$, $\pm[\bar{1}11]$, $\pm[1\bar{1}1]$, $\pm[11\bar{1}]$, i.e. they are oriented along the 3rd order axes of the crystal. This result coincides with the well-known experimental data and with the result following from the Shubnikov-Curie principle.

We have also considered the structure formed by the EFG tensors in $KFeF_4$, where the Mössbauer atoms ^{57}Fe occupy the special position 4(a) (the space group D^{17}_{2h}). In this case there are 9 linearly independent TBF's. Constructing the EFG's by mixing TBF's with the unknown C_λ gives the following result:

$$\varphi(01|1,3)=\begin{pmatrix} c_1 & 0 & c_2 \\ 0 & c_3 & 0 \\ c_2 & 0 & c_4 \end{pmatrix}, \quad \varphi(01|2,4)=\begin{pmatrix} c_1 & 0 & -c_2 \\ 0 & c_3 & 0 \\ -c_2 & 0 & c_4 \end{pmatrix}.$$

The TBF's and the EFG tensors coincide for 1 and 3, 2 and 4 nuclei. It follows from the consideration of eigenvalues and eigenvectors that the EFG tensors are axially symmetric. This result cannot be obtained from Shubnikov-Curie principle because the nuclei's special position symmetry allows the existence of tensor of any kind.

To study PT of crystals with the axially symmetric HFS it is convenient to use its covariant form. The tensor representation based on the PT covariant form was used in [12] to investigate MDP of magnetic crystals.

It was shown earlier that if a combined HFS is present in crystal, the set of reflections corresponding to the quadrupole interaction may differ from a similar set for the magnetic interaction. In this case a reflection essentially corresponding to the combined structure may occur in MDP. Since the symmetry of the combined HFS is described by the group $(P_1 \otimes P_2) \circledS G$, where P_1 and P_2 correspond to the quadrupole and magnetic structures, the TBF's for the combined structure can be represented by products of the TBF's of each structure. This representation is not suitable for investigating PT, because it is essentially a 2nd rank tensor. So we have to construct the TBF's corresponding to the wave vector of magnetic structure. The study of MDP's for various structures with the help of TFR show that MDP's may contain an additional quadrupole ($\chi_m^{(\bar{H})} = 0$, $\chi_q^{(\bar{H})} \neq 0$), magnetic ($\chi_q^{(\bar{H})} = 0$, $\chi_m^{(\bar{H})} \neq 0$) and combained ($\chi_q^{(\bar{H})} = 0$, $\chi_m^{(\bar{H})} = 0$, $\chi_c^{(\bar{H})} \neq 0$) reflections simultaneously. The MDP examination of the above struvtures gives the following results.

The space group O_h^7, the special position of the Mössbauer isotope atoms 16(d). Investigating ferromagnetically ordered structure carried out in [8] gives the representation τ_9, so the MDP have no magnetic reflections. The analysis of TFR for $\bar{k} = 0$ star gives the quadrupole reflections of the $0k\ell$ type: $k = 4n$, $\ell = 4n+2$ but does not give the combined reflections. The examination of mutual orientations of TBF's eigenvectors and the magnetic vectors shows, that there are two different HF structures in the Moessbauer spectrum. It is shown in [13] that in this case the combined reflections may occur because the resonant conditions are not satisfied simultaneously for all nuclei in the unit cell. These reflections have the following indices: $h0l$ ($h = 4n+2$, $l = 4n$), $hk0$ ($h = 4n+2$, $k = 4n$). They are absent in the MDP corresponding to the quadrupole structure.

The space group D_{2h}^{17}, the special position of the Mössbauer isotope atoms 4(a). From the consideration of the TFR it follows that the quadrupole reflections of the hkl type, $h = 2n+1$, must exist. If there is a ferromagnetic structure in crystal, the MDP does not have to contain any extra reflections except quadrupole. But in the antiferromagnetic case ($KFeF_4$ gives the example of a layer crystal) the MDP has to contain magnetic reflections in transitions with $\Delta M = \pm 1$ similar to those observed in magnetic neutronography.

References

1. Kolpakov, A.V., Ovchinnikova, E.N., Kuz'min, E.N. (1975). Crystallography, 20, 221 (Russian).
2. Kolpakov, A.V., Ovchinnikova, E.N., Kuz'min, R.N. (1977). Crystallography, 22, 901 (Russian).
3. Aivazyan, Yu.M., Belyakov, V.A. (1969). JETP, 56, 346 (Russian).
4. Belyakov, V.A., Aivazyan, Yu.M. (1968). Pisma JETP, 7, 577 (Russian).
5. Kolpakov, A.V., Ovchinnikova, E.N., Kuz'min, R.N. (1979). Phys. Stat.Sol.(b), 39, 511.
6. Litvin, D.B. (1982). J.Math.Phys., 23, 337.
7. Berenson, E., Kotzev, J.N., Litvin, D.B. (1982). Phys.Rev.B, 25, 7523.
8. Izyumov, Yu.A., Naysh, V.E., Ozerov, R.P. (1981). Neutronography of Magnetics, Part 2, Atomizdat, Moscow (Russian).
9. Izyumov, Yu.A., Siromyatnikov, B.H. (1984). Phase Transitions and Crystals Symmetry, Nauka, Moscow (Russian).
10. Kolpakov, A.V., Bushuyev, V.A., Kuz'min, R.N. (1978). Russian Phys. Surveys, 16, 480 (Russian).
11. Andreeva, M.A., Kuzmin, R.N. (1982). Moessbauer gam-

ma-optics, Moscow State Univ., Moscow, 226 (Russian).

12. Ovchinnikova, E.N., Kuzmin, R.N. (1982). Phys. of Metals and Metalurgy, 54, 212 (Russian).

13. Kolpakov, A.V., Ovchinnikova, E.N., Kuzmin, R.N. (1978). Vestnik Moscow Univ., Ser. Phys., 2, 28 (Russian).

14. Heger, G., Geller, R. (1972). Phys. Stat. Sol.(b), 53, 227.

Part V

SYMMETRIES IN OPTICS, ATOMIC, MOLECULAR, AND NUCLEAR PHYSICS

GROUP THEORETICAL METHODS IN LIE OPTICS

Kurt Bernardo Wolf

IIMAS, Universidad Autonoma de México Iztapalapa DF, Mexico

1. Historical perspective and introduction. Optics has served as the expermental basis and model for much of science since Galielo. The minds at work in mechanics were so too in optics, in its early geometric aspects through Snell up to Newton, and its wave aspects since Huygens through Fourier and including Sommerfeld. Hamilton's original application of generating functions for canonical transformations was in optics; only later were they recognized as a sharp tool for abstract mechanics through the introduction of the concept of a phase space with a symplectic metric [1]. This concept in embrionary form, served Heisenberg [2] and Weyl [3] for the introduction of quantum mechanics, where the Hamiltonian is quadratic in momentum. The history and scope of Sophus Lie's work is well known to this audience, as is the fact that its manifold applications have been most successful when applied to systems with quadratic Hamiltonians [4]. Almost invariably, the kinetic energy term is $p^2/2m$. This Hamiltonian is indeed also useful in optics, albeit in its Gaussian approximation, i.e., first-order optics which deals with linear transformations of phase space. The associated methods have been under research since the early seventies from the perspective of quantum mechanics [5] and groups of integral transforms [6, 7].

It came to me as a pleasant surprise, upon reading the work of Nazarathy and Shamir [8], that the theory of canonical transformations had a very natural model in paraxial wave-optical systems. The elements of such systems are usually slabs of homogeneous material separated by "almost flat' refracting surfaces. There is an association between these optical elements and elements of a group with the structure W_NSp(2N,R), where Sp(2N,R) is the symplectic group of linear transformations of 2N-dimensional phase space, in semidirect product with W_N, the Heisenberg-Weyl group of phase-space translations with a one-dimensional center.

From a different direction, the work of Alex Dragt [9] and coworkers [10, 11] in nonlinear orbit analysis for accelerators and electron microscopes, was based on the theory of Lie series [12, 13] applied to classical phase-space variables. This allows the calculation -by electronic computer - of particle or light ray trajectories up to an in principle arbitrarily high aberration order. Upper practical limits for this order are seven or nine. MARYLIE [14] is a program which models magnetic-lens optical elements, concatenates them into an operator description of the system, and applies the latter to the object space to produce the image space result of the system. It works in three dimensions in coordinate space, since magnetic elements are in general not axially-symmetric - they need two transverse coordinates, and one z-axis along which chromatic dispersion of the electron packet occurs. This calls for Sp(6,R).

The phase-space translation group, W_N, is involved, moreover, in a very direct way in radar signal analysis [15, 16, 17]. This is because the aim is to determine the distance through the echo time-lag (space translation). The optimum resolution problem, for instance, has been solved using harmonic oscillator functions. The group of linear transformations pertains the detection of acceleration, and higher-order aberrations to higher-order derivatives of velocity.

The bag of nearby applications of Lie optics grows

with the orderly account it makes of coherent states [18, 19] in their behavior in optical fibers with inhomogeneities and interfaces, and the measurability of phases in microwave systems. This should be germane to the design of optical computers. As it will become clear at the end of this paper, the purely group-theoretical aspects of optical systems, or any other means of image processing, are also of interest in themselves.

Section 2 contains a brief recount of the hamiltonian formulation of free-flight optics and Section 3 of refracting surfaces. Their concatenation in Section 4 is done through Lie group multiplication. We present the symplectic classification of higher-order aberrations, in particular of order three aberrations. In Section 5 we outline their effect on the phase space of geometrical optics and on wavefunctions in wave optics. The handling of inhomogeneous systems is sketched in Section 6, and in Section 7 we offer some open questions.

2. The hamiltonian formulation. The application of Fermat's minimal action principle leads to the hamiltonian formulation of optics. The system is given a reference optical axis z and, in a z=constant plane perpendicular to this axis, we set our coordinate system $\underline{q}$. This 'position' space is two-dimensional for actual optical systems, but is reduced to one dimension if further symmetries exist (cylindrical and axis-symmetric lenses). A light ray crossing the coordinate plane at $\underline{q}$ has its direction given by a second coordinate system of angular variables. If $n(\underline{q},z)$ is the value of the refraction index at the point $(\underline{q},z)$, and $\underline{n}$ is a vector of magnitude n in the z=constant plane, the projection of the ray on the plane, then the coordinate canonically conjugate to $\underline{q}$ is shown [9,20] to be

$$\underline{p} = \underline{n} \sin\theta \qquad (1)$$

where θ is the angle between the ray and the optical

axis. In these terms, the optical Hammiltonian is

$$H = -n\cos\theta = -\sqrt{n^2 - p^2} \tag{2}$$
$$= -n + \frac{1}{2n}p^2 + \frac{1}{8n^3}p^4 + \frac{1}{16n^5}p^6 + \ldots,\ p < n$$

From this starting point, the well-known hamiltonian machinery [21] defines Poisson brackets, equations of motion, canonical transformations of $(\underline{p},\underline{q})$ phase space, and can be used to search for Lie algebras [22]. Free propagation $\hat{F}_z$ in a z-homogeneous medium, in particular, belongs to the one-parameter evolution group

$$\frac{df}{dz} = -\{H,f\} =: \hat{H}f \Rightarrow f(\underline{p},\underline{q};z) = e^{-z\hat{H}}f(\underline{p},\underline{q},0) =: \hat{F}_z f_\circ \tag{3a}$$

$$\hat{F}_z\begin{pmatrix}\underline{p}\\ \underline{q}\end{pmatrix} = \begin{pmatrix}\underline{p}\\ \underline{q} + z\tan\underline{\theta}\end{pmatrix} = \begin{pmatrix}\underline{p}\\ \underline{q} + z\underline{p}/\sqrt{n^2 - p^2}\end{pmatrix}$$
$$= \begin{pmatrix}\underline{p}\\ \underline{q} + z\underline{p}/n\end{pmatrix} + \mathcal{O}^3(\underline{p},\underline{q}) \simeq \begin{pmatrix}1 & 0\\ z/n & 1\end{pmatrix}\begin{pmatrix}\underline{p}\\ \underline{q}\end{pmatrix}. \tag{3b}$$

Equation (3a) is the usual evolution-operator construction which for (2) yields (3b), where the first-order or Gaussian terms are made explicit in matrix form. The Gaussian part is readily wavized when we replace the matrices (finite-dimensional non-unitary representations of Sp(2,R)) by integral canonical transformations [7,8,23] (infinite-dimensional unitary representations of the double cover - symplectic spin - of the same group).

We should be careful in noting that this treatment of Gaussian optics, and that of optical aberrations which will be presented below, disregards the fact that the geometry of optical phase space is not that of a plane, but that, due to (1) and (2), has momentum coordinates which range over a sphere. For N=1 dimension, this has been taken into account in a manner very similar to that of quantum mechanics on a compact space (a circle S^1) in [23]. We have then a mixed Heisenberg-Weyl group, with two infinitesimal generators, $\hat{Q} = i\lambda\!\!\!^{-} d/d\theta$ and $\hat{I} = \lambda\!\!\!^{-} 1$ (where $\lambda\!\!\!^{-} = \lambda/2\pi$ is the reduced wavelenght in the system), and one finite generator $\hat{E}$ (which in the Schrödinger case is exp $i\hat{P}$), whose action is multiplication by exp $i\theta$. The two former generators are self-adjoint in $L^2(S_1)$ and the latter is unitary. Out of this one, nevertheless, we may build the two self-adjoint combinations $\hat{P}:=n(\hat{E}-\hat{E}^{\dagger})/2i$ and $\hat{H} := -n(\hat{E}+\hat{E}^{\dagger})/2$, which in [24] are identified with the momentum and Hamiltonian, (1) and (2). Perhaps surprisingly, these three operators ($\hat{Q}$, $\hat{P}$, and $\hat{H}$) close into a euclidean iso(2) algebra. For paraxial optics, the matrix elements between forward-concentrated wavefunctions [20] of the generators of this euclidean algebra contract to the usual Heisenberg-Weyl algebra W_1. Free propagation belongs to a one-parameter subgroup of the associated Lie group ISO(2), which is thus the dynamical [4] group of free-propagation optics. One of the consequences of this construction, however, is that configuration space must be discrete, with points separated by $\text{\char"0127}$ ...

$$F_z(R,R') = e^{i\pi(T'-R)/2}\, J_{R'-R}(zn), \quad q=\lambda\!\!\!^{-} R, \quad R\in\mathbb{Z} \tag{4}$$

The system is thus related to that of an infinite lattice of masses joined by springs [26]. (The lattice Hamiltonian is $\hat{P}$ instead of $\hat{H}$ as here, i.e. the system is 'rotated' by 90°).

Fundamental Gaussian functions for the system are defined [24] as diffused Dirac deltas, $G_w(q) := F_{iw}(k=q/\lambda\!\!\!^{-},0) = I_k(wn)$, and given in terms of the modified Bessel functions which, indeed, look Gaussian. Finally, we may deform the euclidean algebra (where $P^2 + H^2 = n^2$) to so(2,1) = sp(2,R) or to infinite-dimensional algebras [22].

The state of this study is preliminary, however. More than delving into the fundamentals of the hamiltonian description of optics, here we would like to apply the well-tested Schrödinger formalism on a flat phase space, i.e. on $L^2(R^N)$ in spite of some of its difficulties, since the aim is to calculate efficiently the behaviour of sp(2,R)-defined objects under increasing aberration order. In [21] we pursued this objective for the study of Gaussian beams - coherent, discrete, and correlated - under the aberration of free propagation.

3. Refracting surfaces. Optics contains one very basic phenomenon which does not have a counterpart in mechanics: refracting surfaces. The closest mechanical model is that of a sudden change in the potential, but the truth is that the latter is by force continuous [1]. Besides, the instant of change of the potential should depend on the position, if curved surfaces $z = \zeta(\underline{q})$ are to be subject to analogy. Refracting surfaces may be seen better as 'sudden' finite transformations of phase space which take place at some reference plane (made to conicide with the optical center $\zeta(\underline{0}) = 0$ for convenience). Thus, we search to define an operator $\hat{S}(n,n';\zeta)$ accounting for the refraction transformation at ζ, between media n and n'. When the two media are homogeneous, we showed in [28] a rather neat result. First, this transformation may be factorised as

$$\hat{S}(n,n';\zeta) = \hat{R}(n;\zeta)\ \hat{R}(n';\zeta)^{-1} \tag{5a}$$

Second, the root transformation $\hat{R}(n;\zeta)$ may be written as

$$\bar{\underline{p}} := \hat{R}(n;\zeta)\underline{p} = \underline{p} + \sqrt{n^2 - p^2}\ \ (\underline{\nabla}\zeta)(\bar{\underline{q}}), \tag{5b}$$

$$\bar{\underline{q}} := \hat{R}(n;\zeta)\underline{q} = \underline{q} + \zeta(\underline{q})\underline{p}/\sqrt{n^2 - p^2}, \tag{5c}$$

Third, this root transformation is canonical; it is an implicitly defined transformation, since $\bar{q}$ appears on both sides of (5c), but may be self-replaced by symbolic computer algorithms [28] to any desired aberration order. Fourth, the above mapping is not globally one-to-one, but must have caustic singularities.

The validity of this factorisation property for surfaces of discontinuity in the refraction index has been shown

[29] to extend to the case where ζ is the interface between two continuous, inhomogeneous media $n(\underline{q},z)$ and $n'(\underline{q},z)$. This seems to be thus a rather fundamental property of things refracting worthy of deeper inquiry.

Refracting surfaces, moreover, posess <u>invariants</u>. This is obvious in the form of Snell's law:

$$n \sin\phi = n' \sin \phi', \tag{6}$$

where ϕ and ϕ' are the angles of the incomming and refracted rays with respect to the surface normal at the point of incidence. In terms of the system's specifications, this may be written [30] vectorially as

$$\underline{R} := \bar{\underline{p}}/\sqrt{1+\zeta'^2}, \quad \zeta'^2 := [\underline{\nabla}\zeta(\bar{\underline{q}})]^2 \tag{7}$$

This <u>Snell</u> (vector) invariant has one value in medium n, and the same value in medium n'; it is written in terms of the root-transformed variables $\bar{\underline{p}}$ and $\bar{\underline{q}}$ in (5). In fact, <u>any</u> function of $\bar{\underline{p}}$ and $\bar{\underline{q}}$ alone is an invariant. The Snell invariant in two dimensions has the following interesting property:

$$\{R_1, R_2\} = \tfrac{1}{2}(1+\zeta'^2)^{-2}(\bar{\underline{p}} \times \bar{\underline{\nabla}})\,\zeta'^2 \quad \text{(general surface)} \tag{8a}$$

$$= \left[\frac{d}{d(\bar{\underline{q}}^2)}\left(\frac{-1}{1+\zeta'^2}\right)\right]\bar{\underline{q}} \times \bar{\underline{p}} \quad \text{(axis-symmetric surface)} \tag{8b}$$

$$= \frac{1}{\gamma^2}\, m \qquad \text{(spherical, } \gamma\text{, surface)} \tag{8c}$$

The last expression is

$$m := \underline{q} \times \underline{p} = \bar{\underline{q}} \times \bar{\underline{p}} = \underline{q}' \times \underline{p}', \tag{9}$$

the well-known <u>skewness</u> or <u>Petzval</u> invariant [31, 32] of an axis-symmetric surface. For a spherical surface, we see above, we have an so(3) <u>algebra</u> of invariants which can be proven, predictably enough, to be the generators of phase-space transformations which stem from rotations in space which leave the surface invariant.

The Snell invariants allow for an economical description of aplanatic point pairs [32] and - for spherical surfaces - is explicitly expressible as a function of the object and image coordinates as

$$\underline{R} = \underline{p} + \underline{q}\sqrt{n^2-p^2}/\gamma = \underline{p}' + \underline{q}'\sqrt{n'^2-p'^2}/\gamma = \underline{R}' \qquad (10)$$

Through series expansion $\underline{p}'(\underline{p},\underline{q})$, $\underline{q}'(\underline{p},\underline{q})$, this formula allows the recursive computation of the expansion coefficients by comparison of the independent powers of $\underline{p}$ and $\underline{q}$. This algorithm has been pursued to ninth order [30,31] to yield the analytic expressions for the aberration coefficients (to be detailed below) of a spherical surface. Certain 'selection rules' which determine zeros for some coefficients (generalized spherical aberration, coma, and astigmatism), which have been observed by Dragt [34] and Forest [11] are explained in terms of optical center conditions of the surface. Similar analyses should be applicable to other revolution conics.

4. Concatenation of optical elements. Having described homogeneous free-space propagation and refracting surfaces, we concatenate them to form lens optical systems. Here we have to use the enveloping algebra of the classical Poisson bracket Heisenberg-Weyl algebra. There is a nesting in this infinite-dimensional algebra, however, by polynomial order [35]. The Poisson bracket between an Nth order homogeneous polynomial in $\underline{p}$ and $\underline{q}$, and an N'^{th} order one, is of order $N + N' - 2$. Second-order polynomials thus close into sp(2,R). Under this, N^{th} order polynomials transform (by adjoint action) asamodule. Since polynomials of order larger than $N+1 > 2$ form an (infinite-dimensional) subalgebra, we may build the quotient algebra of the full enveloping algebra modulo the latter [36]. This is the (finite-dimensional) N^{th}-order aberration algebra. It is generated by the quadratic monomials

$$^2\chi^1_1 := p^2,\quad ^2\chi^1_0 := \underline{p}\cdot\underline{q},\quad ^2\chi^1_{-1} := q^2, \qquad (11)$$

and by monomials $^k\chi$: of order $k = 3, 4, \ldots, N+1$. The Poisson brackets among the latter are set congruent with zero whenever the order exceeds $N + 1$.

The group-theoretical classification of aberrations

[30,36,37] is done in terms of finite-dimensional (non-hermitian) irreducible representations of the Gaussian-optics linear symplectic algebra (11). When the optical system has a common axis of rotational symmetry (the optical axis), the aberration algebra will contain only monomials of even order in $\underline{p}$ and $\underline{q}$, being thus a function of p^2, $p\cdot q$, and q^2 only. We may then define the polynomials ${}^k\chi^j_m$, $k=2,4,6,\ldots$, $j=k/2, k/2-2,\ldots$, 1 or 0, $m=j,j-1,\ldots,-j$, as homogeneous polynomials of <u>order</u> k, <u>spin</u> j, and <u>projection</u> m. The last label is the eigenvalue under the sp(2,R) weight operator $(p\cdot q/2)$. These symplectic-classified functions can be written in terms of the ordinary solid spherical harmonics $Y^j_m(\xi)$ in the angular variables

$$\xi_0 := \underline{p}\cdot\underline{q},\quad \xi_{\pm} := \pm\frac{1}{\sqrt{2}}(\xi_1 \pm j\xi_2) = \frac{1}{\sqrt{2}}\begin{cases} p^2 \\ q^2 \end{cases}, \tag{12}$$

with invariant radius

$$\underline{\xi}^2 = -(p^2q^2 - [\underline{p}\cdot\underline{q}]^2) = -(\underline{q}\times\underline{p})^2 =: -2\Xi \tag{13}$$

which is related to the Petzval invariant. The expression for the maximum j polynomials is [30]

$$^{2j}\chi^j_m = \frac{\sqrt{4\pi(2j+1)(j+m)!(j-m)!}}{(2j+1)!!} Y^j_m\,\underline{\xi}$$

$$= \frac{(j+m)!(j-m)!}{(2j-1)!!} \sum_R \frac{(\frac{1}{2}p^2)^{R+m}(\underline{p}\cdot\underline{q})^{j-2R-m}(\frac{1}{2}q^2)^R}{(R+m)!(j-2R-m)!\;R!} \tag{14}$$

For the less-than-maximum j polynomials, they are obtained through multiplication of the order-four invariant (13), to form

$$^{2N}\chi^j_m := \Xi^{\nu}\,{}^{2j}\chi^j_m,\quad j+2\nu = N \text{ and } \nu \text{ integers.} \tag{15}$$

The elements of the $(2n-1)^{\text{th}}$ order aberration algebra may be thus written as

$$S_{2n}(p^2,\underline{p},\underline{q},q^2) = \sum_{N=0}^{n} \sum_{j=\frac{1}{2}N(-2)}^{1\,\mathrm{or}\,0} \sum_{m=-j}^{j} V^{2N-1}_{jm}\,{}^{2N}\chi^j_m, \tag{16}$$

with symplectic aberration coefficients V^{2N-1}_{jm}.

To simplify matters, let us work henceforth in <u>third</u> aberration order, which yields compact, explicit results. Higher orders are as yet under development to find the Berezin bracket [38] for the sp(2,R)-based phase space description. By use of REDUCE [39] symbolic computation algorithms, Miguel Navarro Saad has developed [40] various programs which are being integrated as an interactive master program for the design of optical lens systems to aberration order up to nine; the expressions are all analytic and may be printed in explicit form, spot diagrams may be generated, and other forms of graphic display are being contemplated. Group-theoretically, the bracket algorithm is still incomplete.

For aberration order three [36] the generators of the aberration algebra of third order are (11) and those of fouth order: the symplectic <u>quintuplet</u> is

$$ {}^4\chi^2_2 = p^4,\ {}^4\chi^2_1 = p^2\,\underline{p}\cdot\underline{q},\ {}^4\chi^2_0 = \tfrac{1}{3}\left[p^2q^2 + 2(\underline{p}\cdot\underline{q})^2\right],\ {}^4\chi^2_{-1} = \underline{p}\cdot\underline{q}\,q^2,\ {}^4\chi^2_{-2} = q^4, \quad (17) $$

and the symplectic singlet,

$$ {}^4\chi^0_0 = \tfrac{1}{2}\left[p^2q^2 - (\underline{p}\cdot\underline{q})^2\right] = \Xi \quad (18) $$

The Poisson brackets between fourth-order elements ${}^4\chi^2_m$, ${}^4\chi^0_0$ is zero, while the sp(2,R) generators raise and lower the m eigenvalue, thus:

$$ \hat{\chi}^1_0\ {}^N\chi^j_m = 2m\ {}^N\chi^j_m, \quad (19a) $$

$$ \hat{\chi}^1_{\pm 1}\ {}^N\chi^j_m = 2(m \mp j)\ {}^N\chi^j_m, \quad (19b) $$

(valid for all N and j integer, and j half integer as well).

The Lie group corresponding to the aberration Lie algebra seen above is the 3rd-order <u>aberration group</u>. We choose its <u>presentation</u> as

$$ \begin{aligned} G\{\underline{v},\bar{w};\underline{M}\} &= G\{\bar{\underline{v}},\bar{w};\underline{1}\}\ G\{\underline{0},0;\underline{M}(\bar{\underline{u}})\} \\ &= \exp\Big(\sum_{m=-2}^{2} v_m\,{}^4\hat{\chi}^2_m + w\,{}^4\hat{\chi}^0_0\Big)\ \exp\Big(\sum_{m=-1}^{1} u_m\,{}^2\chi^1_m\Big), \end{aligned} \quad (20) $$

where $\underline{M}(\underline{u})$ is a 2x2 unimodular matrix, $\underline{v}$ a 5-vector, and w a scalar. The group multiplication law is

$$G\{\underline{v}_1,w_1;\underline{M}_1\}\ G\{\underline{v}_2,w_2;\underline{M}_2\} = G\{\underline{v}_1+\underline{v}_2\underline{D}^2(\underline{M}^{-1}_1),w_1+w_2;\underline{M}_1\underline{M}_2\} \quad (21)$$

where $\underline{D}^j(\underline{M})$ (j=2) is the (2j+1) × (2j+1)-dimensional matrix, nonunitary irreducible representation of $\underline{M}\in Sp(2,R)$,

$$D^j_{mm'}\begin{pmatrix}\alpha & \beta\\ \gamma & \delta\end{pmatrix} = \sum_n \binom{j-m}{j+m'-n}\binom{j+m}{n}\alpha^{n}\beta^{j+m+n}\gamma^{j+m'-n}\delta^{n-m-m'} \quad (22)$$

(also valid for half-integer j). Now, the group elements corresponding to the optical elements of the lens system are: Free propagations, given by the one-parameter subgroup line

$$\hat{F}_z=\exp(-z\hat{H})=G\{(-\frac{z}{8n^3},0,0,0,0),0;\ \begin{pmatrix}1 & 0\\ -z/n & 1\end{pmatrix}\} \quad (23)$$

An up-to-quartic refracting interface

$$z = \zeta(\underline{q}) = \alpha q^2 + \beta q^4, \quad (24)$$

for rays passing from a medium n to another n', is associated to

$$\hat{S} = G\{(0,0,\frac{\alpha}{2}[\frac{1}{n'}-\frac{1}{n}],\frac{2\alpha^2}{n'}[n-n'],\frac{2\alpha^3}{n'}[n-n']^2+\beta[n-n']), \quad (25)$$

$$\frac{2\alpha}{3}[\frac{1}{n'}-\frac{1}{n}];\ \begin{pmatrix}1 & -2\alpha(n-n')\\ 0 & 1\end{pmatrix}\}.$$

The names of the third-order (Seidel) aberrations are [34]: v_2, spherical aberration; v_1, coma; $C=v_0=3w/4$, curvature of field; $A=v_0-3w/2$, astigmatism; and v_{-1}, distortion. Suggested by group theory are: [36] v_0, curvatism; w, astigmature; and the last (not counted by Seidel) aberration, v_{-2}, dubbed pocus, since it is the Fourier conjugate of spherical aberration, and accounts for p-unfocusing; this produces a decreased depth of field for large angles.

To concatenate these elements, we place our optical elements from left to right, write under them their corresponding group elements (23) and (25), and multiply using (21). The algebra is manageable by hand for up to around five optical elements. Thereafter, as well as for higher aberration orders, we resort to symbolic computation. The aberration coefficients for eighth and tenth-order surfaces have been obtained by Forest [11] and Navarro-Saad. We note that, due to this symplectic-multiplet classification, the six third-order aberrations transform under Gaussian optics through a block diagonal matrix (5+1=6). For higher aberration orders this reduction of computation may be significant. For aberration order five, for example, dimension are 7+3=10; for orden seven, 9+5+1=15; and for nine, 11+7+3=21. The compounding of aberration polynomials under multiplication, for orders higher than third, however, is still under development. Perhaps predicably, Clebsch-Gordan coefficients and triangle selection rules appear.

5. Aberrations of phase space. We have the aberration group elements and their multiplication. The objects on which this group acts depends on the realization of optics we want. In terms of the geometric phase-space operators introduced in section 2 we must act with the exponential series [13] in (20) on $\underline{p}$ and $\underline{q}$, thereby generating non-linear terms for $j>1$; these terms may be molded into symplectic polynomials which may be classified in a way similar to (14), but in terms of sp(2,R) spinor spherical harmonics. For aberration order three, this is done in detail in [36]. Higher orders, as may be expected, introduce the sp(2,R) spinor $\binom{p}{q}$, each of two so(2)-components, coupled with Clebsch-Gordan coefficients to the harmonics (14). In this way one produces a canonical map, the map is nonlinear as $\underline{p}'(\underline{p},\underline{q})$, $\underline{q}'(\underline{p},\underline{q})$, and linear in the full basis $\{{}^{n}\chi^{s}_{m}\}$.

In [36] and [41] we have cut the phase-space coordinates after the term with the aberration order. This means for aberration order three to have the two linear terms ${}^{1}\chi^{1/2}_{m} = \underline{p},\underline{q}$, a symplectic quarter ${}^{3}\chi^{3/2}_{m}$ and a doublet ${}^{3}\chi^{1/2}_{m}$; the last six polynomials are cubic. We cut after that, since the essential group action is contained in this eight-dimensional space. The representation matrix in this space is minimal in dimension, and contains linear transformations in $\underline{pq}$ space, linear transformations among the six cubic terms (decomposed into a quadruplet and doublet), and finally, mixing cubic terms into the linear ones, a 2 x 4 off-diagonal submatrix, and a scalar. We need not present the details here, since they appear in the references, but only remark that spot diagrams for systems or for individual aberrations up to third order appear in [28].

The aberration group to third order has nine parameters, three Gaussian ones and six aberration parameters. This group may be made to act on other spaces to model wave optics. At the outset we should ward the reader that this matter is not fully controlled yet, but several open avenues and a few caution marks will be sketched below.

The Schrödinger approach consists in using $L^2(R)$ functions and the usual realization of the canonical generators P, Q, in place of the classical Poisson generators $\hat{p}$, $\hat{q}$, replaced into the aberration polynomials (14). Indeed, this works for the Gaussian part [42]. For the aberration part, involving the exponentials of fourth- and higher-order polynomials, this is not so straightforward. All higher-order polynomials are not self-adjoint [43] in $L^2(R)$, nor do they have a unique quantization [13]; $(P\cdot Q)^2$, P^2Q^2 fail on both accounts. The first objection may be circumvented working with matrix elements between forward-concentrated beams [20] or by working within $L^2(S_1)$ [24] as we did in section 2. For the latter, the uniqueness-of-quantization problem, we should demand the Weyl quantization rule, since only [42] by this rule will equations (19) continue to hold.

It does not seem to be possible to build an integral transformation realization of the finite-parameter aberration group, which would act as a Huygens-Fresnel integral to third order. Nevertheless, we may try to cut the space in some way if we consider only the first power term in the exponential series of the aberration part; this means to act with differential operators of fourth order for spherical aberration, third-order for coma, etc., in Weyl order. Very preliminary results indicate [44] that when these operators are made to act on Gaussian spot functions, the resulting pattern indeed models the known diffraction patterns in aberration ([32] , Fig. 9.6), for small coefficients. The effect on Gaussian spots of going to higher orders in the cut-off of the exponential series of the fourth-order terms has been started in collaboration with Wolfgang Lassner (NTZ, Leipzig), but are not yet available. (Our computer at IIMAS had a breakdown -a real crash!)

As may be gathered from the discussion in this section, the algorithms provided by Lie methods and those by traditional ray tracing, are quite different. Once the representing group element has been obtained, it serves for any ray in phase space (or outside it, if we count on it to be in S_2^1), and may be subject to a wave-optics translation. As we now torn to describe inhomogeneous-space optics, the effect of this difference may be appreciated.

6. Inhomogeneous systems and fiber optics with aberration.

When the refraction index depends on $\underline{q}$ or z, we term the optical system inhomogeneous. The first case includes models for optical fibers with a quartic index profile

$$n(q,z) = n_0(z) - \nu(z)q^2 - \rho(z)q^4. \qquad (26)$$

A Hamiltonian depending on z will generate a line which twists within the group, and is not coincident with a subgroup. Our results here are explicit only to third aberration order [41]. Higher orders should follow suit, but there aberrations compound.

The tangent to a line in the group $G(v(z),w(z);\begin{pmatrix}\alpha(z) & \beta(z)\\ \gamma(z) & \delta(z)\end{pmatrix})$ is given by

$$\frac{d}{dz}G\{\underline{v},w;\underline{M}\} = \left[A\hat{X}^1_1+B\hat{X}^1_0+C\hat{X}^1_{1}+\sum_{m=-2}^{2}\lambda_m\hat{X}^2_m+\mu\hat{X}^0_0\right] G\{\underline{v},w;\underline{M}\}, \tag{27a}$$

$$A= (\alpha\dot{\gamma}+\dot{\alpha}\gamma+\gamma^2[\dot{\alpha}\beta-\alpha\dot{\beta}])/2\alpha^2,\ B=\dot{\beta}\gamma-\dot{\alpha}\delta,\ C=\dot{\alpha}\beta-\alpha\dot{\beta}, \tag{27b}$$

$$\underline{\lambda}= \dot{\underline{v}}+\underline{v}\underline{F},\ \mu=\dot{w} \tag{27c}$$

$$\underline{F}:= \dot{\underline{D}}^2(\underline{M})\underline{D}^2(\underline{M}^{-1}) = \left\|\begin{matrix} -4B & -8C & 0 & 0 & 0 \\ 2A & -2B & -6C & 0 & 0 \\ 0 & 4A & 0 & -4C & 0 \\ 0 & 0 & 6C & 2B & -2C \\ 0 & 0 & 0 & 8A & 4B \end{matrix}\right\| \tag{27d}$$

This may be inverted [46] to a set of sequential differential operators for A, B, C, with variable coefficients, or subjected to numerical integration.

When the system is purely q-inhomogeneous, as fibers are, the Hamiltonian stemming from (26) has the form

$$H(\underline{p},\underline{q}) = \frac{1}{2n_0}p^2+\nu q^2+\frac{1}{8n_0^3}p^4 + \frac{\nu}{2n_0^2}p^2q^2 + pq^4 +\mathcal{O}^6(\underline{p},\underline{q}) \tag{28}$$

The generated one-parameter group may be found in general closed form [35,41] as

$$\exp\left(z[\underline{u}\cdot\hat{\underline{\chi}}^1+\underline{\lambda}\cdot\hat{\underline{\chi}}^2+\mu\hat{\chi}^0_0]\right)=G\{z\underline{\lambda}\underline{E}(z\underline{u}),\mu z;\ \underline{M}(z\underline{u})\}, \tag{29a}$$

$$\underline{M}(\underline{u}):=\begin{pmatrix} \cosh\omega - u_0\operatorname{sinch}\omega & -2u_{-1}\operatorname{sinch}\omega \\ 2u_1\operatorname{sinch}\omega & \cosh\omega+ u_0\operatorname{sinch}\omega \end{pmatrix}, \tag{29b}$$

$$\omega:= \pm\sqrt{u_0^2 - 4u_1u_1},\quad \operatorname{sinch}\omega = \omega^{-1}\sinh\omega\ , \tag{29c}$$

$$\underline{E}(\underline{u}) = \underline{D}^2(\underline{B}(\underline{u}))\underline{E}^d(\omega)\ \underline{D}^2(\underline{B}(\underline{u})^{-1}) \tag{29d}$$

$$\underline{B}(\underline{u}) = \frac{1}{\sqrt{2}}\begin{pmatrix} \frac{\omega+u_0}{\omega} & \frac{\omega-u}{2u_1} \\ -\frac{u_1}{2\omega} & 1 \end{pmatrix},\ E^d_{mm'}(\omega) = \delta_{m,m'}\frac{e^{2m\omega}-1}{2m\omega}. \tag{29e}$$

This general exponential relation is closed, and solves analitically every Baker-Campbell-Hausdorff relation within the group. Partial results exist for aberration orders higher than third, and certain particular cases (such as elliptic $\underline{M}(z)$ produced from (28) for fiber optics, and triangular $\underline{M}$ as from refracting surfaces) have been solved, the general case remains to be worked out.

The case for quartic fibers (26) has been given explicitly in [41], and shown that all aberrations but astigmature may be made to vanish by appropriate choices of the rations $n_o:\nu:\rho$ and the fiber lenght z. The calculation is performed through a Bargmann matrix transform $\underline{B}(\underline{u})$ in (29d) which diagonalizes the Gaussian part of the system. A detailed study of this system, with experimental predictions, however, remains to be performed.

7. Outlook. Only the barest essentials of the current program on Lie optics have been outlined here, and much remains to be done. Careful comparison of cost with existing systems from more traditional methods of ray tracing and Fourier optics must be performed before we can claim true improvement in this old science. The work of Dragt [9,14] on accelerator design indicates this is probably the case. Presently, with Navarro Saad, we are developing spot diagrams for spherical surfaces, where aberration orders 3,5,7 and 9 may be compared with the exact result over the full surface. The movement of Gaussian coherent beams within fibers [46] with inhomogeneous necks or kinks [47] is also being readied for graphic display [48] as well as their travel under free-space spherical aberration. This problem has recently been examined together with V.I. Man'ko [20].

We should note, as an unsatisfactory feature of the 'Schrödinger' description, that only wavefront propagation to the right is described. Reflection phenomena, partial and total, must be accounted for, probably through an à la Dirac [49] particle-antiparticle space doubling, see also [20]. This should take into account the two signs of the root function in the Hamiltonian. The off-diagonal terms are expected to involve the refraction index gradient or discontinuity.

Finally, the more fundamental description of optical phase-space and its associated euclidean Heisenberg-Weyl group may yet offer some pleasant surprises when applied to aberration optics. I should offer my apologies for showing only the scaffold of these unfinished matters.

Acnowledgement. I would like to thank the organizers of the Yurmala Seminar for inviting me to present this contribution.

References

1. Guillemin, V., Sternberg, S. (1984). Symplectic Techniques in Physics, Cambridge Univ. Press
2. Heisenberg, W. (1925). Über quantentheoretische Undeutung Kinematischer und mechanischer Beziehungen, Z. Phys. 33, 879-893
3. Weyl, H. (1928). Quantenmechanik und Gruppentheorie, Z. Phys. 46, 1-46 id. (1930). The theory of groups and quantum mechnics. 2nd ed., Dover.
4. Malkin, I.A., Man'ko, V.I. (1979). Dynamical symmetries and coherent states of quantum systems, Nauka, Moscow (Russian)
5. Moshinsky, M., Quesne, C. (1974). Oscillator systems, in: Proceedings of the 15th Solvay Conference in Physics (1970). Gordon and Breach, id. (1971). Linear canonical transformations and their unitary representations. J. Math. Phys. 12, 1772-1780. id. (1971). Canonical transformations and matrix elements. J. Math. Phys. 12, 1780-1783
6. Wolf, K.B. (1974). Canonical transforms I. Complex linear transforms. J. Math. Phys. 15, 1295-1301; II. Complex radial transforms. 2101-2111; (1980). IV. Hyperbolic transforms: continuous series representations of sl(2,R), 21, 680-688
7. Wolf, K.B. (1979). Integral transforms in science and engineering, part IV. Plenum Press.
8. Nazarathy, M., Shamir, J. (1980). Fourier optics described by operator algebra. J. Opt. Soc. Am. 70, 150-158. id. ibid. (1982). First-order optics a canonical operator representation: lossless systems. 72, 356-364

9. Dragt, A.J. (1982). Lectures on nonlinear orbit dynamics. AIP Conference Proceedings, Vol. 87.
10. Douglas, D.R. (1982). Lie algebraic methods for particle accelerator theory. Ph. D. Thesis, University of Maryland.
11. Forest, E. (1984). Lie algebraic methods for charged particle beams and light optics. Ph. D. Thesis, University of Maryland.
12. Dragt, A.J., Finn, J.M. (1976). Lie series and invariant functions for analytic symplectic maps. J. Math. Phys. 17, 2215-2227
13. Steinberg, S. (1984). Factored product expansions of solutions of nonlinear differential equations. SIAM J. Math. Anal. 15, 108-115
14. Douglas, D.R., Dragt, A.J. (1983). MARYLIE, the Maryland Lie algebraic beam transport and particle tracking program. IEEE Trans. Nucl. Sci. NS-30, 2442
15. Klauder, J.R. (1960). Bell Syst. Techn. J. 39, 809; Glauber, R.J. (1963). Phys. Rev. Lett. 10, 84
16. Schempp, W. (1982). Radar reception and nilpotent harmonic analysis. I-V. C.R. Math. Rep. Acad. Sci. Canada 4, 43-48, 139-144, 219-255, 287-292; ibid. (1983) 5. 5, 3-8,35-40 ; Raszillier, H., Schempp, W. Fourier optics from the perspective of the Heisenberg group. Preprint Univ. Bonn HE 84-34, to appear in Lie methods in optics, Proceedings of the CIFMO-CIO workshop on Jan 7-10, 1985 Leon, Mexico, ed. by J. Sanchez Mondragon and K.B. Wolf; Lecture Notes in Physica, Springer Verlag
17. Schmidt, M. (1985). Die reele Heisenberg-Grouppe und einige ihrer Anwendungen in Radarortung und Physik. Diplomarbeit Universität-Hochshule Siegen, DBR, April
18. Dodonov, V.V., Kurmyshev, E.V., Man'ko, V.I. (1980). Phys. Lett. 79A, 150
19. Klauder, J.R., Sudarshan, E.C.G. (1968). Fundamentals of quantum optics. Benjamin

20. Man'ko, V.I., Wolf, K.B. (1985). The influence of aberrations in the optics of gaussian beam propagation. Preprint Univ. Metropolitana April; to appear in shortened version in Lie methods in optics, op.cit.

21. Goldstein, H. (1959). Classical mechanics. Addison Wesley

22. Exner, P., Havlicek, M; Lassner, W. (1976). Canonical realizations of classical algebras. Czech. J. Phys. B26, 1213-1228; Lassner, W. Noncommutiative algebras prepared for computer calculations. In: Proceedings of the International Conference on Systems and Techniques of Analytical Computing and Their Applications in Theoretical Physics. Dubna report D11-80-13

23. Bacry, H., Cadilhac, M. (1981). Metaplectic group and Fourier optics. Phys. Rev. A. 23, 2533-2536

24. Wolf, K.B.(1975). The Heisenberg-Weyl ring in quantum mechanics. In: Group theory and its applications, E.M. Loebl (ed.) 3, Section VI, Acad. Press.

25. Wolf, K.B. (1985). A euclidean algebra of hamiltonian observables in Lie optics. Preprint. Universidad Metropolitana. May

26. Goodman, J.W. (1968). Introduction to Fourier optics. Mc Graw-Hill, Section 2.3

27. Navarro-Saad, M., Wolf, K.B. (1984). Factorization of the phase-space transformation produced by an arbitrary refracting surface. Preprint CINVESTAV April (to appear in J. Opt. Soc. Am.)

28. Navarro-Saad, M. (1985). Cálculo de aberraciones en sistemas ópticos con teoría de grupos. Tesis, Facultad de Ciencias, Universidad Nacional Autónoma de México, Mexico DF, Jan.

29. Dragt, A.J;, Forest, E., Wolf, K.B. to appear in Lie methods in Optics, op. cit.

30. Wolf, K.B. (1985). Symmetry in Lie optics, Preprint Universidad Metropolitana, April

31. Stavroudis, O.N. (1972). The optics of rays, wavefronts, and caustics. Academic Press.
32. Born, M., Wolf, E. (1959). Principles of optics.Pergam
33. Navarro-Saad, M;, Wolf, K.B. (1985). Applications of a factorization theorem for ninth-order aberration optics. Comunicaciones Técnicas IIMAS preprint Desarrollo No. 41, Feb. to appear in Journal of Symbolic Computation.
34. Dragt, A.J. (1982). Lie-algebraic theory of geometrical optics and optical aberrations. J. Opt. Soc. Am. 72, 372-379
35. Wolf, K.B. (1983). Approximate canonical transformations and the treatment of aberrations. One-dimensional simple Nth order aberrations in optical systems. Comunicaciones Técnicas IIMAS preprint No. 352
36. Navarro-Saad, M., Wolf, K.B. (1984). The group-theoretical treatment of aberrating systems. I. Aligned lens systems in third aberration order. Comunicaciones Técnicas IIMAS preprint No. 363
37. Buchdahl, H.A. (1954). Optical aberration coefficients Oxford Univ.
38. Berezin, F.A. (1966). The method of second quantization. Acad. Press.
39. Hearn, A.C. REDUCE-2 User's Manual. University of Utah.
40. Ref. 28 and work in progress.
41. Wolf, K.B. (1984). The group-theoretical treatment of aberrating systems. II. Axis-symmetric inhomogeneous systems and fiber optics in third aberration order. Comunicaciones Técnicas IIMAS preprint No.366
42. García-Bullé, M., Lassner, W., Wolf, K.B. (1985). The metaplectic group within the Heisenberg-Weyl ring. Universidad Metropolitana preprint II, No. 20
43. Klauder, J.R. Wave theory of imaging systems. To appear in Lie methods in optics, op. cit.

44. Wolf, K.B., work in progress, to appear in Lie methods in optics, op. cit.

45. Farnell, G.W. (1958). Canadian J. Phys. 36, 935, Figs. 3

46. Wolf, K.B. (1981). On time-dependent quadratic quantum Hamiltonians. SIAM J. Appl. Math. 40, 419-431

47. Castaños, O., López-Moreno, E., Wolf, K.B. The group-theoretical formulation of gaussian optics. Work in progress.

48. Barral, J.F., Castaños, O., Cuevas, S., Wolf, K.B., work in progress.

49. Plebański, J., private communication.

UNIVERSAL INVARIANTS OF PARAXIAL OPTICAL BEAMS

V.V.Dodonov, O.V.Man'ko
Moscow Physical-Technical Institute

It is well-known that under certain conditions (e.g. the validity of the paraxial approximation) the Helmholtz equation $\Delta E+k^2n^2E=0$ describing propagation of the harmonic wave fields can be reduced by substituting

$$E(x,y,z)=n_0^{-1/2}\psi(x,y,z)\exp\left(+ik\int_0^z n_0(\xi)d\xi\right)$$

to the parabolic equation [1,2]

$$i\lambda\frac{\partial\psi}{\partial z}=-\frac{\lambda^2}{2n_0}\left(\frac{\partial^2\psi}{\partial x^2}+\frac{\partial^2\psi}{\partial y^2}\right)+\frac{1}{2n_0}\left(n_0^2(z)-n^2(x,y,z)\right)\psi. \quad (1)$$

Here $\lambda=\lambda/2\pi=k^{-1}$ is the reduced wavelength in vacuum, $n_0\equiv n(0,0,z)$ is the refractive index of the medium at the axis of the beam propagating in z-direction. Equation (1) can be considered as the Schrödinger equation provided one identifies λ with Planck's constant, and longitudinal coordinate z with "time". Therefore the methods and results developed in quantum mechanics can be effectively applied to problems of paraxial beams propagation, e.g. in optical waveguides.

Consider so-called parabolic waveguides, i.e.media with the quadratic dependence of dielectric permittivity n^2 on the transverse coordinates $x_1=x$ and $x_2=y$: $n^2-n_0^2=$ $=a_{ik}(z)x_ix_k+b_i(z)x_i$, $i=1,2$, where the parameters a_{ik} and b_i may be quite arbitrary real functions of the longitudinal coordinate z. By analogy with quantum mechanics introduce the operators $\hat{x}=x$, $\hat{p}=-i\lambda\,\partial/\partial x$ combined into the four-dimensional vector $\hat{q}=(\hat{x},\hat{y},\hat{p}_x,\hat{p}_y)$. Then the solutions of the Heisenberg equations of motion for the operators $\hat{q}_\alpha$ for all the systems under study ("quadratic media") obviously have the following form (since these equations are linear):

$$\hat{q}(z)=\Lambda(z)\hat{q}(0)+\delta(z) \tag{2}$$

where $\Lambda(z)$ is a 4×4 matrix, $\delta(z)$ a 4-dimensional vector. It is important that the transformation $\hat{q}(0)\rightarrow\hat{q}(z)$ is cannonical, i.e. it preserves the commutation relations $[\hat{q}_j,\hat{q}_k]=i\lambda\Sigma_{jk}$. Therefore Λ is symplectic ($\tilde{\Lambda}$ stands for the transposed matrix):

$$\Lambda(z)\Sigma\tilde{\Lambda}(z)=\Sigma\,,\quad \Sigma=\|\Sigma_{jk}\|. \tag{3}$$

Consider the covariance matrix $Q(z)$ consisting of the second-order central moments $Q_{jk}\equiv\overline{q_j q_k}=\frac{1}{2}\langle\hat{q}_j\hat{q}_k+\hat{q}_k\hat{q}_j\rangle-\langle\hat{q}_j\rangle\langle\hat{q}_k\rangle$, where $\langle\ldots\rangle$ means an averaging. This matrix satisfied the following equation, a consiquence of (2):

$$Q(z)=\Lambda(z)Q(0)\tilde{\Lambda}(z). \tag{4}$$

Comparing (3) and (4) one sees that

$$D(\mu)=\det\left(\Sigma^{-1}Q(z)-\mu\right)=\mu^4+\mu^2 D_2+D_0 \tag{5}$$

does not depend on z for arbitrary values of the parameter μ (it follows from (3) that $\det\Lambda=1$ and the odd powers of μ vanish because $\tilde{\Sigma}=-\Sigma$, $\tilde{Q}=Q$, $D(\mu)=D(-\mu)$). Therefore the quantities D_0 and D_2, which are nothing but the determinant and the sum of the second-order principal minors of $\sum^{-1}Q(z)$, can be called universal invariants of paraxial beams in parabolic waveguides, because they do not depend on z and the explicit form of the function $n^2(x_1,x_2,z)$. The existance of similar invariants in quantum mechanical problems described by means of multidimensional quadratic Hamiltonians, as well as their connection with Poincaré-Cartan's universal invariants of classical mechanics, was discovered in [3]. The explicit forms of the invariants D_2 and D_o are

$$D_2 = (\overline{p_x^2})(\overline{x^2}) + (\overline{p_y^2})(\overline{y^2}) + 2(\overline{xy})(\overline{p_x p_y}) - 2(\overline{xp_y})(\overline{yp_x}) - (\overline{yp_y})^2 - (\overline{xp_x})^2 , \tag{6}$$

$$\begin{aligned} D_o = {} & (\overline{x^2})(\overline{y^2})(\overline{p_x^2})(\overline{p_y^2}) - (\overline{x^2})(\overline{y^2})(\overline{p_x p_y})^2 - \\ & - (\overline{xy})^2(\overline{p_x^2})(\overline{p_y^2}) + (\overline{xy})^2(\overline{p_x p_y})^2 + \\ & + (\overline{xp_x})^2(\overline{yp_y})^2 + (\overline{yp_x})^2(\overline{xp_y})^2 - \\ & - 2(\overline{xp_x})(\overline{yp_y})(\overline{xp_y})(\overline{yp_x}) - 2(\overline{xy})(\overline{p_x p_y})(\overline{xp_x})(\overline{yp_y}) \\ & - 2(\overline{xy})(\overline{p_x p_y})(\overline{xp_y})(\overline{yp_x}) + \\ & + 2(\overline{xy})\left[(\overline{p_x^2})(\overline{yp_y})(\overline{xp_y}) + (\overline{p_y^2})(\overline{xp_x})(\overline{yp_x})\right] + \\ & + 2\,(\overline{p_x p_y})\left[(\overline{y^2})(\overline{xp_y})(\overline{xp_x}) + (\overline{x^2})(\overline{yp_x})(\overline{yp_y})\right] - \\ & - (\overline{y^2})(\overline{p_x^2})(\overline{xp_y})^2 - (\overline{x^2})(\overline{p_y^2})(\overline{yp_x})^2 - \\ & - (\overline{y^2})(\overline{p_y^2})(\overline{xp_x})^2 - (\overline{x^2})(\overline{p_x})^2(\overline{yp_y})^2 . \end{aligned} \tag{7}$$

Suppose that n^2 depends only on x^2+y^2. Then (1) is invariant with respect to rotations in the xy-plane. If the initial state of the beam at z=0 possesses the same property, then expressions (6) and (7) can be simplified:

$$D_2^{(2)} = (\overline{xp_y})^2 - (\overline{xp_x})^2 + (\overline{x^2})(\overline{p_x^2}), \tag{8}$$

$$D_0^{(2)} = \left((\overline{xp_x})^2 + (\overline{xp_y})^2 - (\overline{x^2})(\overline{p_x^2}) \right)^2 . \tag{9}$$

In the case of a planar waveguide, when the dependence on y-variable can be omitted, one gets the invariant

$$\Delta = (\overline{x^2})(\overline{p^2}) - (\overline{xp})^2. \tag{10}$$

Besides the invariants (6)-(10) depending on the second-order moments there are invariants depending on the fourth-order ones. To construct them let us note that the Hamiltonian of eq. (1) in the case of quadratic medium (for simplicity without linear terms $b_i x_i$) is a linear combination of operators $\hat{x}^2, \hat{y}^2, \hat{x}\hat{y}, \hat{p}_x^2, \hat{p}_y^2$, which together with operators $\hat{p}_x\hat{p}_y$, $\hat{x}\hat{p}_y$, $\hat{y}\hat{p}_x$, $\hat{x}\hat{p}_x+\hat{p}_x\hat{x}$, $\hat{y}\hat{p}_y+\hat{p}_y\hat{y}$ are the generators $\hat{r}_k$, k = 1,2,...,10 of the algebra with structure constants satisfying the conditions $C^{\alpha}_{\alpha\beta} = 0$. Such systems possess the sets of universal invariants G_m, which are coefficients of the expansion [3]:

$$G(\mu) = \det(R(z) - \mu g) = \sum_m \mu^m G_m , \tag{11}$$

where the matrix R consists of the average values $R_{\alpha\beta} \equiv \overline{r_\alpha r_\beta} = \frac{1}{2}\langle \hat{r}_\alpha \hat{r}_\beta + \hat{r}_\beta \hat{r}_\alpha \rangle$, and g is Killing-Cartan's matrix: $g = \| g_{\alpha\beta} \|$, $g_{\alpha\beta} = C^{\varrho}_{\alpha\delta} C^{\delta}_{\beta\varrho}$. In the case under consideration one of the most simple invariants, namely G_8, has the following explicit form

$$
\begin{aligned}
G_8 = {} & \overline{(p_x^4)}\,\overline{(x^4)} + \overline{(p_y^4)}\,\overline{(y^4)} + 3\overline{(p_x^2x^2)}^2 + 3\overline{(p_y^2y^2)}^2 + \\
& + 12\overline{(p_xp_yxy)}^2 - 12\overline{(xy^2p_x)}\,\overline{(xp_y^2p_x)} - \\
& - 12\overline{(yx^2p_y)}\,\overline{(yp_x^2p_y)} - 12\overline{(xp_yp_x^2)}\,\overline{(p_xyx^2)} - \\
& - 12\overline{(yp_xp_y^2)}\,\overline{(p_yxy^2)} + 12\overline{(xyp_x^2)}\,\overline{(x^2p_yp_x)} + \\
& + 12\overline{(yxp_y^2)}\,\overline{(p_yp_xy^2)} - 4\overline{(xp_x^3)}\,\overline{(p_xx^3)} - 4\overline{(yp_y^3)}\,\overline{(p_yy^3)} - \\
& - 4\overline{(x^3p_y)}\,\overline{(p_x^3y)} - 4\overline{(xp_y^3)}\,\overline{(p_xy^3)} + 4\overline{(p_x^3p_y)}\,\overline{(x^3y)} + \\
& + 6\overline{(p_x^2y^2)}\,\overline{(p_y^2x^2)} + 2\,\overline{(p_x^2p_y^2)}\,\overline{(x^2y^2)} + \\
& + 3\lambda^2\left[\overline{(p_x^2x^2)} + \overline{(p_y^2y^2)} + 2\overline{(p_xp_yxy)}\right] - 3\lambda^4.
\end{aligned}
\tag{12}
$$

We take into account the relations

$$\overline{(p_xx + xp_x)^2} = 4\overline{(p_x^2x^2)} + 3\lambda^2 ,$$

$$\overline{(p_xx + xp_x)(p_yy + yp_y)} = 4\overline{(p_xp_yxy)} + \lambda^2 ,$$

$$\overline{(p_xyp_yx)} = \overline{(p_xp_yxy)} + \lambda^2/2 .$$

The explicit expressions of other invariants are much more complicated. However, the formulas are simplified in two cases:

1) Axial symmetric waveguides (and initial values). It suffices to consider only four generators: $\hat{x}^2+\hat{y}^2 = \hat{q}$, $\hat{p}_x^2+\hat{p}_y^2=\hat{K}$, $\hat{x}\hat{p}_y-\hat{y}\hat{p}_x=\hat{L}$, $\hat{x}\hat{p}_x+\hat{p}_y\hat{y}=\hat{S}$. Thus one gets the universal invariants (besides obvious $\langle\hat{L}^2\rangle$):

$$G_2^{(2)} = \overline{(SL)}^2 + \overline{(qK)}\,\overline{(L^2)} - \overline{(S^2)}\,\overline{(L^2)} - \overline{(KL)}\,\overline{(qL)},$$

$$
\begin{aligned}
G_1^{(2)} = {} & 4\overline{(S^2)}\,\overline{(L^2)}\,\overline{(qK)} - 4\overline{(qK)}\,\overline{(SL)}^2 + \overline{(L^2)}\,\overline{(q^2)}\,\overline{(K^2)} - \\
& - \overline{(qK)}^2\,\overline{(L^2)} + 4\overline{(KS)}\,\overline{(SL)}\,\overline{(qL)} + 2\overline{(qL)}\,\overline{(KL)}\,\overline{(qK)} - \\
& - 2\overline{(qL)}\,\overline{(KL)}\,\overline{(S^2)} - \overline{(q^2)}\,\overline{(KL)}^2 - 2\overline{(KS)}\,\overline{(L^2)}\,\overline{(qS)} + \\
& + 2\,\overline{(KL)}\,\overline{(qS)}\,\overline{(SL)} - \overline{(qL)}^2\,\overline{(K^2)},
\end{aligned}
$$

$$
\begin{aligned}
G_0^{(2)} = & - \overline{(qK)}^2\,\overline{(S^2)}\,\overline{(L^2)} - \overline{(q^2)}\,\overline{(K^2)}\,\overline{(SL)}^2 + \\
& + \overline{(qK)}^2\,\overline{(SL)}^2 + \overline{(qS)}^2\,\overline{(KL)}^2 - \overline{(q^2)}\,\overline{(KS)}^2\,\overline{(L^2)} - \\
& - \overline{(q^2)}\,\overline{(KL)}^2\,\overline{(S^2)} + \overline{(qL)}\,\overline{(KL)}^2\,\overline{(S^2)} - \\
& - \overline{(L^2)}\,\overline{(qS)}^2\,\overline{(K^2)} - \overline{(qL)}^2\,\overline{(K^2)}\,\overline{(S^2)} - \\
& - 2\,\overline{(qL)}\,\overline{(KS)}\,\overline{(qS)}\,\overline{(KL)} + \overline{(KS)}^2\,\overline{(qL)}^2 + \\
& + \overline{(q^2)}\,\overline{(K^2)}\,\overline{(S^2)}\,\overline{(L^2)} + 2\,\overline{(qS)}\,\overline{(qK)}\,\overline{(KS)}\,\overline{(L^2)} + \\
& + 2\,\overline{(SL)}\,\overline{(KL)}\,\overline{(q^2)}\,\overline{(KS)} - 2\,\overline{(qS)}\,\overline{(qK)}\,\overline{(SL)}\,\overline{(KL)} + \\
& + 2\,\overline{(qS)}\,\overline{(K^2)}\,\overline{(SL)}\,\overline{(qL)} + \overline{(qK)}\,\overline{(KL)}\,\overline{(qL)}\,\overline{(S^2)} - \\
& - 2\,\overline{(qL)}\,\overline{(qK)}\,\overline{(KS)}\,\overline{(SL)}. \qquad (13)
\end{aligned}
$$

2) Planar waveguides. We have three generators $\hat{x}^2$, $\hat{p}^2$, $\hat{x}\hat{p} + \hat{p}\hat{x}$, but only two nontrivial universal invariants:

$$
\begin{aligned}
G_1^{(1)} = & \overline{(p^4)}\,\overline{(x^4)} + 3\,\overline{(x^2p^2)}^2 - 4\,\overline{(p^3x)}\,\overline{(px^3)} + \\
& +3\,\chi^2\,\overline{(x^2p^2)}, \qquad (14)
\end{aligned}
$$

$$
\begin{aligned}
G_0^{(1)} = & \overline{(p^4)}\,\overline{(x^4)}\,\overline{(p^2x^2)} + 2\,\overline{(p^2x^2)}\,\overline{(p^3x)}\,\overline{(xp^3)} - \\
& - \overline{(p^4)}\,\overline{(x^3p)}^2 - \overline{(x^4)}\,\overline{(p^3x)}^2 - \overline{(x^2p^2)}^3 + \\
& + \tfrac{3}{4}\chi^2\left[\overline{(p^4)}\,\overline{(x^4)} - \overline{(p^2x^2)}^2\right].
\end{aligned}
$$

In conclusion let us discuss the meaning of the averaging $\langle \ldots \rangle$. Usually it is defined in terms of the mutual coherence function ρ (an analogue of the density matrix in quantum mechanics) as follows:

$$\langle \hat{q}_\alpha \hat{q}_\beta \rangle = N^{-1} \mathrm{Tr}(\hat{\varrho}\, \hat{q}_\alpha \hat{q}_\beta) = N^{-1} \int \hat{q}_\alpha \hat{q}_\beta\, \varrho(x,x')\Big|_{x'=x} dx ,$$

$$N = \mathrm{Tr}\, \hat{\varrho} \quad . \tag{15}$$

However, other definitions are also possible, e.g. set

$$\langle\langle \hat{q}_\alpha \hat{q}_\beta \rangle\rangle = \tilde{N}^{-1} \mathrm{Tr}(\hat{\varrho}^2 \hat{q}_\alpha \hat{q}_\beta) = \tilde{N}^{-1} \int \varrho(x',x) \hat{q}_\alpha \hat{q}_\beta \varrho(x, x')\, dx\, dx' \tag{16}$$

$$\tilde{N} = \mathrm{Tr}\, \hat{\varrho}^2 ,$$

or

$$\langle\langle\langle q_\alpha q_\beta \rangle\rangle\rangle = \tilde{N}^{-1} \mathrm{Tr}(\hat{\varrho}\, \hat{q}_\alpha \hat{\varrho}\, \hat{q}_\beta) = \tilde{N}^{-1} \int \hat{q}_\alpha \varrho(x,x') \hat{q}_\beta \cdot \varrho(x',x)\, dx\, dx' \quad . \tag{17}$$

As an example we consider gaussian beam with the mutual coherence function of the form $\varrho(x_1,x_2) = A\exp(-ax_1^2 - a^* x_2^2 + 2bx_1x_2)$, $b \geqslant 0$. Then three definitions (15)-(17) lead to the following explicit expressions for covariances and universal invariants of the type (10):

$$\langle x^2 \rangle = \frac{1}{2} \, \frac{1}{a^*+a-2b} , \quad \langle p^2 \rangle = 2\lambda^2 \frac{aa^* - b^2}{a+a^* - 2b} ,$$

$$\langle xp \rangle = \frac{i\lambda}{2} \, \frac{a - a^*}{a+a^* - 2b} , \qquad \Delta_1 = \frac{\lambda^2}{4} \, \frac{a+a^* + 2b}{a+a^* - 2b} .$$

$$\langle\langle x^2 \rangle\rangle = \frac{a + a^*}{2((a^* + a)^2 - 4b^2)} , \quad \langle\langle p^2 \rangle\rangle = \frac{2\lambda^2 (a+a^*)(aa^* - b^2)}{(a+a^*)^2 - 4b^2} ,$$

$$\langle\langle xp \rangle\rangle = \frac{i\lambda (a^2 - a^{*2})}{2((a+a^*)^2 - 4b^2)} , \quad \Delta_2 = \frac{\lambda^2}{4} \, \frac{(a+a^*)^2}{(a^* + a)^2 - 4b^2} ,$$

$$\langle\!\langle x^2\rangle\!\rangle = \frac{b}{(a+a^*)^2-4b^2}, \quad \langle\!\langle p^2\rangle\!\rangle = 4\lambda^2 b\,\frac{aa^* - b^2}{(a+a^*)^2-4b^2},$$

$$\langle\!\langle xp\rangle\!\rangle = \frac{i\lambda b(a-a^*)}{(a+a^*)^2-4b^2}, \quad \Delta_3 = \frac{\lambda^2 b^2}{(a+a^*)^2-4b^2}.$$

All the three forms of the invariants (10) are equivalent in the case under study to the relation $b/(a+a^*) = \text{const}$. It means that for any gaussian beam propagating in an arbitrary quadratic medium the ratio of the correlation radius to the width of the beam remains constant.

References

1. Leontovich,M.A., Fock,V.A.(1946). ZhETPh 16, 557,

2. Marcuse,D. (1972). Light Transmisstion Optics. Van Nostrand Reinhold, N.Y.

3. Dodonov,V.V., Man'ko,V.I. (1985). In: Group Theoretical Methods in Physics, Markov,M.A. et al. (Eds.). Harwood Academic Publishers, Chur-London-Paris-N.Y., 1, p. 591.

CORRELATED COHERENT STATES AND WAVE PROPAGATION THROUGH OPTICAL ACTIVE QUADRATIC-INDEX MEDIA

S.G. Krivoshlykov, N.I. Petrov, I.N. Sisakyan
Institute for General Physics, the USSR Academy of Sciences, Moscow

1. Introduction. The choice of the representation (the complete set of functions) is very important for describing a field in many problems of wave propagation. A representation proved to be convenient here depends on excitation conditions and waveguide properties. It was shown [1] that the coherent states (CS) representation and the integrals of motion method turn to be extremely convenient for investigating Gaussian beam propagation in quadratic-index media. In [2] the correlated coherent states (CCS) were used to describe arbitrary Gaussian beam propagations in such media. The coherent state representation turns to be also convenient for description of nonparaxial propagation of both coherent and partially coherent radiation in the gradient medium [3,4]. Here we show that CCS turn to be also more convenient representation for investigating wave propagation in optic active quadratic media. CCS in such media describe a Gaussian wave packets with constant parameters during beam propagation. These states permit the concept of geometric ray width and wave front curvature to be introduced by natural manner and the relationship between wave and ray optics description to be clearly visualized for quadratic media with loss or gain.

2. Arbitrary Gaussian beams, CCS, rays and modes in the quadratic media with loss or gain. It is known (see, for example, ref.5) that in the weakly inhomogeneous medium

(i.e., such that $\Delta n/(2\pi n) \ll 1$ for distances of the order of λ, where n is the refractive index of the medium and λ is the wavelength characteristic of tree space) the Maxwell equations for the monochromatic electric field component in the Fock-Leontovith paraxial approximation can be reduced to the equivalent Schrodinger equation. Thus, if the z axis of a rectangular coordinate system $x_1 x_2 z$ is chosen in the direction of light propagation, we get the reduced field

$$\Psi(x_1,x_2,z)=n_0^{\frac{1}{2}} E(x_1,x_2,z)\exp\{-ik\int_0^z n_0 dz\},$$

obeying the equation

$$\frac{i}{k}\frac{\partial\Psi}{\partial\xi} = -\frac{1}{2k^2}\left(\frac{\partial^2\Psi}{\partial x_1^2}+\frac{\partial^2\Psi}{\partial x_2^2}\right)+\frac{1}{2}(n_0^2-n^2)\Psi = H\Psi, \quad (1)$$

where $E(x_1,x_2,z)$ is the monochromatic electric field component, $n_0 = n(0,0,z)$, $k=\frac{2\pi}{\lambda}$, $\xi=\int_0^z\frac{dz}{n_0}$. Here the Hamiltonian is non-Hermitian since the refractive index is complex value.

Investigate now a propagation of the arbitrary beams through the longitudinally inhomogeneous quadratic medium with loss or gain. Consider for simplicity a two-dimensional parabolic index medium

$$n^2(x,\xi)=n_0^2-\omega^2(\xi)x^2, \quad n_0=n_{0r}-in_{0i}, \quad (2)$$

where $\omega(\xi)=\omega_r(\xi)-i\omega_i(\xi)$ is the complex gradient parameter, which describes the transverse parabolic distribution. It was shown in [8] that wave packets with constant width and radius of the wave front curvature can propagate for $\omega(\xi)=const$ in the optic active medium (2). It turns out that such wave packets correspond to CCS which was introduced in [9]. Since the trajectories of considered wave packets coincide with those whose radiation energy propagates, then such beams can be considered as a rays in the quadratic optic active media.

CCS which are eigenstates of the operator $\hat{a}(\rho,\chi,\varphi)= = e^{i\varphi}\left(\frac{k}{2\rho\cos\chi}\right)^{1/2}(\hat{x}+i\rho e^{i\chi}\hat{p})$

$\hat{a}(\rho,\chi,\varphi)\Psi_\alpha(x)=\alpha\Psi_\alpha(x)$

in the coordinate representation have the form [2]:

$$\Psi_\alpha(x)=\left(\frac{k\cos\chi}{\pi\rho}\right)^{1/4}\exp\left\{e^{-i\chi}\left(x\sqrt{\frac{k}{2\rho}}-\sqrt{\cos\chi}\,e^{-i\varphi}\alpha\right)^2+\frac{\alpha^2}{2}e^{-2i\varphi}-\frac{|\alpha|^2}{2}\right\}, \quad (3)$$

where

$$\rho>0, -\pi/2<\chi<\pi/2, 0<\varphi<2\pi, \alpha=e^{i\varphi}\left(\frac{k}{2\rho\cos\chi}\right)^{1/2}\left(\langle\hat{x}\rangle+i\rho e^{i\chi}\langle\hat{p}\rangle\right)$$

Parameters ρ, χ, φ of the CCS which are adjusted with the parameters ω_z and ω_i of medium (2) with $\omega(\xi)=const$ are connected with these parameters by formulas:

$$\rho=\frac{1}{|\omega|}=\frac{1}{\sqrt{\omega_z^2+\omega_i^2}}, \quad \chi=\mathrm{arctg}\,\frac{\omega_i}{\omega_z}, \quad \varphi=-\frac{1}{2}\mathrm{arctg}\,\frac{\omega_i}{\omega_z}. \quad (4)$$

Average values of coordinates operators in representation (3) give the ray trajectories. For these states operators $\hat{x}$ and $\hat{p}$ obey the uncertainty relation (equality) in an arbitrary plane $z=const$:

$$\langle(\Delta x)^2\rangle_\alpha\langle(\Delta p)^2\rangle_\alpha\cos^2\chi=\frac{1}{4k^2}. \quad (5)$$

In this sense these states remain maximum classical along all longitudinal coordinate z and are similar to the usual coherent states in the medium without loss or gain. Therefore, it is the most convenient representation for describing Gaussian beams propagation through the quadratic-index media with loss or gain.

CCS $\Psi_\alpha(x,\xi)$ are generating functions of the new class of modes

$$\Psi_n(x,\xi) = \left(\frac{k\cos\chi}{\pi\mu}\right)^{\frac{1}{4}} \frac{(e^{-i\chi})^{\frac{n}{2}}}{(2^n n!)^{1/2}} e^{-i(n+\frac{1}{2})\omega\xi} \cdot$$

$$\cdot \exp\left(-\frac{k\omega}{2}x^2\right) \mathcal{H}_n\left(\sqrt{\frac{k\cos\chi}{\mu}}\, x\right). \tag{6}$$

These modes are described by the Gauss-Hermite functions with spherical wavefronts.

3. Operator method for calculation of the paraxial Gaussian beams parameters in the quadratic-index medium with loss or gain. The existence of a dynamic symmetry group whose generators create rays and modes in medium permits to study all the dynamics of the system looking at the operators of this group. This approach permits the parameters of the arbitrary Gaussian beams in the quadratic media with loss or gain to be calculated using an algebraic procedure only, i.e. without using evident form of a field and without calculating the correspondings integrals. Trajectory, slope, width, ray-slope dispersion and the complex radius of curvature can be obtained calculating the corresponding average values of coordinate and position operators [8]. As an example we apply this method to quadratic Hamiltonian corresponding to focusing or defocusing media:

$$n^2(x,\xi) = n_0^2 \mp \omega^2(\xi)x^2. \tag{7}$$

Let us consider the case of excitation at the plane $z = 0$ by a Gaussian beam with arbitrary width and radius of wavefront curvature, which is described by CCS [9] and is an eigenfunction of $\hat{a}(\mu, \chi, \varphi)$. The wavefront curvature z and the beam radius w_0 are connected with the parameters μ, χ, φ as follows:

$$z = \frac{\mu}{\sin\chi},\ w_0^2 = \frac{\mu}{k\cos\chi}. \tag{8}$$

The evolutions of the operators $\hat{a}(\mu,\chi,\varphi)$ and $\hat{a}^{+}(\mu,\chi,\varphi)$ in the paraxial approximation may be described by equations [12]:

$$\hat{A} = \xi(\xi)\hat{a}_{o} + \eta(\xi)\hat{a}_{o}^{+};$$

$$\hat{A}_{1} = \eta_{1}(\xi)\hat{a}_{o} + \xi_{1}(\xi)\hat{a}_{o}^{+}, \tag{9}$$

where the functions ξ, η, η_1, ξ_1 satisfy the conditions:

$$\ddot{\xi} \pm \omega^{2}(\xi)\xi = 0,\ \xi(0) = 1;\ \ddot{\xi}_{1} \pm \omega^{2}(\xi)\xi_{1} = 0,\ \xi_{1}(0) = 1;$$

$$\ddot{\eta} \pm \omega^{2}(\xi)\eta = 0,\ \eta(0) = 0;\ \ddot{\eta}_{1} \pm \omega^{2}(\xi)\eta_{1} = 0,\ \eta_{1}(0) = 0. \tag{10}$$

Calculating the average values describing the beam parameters in the medium we get

$$x_{\alpha}(\xi) = \sqrt{\frac{\mu}{2k\cos\chi}}\left\{\frac{\eta e^{-i\chi/2}\alpha^{*}}{|\xi_{1}|^{2}-|\eta|^{2}} + \frac{\xi_{1}^{*}e^{-i\frac{\chi}{2}}\alpha}{|\xi_{1}|^{2}-|\eta|^{2}} + \text{c.c.}\right\} \tag{11a}$$

$$p_{\alpha}(\xi) = -i\sqrt{\frac{1}{2\mu k\cos\chi}}\left\{\frac{\eta e^{i\frac{\chi}{2}}\alpha^{*}}{|\xi_{1}|^{2}-|\eta|^{2}} + \frac{\xi_{1}^{*}e^{i\frac{\chi}{2}}\alpha}{|\xi_{1}|^{2}-|\eta|^{2}} - \text{c.c.}\right\} \tag{11b}$$

$$\Delta x_{\alpha}^{2}(\xi) = \frac{\mu}{2k\cos\chi} + \frac{\mu}{2k\cos\chi}\left\{\frac{\xi_{1}^{*}\eta e^{-i\chi}}{|\xi_{1}|^{2}-|\eta|^{2}} + \frac{\xi_{1}\eta^{*}e^{i\chi}}{|\xi_{1}|^{2}-|\eta|^{2}}\right\} + \frac{\mu}{k\cos\chi}\frac{|\eta|^{2}}{|\xi_{1}|^{2}-|\eta|^{2}} \tag{11c}$$

$$\Delta p_{\alpha}^{2}(\xi) = \frac{1}{2\mu k\cos\chi}\left\{1 - \frac{\xi_{1}^{*}\eta e^{i\chi}}{|\xi_{1}|^{2}-|\eta|^{2}} - \frac{\xi_{1}\eta^{*}e^{-i\chi}}{|\xi_{1}|^{2}-|\eta|^{2}} + \frac{2|\eta|^{2}}{|\xi_{1}|^{2}-|\eta|^{2}}\right\} \tag{11d}$$

$$\frac{1}{q_{\alpha}} = \frac{-i}{2k\Delta x_{\alpha}^{2}\cos\chi}\left\{\left[\frac{\xi_{1}^{*}\eta}{|\xi_{1}|^{2}-|\eta|^{2}} + \frac{|\eta|^{2}e^{i\chi}}{|\xi_{1}|^{2}-|\eta|^{2}} - \text{c.c.}\right] - e^{-i\chi}\right\} \tag{11e}$$

Thus the operator method permit evident expressions for the parameters of arbitrary Gaussian beam in the focusing and defocusing media to be obtained in the most general case.

One can see from eq.(11) that the width, ray-slope dispersion and the complex radius of the beam curvature are the same for all rays α . As is shown in [13] the beam parameter behaviour when the gradient parameter in (2) satisfies $\omega(\pm\infty)=\omega_\pm$ can be expressed via ε such that $\ddot{\varepsilon}+\omega^2(\xi)\varepsilon=0$. The solution of this equation formally coincides with that of the Schrodinger equation and describes a wave transmission through the barrier with the complex potential. Hence the beam parameter behaviour after the barrier may be completely determined by the parameters Γ and φ , which determine the quantum mechanical reflection coefficient off the barrier $\omega^2(\xi)$ and phase of the wave, respectively.

For example, a beam width after the barrier is given by (11c)

$$\Delta x_\alpha^2=\frac{1}{2k\omega_2}\,\frac{e^{-2q_i z}+\rho e^{2q_i z}+2\sqrt{\rho}\cos(2q_2 z+\varphi)}{e^{-2q_i z}-\rho e^{2q_i z}+2\frac{\omega_i}{\omega_2}\sqrt{\rho}\sin(2q_2 z+\varphi)}, \qquad (12)$$

where $\rho=|\Gamma|^2$, $q_i=\mathrm{Im}\,\omega/n_0$, $q_2=\mathrm{Re}\,\omega/n_c$.

The complex reflection coefficient Γ for some barriers $\omega(\xi)$ can be calculated exactly. For example, for the barrier $\omega(\xi)=\omega_-$, when $\xi<0$ and $\omega(\xi)=\omega_+$ when $\xi\geqslant 0$ (two media with different gradient parameters) $\Gamma=(\omega_+-\omega_-)/(\omega_++\omega_-)$.

It is obvious from (12) that a beam width is constant $\Delta x_\alpha^2=1/2k\omega_2$ as $z\to\infty$ for all values of parameter ρ and for all excitation conditions. For $\rho\neq 0$ the distance at which a beam width takes a constant value is determined by q_i . However for $\rho=0$ such a picture takes place immediately after the barrier, i.e. a beam turns to be adjusted with the medium. This phenomenon may be used e.g. to achieve matching when waveguides with different gradient parameters ω_- and ω_+ are to be connected.

For this purpose we can put between them a section of waveguide with the gradient parameter ω_0 and the length L, the values of which are found from the condition $\varrho = 0$.

The parameter ϱ for such barrier is given by

$$\varrho = \left| \frac{(\omega_- - \omega_0)(\omega_0 + \omega_+)e^{-2i\omega_0 L} + (\omega_- + \omega_0)(\omega_0 - \omega_+)e^{2i\omega_0 L}}{(\omega_- + \omega_0)(\omega_0 + \omega_+)e^{-2i\omega_0 L} + (\omega_- - \omega_0)(\omega_0 - \omega_+)e^{2i\omega_0 L}} \right|^2 \qquad (13)$$

Note that the developed algebraic approach proves also effective for investigating nonparaxial wave propagation in quadratic-index media with loss or gain [8].

References

1. Krivoshlykov, S.G., Sisakyan, I.N. (1980). Quantum Electron., 7, 553. (Russian).
2. Krivoshlykov, S.G., Kurmyshev, E.U., Sisakyan, I.N. (1983). Preprint PhIAN N 166.
3. Krivoshlykov, S.G., Sisakyan, I.N. (1983). Quantum Electron., 10, 735.
4. Krivoshlykov, S.G., Petrov, N.I., Sisakyan, I.N. (1985). Preprint IOFAN, N10.
5. Marcuse, D. (1972). Light Transmission Optics. Van Nostrand Reinhold Company, N.Y.
6. Leontovich, M.A., Fock, V.A. (1946). Zh. El. Ph., 16, 557.
7. Arnaud, I.A. (1969). Appl. Opt., 8, 1909.
8. Krivoshlykov, S.G., Petrov, N.I., Sisakyan, I.N. (1985). Preprint IOFAN N127.
9. Dodonov, V.V., Kurmushev, E.V., Man'ko, V.I. (1980). In: Group Theoretical Methods in Physics, 1, p. 227, Nauka, Moscow. (Russian).

10. Malkin, I.A., Man'ko, V.I. (1979). Dynamical Symmetries and Coherent States of Quantum Systems, Nauka, Moscow. (Russian).

11. Nazarathy, M., Hardy, A., Shamir, I. (1982). I. Opt. Soc., Amer. 72, 356, 1398, 1409.

12. Dodonov, V.V., Man'ko, V.I. (1983). Proceedings of P.N. Lebedev Physical Institute 152, 145. (Russian).

13. Krivoshlykov, S.G., Petrov, N.I., Sisakyan, I.N. (1985). In: Group Theoretical Methods in Physics, Nauka, Moscow. (Russian).

A MATRIX METHOD FOR ANALYSIS OF OPTICAL SYSTEMS WITH ABERRATIONS AND MISALIGNMENTS

V.P. Karassiov, Z.S. Sazonova

Physical Institute of the USSR Academy of Sciences, Moscow

1. Introduction. Matrix methods were introduced in optics comparatively long ago [1,2]. Now they have been used widely and fruitfully both in research work and engineering since Kogelnik's papers [3,4]. Kogelnik's ABCD-technique provides the calculation of light propagation through composed optical systems within the linear approximation of geometric optics (GO) and the paraxial wave optics. This universality results from the known analogy between the classical and quantum mechanics [5,6] and it is intimately connected with the theory of canonical transformations and the representation theory of groups Sp (2n) for n=1,2 [6-8].

The ABCD-technique, however, is well adapted only for calculating ideal optical systems.

In real devices we have different misalignments of optical elements and aberrations. Therefore in optical design we need to go over the linear approximation of GO. There are standard perturbation procedures for calculating the aberrations within a nonlinear GO [5,6,9]. But these procedures are based on using irregular nonlinear operations and are labor consuming. Therefore during the last years many authors have been examining different possibilities to improve calculating schemes within nonlinear GO [10-14]. These developments exploit that within the paraxial GO the light propagation is described by a symplectic mapping M on the four (or two)-dimensional phase space; in the linear approximation the mapping M reduces to ordinary matrices of the ABCD-method. Along this line especially interesting are the works by Wolf [13, 14] where he proposes a linear group-theoretical scheme for calculating aberrations of third-order for rotationally symmetric lens systems. On this way of "linearization" of calculations within GO of any order the alternative (matrix) approach was independently suggested in [15]. In this article we give a modified (algebraically closed) version of this method with application to analysis of composed optical systems with space axial contours and misalignments. We also explicitly give matrix operators for calculating second-order aberrations for optical systems with inclined axis of symmetry (ring resonators, waveguides, etc.).

2. General scheme of matrix method for analysis of ideal (aligned) systems. We consider a nonlinear quasiparaxial optics of real multielement systems, composed from a number of optical (refractive or reflective) surfaces Σ_i and homogeneous medium spaces between Σ_i and Σ_{i+1} ("in-between" spaces $\Delta\Sigma_i$).

We use the Lagrangian formulation for describing light rays by four-dimensional vector-column

$$W=(W_1,W_2,W_3,W_4)^T=(X,X',Y,Y')^T,\ X'=\frac{dx}{dz}\cdot Y'=\frac{dy}{dz}$$

where T is the transposition, the axis Z is directed along the axial ray. We associate i-th frame of reference $(X_i\ Y_i\ Z_i)$ with the surface Σ_i, where the axis z_i is directed along the axial ray, and axes Y_i, X_i are in the perpendicular plane, components of vectors in the i-th and (i+1)-th frames of reference are connected by transformations from the three-dimensional euclidean motion group E (3). In a medium space $\Delta\Sigma_i$ light rays are described in the i-th frame of reference.

We describe light rays through an ideal optical system by a nonlinear mapping M_i of ray vectors W on the i-th optical element in the form

$$W \to W_i = M_i \cdot W = \sum_{i=1}^{\infty} M_i^{(r)} \cdot W^{[r]} \quad (1)$$

where $W^{[r]}$ is an r-th rank symmetric tensor of the dimension $d(r) = 1/6\,(r+1)(r+2)(r+3)$, $(W^{[1]} = W, W^{[2]} = (W_1^2, W_1 W_2, \ldots, W_4^2), \ldots)$ [16]; $M_i^{(r)}$ are $4 \times d(r)$ -matrix operators which describe transformations of the r-th order GO (specifically, for r=1, $M^{(1)}$ are just matrices of the ABCD-technique). For ideal optical systems we have two kinds of operators $M_i^{(r)}$: 1) operators $M_i^{(r)} = T_i^{(r)}$ describe a light ray tracing through free space ; 2) operators $M_i^{(r)} = R_i^{(r)}$ specified by Snell's reflection (or refraction) law on the i-th surface.

Transformations of ray vectors W through a complete optical system are described by an equation similar to (1) with operators $M^{(r)}$ specified by composition rule of transformations M_i and M_{i+1} on two consequent optical elements.

Consider GO of 2nd order when (1) reduces to

$$W \to M_i \cdot W = M_i^{(1)} \cdot W + M_i^{(2)} W^{[2]} + O(W^{[3]}) \quad (1a)$$

then the composition rule has comparatively simple form:

$$(M_{i+1}^{(1)}, M_{i+1}^{(2)}) \cdot (M_i^{(1)}, M_i^{(2)}) = (M^{(1)}, M^{(2)}),$$
$$M^{(1)} = M_{i+1}^{(1)} \cdot M_i^{(1)}, \; M^{(2)} = M_{i+1}^{(1)} \cdot M_i^{(2)} + M_{i+1}^{(2)} \cdot D^{[2]}(M_i^{(1)}), \quad (2)$$

where $D^{[2]}(M_i^{(1)})$ are matrices of transformations of symmetric tensors of the 2nd rank determined by group-theoretical methods [16]. Notice some similarity of the composition rule (2) with the structure of the composition for semidirect product of groups. Similarly, we can obtain generalizations of the rule (2) for transformations within GO of arbitrary order. Explicit expressions for $M^{(r)}$ include $M_\alpha^{(k)}$ and $D_\alpha^{[k]}$, $\alpha = i$, $i+1$, with $1 \leqslant k \leqslant r$, and some group-theoretical quantities: one-line Young symmetrizers or Clebsch-Gordan coefficients of the group $G' = \{M^{(1)}\} \subset SL(4, R)$.

But in the case of axially- symmetric optical systems when $M^{(2)} = 0$ we obtain a composition rule for the GO of the 3rd order from (2) by substituting $D^{[2]} \to D^{[3]}$, $M^{(2)} \to M^{(3)}$. Now derive explicit expressions for operators $M^{(2)} = T^{(2)}$ and $M^{(2)} = R^{(2)}$.

3. Matrix operators $M^{(2)} = T^{(2)}$ and $R^{(2)}$ [15].

3.1. Determining $M^{(2)} = T^{(2)}$. Suppose that a light ray traces from the surface Σ_i to the (i+1)-th spherical surface Σ_{i+1}, defined in the i-th frame of reference by

$$z_i = d_i - R_{i+1} \cdot \cos \varphi_i + \mathrm{sgn}\, R_{i+1} \left[R_{i+1}^2 - y_i^2 - (x_i + R_i \cdot \sin \varphi_i)^2 \right]^{1/2} \qquad (3)$$

where d_i is the distance between Σ_i and Σ_{i+1} along the optical axis, φ_i the incidence angle of the axial ray on the sphere Σ_{i+1}; $\mathrm{sgn}\, R_{i+1} = 1$ if the curvature centre of Σ_{i+1} is to the left of Σ_{i+1} and -1 otherwise.

The components $W_{i\alpha}$ $(\alpha = 1, 2 \ldots 4)$ of the ray vector on the input of the surface Σ_{i+1} are defined in terms components W_α of the initial vector W by the formulas

$$\begin{aligned} W_{i1} &= W_1 + W_2 \cdot z(W); \quad W_{i2} = W_2 \\ W_{i3} &= W_3 + W_4 \cdot z(W); \quad W_{i4} = W_4 \end{aligned} \qquad (4)$$

where the function $z(W)$ is a common solution of eqs. (3), (5) (with $W_{i1} = X_1$, $W_{i3} = Y_1$) and the equation $Z_i = z + Z_1$ (Z_1 parametrises ray's position on Σ_i). Hence we obtain explicit expressions for $T^{(2)}$ by expanding the function $z(W)$ in a power series in the variables W_α. In particular, we have the following expressions for the operators $T^{(1)} \in Sp(4, R)$ and $T^{(2)}$: (5)

$$T^{(1)} = \begin{pmatrix} 1 & \ell & 0 & 0 \\ 0 & 1 & 0 & 0 \\ 0 & 0 & 1 & \ell \\ 0 & 0 & 0 & 1 \end{pmatrix}; \qquad T^{(2)} = -\tan\varphi \begin{pmatrix} 0&1&0&0&\ell&0&0&0&0&0 \\ 0&0&0&0&0&0&0&0&0&0 \\ 0&0&0&1&0&0&\ell&0&0&0 \\ 0&0&0&0&0&0&0&0&0&0 \end{pmatrix}$$

3.2 Determining $M^{(2)} = R^{(2)}$. Consider the ray's reflection on the spherical surface Σ_{i+1} defined by eq.(3). We define components of the incident ray vector in the i-th frame of reference and those of the reflected ray in (i+1)-th frame of reference.

Then the "space" components (W_1, W_3) of the reflected and the incident rays are connected by

$$W_{i+1,1} = U_{11} \cdot W_{i1} + U_{12} \cdot W_{i2} + U_{13} \cdot \rho(W_{i1}; W_{i3}) \qquad (6)$$

$$W_{i+1,3} = U_{21} \cdot W_{i1} + U_{22} \cdot W_{i2} + U_{23} \cdot \rho(W_{i1}; W_{i3})$$

where $U_{\alpha\beta}$ are elements of a three-dimensional orthogonal matrix U, $W_{i1} = X_i$ $W_{i3} = Y_i$, and $\rho(W_{i1}, W_{i3}) = (Z - d_i)$ with Z_i defined by (3). The "angle" components $W_{i+1,2}$, $W_{i+1,4}$ of the reflected ray vector are given by

$$W_{i+1,2} = \frac{k_{i+1,1}}{k_{i+1,3}}; \quad W_{i+1,4} = \frac{k_{i+1,2}}{k_{i+1,3}} \qquad (7)$$

where

$$k_{i+1,\alpha} = \sum_{\beta=1}^{3} U_{\alpha\beta} \cdot k_{i\,refl,\beta}, \quad \alpha = 1, 2, 3 \qquad (8)$$

Here components $k_{i\,refl,\beta}$ are determined from the reflection law given in the frame of reference by

$$(\vec{K}_{i\,refl} - \vec{K}_i) \times \vec{N}^{(i+1)} = 0 \qquad (9)$$

where $\vec{k}_{i\,refl} = (k_{i\,refl,\beta})$, $\vec{k}_i = (k_{i\alpha}) = (1 + W_{i2}^2 + W_{i4}^2)^{-1/2} \cdot (W_{i2}, W_{i4}, 1)$ is the unit vector along the incidental ray, $N^{(i+1)} = (N_\alpha^{(i+1)})$ is that along the normal to the surface Σ_{i+1},

$$(N_\alpha^{(i+1)}) = R_{i+1}^{-1} \cdot \{W_{i+1} + R_{i+1} \cdot \sin\varphi_i; W_{i3}; z_i - d_i + R_{i+1} \cdot \cos\varphi_i\}.$$

The matrix U in (6), (8) is the product $U = U'' \cdot U'$ of the two matrices,

$$U' = \begin{pmatrix} -\cos 2\varphi_i & 0 & \sin 2\varphi_i \\ 0 & 1 & 0 \\ -\sin 2\varphi_i & 0 & -\cos 2\varphi_i \end{pmatrix} \qquad (10)$$

(determines the position of the axial ray in the (i+1)-th frame of reference; the rotation in (xz)-plane by the angle $\pi - 2\varphi_i$) and

$$U'' = \begin{pmatrix} v_{11} & v_{12} & 0 \\ v_{21} & v_{22} & 0 \\ 0 & 0 & 1 \end{pmatrix} \qquad (11)$$

(fixes the (i+1)-th frame of reference on the surface Σ_{i+1} ; arbitrary orthogonal transformation in the (xy)-plane).

Combining (6)-(11) we get transformation of ray vectors W by reflection on the sphere (3). (We may similarly get transformation of vectors W by a refractive surface). Expanding the result in power series in $W_{i\alpha}$ we get explicit expressions for $R^{(i)}$. Specifically, for $R^{(1)}$ and $R^{(2)}$ we have the following expressions:

$$R^{(1)}=\begin{pmatrix} -v_{11} & 0 & v_{12} & 0 \\ av_{11} & -v_{11} & -b\cdot v_{12} & v_{12} \\ -v_{21} & 0 & v_{22} & 0 \\ av_{21} & -v_{21} & -bv_{22} & v_{22} \end{pmatrix} \tag{12a}$$

where $a^{-1}=\frac{R_{i+1}}{2}\cdot\cos\varphi_i$; $b^{-1}=\frac{R_{i+1}}{2}\cdot\cos^{-1}\varphi_i$, $R^{(1)}\in SL(4,R)$ and $R^{(1)}\in Sp(4,R)$ if $U''=I$ (the unit matrix);

$$R^{(2)}=\frac{\tan\varphi_i}{4}\begin{pmatrix} -2av_{11} & 0 & 0 & 0 & 0 & 0 & 0 & -2v_{11}\cdot b & 0 & 0 \\ v_{11}a^{-1} & 0 & 2v_{12}\cdot ab & 0 & 0 & -4v_{12}b & 0 & v_{11}\cdot(3+\tan^2\varphi_i)b^{-1} & 4v_{12}a & 0 \\ -2av_{21} & 0 & 0 & 0 & 0 & 0 & 0 & -2b\cdot v_{21} & 0 & 0 \\ v_{21}a^{-1} & 0 & 2v_{22}ab & 0 & 0 & -4v_{22}b & 0 & v_{21}\cdot(3+\tan^2\varphi_i)b^{-1} & -4v_{22}b & 0 \end{pmatrix} \tag{12b}$$

The explicit expressions for matrices $R^{(2)}$ and $T^{(2)}$ of the higher order ($r\leqslant 5$) are given in [15].

Thus we have formulated a simple algebraically closed matrix method for calculating and designing composed optical systems with aberrations. Now we briefly discuss its modification in applications to optical systems with misalignments.

4. <u>Optics of misaligned systems</u>. In real devices we have different misalignments of optical surfaces and other optical elements: small displacements and rotations. These misalignments give rise to a modification of light ray vector transformation and to establishing a new optical axis in the considered system, the axis obtained as a solution of an eigenfunction problem [15].

First calculate ray vector transformation in systems with small misalignments within the linear GO, and then discuss the general case. Let

$$Z_{i-1}=d_{i-1}+g^{(i)}(x_{i-1,1};x_{i-1,2});$$

$$x_{i-1,1} = x_{i-1} = W_1, \quad x_{i-1,2} = y_{i-1} = W_3 \tag{13}$$

be the equation of the i-th surface in an ideal optical system, given in the (i-1)-th frame of reference. Then an equation of this surface with misalignments has the following form up to third order terms.

$$Z_{i-1} = d_{i-1} + \Delta z_i + (N^{(i)}_{\xi} + \Delta N^{(i)}_{\xi}) \cdot x_{i-1,\xi} + \frac{1}{2} S^{(i)}_{\xi 2} x_{i-1,\xi} \cdot x_{i-1,2} \tag{14}$$

where d_{i-1} is the distance between Σ_{i-1} and Σ_i along the optical axis, $N^{(i)}_{\alpha} = \dfrac{\partial g^{(i)}}{\partial x_{i-1,\alpha}}$; $\alpha = 1, 2,$

$$(S^{(i)}_{2\xi}) = \left(\frac{\partial^2 g^{(i)}}{\partial x_{i-1,\xi} \, \partial x_{i-1,2}} \right)_{\xi, 2 = 1,2},$$

quantities Δz_i, and $\Delta N^{(i)}_{\xi}$, $\xi = 1,2$ are determined by misalignments. (Here and in what follows we use Einstein's summation rule). We give explicit expressions for Δz_i and $\Delta N^{(i)}_{\xi}$ in two cases.

1°. Displacement of the surface (13) by a constant vector $\vec{u} = (u_{i1}; u_{i2}; t_i)$

$$\begin{aligned} \Delta z_i &= t_i - N^{(i)}_{\xi} \cdot u_{i,\xi} + \frac{1}{2} S^{(i)}_{\xi 2} \cdot u_{i,\xi} \cdot u_{i,2} \\ \Delta N^{(i)}_{\xi} &= - S^{(i)}_{\xi 2} \cdot u_{i2} \, . \end{aligned} \tag{15}$$

2°. Rotation of the surface (13) around a fixed point $\{x^{\circ}_{i-1,\alpha}\}_{\alpha = 1,2,3}$ described by infinitesimal orthogonal matrix $V = (V_{\alpha\beta})$:

$$V_{\alpha\beta} = \delta_{\alpha\beta} + v_{\alpha\beta}; \; v_{\alpha\beta} = -v_{\beta\alpha}, \; |v_{\alpha\beta}| = O(1):$$

$$\begin{aligned} \Delta z_i &= v_{\xi 3} \cdot N^{(i)}_{\xi} \cdot d_{i-1} - N^{(i)}_{\alpha} \cdot v_{\alpha\beta} \cdot x^{\circ}_{i-1,\beta} \\ \Delta N^{(i)}_{\xi} &= N^{(i)}_{\alpha} \cdot v_{\alpha\xi} + N^{(i)}_{\xi} \cdot N^{(i)}_{\zeta} \cdot v_{\zeta 3} - S^{(i)}_{2\xi} \cdot (v_{2\gamma} \cdot x^{\circ}_{i-1,\gamma} - v_{23} d_{i-1}). \end{aligned} \tag{16}$$

The transformation of light rays on the output of the misaligned surface (14) is unlike (1), given by

$$W_i = M^{(1)} \cdot W_{i-1} + \Delta W^{(i)} \tag{17}$$

where $M^{(1)}$ is "unperturbed" linear matrix operator (see Sect. 2, 3), and "displacement vector" $\Delta W^{(i)}$ is of the form

$$\Delta W^{(i)} = \begin{pmatrix} \Delta z_i \cdot U_{13} \\ \Delta z_i \cdot U_{23} \\ [(n_1/n_2)-U_{33}]\cdot[A^{-1}]_{1,2}\cdot \Delta N_2^{(i)} \\ [n_1/n_2-U_{33}]\cdot[A^{-1}]_{2,2}\cdot \Delta N_2^{(i)} \end{pmatrix} \tag{18}$$

where n_1 and n_2 are refractive indices of media divided by the surface (14), the matrix A is $(A_{\xi 2}) = (U_{\xi 2} + U_{\xi 3} \cdot N_2^{(i)})_{\xi,2=1,2}$ with matrix elements $U_{\alpha\beta}$ defined by (10), (11). A composition of two transformations (17) is given by the product of the 5 x 5 - matrices $\begin{pmatrix} M^{(i)} & \Delta W^{(i)} \\ 0 & 1 \end{pmatrix}$. In the general case of the GO of arbitrary order we obtain the light ray transformation law by substituting $M^{(1)} \to M$ (general optical mapping that accounts aberrations) in (17) with a suitable modification (18) for $\Delta W^{(i)}$; the composition law of the two transformations is also modified. The work along this line is now in the progress.

5. Conclusion. Point out some possibilities of developments in the framework of the above approach. Our method is well-adapted for calculating composed optical systems with homogeneous medium spaces between optical surfaces. Combining it with the techniques [10, 12-14] we can obtain its generalization for inhomogeneous media. Another line of developments is further examination of the algebraic properties of our technique, particularly, its relations with the theory of representations of Sp (2n,R). It is also interesting to investigate possibilities to apply results obtained in the quasiparaxial wave optics (particularly, along the line of Kogelnik's papers [3,4]). The coherent states technique [17] appears to be useful for this analysis (cf. [18]). The authors are grateful to Prof. V.P. Bykov for helpful discussions.

References

1. Gerrard, A., Burch, J.M. (1975). Introduction to Matrix Methods in Optics John Wiley & Sons, London et al.

2. Sampson, R.A. (1913). Phil. Trans. Roy. Soc., 212,149.
3. Kogelnik, H. (1965). Appl. Opt. 4, 1562.
4. Kogelnik, H., Li, T. (1966). Appl. Opt. 5, 1550.
5. Ghatak, A.K., Thyagarajan, K. (1978). Contemporary Optics,Plenum, N.Y.
6. Guillemin, V., Sternberg, S. (1977). Geometric asymptotics. AMS, Providence.
7. Stoler, D. (1981). J. Opt. Soc. Am., 71, 334.
8. Bacry, H., Cadilhac, M. (1981). Phys. Rev. A23, 5, 2533
9. Stavroudis, O.N. (1972). The Optics of Rays, Wavefronts and caustics. Academic Press, N.Y.
10. Dragt, A.J. (1982). J. Opt. Soc. Am. 72, 372.
11. Nazarathy, M., Shamir, F. (1982). J. Opt. Soc. Am. 72, 356.
12. Steinberg, S. (1984). SIAM J. Math. Anal. 15, 108.
13. Wolf, K.B. (1984). Commun. Tecnicas IIMAS, N 366, Mexico.
14. Saad, M.N., Wolf, K.B. (1984). Preprint CINVESTAV, Mexico.
15. Sazonova, Z.S. (1982). Thesis, Moscow. (Russian).
16. Weyl, H. (1946). The Classical Groups Princeton Univ.,Princeton.
17. Malkin, I.A., Manko, V.I. (1979). Dynamical Symmetries and Coherent States of Quantum Systems. Nauka, Moscow. (Russian).
18. Krivoshlykov, S.G., Sissakian, I.N. (1980). Opt. and Quant. Electronics, 12, 463.

ON CALCULATING EINSTEIN COEFFICIENTS FOR TIME-PERIODIC SYSTEMS. INTEGRAL OF MOTION METHOD

E.V.Ivanova

S.Ordgonikidse Moscow aviation institute

When systems with time-periodic Hamiltonians are involved, representation by quasi-energy (QE) and quasi-energetic states (QES) proves to be most convenient, see the survey [1] and references therein. Quasi-energetic wave functions are solutions of the Schrödinger equation satisfying the condition

$$|\vec{n};t+T\rangle = e^{-i\varepsilon T}|\vec{n};t\rangle, \qquad (1)$$

where QE ε is determined up to an integral number of oscillation quanta. T is the period of the system. Throughout the paper we use the system of units in which $c = \hbar = 1$.

The integrals of motion method and the coherent states method prove to be effective for constructing QES for quadratic and some non-quadratic systems [2,3] . The integrals of motion method is useful also for calculating the radiation power of systems [4,5] .

In this work the Einstein coefficients are obtained for time-periodic quadratic systems. Some aspects of this problem for arbitrary time-periodic systems have been considered in [1,6] . We use here the results obtained in [7-9] .

For stationary systems the Einstein coefficient $\mathcal{A}$ is known to characterize the probability of spontaneous transition per unit time between stationary states $|\vec{n}\rangle$ and $|\vec{m}\rangle$. The Einstein coefficient $\mathcal{B}$ corresponds to the probability of induced transitions per unit time between these states. If the system obeys the Boltzmann distribution law, the connection between the coefficients $\mathcal{A}$ and $\mathcal{B}$ in thermodynamic equilibrium is defined by the relation

$$\mathcal{A} = \frac{\hbar\omega^3}{\pi^2 c^3}\,\mathcal{B}$$

or

$$\mathcal{A} = \frac{\omega^3}{\pi^2}\,\mathcal{B} \tag{2}$$

if $c = \hbar = 1$. Here ω is the frequency of the transition between the states $|\vec{n}\rangle$ and $|\vec{m}\rangle$.

As in the stationary case, for time-periodic systems we assume that the Einstein coefficient $\mathcal{A}$ characterizes the probability per unit time averaged over period for spontaneous QES transitions, and the Einstein coefficient $\mathcal{B}$ characterizes the same for induced QES transitions.

We give an account of the common scheme for calculating Einstein coefficients for time-periodic systems by the integrals of motion method and illustrate it by a non-relativistic particle with charge e, mass m in a time-dependent homogeneous magnetic field $\vec{\mathcal{H}} = \{0, 0, \mathcal{H}(t)\}$ with the vector potential

$$\vec{\mathcal{A}} = \{-\mathcal{H}y, 0, 0\}; \quad \mathcal{H}(t+T) = \mathcal{H}(t); \quad T = \frac{2\pi}{\omega} \tag{3}$$

and in a time-dependent inhomogeneous electric field with the scalar potential

$$\varphi = \frac{U(t)}{2R^2}(y^2 - z^2); \quad U(t+T) = U(t); \quad R = const, \tag{4}$$

which can be created by means of a quadrupole condenser

[8].

The Hamiltonian of a charged particle in the fields (3) and (4) can be written as

$$H = H_{X,Y} + H_Z\ ;\ P_x = -i\frac{\partial}{\partial x}\ ;\ \omega_o(t) = \frac{e}{m}\mathcal{H}(t);$$

$$H_{X,Y} = \frac{P_x^2}{2m} + \frac{P_y^2}{2m} + \omega_o(t)\, y P_x + \left[\frac{m}{2}\omega_o^2(t) + \frac{e}{2R^2}U(t)\right]y^2; \tag{5}$$

$$H_Z = \frac{P_z^2}{2m} - \frac{e}{2R^2}U(t)\, z^2.$$

It is easy to see that the operators

$$A = \frac{i}{\sqrt{2m}}(\sigma P_x + \varepsilon P_y - m\dot{\varepsilon} y); \qquad B = \frac{i}{\sqrt{2m}}(\mu P_x + S P_y - imx - m\dot{s} y) \tag{6}$$

are integrals of motion, i.e. they commute with $H_{X,Y} - i\,\partial/\partial t$. In the relations (6)

$$\sigma = \int_{t_o}^{t} \omega_o(\tau)\varepsilon(\tau)d\tau; \qquad S = \frac{1}{2}(\sigma\varepsilon^* - \sigma^*\varepsilon); \qquad \mu = 1 + \int_{t_o}^{t}(\omega_o s + i)d\tau, \tag{7}$$

and $\varepsilon(t)$ is the solution of the equation

$$\ddot{\varepsilon} + \frac{\Omega^2(t)}{4}\varepsilon = 0; \quad \frac{\Omega^2(t)}{4} = \omega_o^2(t) + \frac{e}{mR^2}U(t), \tag{8}$$

for which the condition

$$\varepsilon(t)=|\varepsilon(t)|\exp\left[i\int_0^t|\varepsilon(\tau)|^{-2}d\tau\right] \tag{9}$$

holds.

The operators A and B satisfy Bose commutation relations for the lowering and raising operators:

$$[A,A^+]=[B,B^+]=1;\ [A,B]=[A,B^+]=0. \tag{10}$$

Note that the integrals of motion for a nonrelativistic charge in a time -dependent magnetic field of type (3) and in a time-dependent electric field of the oscillatory type close to type (4) have been constructed by I.A.Malkin and V.I.Man'ko [2] .

It is easy to obtain the expression for coherent states of a charge in the fields (3), (4) which are the eigenstates of the operators A and B:

$$\begin{gathered} A|\alpha,\beta;t\rangle=\alpha|\alpha,\beta;t\rangle;\ B|\alpha,\beta;t\rangle=\beta|\alpha,\beta;t\rangle; \\ i\frac{\partial}{\partial t}|\alpha,\beta;t\rangle=H_{X,Y}|\alpha,\beta;t\rangle; \\ \pi^{-2}\int d^2\alpha\, d^2\beta\,|\alpha,\beta;t\rangle\langle\alpha,\beta;t|=1, \end{gathered} \tag{11}$$

where α and β are arbitrary complex numbers. We can obtain also the expression for Fock states of this system, which are eigenstates of the operators A^+A and B^+B:

$$\begin{gathered} A^+A|n_1,n_2;t\rangle=n_1|n_1,n_2;t\rangle; \\ B^+B|n_1,n_2;t\rangle=n_2|n_1,n_2;t\rangle, \\ i\frac{\partial}{\partial t}|n_1,n_2;t\rangle=H_{X,Y}|n_1,n_2;t\rangle; \end{gathered} \tag{12}$$

$$\langle m_1, m_2; t | n_1, n_2; t\rangle = \delta_{m_1 n_1} \delta_{m_2 n_2};$$

$$n_{1,2}, m_{1,2} = 0, 1, 2, \ldots \; .$$

We confine ourselves to the case of stable classical charge motion. Note that the quasi-energy spectrum is discrete in this case. Then, according to Flocke's theorem the solution of equation (8) is

$$\varepsilon(t+T) = e^{i\Omega_1 T} \varepsilon(t); \; \Omega_1 = \frac{1}{T} \int_0^T |\varepsilon(\tau)|^{-2} d\tau . \tag{13}$$

For the operator A in (6) using (13), we find

$$A(T) = e^{i\Omega_1 T} A(0). \tag{14}$$

From this we have

$$|\alpha; T\rangle = e^{-i\frac{\Omega_2}{2}T} |e^{-i\Omega_1 T} \alpha; 0\rangle . \tag{15}$$

Since the coherent states $|\alpha, \beta; t\rangle$ are the generating functions for the Fock states $|n_1, n_2; t\rangle$ [2], the eigenstates of the operator A^+A are QES [3] , i.e. they satisfy

$$|n_1; t+T\rangle = \exp[-i\Omega_1 T (n_1 + \tfrac{1}{2})] |n_1; t\rangle;$$

$$\mathcal{E}_{n_1} = \Omega_1 (n_1 + \tfrac{1}{2}). \tag{16}$$

Now let us express the Hamiltonian of the interaction of the system (3),(4) with radiation field in terms of integrals of motion (6). In the dipole approximation the interaction Hamiltonian is

$$H_{int} = -\frac{e}{\sqrt{2m}} \sum_{\lambda,\varrho} \sqrt{\frac{2\pi}{V\omega_\lambda}} (C_{\lambda\varrho} + C^+_{\lambda\varrho}) \times$$
$$\times (\sigma_1 A + \sigma_1^* A^+ + \sigma_2 B + \sigma_2^* B^+), \tag{17}$$

where $C_{\lambda\varrho}$ ($C^+_{\lambda\varrho}$) are annihilation(creation) Bose operators of a photon with frequency ω_λ, wave vector K_λ and polarization vector $\vec{e}_{\lambda\varrho}$ and V is the normalized volume. The variables $\sigma_{1,2}$ are determined as

$$\sigma_1 = \omega_o \varepsilon^* (\vec{e}_{\lambda\varrho})_x + \dot{\varepsilon}^* (\vec{e}_{\lambda\varrho})_Y ;$$
$$\sigma_2 = (\omega_o s^* - i)(\vec{e}_{\lambda\varrho})_x + \dot{s}^* (\vec{e}_{\lambda\varrho})_y ; \tag{18}$$
$$(\vec{e}_{\lambda\varrho})_z = 0.$$

Taking into account the relations

$$A|n_1,n_2;t\rangle = \sqrt{n_1}\,|n_1-1,n_2;t\rangle;$$
$$B|n_1,n_2;t\rangle = \sqrt{n_2}\,|n_1,n_2-1;t\rangle;$$
$$A^+|n_1,n_2;t\rangle = \sqrt{n_1+1}\,|n_1+1,n_2;t\rangle; \tag{19}$$
$$B^+|n_1,n_2;t\rangle = \sqrt{n_2+1}\,|n_1,n_2+1;t\rangle,$$

it is easy to obtain [8] the expressions for the probabilities per unit time averaged over period for spontaneous and induced QES transitions.

From this we obtain the Einstein coefficients $\mathcal{A}$ and $\mathcal{B}$ for the transition frequency $\omega_\lambda = \omega\ell - \Omega_1 > 0$:

$$\mathcal{A} = \frac{e^2}{3m}(\omega\ell - \Omega_1)^2 (n_1+1)\,|\tau_1(\ell)|^2; \tag{20}$$

$$\mathcal{B} = \frac{\pi^2 e^2}{3m} \frac{(n_1+1)}{\omega l - \Omega_1} |\eta_1(l)|^2;$$
$$\Omega_1 \neq \nu \frac{\omega}{2}; \quad \nu = 0, \pm 1, \pm 2, \ldots,$$

where $\eta(l)$ are the Fourier series coefficients of the periodic function

$$\eta(t) = \eta(t+T); \; \eta(t) = \sigma_1(t) e^{i\Omega_1 t}. \tag{21}$$

In the particular case

$$U(t) = U_o + U_1 \cos \omega t; \; U_o, U_1 = const, \tag{22}$$

assuming $U_1/U_o \ll 1$, $\omega = 2\Omega_2(1+\Delta)$, where $\Omega_2^2 = eU_o/mR^2$; $\Delta \to 0$, we have

$$\Omega_1 = \frac{1}{2}\sqrt{\Delta^2 - \frac{eU_o}{4mR^2}\left(\frac{U_1}{U_2}\right)^2}. \tag{23}$$

Now we find the Einstein coefficients for a time-periodic quantum system with a discrete quasi-energy spectrum whose Hamiltonian is a general quadratic form with respect to coordinates and momenta [7], i.e. we consider a quantum system of N charged particles in a time-periodic electromagnetic field with the Hamiltonian of the form

$$H(t) = \sum_{\alpha,\beta=1}^{6N} B_{\alpha\beta}(t) Q_\alpha Q_\beta + H_o(t) \equiv$$
$$\equiv \vec{Q} B(t) \vec{Q} + H_o(t); \; H(t+T) = H(t), \tag{24}$$

where $B(t+T) = B(t)$ is a $6N \times 6N$ real symmetric matrix, $H_o(t+T) = H_o(t)$ a time-periodic function and

$\vec{Q} = \left\| \begin{matrix} \vec{p} \\ \vec{q} \end{matrix} \right\|$ is the 6N-dimensional vector with $\vec{p} = (p_1, \ldots, p_{3N})$; $\vec{q} = (q_1, \ldots, q_{3N}) \equiv (\vec{q}^{\,1}, \ldots, \vec{q}^{\,b}, \ldots, \vec{q}^{\,N})$, where $\vec{q}^{\,b}$ is the position operator of the b-th charged particle and $\vec{p}^{\,b} = -i\partial/\partial\vec{q}^{\,b}$ is the particle's momentum operator.

I.A.Malkin and V.I.Man'ko got [2] 2N Hermitian integrals of motion for the systems with the Hamiltonian (24) for arbitrary $B(t)$ and $H_o(t)$

$$\vec{I}(t) = \Lambda(t)\vec{Q}. \tag{25}$$

In formula (25) the $6N \times 6N$ symplectic matrix $\Lambda(t)$ is defined so that

$$\dot{\Lambda}(t) = -2\Lambda(t)JB(t); \quad \Lambda(0) = E_{6N}, \tag{26}$$

where

$$J = \left\| \begin{matrix} 0_{3N} & -E_{3N} \\ E_{3N} & 0_{3N} \end{matrix} \right\|; \qquad B = \left\| \begin{matrix} B_1 & B_2 \\ \tilde{B}_2 & B_4 \end{matrix} \right\|. \tag{27}$$

$\sim$ stands for the transposition, and E_{3N} is the $3N \times 3N$ unit matrix.

Let us consider the classical system corresponding to the quantum system (24). The matrix of the fundamental system of solutions of equations determining the path of this classical system is the matrix $\Lambda^{-1}(t)$ (see (26)). In accordance with Flocke's theorem

$$\Lambda^{-1}(t) = U(t)\exp\left[\frac{t}{T}\ln\Lambda^{-1}(T)\right], \tag{28}$$

where $U(t + T) = U(t)$ is the periodic symplectic matrix, $U(l)$ are the coefficients of its Fourier - transform. Suppose that the stable classical system with different eigenvalues of $\Lambda^{-1}(T)$ corresponds to the quantum system (24). Then since $\Lambda(t)$ is symplectic then [7]:

$$\tilde{\Lambda}(T)\vec{f}^{(j)} = \exp(i\omega_j)\vec{f}^{(j)},$$

where (29)

$$\omega_j > 0,\ \vec{f}^{(3N+j)} = \vec{f}^{(j)*},$$
$$\vec{f}^{(j)} J \vec{f}^{(k)*} = \pm\delta_{jk},$$
$$\omega_j \neq \frac{\omega}{2}s,\quad \omega_j \pm \omega_k \neq \omega s \quad \text{for } j \neq k;$$
$$j = 1, \ldots, 3N;\quad s = 0, \pm 1, \pm 2, \ldots .$$

set

$$\mathfrak{K}_j = \operatorname{sign}\left(\vec{f}^{(j)} J \vec{f}^{(j)*}\right)\frac{\omega_j}{T};$$

$$\hat{\mathfrak{K}} = \begin{Vmatrix} \mathfrak{K}_1 & 0 & \ldots & 0 \\ 0 & \mathfrak{K}_2 & \ldots & 0 \\ \cdot & \cdot & \cdot & \cdot \\ 0 & 0 & \ldots & \mathfrak{K}_{3N} \end{Vmatrix};\quad \vec{F}^{(j)} = \begin{cases} \vec{f}^{(j)}, & \mathfrak{K}_j > 0 \\ \vec{f}^{(j)*}, & \mathfrak{K}_j < 0 \end{cases} \tag{30}$$

Then we may set

$$A_j(t) = \vec{I}(t)\vec{F}^{(j)};\ A_j^{+}(t) = \vec{I}(t)\vec{F}^{(j)*}. \tag{31}$$

Now we write the Hamiltonian of interaction of the system (24) with the radiation field in the form

$$H_{int}(t) = -e\sum_{\lambda,\varrho}\sqrt{\frac{2\pi}{V\omega_\lambda}}\,\vec{\sigma}_{\lambda\varrho}\left(c_{\lambda\varrho} + c^{+}_{\lambda\varrho}\right)\vec{V}(t), \tag{32}$$

Here e are charges of particles. We suppose that the

charges coincide. Set

$$\vec{V}(t)=\sum_{l=-\infty}^{+\infty}\Big[e^{i\omega l t}\,\varkappa(l)\,e^{-i\hat{\Omega}t}\,\vec{A}(t)\;+ \tag{33}$$

$$+e^{-i\omega l t}\,\varkappa^{*}(l)\,e^{i\hat{\Omega}t}\,\vec{A}^{+}(t)\Big],$$

where

$$\varkappa(l)=2i\big[B_1(U_2E_1^{+}-U_1F_2^{+})+B_2(U_4F^{+}-U_3F_2^{+})\big]; \tag{34}$$

$$U(l)=\begin{Vmatrix}U_1 & U_2\\ U_3 & U_4\end{Vmatrix};\; F=\begin{Vmatrix}F_1 & F_2\\ F_1^{*} & F_2^{*}\end{Vmatrix};\;\begin{Vmatrix}F_1^{1} & \dots & F_{6N}^{1}\\ \cdot & \cdots & \cdot\\ F_1^{3N} & \dots & F_{6N}^{3N}\end{Vmatrix}=\|F_1\,F_2\|.$$

Taking into account the action of $A_j(t)$ and $A_j^{+}(t)$ on QES (19), it is easy to obtain the expressions for the probabilities per unit time averaged over the period for spontaneous and induced QES transitions. Then we find the Einstein coefficients $\mathcal{A}$ and $\mathcal{B}$ for the transition frequency $\omega_\lambda = \omega l - \Omega_j > 0$ [7] :

$$\mathcal{A}=\frac{4}{3}e^{2}(\omega l-\Omega_j)(n_j+1)\left|\vec{z}^{\,(j)}(l)\right|^{2}; \tag{35}$$

$$\mathcal{B}=\frac{4}{3}\pi^{2}e^{2}\,\frac{n_j+1}{(\omega l-\Omega_j)^{2}}\left|\vec{z}^{\,(j)}(l)\right|^{2}, \tag{36}$$

where $\vec{z}^{\,(j)}(l)$ is the j-th row of the matrix $\tilde{\varkappa}(l)$.

Let us consider the case of quasi-energy continuum [9]. In this case motion of the classical system corresponding to the quantum system (24) is unstable, and therefore the eigenvalues of $\Lambda^{-1}(T)$ for the system (24) are real numbers. We write them in the form

$$\lambda_j = \exp \vartheta_j \; ; \; \lambda_{3N+j} = \exp(-\vartheta_j), \text{where} \qquad (37)$$
$$\vartheta_j > 0 \; ; \; j = 1,2,\dots,3N$$

Assume that $\vartheta_j \neq \vartheta_k$; $j \neq k$; $j,k = 1,2,\dots,3N$. The eigenvectors of $\tilde{\Lambda}(T)$ are

$$\tilde{\Lambda}(T)\vec{f}^{(j)} = \exp(\vartheta_j)\vec{f}^{(j)};$$
$$\tilde{\Lambda}(T)\vec{f}^{(3N+j)} = \exp(-\vartheta_j)\vec{f}^{(3N+j)}. \qquad (38)$$

Since $\lambda_j \cdot \lambda_{3N+j} = 1$, then $\vec{f}^{(3N+j)} I \vec{f}^{(j)} \neq 0$, i.e. $\vec{f}^{(j)}$ and $\vec{f}^{(3N+j)}$ can be normalized so that $\vec{f}^{(3N+j)} I \vec{f}^{(k)} = \delta_{jk}$.

Hence, one may set

$$X_\alpha(t) = \vec{I}(t)\vec{f}^{(\alpha)}; \; \vec{X}(t) = \theta(t)\vec{Q}; \; \alpha = 1,2,\dots, 6N, \qquad (39)$$

where

$$\theta(t) = f\Lambda(t); \; \tilde{f} = \| \tilde{\vec{f}}^{(1)}, \dots, \tilde{\vec{f}}^{(3N)}, \tilde{\vec{f}}^{(3N+1)}, \dots, \tilde{\vec{f}}^{(6N)} \|, \qquad (40)$$

Such that

$$[X_\alpha(t), X_\beta(t)] = iJ_{\alpha\beta} \, . \qquad (41)$$

We obtain the states $|\vec{k}\,;\,t\rangle$ of the system (24) with the quasi-energy continuum, which are eigenstates for the integrals of motion $\vec{X}_\alpha$:

$$X_j(t)|\vec{k};t\rangle = k_j|\vec{k};t\rangle;$$
$$j=1,2,\ldots,3N;\quad |\vec{k};t\rangle = \prod_{j=1}^{3N}|K_j;t\rangle; \tag{42}$$
$$i\frac{\partial}{\partial t}|\vec{k};t\rangle = H(t)|\vec{k};t\rangle;\ \langle\vec{k}';t|\vec{k};t\rangle = \delta(\vec{k}-\vec{k}').$$

The states $|\vec{k};t\rangle$ are generating functions for the QES $|\{\nu_j;\varepsilon_j\};t\rangle$ obtained from $|\vec{k};t\rangle$ by the Mellin transformation [3] ; here k_j , ν_j are real numbers, and ε_j is either 0 or 1 .

As in the above general scheme of calculating Einstein coefficients we find the matrix elements

$$\langle k_j';t|X_j(t)|k_j;t\rangle = k_j\delta(k_j-k_j');$$
$$\langle k_j';t|X_{3N+j}(t)|k_j;t\rangle = -i\frac{\partial}{\partial k_j}\delta(k_j-k_j'); \tag{43}$$
$$j=1,2,\ldots,3N.$$

Since the expressions $\langle k_j';t|X_\alpha(t)|k_j;t\rangle$ are generating functions for the matrix elements $\langle\{\nu_j';\varepsilon_j'\};t|X_\alpha(t)|\{\nu_j;\varepsilon_j\};t\rangle$, we find from (43) and (41) that

$$\langle\{\nu_j';\varepsilon_j'\};t|X_j(t)|\{\nu_j;\varepsilon_j\};t\rangle =$$
$$= 4\pi\delta(\nu_j-\nu_j'+i)\delta_{\varepsilon_j+\varepsilon_j';1}; \tag{44}$$
$$\langle\{\nu_j';\varepsilon_j'\};t|X_{3N+j}(t)|\{\nu_j;\varepsilon_j\};t\rangle =$$
$$= -4\pi\left(\nu_j-\frac{i}{2}\right)\delta(\nu_j-\nu_j'-i)\delta_{\varepsilon_j+\varepsilon_j',1}.$$

The obtained expressions make sense for states with complex values of quasi-energy, i.e. quasi-energetic states with an account of relaxation processes.

Expressions for the matrix elements similar to (44), i.e. containing δ-functions of complex variable, enter the interaction Hamiltonian also in the case of mixed quasi-energy spectrum. Thus, in this case, too, the quasi-energetic states with the relaxation processes should be considered.

Acknowledgements are due to V.I.Man'ko for encouragement and helpful discussions.

References

1. Zeldovich,Ya.B. (1973). Russian Phys. Surveys, 110, 139 (Russian).
2. Malkin,I.A., Man'ko,V.I. (1979). Dynamic Symmetrical and Coherent States of Quantum Systems. Nauka, Moscow (Russian).
3. Malkin,I.A., Man'ko,V.I., Schustov,A.P. (1975). The coherent states and the quasienergy spectra of arbitrary quadratic systems. Preprint Ph IAN USSR, Moscow.
4. Ivanova,E.V., Malkin,I.A., Man'ko,V.I. (1977). Int. J. Theor. Phys., 16, 503.
5. Ivanova, E.V., Malkin,I.A., Man'ko,V.I. (1977). J. Phys., 10A, L75.
6. Berson,I.Ya. (1970). Izv. AN Latv. SSR,3,3 (Russian).
7. Ivanova,E.V., Man'ko,V.I. (1983). Short notes on physics Ph IAN, 1, 13 (Russian).
8. Ivanova,E.V. (1981). Motion and radiation of a charged particle in variable magnetic field and variable electric field of quadrupole condenser. VINITI deposition N 4173 (Russian).
9. Ivanova,E.V. (1983). On radiation of time-periodic quadratic systems with a quasi-energy continuum. VINITI deposition N 5183 (Russian).
10. Ivanova,E.V. (1984). On radiation of time-periodic quadratic systems with mixed quasi-energy spectrum. VINITI deposition N 605 (Russian).

APPLICATION OF FOUR- AND FIVE-DIMENSIONAL ROTATION GROUPS TO DESCRIPTION OF APPROXIMATE SYMMETRIES OF MANY-ELECTRON ATOMS

J.M.Kaniauskas, V.C.Simonis, Z.B.Rudzikas
Institute of Physics, Vilmius, 232600, K.Pozelos 54

1. <u>Group-Theoretical Methods Describing Mixed Configurations</u>. Accounting for correlation effects in many-electron atoms is one of the fundamental problems in modern theory of atomic spectra [1]. The principal part of Coulomb interaction among electrons is usually considered with the help of the Hartree-Fock method in a single-configuration approximation, whereas the rest part of this interaction describes the correlation energy of an atom. Calculating this energy requires to abaudon a single-configuration approximation with a fixed distribution of electrons among atomic shells. Unlike some other quantum-mechanical many-body systems (solid states, atomic nuclei, etc.), atomic spectra, due to correlation effects, have no collective branches of quasi-particle excitations of the system. This fact ensures high accuracy of the central field approximation as a zero approximation for atomic systems.

However in the theory of many-electron atoms one rather often comes across a different situation when the role of one or several admixed configurations is comparable with that the principal configuration. In such cases it is necessary already in the zero approximation to abaudon the single-configuration approximation and to look for the possibilities to adopt new, more adequate models.

One of these possibilities is to use the group-theoretical methods for constructing multi-configurational wave-functions. These methods were first used in the theory of nucleus by Elliott [2] who considered mixed nucleonic configurations $(s+d)^N$ using their degeneracy in the harmonic potential field. Atomic mixed configurations of the type $(l_1+l_2)^N$ were considered from the group-theoretical point of view in [3,4]. Grounding on the fact that all the eigenfunctions of such complex of configurations form a basis of antisymmetric representation $\{1^N\}$ of the unitary group $U\,4(L_1+l_2+1)$, in these papers the problem of classifying atomic states was transformed in to problem of reducing this group with respect to non-canonical chains of subgroups conserving the total orbital L and the spin angular momenta S.

Shortcomings of this approach are difficulties arising when one grounds, with the help of some physical speculations, the reduction chain of the subgroups chosen. As the result the additional quantum numbers, adopted in these cases to classify atomic states, are not "exact" ones the energy matrix, defined in the basis of the corresponding wave-functions, is to a large extent a non-diagonal one. Moreover, the possibilities to make use of these bases are essentially restricted as of the group theory is not developed enough.

Considering the tensorial properties of the wave-functions and operators in the supplementary spaces of quasi-spin and isospin may be used as an alternative to the method of the groups of higher ranks. In the case of a shell of equivalent electrons l^N the values of the z-projection of the quasispin operator Q_z are defined by the difference between the number of particles N and the number of holes N in a shell, i.e. $Q_z = (N-N_h)/4$; whereas the electron creation and annihilation operators are components of the triple tensor of the rank $q=1/2$ in a quasispin space:

$$a^{(q\ell s)}_{\varrho m \mu} = \begin{cases} a^{(\ell s)}_{m\mu} ; & \varrho = 1/2 \\ \tilde{a}^{(\ell s)}_{m\mu} ; & \varrho = -1/2 \end{cases} .$$

The isospin formalism may be applied for the configurations of the type $n_1 l^{N_1} n_2 l^{N_2}$. Moreover, rotations in the isospin space correspond to some transformations of the radial orbitals of these configurations. The quasispin and isospin methods are based on the developed angular momentum theory,and they are described in detail in the monograph [5] . The results of the studies of approximate symmetries of complex atoms with the help of the above-mentioned methods were reported in the previous seminar [6] . Therefore, here we shall consider the approach based on these methods when the groups of four- and five-dimensional rotations are used for the description of the mixed configurations $(l_1+l_2)^N$.

Possibilities to chose various schemes of groups may be deduced from the commutation relations between the irreducible tensorial operators

$$W^{(k)}_{M}(\gamma_1,\gamma_2) = \left[a^{(\gamma_1)} \times a^{(\gamma_2)}\right]^{(k)}_{M} = \sum_{\upsilon_1 \upsilon_2} \begin{bmatrix} \gamma_1 & \gamma_2 & k \\ \upsilon_1 & \upsilon_2 & M \end{bmatrix} a^{(\gamma_1)}_{\upsilon_1} a^{(\gamma_2)}_{\upsilon_2} \quad (1.1)$$

(here and further on the abbreviations for triple quantities are used: $(\gamma) = (q\, l\, s)$ and $(k) = (k_1 k_2 k_3)$ - for ranks; $\upsilon = \varrho\, m\, \mu$ and $M = M_1 M_2 M_3$ - for projections; the algebraic expressions, containing these abbreviations, ought to be understood as multiplicative functions).

In the most general case these relations may be presented in the following form:

$$\left[W^{(k_{12})}_{m_{12}}(\gamma_1,\gamma_2),W^{(k_{34})}_{m_{34}}(\gamma_3,\gamma_4)\right]=\sum_{\varkappa}[k_{12},k_{34}]^{1/2}\begin{bmatrix}k_{12} & k_{34} & \varkappa\\ m_{12} & m_{34} & m_\varkappa\end{bmatrix}\times$$

$$\times\Big((-1)^{\gamma_2+\gamma_3+k_{12}+k_{34}}\delta(\gamma_1,\gamma_4)\begin{Bmatrix}k_{12} & k_{34} & \varkappa\\ \gamma_3 & \gamma_2 & \gamma_1\end{Bmatrix}W^{(\varkappa)}_{m_\varkappa}(\gamma_3,\gamma_2)-$$

$$-(-1)^{\gamma_3+\gamma_4+k_{34}}\delta(\gamma_2,\gamma_4)\begin{Bmatrix}k_{12} & k_{34} & \varkappa\\ \gamma_3 & \gamma_1 & \gamma_2\end{Bmatrix}W^{(\varkappa)}_{m_\varkappa}(\gamma_3,\gamma_1)+ \tag{1.2}$$

$$+(-1)^{\gamma_1+\gamma_4+k_{12}+\varkappa}\delta(\gamma_1,\gamma_3)\begin{Bmatrix}k_{12} & k_{34} & \varkappa\\ \gamma_4 & \gamma_2 & \gamma_1\end{Bmatrix}W^{(\varkappa)}_{m_\varkappa}(\gamma_2,\gamma_4)-$$

$$-(-1)^{\gamma_1+\gamma_4+\varkappa}\delta(\gamma_2,\gamma_3)\begin{Bmatrix}k_{12} & k_{34} & \varkappa\\ \gamma_4 & \gamma_1 & \gamma_3\end{Bmatrix}W^{(\varkappa)}_{m_\varkappa}(\gamma_1,\gamma_4)\Big).$$

In particular, while specifying the type of mixed configurations, this formula defines the commutation relations among generators of the corresponding group.

2. <u>The Group of Fourdimensional Rotations.</u> Let us consider the complex of configurations $(l_a+l_b)^N$ with $l_b=l_a+1$, and construct in the orbital space the following 1st rank tensor

$$A^{(1)}_{\varrho}=\alpha L^{(1)}_{\varrho}(a)+\beta L^{(1)}_{\varrho}(b)+\gamma W^{(010)}_{0\varrho 0}(a,b)+\delta W^{(110)}_{0\varrho 0}(a,b). \tag{2.1}$$

Using (1.2) it is easy to find that this operator with coefficients

$$\alpha = \frac{l_a+2}{l_a+1}; \qquad \gamma = 2i\sqrt{\frac{(2l_a+1)(2l_a+3)}{3(l_a+1)}}\cos\psi;$$
$$\beta = \frac{l_a}{l_a+1}; \qquad \delta = -2\sqrt{\frac{(2l_a+1)(2l_a+3)}{3(l_a+1)}}\sin\psi. \tag{2.2}$$

together with the operator of the angular momentum $\vec{L} = \vec{L}(a)+\vec{L}(b)$ satisfies the usual commutation relations for generators of SO(4). The angle ψ in (2.2) is an arbitrary parameter.

It follows from (2.2) that there is some freedom in choosing such group which, as we will show, is defined by tensorial properties of the secondary quantization operators in the quasispin space. It is known [5] that for these operators the finite rotations in the quasispin space around an arbitrary axis define the special canonical Bogoliubov transformation. If we confine ourselves to the rotations on the angle φ around the axis Z of this space, then we arrive at the phase transformations of these operators

$$a^{(qls)}_{\varrho m\mu} = e^{i\varrho\varphi} a^{(qls)}_{\varrho m\mu} \tag{2.3}$$

or, in other words, at the components of the quasispin operator, defined with respect to the rotated axes ox' and oy'. Considering this rotation in "local" sense,i.e. defining for each of the shells its own angle rotation $\varphi(l_i)$, we find that transformation (2.3) with $\psi = [\varphi(l_b) - \varphi(l_a)]/2$ defines the transition from the operator (2.1) to the operator

$$A_{\rho}^{(1)} = \frac{l_a+2}{l_a+1} L_{\rho}^{(1)}(a) + \frac{l_a}{l_a+1} L_{\rho}^{(1)}(b) + 2i\sqrt{\frac{(2l_a+1)(2l_b+3)}{3(l_a+1)}} W_{0\rho 0}^{(010)}(a,b). \tag{2.4}$$

Due to the fact that this operator is a scalar in the bothspin and quasispin spaces the wave-functions of the corresponding basis as well as the characteristics of the irreducible representations of the group SO(4) will be classified by the quantum numbers of the total spin S and quasispin Q. Since locally SO(4) and SU(2) x x SU(2) are isomorphic it is possible to define the operators

$$_1J_{\rho}^{(1)} = \frac{1}{2}(L_{\rho}^{(1)} + A_{\rho}^{(1)}), \quad _2J_{\rho}^{(1)} = \frac{1}{2}(L_{\rho}^{(1)} - A_{\rho}^{(1)}) \tag{2.5}$$

satisfying the usual commutation relations for the irreducible components of the angular momentum operator, moreover , $[{}_1\vec{J}, {}_2\vec{J}] = 0$. Linear combinations of the secondary quantization operators

$$a_{\rho m_1 m_2 \mu}^{(q j_1 j_2 s)} = \sum_{lm} \begin{bmatrix} j_1 & j_2 & l \\ m_1, & m_2 & m \end{bmatrix} a_{\rho m \mu}^{(q l s)} \tag{2.6}$$

define the components of irreducible tensorial operators of ranks $j_1 = (2l_a+1)/2$, $j_2 = 1/2$ in corresponding spaces.

Using representation (2,6) we get an expression of the secondary-quantized form of the energy operator in terms of the irreducible tensors of SO(4). Analysis of individual terms of such an expression allows us to define necessary conditions and concrete possibilities to realize SO(4) as a group of an approximate symmetry of a many-electron atom.

However, it ought to be mentioned that the wave-functions transforming with respect to the irreducible representations of SO(4) in the general case can not be characterized by a definite parity quantum number, i.e. to ensure a parity condition one has to form some linear combinations of these functions. For two-electron atoms some methods to construct such functions were proposed in [7,8].

3. <u>Groups of Fivedimensional Rotations</u>. Richer are the possibilities of the group theoretical description of the complex of configurations $(n_a l_a + n_b l_b)^N$ with $l_a = l_b = l$. In this case ten scalars in the orbital and spin spaces [5]

$$A^{(00)} = \left[a^{(ls)} \times a^{(ls)}\right]^{(00)}; \qquad B^{(00)} = \left[a^{(ls)} \times \tilde{a}^{(ls)}\right]^{(00)};$$

$$C^{(00)} = \left[\tilde{a}^{(ls)} \times \tilde{a}^{(ls)}\right]^{(00)}; \qquad D^{(00)} = \left[b^{(ls)} \times b^{(ls)}\right]^{(00)};$$

$$E^{(00)} = \left[b^{(ls)} \times \tilde{b}^{(ls)}\right]^{(00)}; \qquad F^{(00)} = \left[\tilde{b}^{(ls)} \times \tilde{b}^{(ls)}\right]^{(00)}; \tag{3.1}$$

$$G^{(00)} = \left[a^{(ls)} \times b^{(ls)}\right]^{(00)}; \qquad H^{(00)} = \left[\tilde{a}^{(ls)} \times \tilde{b}^{(ls)}\right]^{(00)};$$

$$I^{(00)} = \left[a^{(ls)} \times \tilde{b}^{(ls)}\right]^{(00)}; \qquad J^{(00)} = \left[\tilde{a}^{(ls)} \times b^{(ls)}\right]^{(00)}$$

form the algebra of SO(5). If the characteristics of the irreducible representations of this group are adopted to classify the corresponding atomic states, then a set of additional quantum numbers is defined by possible schemes of the reduction of this group.

First let us point out three reduction schemes in which the configuration is conserved as a quantum number.

1) $SO(5) \supset SU(2)^T \times U(1)^N$ in which the isospin group $SU(2)^T$ is generated by the operators

$$T_{-1}^{(1)} = i\sqrt{2\ell+1}\,[a^{(\ell s)} \times \tilde{b}^{(\ell s)}]^{(00)};\quad T_{0}^{(1)} = \frac{i}{2}(\hat{N}_b - \hat{N}_a);$$
$$T_{1}^{(1)} = i\sqrt{2\ell+1}\,[b^{(\ell s)} \times \tilde{a}^{(\ell s)}]^{(00)} \tag{3.2}$$

and U(1) is determined by the particle number operator N.

2) $SO(5) \supset SU(2)^F \times U(1)^{N_a - N_b}$ in which the F-spin group $SU(2)^F$ is generated by the operators

$$F_{-1}^{(1)} = -i\sqrt{2\ell+1}\,[\tilde{a}^{(\ell s)} \times \tilde{b}^{(\ell s)}]^{(00)};\quad F_{0}^{(1)} = \frac{i}{2}(\hat{N} - 4\ell - 2);$$
$$F_{1}^{(1)} = i\sqrt{2\ell+1}\,[a^{(\ell s)} \times b^{(\ell s)}]^{(00)} \tag{3.3}$$

and U(1) is determined by the operator of the difference in the number of electrons on the shells.

3) $SO(5) \supset SU(2)^Q$ with the corresponding quasispin operators of separate shells

$$Q_{1}^{(1)}(\ell_j) = -i\sqrt{\frac{2\ell_j+1}{2}}\,[a^{(\ell_j s)} \times a^{(\ell_j s)}]^{(00)};$$
$$Q_{-1}^{(1)}(\ell_j) = -i\sqrt{\frac{2\ell_j+1}{2}}\,[\tilde{a}^{(\ell_j s)} \times \tilde{a}^{(\ell_j s)}]^{(00)}; \tag{3.4}$$
$$Q_{0}^{(1)}(\ell_j) = \frac{i}{2}(\hat{N}_j - 4\ell_j - 2).$$

Let us note that the first two reduction schemes are non-canonical, whereas the third is canonical one.

The transition to many-configuration wave-functions requires to refuse the reduction schemes described above and to look for new possible schemes which do not fix the number of electrons in separate shells.

It is possible to obatin a fairly simple group-theoretical classification scheme of mixed configurations following the way presented in the previous section and getting to the total quasispin of the configuration with the help of vectorial coupling of the quasispin momenta of separate shells. Other schemes may be found as a result of extending the F-spin group (3.3) to the rotation group of some fourdimensional space. One of such extensions can be obtained introducing the operators

$$B_1^{(1)} = Q_1^{(1)}(a) + Q_1^{(1)}(b);$$

$$B_{-1}^{(1)} = Q_{-1}^{(1)}(a) - Q_{-1}^{(1)}(b); \qquad (3.5)$$

$$B_0^{(1)} = iT_x$$

and cjecking with the help of (1.2) whether they, together with the F-spin operators, make the algebra of SO(4).

Exactly the same algebra is made together with the F-spin operators by the operators

$$C_{-1}^{(1)} = -i\left[Q_{-1}^{(1)}(a) - Q_{-1}^{(1)}(b)\right]; \quad C_0^{(1)} = iT_y;$$

$$C_1^{(1)} = i\left[Q_1^{(1)}(a) - Q_1^{(1)}(b)\right]. \qquad (3.6)$$

The choice of basis, expedient from the physical point of view, requires the analysis of the expression of the energy operator in terms of irreducible tensorial operators of corresponding groups.

References

1. Nikitin,A,A., Rudzikas,Z.B. (1983). Foundations of the theory of spectra of atoms and ions. Nauka, Moscow (Russian).

2. Elliot,I.P. (1958). Proc. Roy. Soc. A245, 61.

3. Feneuille,S. (1967). J. Phys. 28, 315.

4. Butler,P.H., Wybourne,B.G. (1970). J. Math. Phys. 11, 2512.

5. Rudzikas,Z.B., Kaniauskas,J.M. (1984). Quasispin and isospin in atom theory Mokslas, Vilnius (Russian ; extended version in English to be published by Nauka in 1986).

6. Kaniauskas,J.M., Simonis,V.C., Rudzikas,Z.B. (1983). In: Group-theoretical Methods in Physic, Nauka, Moscow, p. 276 (Russian).

7. Wulfman,C. (1973). Chem. Phys. Lett. 23, 370.

8. Nikitin,S.I., Ostrovkij,V.N. (1980). In: Group-theoretical Methods in Physics, Nauka, Moscow, 189 (Russian).

CALCULATIONS FOR TWO-PHOTON IONIZATION OF ATOMIC HYDROGEN ABOVE THE PHOTOELECTRIC THRESHOLD BY ANALYTIC CONTINUATION

E.Karule

Institute of Physics, Latvian SSR Academy of Sciences, 229021 Riga, Salaspils

Multiphoton ionisation of atomic hydrogen is investigated by many authors although experimental data are scarce. The interest of theoreticians can be explained racalling that the well-known wave functions of atomic hydrogen allow one to work out reasonable approximations that can be applied to alkali atoms, negative ions, etc.

Unlike the single-photon processes, multiphoton processes strongly depend on the intensity of the incident radiation. The transition rate for N-photon ionisation of an atomic shell with principal quantum number n_1 is defined by

$$Q^{(N)}_{n_1}/I^{N-1} = n_1^{-2}\sum_{\ell_1=0}^{n_1-1}(2\ell_1+1)Q^{(N)}_{n_1\ell_1}/I^{N-1}. \qquad (1)$$

In evaluating the ionisation rate, the main difficulty is to calculate the radial parts of transition matrix elements. If the intensity of the radiation field is less than or about 10^7 V/cm, then the perturbation theory can be used. Let us consider the interaction between the atom and the radiation field as dipole interaction, then using Green's function formalism in the lowest non-vanishing order of the non relativistic perturbation theory, the radial parts of transition matrix elements can be written in the form

$$T^{(N)}_{rad}(n_1\ell_1, E\ell_2)=\int_0^\infty R_{E\ell_2}(r_1)r_1^3\prod_{j=0}^{N-1}\int_0^\infty G_{L_j}(r_j,r_{j+1};\Omega_j)\times \tag{2}$$

$$\times R_{n_1\ell_1}(r_N)r_{j+1}^3\,dr_1\,dr_{j+1}$$

where $\Omega_j = 2^{-1}n_1^{-2} + j\omega$ is energy, L_j the orbital momentum of the electron in a virtual state. Coulomb Green's function (CGF) can be written as atomic hydrogen energy eigenfunction expansion

$$G_L(r,r';\Omega) = -S_n \frac{R_{nL}(r)R_{nL}(r')}{E_n-\Omega}. \tag{3}$$

The expansion (3) we used in the earliest two-photon ionisation calculations for atomic hydrogen [1-3]. But the authors reduced the problem to solving a 1st order inhomogeneous differential equation. In [4], this method was generalized for multiphoton ionisation. Beb and Gold [5] summed the series (3) directly but they had not taken into account the continuum and had taken into account only partially the contribution from states whose energy is not close to that of the intermediate state.

More convenient for calculations is the CGF Sturmian expansion [6]

$$G_L(r,r';\Omega) = -\rho\sum_{m=L+1}^{\infty}\frac{S_{mL}(r)S_{mL}(r')}{m-\rho}, \quad \Omega<0,\ \rho=(-2\Omega)^{-1/2}. \tag{4}$$

Sturmian functions S_{mL} are the charge eigenfunctions. An advantage of this expansion is that there is only a summation over the discrete spectrum. A method equivalent to the KCF Sturmian expansion was first used by Podolsky [7] for calculating two-photon bound-bound transition rates. The Sturmian expansion (4) can be rewritten also in the integral form

$$G_L(r,r';\Omega) = 2(rr')^{-1/2}\int_0^\infty dt \exp\left(-\frac{r+r'}{p}\cosh t\right) \times \\ \times (\coth t/2)^{2p} I_{2L+1}\left(\frac{2rr'}{p}\sinh t\right), \quad \Omega < 0. \tag{5}$$

Calculations for two-photon ionisation of atomic hydrogen using expression (5), are carried out e.g. in [8, 9]. In the CGF integral form, the radial variables are not separated. Therefore, it is less appropriate for calculating multiphoton processes than the Sturmian expansion (4).

KGF Sturmian expansion (4) allows us to obtain expressions for the radial transition matrix elements for multiphoton processes on atomic hydrogen in closed form, which was first done in [10]. Calculations for multiphoton ionisation $N \leqslant 16$ of atomic hydrogen in the ground state applying expressions derived in [10] are carried out in [11].

Expressions (4), (5) for KGF, the same as expressions for transition matrix elements obtained using these expressions, are convergent only for $\Omega < 0$. Therefore, all the above calculations are carried out for the cases where minimum number of photons necessary for ionisation is absorbed meaning that energies of all intermediate states are negative. But now it is also possible to observe [12] the multiphoton processes where more photons than the minimum number required for the ionisation of an atom are absorbed. Then the energy of one or more intermediate states belong to the continuum. Multiphoton ionisation rates for the cases there more than one intermediate states belong to the continuum are not evaluated. The simplest case for theoretical investigation is the two-photon ionisation above the photoelectric threshold. Klarsfeld [13] was the first to calculate transition rates for the two-photon ionisation of atomic hydrogen in the ground state. He used the KGF integral representation (5) and obtained the expressions for transition matrix elements whose convergence depends on that of Appell's hypergeometric functions F_1. Applying transformation formulas for F_1 and using the new Tailor's double series expansion, Klarsfeld made calculations in a wide energy domain below and above the threshold. In [14], we calculated transition rates for two-photon ionisation of H in the ground state above the photoelectric threshold by analytical continuation of transition matrix elements derived by means of the Sturmian expansion (4).

Calculations for the two-photon ionisation of excited states are more difficult than those for the ground state, as the wave functions of excited states are more complicated and it is necessary to calculate more matrix elements. Different methods have been proposed for calculations. In [15, 16] methods for resumming the initial series are given. But since the new series converge due to the respective compensation of terms in the initial series, precision is lost, especially at the threshold region. In [15, 16] calculations are done for the ionisation from series with n = 2,3,4 above the photoelectric thresholdexcept the region close to the threshold. Above the threshold, as well as below the threshold, inhomogeneous differential equations may be solved. This method is used in [17], where transition rates for two-photon ionisation of ns ($n \leqslant 6$) states by circularly polarised light are calculated above the photoelectric threshold, but it is a labor consuming method.

We have calculated [18-20] ionisation rates for atomic hydrogen in excited states n $\leqslant$ 9 using the method of analytical continuation of series for transition matrix elements, obtained by means of the Sturmian expansion (4) for KGF. The calculations are done for the cases of lineary and circularly polarised light in the λ-region $20n_1^2$Å to $911 \cdot 76n_1^2$Å (photoionisation threshold).

Usually, when the Sturmian expansion (4) is used, the integration over both radial variables is carried out first and the resukt is obtained in a closed form as the infinite sum of Gaussian hypergeometric functions. Finally, all the summations are done on a computer. If we want to continue such expressions analytically bejond the threshold, it is necessary first to reverse the order of summation. Further the new series expressed as the series of Gaussian hypergeometric functions, which can be continued analytically. It is easier to continue double series, if they are expressed as Appell's hypergeometric functions of two variables. Double series in the form of Appell's functiona can be obtained using the KGF integral representation e.g. the Sturmian expansion (4) summed directly over intermediate states. Another way to obtain result in Appell's functions is to use the KGF Sturmian expression (4), and then to integrate over one of the radial variables and to sum directly over intermediate states. At last we have to integrate over the second radial variable. In [14, 18-20] expressions for radial matrix elements, the convergence of which completely depends on convergence of Appell's functions

$$F_1\left(\alpha, \beta, \beta', \gamma; \frac{1}{z_1 z_2}, \frac{z_2}{z_1}\right) \qquad (6)$$

are derived in this way.

Below and above the threshold $|1/z_1 z_2| < 1$ (see

Table 1. Behaviour of variables below $(\Re < 0)$, above $(\Re > 0)$ and at the photoelectric threshold $(\Re = 0)$

Variable	$\Re<0$	$\Re>0$	$\Re=0$
n_1	integer	integer	integery
ω	$0.5 < 2n_1^2\omega < 1$	$2n_1^2\omega > 1$	$2n_1^2\omega = 1$
$\sigma = -i/k_2$	immaginary	imaginary	imaginary
$p = (-2\omega)^{-1/2}$	real	imaginary	∞
$z_1 = \frac{p+n_1}{p-n_1}$	$-6 \leqslant z_1 < -1$	$\exp(i\varphi)$ $\pi < \varphi \leqslant 2\pi$	-1
$z_2 = \frac{p+\sigma}{p-\sigma}$	$\exp(i\Psi)$ $\pi < \Psi \leqslant 2\pi$	$-6 \leqslant z_2 < -1$	-1
$z_3 = \frac{\sigma+n_1}{\sigma-n_1}$	$\exp(i\delta)$ $\pi/2 < \delta \leqslant \pi$		i
$\lvert 1/z_1 z_2\rvert$	<1	<1	1
$z_2 z_1$	$(1/z_1 z_2)^*$	$\lvert z_2/z_1\rvert > 1$	1
z_1/z_2	$\lvert z_1/z_2\rvert > 1$	$(1/z_1 z_2)^*$	1

Table 1) while $|z_2/z_1|>1$ above the threshold and series (6) deverge. Let us write F_1 as the sum of Gaussian hypergeometric functions

$$F_1 = \sum_{n=0}^{\infty} \frac{(\alpha)_n (\beta)_n}{n!(\gamma)_n (z_1 z_2)^n} \, {}_2F_1(\alpha+n, \beta'; \gamma+n; z_2/z_1). \quad (7)$$

By analytical continuation of ${}_2F_1$, we obtain two series: the double series depending on variables $1/z_1 z_2 < < 1$ and a series depending on z_2^{-2}. Moduli of all these variables are above the threshold, hence, the series converge. Series depending on such variables, are used in [14, 18, 19]. But near the threshold $\lambda = 900 n_1^2 Å$

convergence of double series is very slow, as both variables are close to 1. By means of the other analytical continuation formulas for Gaussian hypergeometric functions we got expressions for radial transition matrix elements as double series of powers of $1/z_1z_2$ and $1/z_2z_3$ and series of powers of z_2^{-2}. Contribution of the last series is so small that it can be neglected. Near the threshold, the new double series converge 10 times faster than those of powers of $1/z_1z_2$ and z_1z_2. If $\rho \to \infty$, we can get series convergent at the threshold. Above the threshold, the expressions for the radial transition matrix elements for two-photon ionisation of atomic hydrogen, derived by means of analytical continuation, can be written as

$$T_{rad}(n_1 l_1 ; E l_2) = -2^{-3} C_k k_2^{\,l_2} \{(n_1 - l_1)_{2l_1+1}\}^{1/2} \times$$
$$\times \{(2L+1)!\}^{-1} (2l_1+2)_{L-l_1+2}\, \omega^{-2} (T_1 + T_2) \tag{8}$$

where C_k is the normalisation factor of the Coulomb wave function, the wavenumber of ejected electron,

$$T_1 = 2^{-2}(2l_2+2)_{L-l_2+2}\, n_1^{\,l_2-2}\, \omega^{-1} p(p+L+1)^{-1} \times$$
$$\times (1-z_1)^{l_1-L+2} (1+z_3)^{L+l_2-2}\, z_3^{-q-L} \times$$
$$\times \sum_{m=0}^{n_1-l_1-1} \frac{(-n_1+l_1+1)_m (L+l_1+4)_m}{m!\,(2l_1+2)_m} (1-z_1)^m \times \tag{9}$$
$$\times \sum_{n=1}^{m+l_1-L+2} \frac{(L-l_1-m-2)_n (p+L+1)_n}{n!\,(2L+2)_n} (1+1/z_1)^n \times$$

$$\times \sum_{\beta=0}^{n-1} \frac{(2L+2)_\beta}{(p+L+2)_\beta} (1+1/z_1)^{-\beta} \times$$

$$\times \sum_{\alpha=0}^{\beta} \frac{(-\beta)_\alpha (L+\ell_2+4)_\alpha}{\alpha! (2L+2)_\alpha} \left(\frac{\frac{1}{n_1} - \frac{1}{p}}{\frac{1}{n_1} + ik_2} \right)^\alpha \times$$

$$\times {}_2F_1(i/k_2 + \ell_2 + 1, \ell_2 - L - 2 - \alpha; 2\ell_2 + 2; 1 - z_3).$$

Since (9) is valid for all energies

$$T_2 = n_1^{-L-2} (ik_2)^{L-\ell_2-3} z_1 (1-z_1)^{\ell_1 - L} (i/k_2 + \ell_2 + 1)_{L-\ell_2+2} \times$$

$$\times \omega^{-L} z_2^{-p+L-1} z_3^{i/k_2+1} (z_2 - 1)(1/n_1 + 1/p)(1/n_1 - ik_2)^{-1} \times$$

$$\times \sum_{m=0}^{n_1-\ell_1-1} \frac{(-n_1+\ell_1+1)_m (L+\ell_1+4)_m}{m! (2\ell_1+2)_m} (1-z_1)^m \times \tag{10}$$

$$\times {}_2F_1(L-\ell_1-m-2, p+L+1; 2L+2; 1+1/z_1)$$

$$\sum_{n=0}^{\ell_2-L+2} \frac{(L-\ell_2-2)_n}{n! z_3^n} \sum_{s=0}^{L-\ell_2+2} \frac{(\ell_2-L-2)_s (-i/k_2+\ell_2+1)_s}{s! (-i/k_2 - L - 2)_s z_3^s} \times$$

$$\times (ik_2 + L + 2 - n - s)^{-1} g(n_1, L, p, q, n, s)$$

where

$$g(n_1, L, p, q, n, s) =$$

$$= \sum_{x=0}^{2L} \frac{(p-L)_x}{x!} \left(1 - 1/z_1 z_2\right)^{x} \left(1 - 1/z_2 z_3\right)^{x-2L-1} \times$$

$$\times {}_2F_1\left(x-2L, -i/k_2+n+s-L-2; -i/k_2+n+s-L-1; 1/z_2 z_3\right) + \tag{11}$$

$$+ \sum_{x=2L+1}^{\infty} \frac{(p-L)_x}{x!} \left(1 - 1/z_1 z_2\right)^{x} \times$$

$$\times {}_2F_1\left(1, -i/k_2+n+s+L-1-x; -i/k_2+n+s-L-1; 1/z_2 z_3\right).$$

Convergence of T_2 depends on expression (11), which converges not only in the vicinity of the threshold, but also in all domains above the threshold. Some results of the calculations are presented in Tables 2 and 3.

In [21,22], the KGF Sturmian expansion (4) is continued in the region of positive energies. But instead of a single series expansion as (4), they obtained the double series expansion for KGF. Ionisation rates for two-photon ionisation of atomic hydrogen in excited states are calculated by this method only for $\lambda \lesssim 603 n_1^2$ A above the photoelectric threshold [22].

Analytic continuation of transition matrix elements permitted us to calculate ionisation rates for all energies above the photoelectric threshold.

Table 2. Two-photon transition tates (in $W^{-1}cm^4$) for ionisation of atomic hydrogen in n_1=7,8,9 states by linearly polarised light and ratio of cross-sections for ionisation by circularly and linearly.

λ/n_1^2 A	$n_1=7$		$n_1=8$		$n_1=9$	
	Q_1/I	Q_c/Q_1	Q_1/I	Q_c/Q_1	Q_1/I	Q_c/Q_1
20	3.89-38	0.96	1.06-37	0.98	2.58-37	1.01
50	1.22-35	1.05	3.36-35	1.07	8.23-35	1.09
100	9.57-34	1.11	2.65-33	1.12	6.52-33	1.14
200	7.60-32	1.16	2.11-31	1.17	5.21-31	1.18
300	9.86-31	1.19	2.74-30	1.20	6.77-30	1.20
400	6.09-30	1.21	1.70-29	1.22	4.19-29	1.22
500	2.50-29	1.23	6.97-29	1.23	1.73-28	1.24
600	7.94-29	1.24	2.21-28	1.24	5.50-28	1.25
700	2.11-28	1.25	5.89-28	1.26	1.45-27	1.26
800	4.94-28	1.26	1.38-27	1.27	3.40-27	1.27
900	1.03-27	1.28	2.91-27	1.27	7.18-27	1.28

Table 3. Two-photon transition rates (in $W^{-1}cm^4$) for ionisation of atomic hydrogen in $n_1 \leq 9$ states at the photoelectric threshold (λ=911.76n_1^2Å).

n_1	Q_c/I	Q_1/I	Q_c/Q_1
1	4.34-34	3.63-34	1.22
2	9.45-32	7.50-32	1.26
3	2.15-30	1.70-30	1.27
4	1.97-29	1.55-29	1.27
5	1.09-28	8.58-29	1.27
6	4.43-28	3.48-28	1.27
7	1.45-27	1.13-27	1.28
8	4.24-27	3.30-27	1.29

9	1.10-26	8.49-27	1.29

References

1. Zernik, W. (1964). Phys. Rev., 135, A51.
2. Zernik, W., Klopfenstein, R.W. (1965). J.Math.Phys., 6, 262.
3. Zernik, W. (1968). Phys. Rev., 176, 420.
4. Gontier, Y., Trahin, M. (1968). Phys. Rev., 172, 83.
5. Bebb, H.B., Gold, A. (1966). Phys. Rev., 143, 1.
6. Hostler, L.C. (1970). J.Math. Phys., 11, 2966.
7. Podolsky, B. (1928). Proc.Natl.Acad.Sci.US, 14, 253.
8. Klarsfeld, S. (1969). Lett.Nuovo Cimento, 1, 682; 2, 548.
9. Rapoport, L., Zon, B., Manakov, N. (1969). Sov.Phys. JETP, 28, 480.
10. Karule, E. (1971). J.Phys.B: At.Mol.Phys., 4, L67.
11. Karule, E. (1975). In: Atomic Processes Zinatne, Riga, p.5.
12. Agostini, P., Fabre, F., Mantray, G., Petite, G., Rahman, N.K. (1979). Phys.Rev.Lett., 42, 1127.
13. Klarsfeld, S. (1970). Lett. Nuovo Cimento, 3, 395.
14. Karule, E. (1978). J.Phys.B: At.Mol.Phys., 11, 441.
15. Klarsfeld, S., Maquet, A. (1979). Phys.Lett., 73A, 100.
16. Klarsfeld, S., Maquet, A. (1979).J.Phys.B: At.Mol.Phys., 12, L553.
17. Aymar, M., Crance, M. (1980). J.Phys.B: At.Mol.Phys., 13, L287.
18. Karule, E.M. (1983). Report presented at Section of Photoprocesses of Atomic and Electron Collision Scientific Council of the USSR Academy of Sciences, Uzgorod, May (Russian).
19. Karule, E. (1984). In: Nonlinear processes in two-lectron atoms, M., USSR Acad. Sci. Moscow, 209 (Russian).
20. Karule, E. (1985). J.Phys.B: At.Mol.Phys., 18, 2207.
21. Manakov, N.L., Marmo, S.I., Fainshtein, A.G. (1984). Theor.Math.Phys., 59, 49 (Russian).
22. Fainshtein, A.G., Manakov, N.L., Marmo, S.I. (1984). Phys.Lett., 104A, 347.

THE THRESHOLD PHENOMENA IN THREE-PARTICLE COULOMB SYSTEMS

M.K.Gailitis

Institute of Physics, Latvian SSR Academy of Sciences, Riga-Salaspils

This paper examines the reactions

$$e+H \to e+H,\ p+H \to p+H,\ e^{+}+H \to p+(e^{+}e),$$
$$p+He^{+} \to \alpha+H,\ e+(ee^{+}) \to e+(ee^{+}),\ \mu^{-}+H \to e+(\mu^{-}p) \tag{1}$$

and some other near the threshold where the three-particle system splits into two bodies, a and b, the charged particle $(Z_a = Z_1)$ and the neutral one $(Z_b = Z_2 + Z_3 = 0)$; an excited Coulomb pair (2,3). We discuss effect of the dynamical symmetry of motion with Coulomb forces between the particles in the threshold behaviour pattern.

According to Wigner [1] there are three forms of threshold behaviour of two-body thresholds:

1) At Coulomb repulsion in new channels $(Z_a Z_b > 0)$, the new mode excitation cross-sections near the threshold exponentially tend to zero

$$\sigma_{if} \sim e^{-2\pi Z_a Z_b / v_f}, \tag{2}$$

where v_f is the relative velocity od particles a, b.

2) At Coulomb attraction they vary jumplike, vanishing below the threshold and assuming a finite value immediately above it:

$$\sigma_{if} \sim const. \tag{3}$$

3) Without Coulomb forces $(Z_a Z_b = 0)$ the threshold behaviour is controlled by centrifugal repulsion and the excitation cross-sections increase from zero on the thres-

hold by the power law

$$\sigma_{if} \sim k_f^{2\ell_f+1} , \tag{4}$$

where k_f and ℓ_f are the momentum and the angular momentum of the new mode.

For reactions (1) the pair charge is zero, and the condition for the power law (4) may seem to have been observed. Actually, the threshold behaviour is essentially different, and this paper deals with just this matter.

The $O(4)$ symmetry causes the degeneration of energie and the linear Stark"s effect with excited states of atom H and all Coulomb-bound pairs, It is the physical cause of deviation from the law (4) in reaction (1).

We denote intrinsic coordinates and reduced mass of the pair by $\vec{r}$ and m, coordinates and reduced mass for a cingle particle relative to the pair by $\vec{\rho}$ and μ. At $Z_a = 1$, the single particle acts upon the pair with electric field $\vec{\mathcal{E}} = -\vec{\rho}/\rho^3$ which causes linear Stark's effect, the degenerated excited states of the pair split into components with energy differences proportional to the electric field

$$\Delta E_j = -\left(\vec{\mathcal{E}} \cdot \frac{\vec{d}}{m}\right) = -\frac{3}{2}\frac{NK_j}{m\rho^2} , \quad |K_j| \leqslant N-1 \tag{5}$$

where $\frac{\vec{d}}{m}$ is the pair's dipole momentum. The latter and ΔE_j are inversely proportional to m because so are the pair dimensions. N is the principal quantum number of the pair, and K_j is electric quantum number. We take $Z_2 = 1$ in (5). The splitting (5) means that a single particle interacts with the pair with potential ΔE_j, dependent on the pair state j. Both ΔE_j and the centifugal energy are inversely propor tional to ρ^2.

Therefore, their total

$$\frac{\ell(\ell+1)}{2\mu\rho^2} + \Delta E_j = \frac{a_j}{2\mu\rho^2} \tag{6}$$

determines a threshold behaviour pattern instead of a purely centrifugal term in cases of short-range forces. If in (6)

$$a_j = \lambda_j(\lambda_j + 1)\ , \quad \lambda_j + \frac{1}{2} = \sqrt{a_j + \frac{1}{4}}\ , \tag{7}$$

λ_j replaces ℓ_f in (4), and it can be shown [2,3] that

$$\sigma_{if} \sim \left|\sum_j C^j_{if}\, k_f^{\lambda_j + 1/2}\right|^2, \tag{8}$$

i.e. near the threshold the excitation cross-sections are proportional to the squared absolute value of the linear combination of k_f in various powers.

Determining a_j is slightly impeded by the fact that in the representation with a diagonal dipole part the centrifugal part is slightly nondiagonal. Therefore we temporarily replace the angular momentum ℓ of one single particle by the total angular momentum L to illustrate the qualitative aspects. In this approximation $a_j = L(L+1) - 3wNK_j$ depends on masses only via dimensionless parameter $w = \mu/m$, the ratio of both reduced masses.

For some first partial waves with $L < L_{c2} = \sqrt{3wN(N-1)} - 1/2$ some states j have the dipole attraction stronger than the centrifugal repulsion, with $a_j < -1/4$, and imaginary, $\lambda_j + 1/2$ $|k_f^{\lambda_j + \frac{1}{2}}| = |k_f^{i\,\mathrm{Im}\,\lambda_j}| = 1$, the cross-sections (8) near the threshold do not tend to zero but remain finite contrary to Wigner's law (4).

The imaginary $\lambda_j + 1/2$ cause some other phenomena: series of resonances below the thresholds in transitions between old channels are distributed via a geometri progression

$$\frac{\varepsilon_n}{\varepsilon_{n+1}} = \frac{\Gamma_n}{\Gamma_{n+1}} = R_j = e^{2\pi/\mathrm{Im}\,\lambda_j}, \quad \varepsilon = E - E_t . \tag{9}$$

The phenomenon (9) coexists with an infinite number of boundstates in the potential

$$V(\rho) \sim a_j/(2\mu\rho^2) \tag{10}$$

for large ρ, with $a_j < -1/4$.

For $L > L_{cr}$ the situation changes. The centrifugal repulsion exceeds the dipole attraction, all λ_j are real, the resonance series (9) desappear, the threshold behaviour of excittation cross-sections assumes the form (4) with l_f replaced by the smallest of λ_j :

$$\sigma_{if} \sim K_f^{2\lambda_1+1}, \quad \lambda_1 = \min(\lambda_j) . \tag{11}$$

Thus, the pattern of threshold behaviour (1) lies between laws (3) and (4). For partial waves with large angular momentum $L > L_{cr}$ the behaviour (11) is closer to (4), and with small angular momentum $L < L_{cr}$ both phenomena, the finite excitation cross-sections immediately above the threshold and the infinite number of resonances below the threshold are similar to Coulomb attraction case. Nevertheless, the resonances series are of different forms. In Coulomb attraction, they follow Rydberg's law, while reactions (1) comply with geometric progression law (9).

Specifying the approach, we must consider the variance of angular momentum l in different new modes. The main results remain unchanged, only the eigenvalues of the matrix

$$\Lambda = \ell(\ell+1) + \alpha, \quad \alpha = 2wd \tag{12}$$

must be used as a_j. Here, d is the matrix of a pair's dipole momentum projection on $\vec{\varrho}$ calculated by regarding the reduced mass inside the pair as a unit and replacing by zero the elements between modes with different N. The matrix Λ is cell-diagonal. Each cell depends on L, N and parity P. Its eigenvalues are degenerated with respect to parity, all eigenvalues with $P = (-1)^{L+1}$ are also the eigenvalues with $P = (-1)^{L}$ (but not vice versa). It is due to the existence of an operator commuting with Λ and anticommuting with P [4-6]. On the thresholds with N = 2 and 3 the eigenvalues a_j are roots of square and cubic equations. The order of the equation increases with N. Therefore, it is useful to have approximate expressions of a_j applicable for any N. They can be found by regarding $\ell(\ell+1)$ as perturbation related to α. The 0th order approximation

$$a_j^{(0)} = -3wK_jN \tag{13}$$

follows from (5). In the 2nd order approximation [5]

$$a_{KT}^{(2)} = -3wKN + L(L+1) + (N^2-1-K^2-3T^2)/2$$
$$-K[8L(L+1)+N^2-1-K^2-15T^2]/(12wN) \tag{14}$$

depends on two integer quantum numbers K and T, where $0 \leqslant T \leqslant L$, $N-1-K-|T| \geqslant 0$ and being even T is 0 only if $P = (-1)^{L}$. The lowest eigenvalue a_1 features $K = N-1$, $T = 0$. It follows from (13,14):

$$l_{min}(l_{min}+1)-3wN(N-1)\leqslant a_1\leqslant L(L+1)+(N-1)(1-3wN),$$
$$l_{min}=max\,(0,L+1-N). \tag{15}$$

The left inequality is due to $l(l+1)\geqslant l_{min}(l_{min}+1)$. Therefore each a_j is at least $l_{min}(l_{min}+1)$ times greater than its $a^{(0)}_{KT}$. The right inequality follows from the theorem saying that the first approximation of the perturbation theory yields the upper boundary of the lower eigenvalue. Eq.(15) can be wxitten in the form of two sufficient conditions

$$a_1\leqslant -\frac{1}{4} \quad \text{for} \quad (L+\tfrac{1}{2})^2<(3wN-1)(N-1), \tag{16}$$

$$a_1\geqslant -\frac{1}{4} \quad \text{for} \quad (l_{min}+\tfrac{1}{2})^2>3wN(N-1). \tag{17}$$

They show that number of partial waves with finite cross-sections at the threshold and resonance series (9) below the threshold change with N and w.

The mass effect is contained only in the equation

$$w=\frac{\mu}{m}=\frac{m_1(m_2+m_3)^2}{m_2m_3(m_1+m_2+m_3)}. \tag{18}$$

where $w<1$ ($w=0.0054$) only for the reaction $e(\mu p)$. For $N\leqslant 5$, it has parameters (17) observed for all L and the excitation cross-sections at the first thresholds begin with zero. Nevertheless, even for that reaction eq.(16) is observed at sufficiently high thresholds $(N\geqslant 63)$ and the cross-sections are finite immediately above the threshold. It was found by numerical computation that $a_1<-1/4$ and the cross-sections wer finite for $N\geqslant 9$.

For all other reactions (1) $w > 1$ and (16) is met on all excitation thresholds for a few lower partial waves. Their number increases with w or N. This can be illustrated by computation results, which provide finite cross-sections for the following angular momenta L

$$e\text{-}H: w = 1.000000296, \quad L \leqslant 2\,(N=2), \; L \leqslant 4\,(N=3),$$

$$p\text{-}H: w = 918 \cdot 83, \quad L \leqslant 73\,(N=2), \quad L \leqslant 128\,(N=3).$$

The computation results also show that in reactions with $w \sim 1$ the geometric progression law (9) is manifest, i.e. ratios R_j are large and with increasing n the resonances rapidly tend towards the threshold. As w grows, $R_j \to 1$. In the small interval of n the resonance spectrum is almast equidistant and resembles molecular vibration spectrum, but for the widths and intervals between the resonances decreasing proportionally to the distance from the threshold

$$(\varepsilon_n - \varepsilon_{n+1})/\varepsilon_n = 2\pi / \sqrt{3wKN} + O\,(w^{-1}). \qquad (19)$$

For instance, the rwonance spectrum is like that for p-H scattering.

So far the relativistic splitting δE of the threshold levels was neglected. Therefore the results are applicable for $|\varepsilon| \gg (L+1)\ \delta E$. Very close to the threshold where this condition fails, the excitation cross-sections tend zero and the resonance series break. Furthermore, with removed degeneration there are several closely situated thresholds, and the study of the behaviour of the cross-sections individual near each of them is required. Since Dirac's equation has doubly degenerated levels and Lamb's shift fully removes the degeneration, the changes in the threshold behavioural laws take place in two stages: partly (e.g. R_j and E_t change in (9)) at distances of the order of Dirac's splitting and fully at distances of the order of Lamb'sshift.

References

1. Wigner, E.P. (1948). Phys.Rev., 73, 1002-9.

2. Gailitis, M. (1982). J;Phys., B15, 3423-40.

3. Gailitis, M. (1984). In: Electronic and Atomic Collisions, Elsevier Science Publishers B.V., 731-42.

4. Herrick, D.R. (1975). Phys.Pev., A12, 413-24.

5. Herrick, D.R; (1978). Phys.Rev., A17, 1-10.

6. Nikitin, S.I., Ostrovsky, V.N. (1978). J;Phys., B11, 1681-93.

SYMMETRY AND CRITICAL PHENOMENA IN ROTATIONAL SPECTRA OF ISOLATED MICROSYSTEMS

I.M. Pavlichenkov, B.I. Zhilinskii
I.V. Kurchatov Institute of Atomic Energy, Moscow

Considerable progress has been recently made in the study of rotational excited states of molecules and atomic nuclei. Modern experimental technique enables one to excite rotational states of such many-particle systems with angular momentum up to $100\hbar$. The centrifugal distortion effects are important for these momenta and the microsystem is an asymmetric rotator. (Molecules may be asymmetric even in low rotational states.) The spectrum of the rotational excitation of an asymmetric top consists of rotational multiplets (RM) (RM is a set of states with a given value of the rotational quantum number J.) We study centrifugal distortion effects in an isolated rotational band for a nondegenerate vibrational state. Set

$$H_{eff} = \sum_{\alpha,\beta} c_{\alpha\beta} J_\alpha J_\beta + \sum_{\alpha,\beta,\gamma,\delta} c_{\alpha\beta\gamma\delta} J_\alpha J_\beta J_\gamma J_\delta + \dots , \qquad (1)$$

where J_α $(\alpha = x, y, z)$ are the angular momentum operators in the fixed body frame. The Hamiltonian (1) may be obtained by the generalized density matrix method used to describe collective excitations of atomic nuclei [1]. The operator perturbation theory is widely used to obtain rotational effective operators for molecular systems. Let us expand H_{eff} (1) in the series in irreducible spherical tensor operators:

$$H_{eff}=\sum_{\mathcal{Q}\geq k}^{\infty}\vec{J}^{2\mathcal{Q}}\sum_{k=0}^{\infty}\sum_{m=0}^{2k}\{t_{2\mathcal{Q},2k,m}T_{2k,m}+ +(-1)^m t^*_{2\mathcal{Q},2k,m}T_{2k,-m}\}, \qquad (2)$$

where

$$T_{2k,m}=(-1)^m T^+_{2k,m}=f_{2k,m}(\vec{J}^2,J_z)J_-^m, \qquad (3)$$

and $\vec{J}^2=J_x^2+J_y^2+J_z^2$, $J_\pm=J_x\pm iJ_y$ (J_+ is the lowering and J_- the raising operators in the fixed-body frame), f a real-valued function. It is more convenient to regroup the sum (2) and to write the effective Hamiltonian in the form

$$H_{eff}=\sum_{m=0}^{\infty}\{g_m(\vec{J}^2,J_z)J_-^m+J_+^m g_m^*(\vec{J}^2,J_z)\}. \qquad (4)$$

The functions g_m depend on the coefficients t of the expansion (2). Invariance of (1) with respect to some discrete symmetry group additionally restricts the coefficients.

Classification of the stationary axes of rotation is useful in the study of variation of the RM structure with quantum number J. The directions of stationary rotation axes can be found from the equations

$$\{H_{eff},J_\alpha\}_{P.B.}=0,\ \alpha=x,y,z, \qquad (5)$$

where H_{eff} is a classical analogue of the Hamiltonian (4). Each stable stationary axis of rotation properly oriented in the fixed body frame is associated with the group of levels corresponding to the precession of the angular momentum vector $\vec{J}$ around this axis. An unstable stationary axis of rotation corresponds to a separatrix of the set of classical trajectories of the end of $\vec{J}$ on a sphere of radius J. The separatrix separates the trajectories corresponding to the precession around different stable axes. Some discrete symmetry group results in appearance of equivalent axes. The precession motions around the equivalent axes coincide up to symmetry transformation. A system of the equivalent axes is responsible for a fine structure of the RM. This structure is called cluster in molecular spectroscopy [2]. Thus the structure of the RM may be completely deduced from the totality of stationary rotation axes and their stability.

We distinguish regular and critical changes in the multiplet level structure. In the first case, the variation of the quantum number J, which is supposed to be rather large, results only in a smooth change of the orientation of the stationary axes bringing about monotonous dependence of the energy levels on J. These regular changes do not affect the cluster structure of the multiplet energy levels. The critical phenomena are associated with the change of either the number of stationary axes, or their stability. Near a critical point a redistribution of the part of the energy levels of the RM occurs corresponding to the changed stable stationary axis of rotation. Cluster structure of the multiplet changes passing through critical point [3]. Critical phenomena are connected with the symmetry of the system. The symmetry of the whole system and the local symmetry of rotation axes should be distinguished. The latter is of primary importance since local symmetry may be broken at critical point without any change of the symmetry of the total system, and the number of equivalent axes may increase.

Let us consider different local symmetry elements for an axis z of the body-fixed frame. If the axis z is a symmetry axis C_n, only the terms of the sum (4) having $m = nk$, $k = 0, 1, 2, \ldots$, are nonzero. If z lies in a symmetry plane, for example, in the zx plane, $g^*_m = g_m$. For the C_{nv} axes the terms of the sum (4) must simultaneously satisfy both the requirements. The symmetry axes C_{nh} do not bring about any new properties to the Hamiltonian (4) since $C_{2kh}=C_{2k}$ for this Hamiltonian and $C_{(2k+1)h} = C_{2(2k+1)}$. Note that an arbitrary rotation of the body-fixed frame around the z-axis does not break the local symmetry C_{nv}. An arbitrary rotation around an axis perpendicular to the symmetry plane does not affect the local symmetry C_s. We use the effective Hamiltonian (4) to find the classical rotation energy $E(\theta, \varphi)$ depending on the polar angles θ, φ specifying the orientation of $\vec{J}$ in the body-fixed frame. The other problem is to study the surface of the rotation energy $E(\theta, \varphi)$ near the z-axes depending on the parameter J for various local symmetry groups.

1. <u>The local symmetry group C_1</u>. The axis z is in general position and there are no additional restrictions on the functions g_m. Let us use a series expansion of the rotation energy $E(\theta, \varphi)$ in θ (assuming θ to be small) and introduce Cartesian coordinates $\xi = \theta \cos \varphi$, $\eta = \theta \sin \varphi$ in a vicinity of the north pole of the sphere of radius J centered at the origin of the body-fixed frame. It suffices to take into account cubic terms in ξ and η.

$$E(\xi,\eta) = E_0(J) + a_{10}\xi + a_{01}\eta + a_{02}\eta^2 + a_{11}\xi\eta + \\ + a_{20}\xi^2 + a_{30}\xi^3 + a_{21}\xi^2\eta + a_{12}\xi\eta^2 + a_{03}\eta^3, \tag{6}$$

where E_0 is the energy of rotation around the z axis. The coefficients a_{mn} depend on the angular momentum J and, in general, on the Euler angles Φ, ϑ, Ψ, specifying an arbitrary rotation of the body-fixed frame. This enables one to simplify the expansion (6). We eliminate the nondiagonal in ξ and η cubic terms by the transformation $\xi = \xi' + p\eta'^2$, $\eta = \eta' + q\xi'^2$. Then we choose the angles φ, ϑ, ψ so that $a_{01} = a_{20} = a_{11} = 0$. We define a critical point J_c (if there is any) from the condition $a_{10}(J_c) = 0$. Near a critical point the rotation energy surface is of the form

$$E(\xi,\eta) = E_o(J) + a_{10}\xi + a_{02}\eta^2 + a_{30}\xi^3, \tag{7}$$

and $a_{10}(J) = -\alpha(J - J_c)$.

The investigation of the surface (7) shows that for $J > J_c$, if α and $a_{30}(J_c)$ are of the same sign, and for $J < J_c$, if the signs of α and $a_{30}(J_c)$ are different, there exist two stationary axes (stable and unstable). These two axes coincide at J_c. The rotation energy around the stable axis is smaller than that around the unstable one for $a_{02}(J_c) > 0$ and vice versa for $a_{02}(J_c) < 0$. Thus, the critical phenomenon considered is connected with the appearance of two stationary axes in general position. Near a critical point we have

$$\frac{d^2E}{dJ^2} \approx -\frac{1}{2}\sqrt{\frac{\alpha^3}{3a_{30}(J_c)}}\,(J - J_c)^{-1/2}, \tag{8}$$

where E(J) is the energy of rotation around the stable axis.

2. <u>The local symmetry group C_s</u>. Let the symmetry plane C_s coincide with the plane xz of the body-fixed frame. According to Hamiltonian (4) the rotation energy surface close to the axis lying in the symmetry plane has the form

$$E(\xi,\eta) = E_o(J) + a_{10}\xi + a_{20}\xi^2 + a_{02}\eta^2 + a_{30}\xi^3 \tag{9}$$

$$+a_{12}\xi\eta^2 + a_{40}\xi^4 + a_{22}\xi^2\eta^2 + a_{04}\eta^4 + \dots ,$$

where $E_0(J)$ is the energy of rotation around the z-axis. If $a_{02} \neq 0$ for any J, the stationary axis regularly varies its orientation remaining on the symmetry plane. At a critical point $a_{10} = 0$ similar to the case of the local symmetry group C_1. This corresponds to the appearance of two stationary axes lying on the plane C_s. The critical point J_c defined by the equation $a_{02}(J_c) = 0$ is a new one. Let the initial stationary rotation axis coincide with the z-axis of the body-fixed frame. Then the energy surface (9) close to J_c can be transformed up to the required accuracy into the form

$$E(\xi,\eta) = E_0(J) + a_{20}\xi^2 + a_{02}\eta^2 + a_{04}\eta^4, \tag{10}$$

with $a_{02}(J) = -\alpha(J - J_c)$.

The critical point J_c is connected with the appearance of two additional equivalent stationary rotation axes mirror symmetric with respect to the plane C_s. The rotation energy for these two axes is $E_1(J) = E_0(J) - \alpha^2(J - J_c)^2/4a_{04}(J_c)$. Assume that the equivalent rotation axes appear for $J > J_c$. If $a_{20}(J_c)$ and $a_{04}(J_c)$ are of the same sign, two equivalent stable axes with the energy $E_1(J)$ appear for $J > J_c$ instead of one stable rotation axis with the energy $E_0(J)$ for $J < J_c$. If the above coefficients are positive and $\alpha > 0$, the stable axes correspond to the minimum of the rotational energy. If the coefficients are negative and $\alpha < 0$, the stable axes correspond to the maximum. In both cases the second derivative of the maximal or minimal rotation energy with respect to J has a discontinuity at J_c. Moreover, the number of states in the cluster doubles for $J > J_c$. If $a_{20}(J_c)$ and $a_{04}(J_c)$ have opposite signs, the unstable for $J < J_c$ rotation axis with the energy $E_0(J)$ becomes stable for $J > J_c$. The energy $E_0(J)$ is maximal for $d > 0$, $a_{20}(J_c) < 0$, $a_{04}(J_c) > 0$ and minimal for the signs of these coefficients opposite to the indicated.

3. <u>The local symmetry group C_2 and C_{2v}</u>. If the symmetry axis C_2 coincides with the z-axis of the body-fixed frame, then according to (4) the rotation energy surface close to this direction is of the form

$$E(\xi,\eta) = E_0(J) + a_{20}\xi^2 + a_{11}\xi\eta + a_{02}\eta^2 + a_{40}\xi^4 + a_{31}\xi^3\eta + a_{22}\xi^2\eta^2 + a_{13}\xi\eta^3 + a_{04}\eta^4 + \ldots \quad (11)$$

All the terms containing odd powers of η should be omitted from eq.(11) for the C_{2v}-axis. In both cases the axis z is a stationary one with the rotation energy $E_0(J)$. The critical point is defined equating to zero either $a_{20}(J_c)$ or $a_{02}(J_c)$ (but not simultaneously). In these cases the expansion (11) can be easily reduced to (10) or to a similar expression where ξ and η are interchanged. Thus the same type of the critical phenomena takes place for the symmetry groups C_2 and C_{2v} as for the C_s group.

Another type of the critical phenomena exists for the local symmetry considered when all the terms of Hamiltonian (4) containing $J_\pm^4$, $J_\pm^6$, etc., may be omitted. Let us consider e.g. the rotation of a nonrigid asymmetric rotator described by the Hamiltonian

$$H = AJ_x^2 + BJ_y^2 + CJ_z^2 + d(\vec{J}^2, J_z^2) + (1/2)(J_+^2 + J_-^2)\, e(\vec{J}^2, J_z^2), \quad (12)$$

where $A < B < C$ are rotational constants. The energy of the rotation around the axes x and y equals

$$E_x(J) = AJ^2 + d(J^2, 0) + J^2 e(J^2, 0),$$
$$E_y(J) = BJ^2 + d(J^2, 0) - J^2 e(J^2, 0). \qquad (13)$$

respectively. The critical point is defined by the condition $2e(J_c^2, 0) = B - A$. Close to J_c we have $e(J^2, 0) = (B-A)/2 + \alpha(J - J_c)$ and the rotation energy surface may be approximated as

$$E(J,\varphi) = E_x(J)/2 + E_y(J)/2 + \alpha J^2 (J - J_c) \cos 2\varphi, \qquad (14)$$

if $C - (A+B)/2 \pm e(J_c^2, 0) > 0$. Two rotational levels $E(J, 0)$ and $E(J, \pi/2)$ corresponding to different orientations of the angular momentum vector intersect at J_c. If $\alpha > 0$, low branches of the intersecting curves correspond to stable rotation axes; high branches correspond to unstable ones. Thus, $\vec{J}$ rotates at J_c by $\pi/2$ with respect to the body-fixed frame. The first derivative with respect to J of the energy of rotation around the stable axes has a jump discontinuity at J_c [4].

4. <u>The local symmetry group C_3 and C_{3v}</u>. Let the axis z of the body-fixed frame be the symmetry axis C_3 or C_{3v}. The difference between energy surfaces for these two axes is not essential. Near the C_{3v} axis the rotation energy surface has the form

$$E(\theta, \varphi) = E_0(J) + a\theta^2 + b\theta^3 \cos 3\varphi, \qquad (15)$$

where $E_0(J)$ is the energy of rotation around the z axis which is a stationary rotation axis. A critical point is determined from $a(J_c) = 0$ $(a(J) = -\alpha(J - J_c)$ near J_c). The type of stability of the rotation axis changes at J_c. $E_0(J)$ is transformed from minimum to maximum for $\alpha < 0$ and vice versa for $\alpha > 0$. This transition is due to three equivalent unstable rotation axes whose directions are close to that of z . A cardinal rearrangement of the cluster structure takes place at J_c. The energy of the clusters depends differently (increases or decreases) on the angular momentum projection M on the z-axis for $J < J_c$ and for $J > J_c$.

5. <u>The local symmetry group C_4 or C_{4v}</u>. The critical phenomena for these two groups are the same. They are due to the behaviour of the rotation energy surface near the z-axis, coinciding with the C_{4v} symmetry axis. The energy surface has the form

$$E(\theta,\varphi) = E_o(J) + a\theta^2 + (b + c\cos 4\varphi)\theta^4. \tag{16}$$

The critical point can be found from the condition $a(J_c)=0,\ a(J)=-\alpha(J - J_c)$. Without loss of generality we may set $c(J_c) > 0$. Two types of transitions are possible. If $|b(J_c)| > c$, for $b > c$ and $\alpha > 0$ the minimum $E_o(J)$ turns for $J > J_c$ into maximum and a new minimum with the energy $E_o(J) - \alpha^2(J - J_c)^2/4(b - c)$. If $|b(J_c)| > c$, $b < -c$ and $\alpha > 0$, the maximum $E_o(J)$ turns for $J > J_c$ into minimum and a new maximum with the energy $E_o(J) - \alpha^2(J - J_c)^2/4(b+c)$. The additional minimum (maximum) corresponds to four equivalent rotation axes close to the z-axis and rotated with respect to each other by $\pi/2$. The second derivative of the minimal (maximal) rotation energy with respect to J has a jump discontinuity at the critical point. Another evidence of this critical phenomenon is the appearance of the eight-fold (or four-fold) degenerate clusters beyond the transition point. If $|b(J_c)| < c$, the critical phenomenon is similar to that discussed for the C_{3v} group. For $\alpha > 0$ the maximum $E_o(J)$ changes for $J > J_c$ into the minimum $E_o(J)$. For $\alpha < 0$ the inverse transition takes place.

6. The local symmetry group C_n and C_{nv}, $n \geq 5$. These axes may exist only for large and heavy molecules. The energy surface close to the z-axis which coincides with the symmetry axis C_n or C_{nv} may be written in the form

$$E(\theta,\varphi) = E_0(J) + a\theta^2 + b\theta^4 + c\,\theta^n \cos n\varphi . \qquad (17)$$

A critical point can be determined from the condition $a(J_c) = 0$. Only one type of critical phenomenon takes place for all the above-mentioned axes. This phenomenon is similar to the transition for the C_4 or C_{4v} axes for $|b(J_c)| > c$, due to the smallness of the last term in the expression (17) defining the number of the equivalent axes.

All the considered critical phenomena that may exist for rotational excitations of finite-particle systems are presented in Table. A graphic representation of the critical phenomenon in the plane (E,J) is given. The solid lines correspond to the energy of rotation around the stable axes. A stable axis corresponds to a minimum if the solid line is lower than the dotted one and vice versa. A dotted line shows the energy of rotation around the unstable axis associated with a saddle point of the energy surface $E(\theta, \varphi)$. The lines in the figures are numbered according to the multiplicity of equivalent axes.

References

1. Belayev, S.T., Zelevinskii, V.G.(1973). Sov. J. Nucl. Phys. 16, 657; 17, 269.
2. Harter, W.G., Patterson, C.W. (1984). J. Chem. Phys. 80, 4241.
3. Zhilinskii, B.I. (1979). J. Mol. Spectrosc. 78, 203.
4. Pavlichenkov, I.M. (1982). Sov. Phys. JETP, 82, 5.

Symmetry axis	Energy surface $E(\theta, \varphi)$ close to symmetry axis z	Critical point	Graphic representation of singularity (E, J)	Singularity at critical point
C_1	$E_0(J) + a_{10}\xi + a_{02}\eta^2 + a_{30}\xi^3$	$a_{10}(J_c)=0$		$E''(J_c) = \infty$
C_s C_2, C_{2v}	$E_0(J) + a_{20}\xi^2 + a_{02}\eta^2 + a_{04}\eta^4$	$a_{02}(J_c)=0$		$\Delta E''(J_c) = -\dfrac{\alpha^2}{2a_{04}(J_c)}$
C_3, C_{3v}	$E_0(J) + a\theta^2 + b\theta^3\cos 3\varphi$	$a(J_c)=0$		—
C_4, C_{4v}	$E_0(J) + a\theta^2 + \theta^4(b + c\cdot\cos 4\varphi)$	$a(J_c)=0$		$\Delta E''(J_c) = \dfrac{-\alpha^2}{2\{b(J_c) - c(J_c)\}}$, —
C_n, C_{nv} $n \geqslant 5$	$E_0(J) + a\theta^2 + b\theta^4 + c\,\theta^n\cos n\varphi$	$a(J_c)=0$		$\Delta E''(J_c) = -\dfrac{\alpha^2}{2b(J_c)}$

STRUCTURE AND FEATURES OF THE CONSTANT NON-ABELIAN FIELD HAMILTONIANS GENERATED BY CARTAN ALGEBRAS

V.Vanagas

Institute of Physics, the Lithuanian SSR Academy of Sciences

<u>1. Introduction and general definitions.</u> We discuss the features of the Hamiltonians published in [1] and reported in [2], generated by the infinitesimal operators of Cartan algebras, acting in the irreducible space $R^{(\lambda)}$ of compact matrix groups. In case of the unitary group algebra in some approximation (see for details [1-10] in [1]), such Hamiltonians have been used for the gluon vacuum model.

The expression of the constant non-abelian field Hamiltonian H in terms of infinitesimal operators of the SU(N) algebra have been obtained and discussed in [3-5]. It consists of the kinetic energy H_k and the potential energy H_p terms

$$H_k = \mathrm{tr} \sum_{s=1}^{r_o} \frac{\partial}{\partial \bar{a}^{+s}} \frac{\partial}{\partial a^s}$$

$$H_p = \mathrm{tr} \sum_{s\neq s'=1}^{r_o} \left[a^s, a^{s'}\right]^+ \left[a^s, a^{s'}\right], \qquad (1)$$

where $\left[a^s, a^{s'}\right]$ is the commutator of the following linear forms of the SU(N) infinitesimal operators $\hat{I}_k$

$$a^s = \sum_{k=1}^{p(N)} x_k^s \hat{I}_k. \qquad (2)$$

Operators $\hat{I}_k$ act in the SU(N) irreducible space $R^{[1-1]}$ of the ajoint representation $[1-1] \equiv [1\ 0 \ldots 0 - 1]$,

x_k^s are real variables, $k = 1, 2, \ldots, n(N)$, where $n(N)^k$ is the number of $\hat{I}_k$ and r_o is the space dimension. For the pilot studies $r_o = 2$, the realistic case is $r_o = 3$. In case of the SU(2) algebra with $r_o = 2, 3$ the explicit expression of H in terms of x_k^s has been obtained in [6,7], where the features of the states of H in the hyperspherical basis have also been discussed.

Let us start with some generalization of the Hamiltonian described. In addition to the SU(N) infinitesimal operators acting in the space $R^{(U)}$ of the ajoint representation U, we will also consider the operators of the unitary U(N), special unitary SU(N), orthogonal O(N) and symplectic $S_p(N)$ group algebras, acting in the space $R^{(\lambda)}$ of an arbitrary irreducible representation λ. Instead of real variables x_k^s we will use complex ones z_k^s. Our aim is to obtain H depending on z_k^s in an explicit form and to examine some properties of the operators H and their states.

2. Explicit expressions of the Hamiltonians. Denote the algebras mentioned above as G(N). If we take (2) in the matrix representation for $\hat{I}_k$ acting in $R^{(\lambda)}$ then a^s also turns into a matrix operator in $R^{(\lambda)}$. Using the expressions of the structure constants of G(N) in terms of Clebsch-Gordon coefficients [8,9], the Hamiltonian H_k can be presented in the following form [1]

$$H_k(G(N)) = \frac{d_u}{d_\lambda} (K^{(2)}_{G(N)}(\lambda))^{-1} \sum_{k,s} \frac{\partial}{\partial \overset{*}{z}{}_k^s} \frac{\partial}{\partial z_k^s} , \qquad (3)$$

suitable for SU(N), O(N) and $S_p(N)$; the expression of H_k for U(N) see in [1,2]. In (3) d_u and d_λ denote the dimensions of the G(N)-irreducible representations u and λ, $K^{(2)}_{G(N)}(\lambda)$ - the eigenvalues of Casimir operators. We use the infinitesimal operators of G(N), defined so that the eigenvalues of $K^{(2)}_{G(N)}(\lambda)$ have the expressions

(18.29), (18.34) and (18.35) of [10]. As shown in [1] in the cases SU(N), O(N) and $S_p(N)$ the potential energy has the following form

$$H_p(G(N)) = \frac{d_\lambda}{d_u} \frac{K^{(2)}_{G(N)}(\lambda)}{(m(N))^2} \sum_{s\neq s'=1}^{r_o} \sum_{k_1k_2k_1'k_2'} \overset{*}{z}{}^{s}_{k_1} \overset{*}{z}{}^{s'}_{k_2} z^{s}_{k_1'} z^{s'}_{k_2'} \times \tag{4}$$

$$\times \sum_k C^{u\ u\ u}_{k_1 k_2 k} C^{u\ u\ u}_{k_1' k_2' k},$$

where m(N) equals $\sqrt{2N}$, $\sqrt{N-2}$ and $\sqrt{N+2}$ for SU(N), O(N) and $S_p(N)$ respectively. The explicit formulas of the quantity defined in (4) as the sum over k follows from the results presented in [8,9]. IN [1] it has been proved, that for SU(N) case (4) implies

$$H_p(SU(N)) = \frac{d_\lambda}{2N^2(N^2-1)} K^{(2)}_{SU(N)}(\lambda) \sum_{s\neq s'=1}^{r_o} (V^{ss'}_p + \overset{*}{V}{}^{ss'}_p). \tag{5}$$

The expression of $V^{ss'}_p$ is given in [2]. For an SU(2) - irreducible representation denoted by J, (5) reduces to

$$H_p(SU(2)) = \frac{J(J+1)(2J+1)}{24} \sum_{s\neq s'=1}^{r_o} \sum_{k\neq k'} \overset{*}{z}{}^{s}_{k} \overset{*}{z}{}^{s'}_{k} (z^{s}_{k} z^{s'}_{k'} - z^{s'}_{k} z^{s}_{k'}), \tag{6}$$

where k = 11, 21, 12 and similarly for k'; the last expression for the real variables x^s_k, J = 1/2 and r_o = 2, 3 has been obtained in [6].

The complexity of the potential energy (5) originates from the properties of the SU(N) algebra. The potential energy, derived using the structure constants of the U(N) algebra is more simple. In [1] it is proved that in this case

$$H_p(U(N)) = \frac{d_\lambda}{d_u} K^{(2)}_{SU(N)}(\lambda) \sum_{s\neq s'=1}^{r_o} \sum_{p',q'=1}^{N} \overset{*}{Y}{}^{ss'}_{p'q'} Y^{ss'}_{p'q'} . \tag{7}$$

The expression of $Y^{ss'}_{p'q'}$ is presented in [1,2].

3. Irreducible components and decomposition of H. In the following discussion we will treat z_k^s as real variables x_k^s. All Hamiltonians considered are scalars with respect to the colour group SO_{r_o}, acting on the indices s of x_k^s. Thus the SO_{r_o}-irreducible representation ω_o is the integral of motion; the SO_{r_o}-scalar states correspond to $\omega_o = (0)$ (in the case $r_o = 3$, ω_o means the angular momentum L_o). Due to the polynomial-type structure of (4) the algebraic methods, developed in the microscopic nuclear theory, can be used to study the properties of H. We will sketch some results, obtained in [1] in this direction.

A remarkable parallel between the microscopic models of n-particle nucleus, introduced in the framework of the restricted dynamics [10-15], and the states of the non-abelian field Hamiltonians can be traced. In particular, the following O(N) irreducible decomposition has been obtained [1]

$$H(G(N)) = H^{(0)}_{coll}(G(N)) + H'(G(N)), \tag{8}$$

where

$$H'(G(N)) = \sum_{W\neq(0),\nu_W} H^{(W)}_{\nu_W}(G(N)). \tag{9}$$

In (8) and (9), (0) and (W) denote the scalar and non-scalar O(N)-irreducible representations respectively contained in the direct product $(1)_n \times (1)_n \times (1)_n \times (1)_n$

and ν_W is a basis of (W). The eigenfunctions Ψ of H can be decomposed in terms of the $U(r_o n)$-irreducible basis, labelled by the irreducible representations of the unitary scheme chain groups, orthogonal scheme chain groups, or by the symplectic scheme chain groups, equivalent to the unitary scheme (see [1,2,16,17]). The hyperspherical basis [18-20] closely related to the orthogonal scheme basis can be also used. Combining the decomposition (8) with the algebraic properties of the basis, just mentioned above both Hamiltonians H of the constant non-abelian fields and the properties of their states can be investigated.

In order to explain the physical meaning of the first term in (8) make use of the following change of variables

$$x_k^s = \sum_{s_o=1}^{r_o} \rho^{(s_o)} \, {}^{(1)}D^{r_o}_{s_o,s}(g_o) \, {}^{(1)}D^{n}_{n-r_o+s_o}, k^{(g)}, \qquad (10)$$

suitable for $r_o \leq n$, where g_o and g denote some sets of angle variables and $\rho^{(s_o)}$ the radial type variables (see for details [21,11-13]; in [12] the case of $r_o > n$ is also discussed). Using the following chains of groups: $U(r_o n) \supset U(r_o) \times U(n)$, $U(r_o) \supset SO(r_o)$, $U(n) \supset O(n)$ and $O(n) \supset O(N) \times O(N)$ in the case of $U(N)$ algebra, when $n = N^2$, in [1] it has been proved, that (9) for the potential energy can be presented in the form

$$H_p'(U(N)) = H_{p(11)}(U(N)) + H_{p(2)}(U(N)), \qquad (11)$$

where (11) and (2) denote $O(N)$-irreducible properties of H_p'. The first term in (8) for H_p in case of $U(N)$ has the following expression

$$H_{pcoll}(U(N)) \equiv H^{(0)}_{pcoll}(U(N)) =$$

$$= \frac{2\, d_\lambda}{N^3} \cdot K^{(2)}_{SU(N)}(\lambda)\, \left(\frac{r_o - 1}{r_o}\, \rho^4 - \beta^2\right),$$

where ρ and β are the collective type variables, used in the microscopic nuclear theory (see [1,2]). According to the notion of microscopic collectivity introduced in [22, 23], (12) is the collective potential. The dependence of (12) only on ρ and β proves that H_{pcoll} is the $O(N^2)$-scalar part of $H_p(U(N))$, thus (12) added to the $O(N^2)$-scalar part of the kinetic energy operator gives the first term in (8), defined as the collective part $H^{(0)}_{coll}$ of H. The states of $H^{(0)}_{coll}$ exhibit collective properties of the system and, due to the symmetry properties of $H^{(0)}_{coll}$, are characterized by the additional integral of motion, labelled by the O(n)-irreducible representations ω. The Hamiltonians (11) violates the integral of motion ω. We will show, that this violation may vanish, i.e. tnat in some cases, $H'(G(N)) = 0$.

4. Particular cases. Despite seeming complexity of the Hamiltonian H(U(N)), we disclosed only two O(N)-irreducible terms, presented in (11), and have proved, that the O(n) scalar term (12) depends only on the two collective variables. This result, obtained in [1], indicates, that the collective degrees of freedom and thus O(n)-irreducible representations ω, may be important for describing the states of H(G(N)). We will confirm this prediction by the SU(2) Hamiltonian. Let us take the expression (6) treating z^s_k as real variables x^s_k and change them into the new ones using (10). In [1] it is proved, that in the case $r_o = 2, 3$

$$H_p(SU(2)) = \frac{J(J+1)(2J+1)}{12}\left(\frac{r_o-1}{r_o}\rho^4 - \beta^2\right). \qquad (13)$$

Taking into account, that for the SU(N) algebra, H_k in the variables X_k^s is $O(N^2-1)$-scalar operator, we conclude, that in both 2+1 and 3+1 case (i.e. r_o dimensionality + time)

$$H(SU(2)) = H_k(SU(2)) + H_p(SU(2)) = H^o_{coll}(SU(2)), \qquad (14)$$

i.e. in this case in (8) only the first term survives. The eigenvalues for Schrodinger equation of the Hamiltonian (14) depend on L_o, ω, $E_{coll} = E_{coll}(L_o, \omega)$.

The purely collective nature of the Hamiltonian (14) indicates that the collective effects may play important role in the description of the features of the gluon vacuum states in the constant field approach. Many methods, including the operatorial decomposition as well as the calculation technique, developed in the nuclear theory and described in [10-15], can be adapted for the investigations in this direction.

References

1. Vanagas, V. (1985). Sov. J.Nucl.Phys., 41, 1474.
2. Vanagas, V. (1985). The Hamiltonians of the constant non-abelian Fields. In: Proceedings of the Ninth European conference on several-body problems in physics, Tbilisi, Georgia, USSR, - 1984. Ed. Fadeev, L. D., Kopaleishvili, T.I. World Scientific, Singapore-Philadelphia, pp. 28-36.
3. Simonov, Yu.A. (1983). Preprint ITEP-14.
4. Simonov, Yu.A. (1984). Phys. Lett., 136B, 105.
5. Simonov, Yu.A. (1985). Sov.J;Nucl.Phys., 41, 1311.
6. Badalyan, A.M. (1983). Sov.J.Nucl. Phys., 38, 779.
7. Badalyan, A.M. (1984). Sov.J.Nucl.Phys. 39, 947.
8. Taurinskas, M., Vanagas, V. (1975). Lief.fiz.rink.,15, 3, 329.

9. Vanagas, V., Taurinskas, M. (1976). Lief.fiz.rink., 16, 3, 341.

10. Vanagas, V. (1971). Algebraic methods in nuclear theory, Mintis, Vilnius.

11. Vanagas, V. (1976). Sov.J.Part.Nucl., 7, 309; (1980). 11, 454.

12. Vanagas, V. (1977). The microscopic nuclear theory wit the framework of the restricted dynamics. Lecture Notes, Univ. of Toronto.

13. Vanagas, V. (1981). The microscopic theory of the collective motion in Nuclei. Lecture Notes in: Group theory and its applications in physics - 1980. Latin American school in physics, Maxico City. American Institute of Physics, AIP Conference Proceedings, N.Y. 71, p.220.

14. Vanagas, V. (1982). Introduction to the microscopic theory of the collective motion in nuclei. Lecture Notes in: Proc. V-th International School on Nuclear and Neutron Physics and Nuclear Energy, Varna - 1981. Bulg.Acad.Sc., Sofia, pp.185-229.

15. Vanagas, V. (1984). The symplectic models of nucleus. In: Group-theoretical methods in Physics. Proc. Intern Seminar, Zvenigorod, 1982,

Ed. Markov, M.A., Man'ko, V.I., Shabad, A.E. N.Y., Gordon and Breach, A, p. 259.

16. Kretzschmar, M. (1960). Z.f.Phys., 157, 433; (1960). 158, 284.

17. Bargman, V., Moshinski, M. (1960). Nucl.Phys., 18, 697; (1961). 23, 177.

18. Simonov, Yu.A. (1966). Sov.J.Nucl.Phys., 3, 630.

19. Badalyan, A.M., Simonov, Yu.A. (1966). Sov.J.Nucl. Phys., 3, 1032; (1969). 9, 69.

20. Surkov, E.L. (1967). Sov.J.Nucl.Phys., 5, 908.

21. Dzyublik, A.Ya. (1971). Preprint ITP-71-122R, Inst. Theoret.Phys., Kiev.

22. Vanagas, V. (1976). The microscopic theory of the nuclear collective motion. In: Proc. Intern. Symposium on Nuclear Structure, Balatonfüred, Hungary, 1975. Ed. I.Fodor-Lovas, G.Palla, Budapest,1, 167.

23. Vanagas, V. (1976). Sov.J.Nucl.Phys., 23, 950.

LOW-LYING LEVELS OF SPHERICAL NUCLEI, BOSON MODELS AND A NEW TYPE OF SYMMETRY

O.Vorov, V.Zelevinsky
Institute of Nuclear Physics, Novosibirsk 630090

Collective spectra of low-lying excited states in complex atomic nuclei are not completely understood [1]. Magic nuclei (those with occupied nucleon shells) have a stable spherical shape and multipole vibrations around it with small amplitudes and rather high frequencies ($\omega \simeq 1.5 - 2$ MeV). Ground states of all even-even nuclei have quantum numbers $J^{\pi} = 0^{+}$ of the angular momentum J and parity π. This is connected with the condensate of Cooper pairs due to strong superconducting pair correlations and does not imply the sphericity of the nuclear shape. It is well known that the interaction between valence nucleons and the polarization of the magic core lead to the instability of the spherical shape. As a result, nuclei with many nucleons in partially occupied outer shells have a stable quadrupole deformation. The pattern of the low-lying spectrum of a well deformed nucleus is again relatively simple being that of distinctly separated rotational bands built on various intrinsic configurations. The most difficult problem is that of description of pretransitional soft nuclei which are intermediate between magic and deformed ones. For definiteness, below we confine ourselves to even-even nuclei.

Experimental data manifest clearly [2] the collective character of low-lying states in nuclei under consideration. Almost all observed levels can be classified according to the multiplet scheme corresponding to irreducible representations of SU(5). This group is generated by operators $d_\mu^+ d_{\mu'}$, where d_μ^+ and d_μ are operators of creation and annihilation of a quadrupole quantum with the angular momentum $l = 2$ and projection μ. Unlike the case of magic or nearmagic nuclei, typical energy intervals ω between the multiplets are small here ($\omega \approx$ 0.5 MeV) as compared with breaking energies $2E \approx 2$ MeV of Cooper pairs. Among the electric quadrupole transitions between multiplets one can find significantly enhanced ones their probabilities being one or two orders higher than it follows from single-particle estimates.

Thus, this part of the spectrum is dominated by the soft quadrupole collective mode. At small ω, the vibration amplitude is large ($\sim \frac{1}{\sqrt{\omega}}$) so that other (non-collective) degrees of freedom become aware of the slowly changing field of the quadrupole symmetry.Therefore nonlinear phenomena are essential resulting in the effective strong anharmonicity. Along with the virtual deformation, the possibility of the collective rotation around the axis perpendicular to that of quadrupole motion appears. This is the origin of the difficulties of the development of the consistent microscopic theory of soft spherical nuclei.

Phenomenological approaches to the problem can be divided into two types. The classical Bohr-Mottelson description [1] assumes that the collective Hamiltonian H_c exists covering the whole subspace of states generated by the quadrupole collective motion. This Hamiltonian can be expressed in terms of coordinates $\alpha_\mu =$

$$= \frac{1}{\sqrt{2\omega}} [d_\mu + (-1)^\mu d_{-\mu}^+] \equiv \frac{1}{\sqrt{2\omega}} d_\mu^{(+)}$$ and momenta

$$\pi_\mu = -i\sqrt{\frac{\omega}{2}} [d_\mu - (-1)^\mu d_{-\mu}^+] \equiv -i\sqrt{\frac{\omega}{2}} d_\mu^{(-)}$$ of quadrupole phonons. Since the motion is adiabatic, one should

expect that the terms of higher order with respect to π are small so that

$$H_c = U(\alpha) + \tfrac{1}{2} \sum_{\mu\mu'} \left[B_{\mu\mu'}(\alpha), \pi_\mu \pi_{\mu'} \right]_+ \qquad (1)$$

where the potential energy $U(\alpha)$ and the inertia tensor $B_{\mu\mu'}(\alpha)$ depend on the coordinates α_μ only. The general Hamiltonian (1) contains seven independent tensor structures which are arbitrary functions of the rotational invariants $(\alpha^2)_{oo}$ and $(\alpha^3)_{oo}$. Here $(\ldots)_{LM}$ means the vector coupling to the total angular momentum L with the projection M. One actually truncates in some way the expansions of $U(\alpha)$ and $B_{\mu\mu'}(\alpha)$ in these invariants. Then it is necessary to diagonalize the Hamiltonian containing many fitted parameters [3]. Since the anharmonicity is not weak, the number $N_d = \sum_\mu d^+_\mu d_\mu$ of bare quanta is not conserved.

In more recent alternative approaches (the interacting boson approximation, IBA [4,5])bosons are identified with the images of the fermion pairs in the limited subspace of collective states. Then the total boson number N_B is uniquely determined by the nucleon number in the outer shells and should be conserved. In practical calculations one takes into account usually s- and d-bosons with the angular momentum $\ell = 0$ and $\ell = 2$ respectively. For fixed $N_B = N_s + N_d$ such a model corresponds to SU(6). The effective nonconservation of N_d arises due to the excitation $s \rightarrow d$ of the condensate s-bosons into the d-state.

The phenomenological schemes of both types give, in general, a reasonable description of the data. But the abundance of free parameters together with the lack of the selection principles cause dissatisfaction. The attempts to obtain the collective Hamiltonian (1) microscopically from the nucleon interaction meet extremely complicated calculations [6,7] using the boson expansion of fermion operators [6,8] . The boson expansion procedures are rapidly convergent at the weak anharmonicity. This is not the case in real nuclei. Apart from that, it is necessary to take into account the coherent responce of noncollective degrees of freedom [9,10] particularly the virtual rotation [11] . As for the different versions of IBA the serious shortcomings of its microscopic justification are not overcome [2,12] . Specific predictions of this model (e.g. the cut-off of rotational bands) have no experimental support.

The analysis [13,14] with the aid of the microscopic estimates of the main features of the quadrupole motion (collectivity and adiabaticity) makes it possible to establish the most important contributions to the collective Hamiltonian, namely the quartic anharmonicity and the virtual rotation. Hence, we are able to formulate a simple phenomenological scheme taking into account the quartic anharmonicity and angular momentum effects (QAAM [13]).

It can be shown that the QAAM scheme gives a reasonable agreement with the vast set of data for the typical soft spherical nuclei. The attractive advantages of the method are the small number of free parameters and the simplicity of the computations (the essential part of it can be carried out analitically).

Thus , we find that the new type of symmetry is approximately realized in the soft nuclei: the symmetry of the quartic five-dimensional oscillator. As it was shown [14] the ground band of ^{100}Pd gives a good example of such a symmetry. The angular momentum effects as well as minor corrections due to other anharmonic terms are superimposed on the main symmetry.

In accordance with the microscopic analysis and with the experimental data, the square of the five-dimensional angular momentum, i.e. the Casimir operator of O(5)

$$C\,[O(5)] = \tfrac{1}{2} \sum_{\mu\mu'} [d^+_\mu d_{\mu'} - (-1)^{\mu+\mu'} d^+_{-\mu'} d_{-\mu}]\,[d^+_{\mu'} d_\mu - (-1)^{\mu+\mu'} d^+_{-\mu}\, d_{-\mu'}], \qquad (2)$$

is an approximate constant of motion in spite of the strong anharmonicity, This operator has quantized eigenvalues:

$$C\,[O(5)] = v(v+3) \qquad (3)$$

where the integer v is seniority number of nonpaired bosons. The $O(5)$-symmetry is confirmed by the multiplet structure as well as by the selection rule $|\Delta\, v| = 1$ for the enhanced collective E2-transitions. In our scheme this symmetry follows naturally from the dominance of the quartic anharmonicity. Let us consider an arbitrary quartic phonon Hamiltonian $H^{(4)}$ with the only restriction being the time reversal invariance:

$$H^{(4)} = \sum_{L=0,2,4} \{ \gamma_L((\alpha^2)_L(\alpha^2)_L)_{00} + \sigma_1([\,(\alpha^2)_L(\pi^2)_L]_+)_{00} + \gamma'_L((\pi^2)_L(\pi^2)_L)_{00} \} \qquad (4)$$

This Hamiltonian is rotationally invariant in the five-dimensional space.

The proof is based on the isolation of photon pairs coupled to the three-dimensional angular momentum $L=0$ (the pair number is $n = (N_d-v)/2$ and the construction of $O(5)$-invariant operators P^+ and P creating and annihilating these condensate pairs without changing the seniority v:

$$P^+ = \tfrac{1}{2} \sum_\mu (-1)^\mu\, d^+_\mu\;\; d^+_{-\mu}\;, \quad P = \sum_\mu (-1)^\mu\, d_\mu\, d_{-\mu}. \qquad (5a)$$

Together with the operator

$$P_o = \quad (N_d + 5/2)/2 \qquad (5b)$$

the operators (5a) generate the noncompact group SU(1,1) with the Casimir operator C [SU(1,1)] connected with (3):

$$C\,[SU(1,1)] = P_o^2 - \tfrac{1}{2}(PP^+ + P^+P) = \tfrac{1}{4}C\,[O(5)] + 5/16 \quad (6)$$

Now one can transform the Hamiltonian (4) to an explicitly O(5)-invariant form

$$H^{(q)} = \lambda(2P_o+P+P^+)^2 + \lambda'(2P_o-P-P^+)^2 + \quad (7)$$
$$+ \alpha[2P_o-P-P^+, 2P_o+P+P^+]_+ + \beta C\,[SU(1,1) + \gamma J(J+1) + \zeta$$

where J stands for the O(3)'s angular momentum and coefficients are as follows:

$$\lambda = \frac{1}{5}\sum_{L=0,2,4}\sqrt{2L+1}\,\varphi_L\gamma_L\,, \quad \lambda' = \frac{1}{5}\sum_{L=0,2,4}\sqrt{2L+1}\,\varphi_L\gamma_L'\,,$$

$$\alpha = \frac{1}{5}\sum_{L=0,2,4}\sqrt{2L+1}\,\varphi_L\sigma_L\,, \quad (8a)$$

(8b)

$$\beta = \frac{8}{7}(\sigma_4 + \frac{4}{\sqrt{5}}\sigma_2), \quad \gamma = \frac{2}{7}(\frac{\sigma_2}{\sqrt{5}} - \frac{\sigma_4}{\sqrt{3}}), \quad (8c)$$

$$\zeta = -\frac{5}{14}(11\,\sigma_4 + \frac{8}{\sqrt{5}}\sigma_2)$$

$$\varphi_o = 1\,, \quad \varphi_2 = \varphi_4 = \frac{2}{7} \quad (8d)$$

The Hamiltonian (7) together with the harmonic term $H^{(2)}$ containing a bare photon frequency Ω can be solved analytically. The appropriate method of solution is the v-dependent canonical transformation [15] which chooses optimum parameters of the boson pair condensate and renormalized phonon frequency ω_v for each subspace with a fixed v . For the lowlying states, such an approximate procedure guarantees the high precision of results avoiding in many cases the numerical diagonalization. The stationary states can be labelled with the number $\tilde{N} = 2\tilde{n}+v$ of new (renormalized) quanta and the exact constants of motion v, J and M.

The most interesting case is that of the soft collective mode (the adiabatic limit). Here the main terms in (7) are the first one and the quasirotational term $\gamma J(J+1)$. The physical meaning of these terms was discussed above. Other terms can be readily taken into account as corrections. Note that the quasirotational correction should be considered for all states except the single-phonon one ($J=2$, $\tilde{n} = 0$, $v = 1$). By definition, this is the pure collective state which serves as a reference point for the calculation of the responce of noncollective degrees of freedom.

In the limit adiabatic situation one can neglect the harmonic term $H^{(2)}$ $(\Omega \rightarrow 0)$ so the model has only one parameter γ to calculate energy ratios. Transition probabilities for the enhanced (allowed in the harmonic approximation [1]) E2-transitions with $|\Delta N_d| = |\Delta v| = 1$ are determined by the operator

$$T_\mu^{(E2)} = d_\mu^{(+)} + \mathcal{K}\,(d^{(+)2})_o\, d_\mu^{(+)} \tag{10}$$

Here the second term arises from the boson expansion of the fermion quadrupole operator when the dominant role of the quartic anharmonicity is properly taken into account. We have neglected in (10) small terms $\sim (d^2)_\mu$ responsible for the weak (forbidden in the harmonic approximation) transitions and for the quadrupole moments of excited states.

The probabilities of intraband E2-transitions inside the yrast band (quasirotational band of states with aligned phonons, $J = 2v$, $\tilde{n} = 0$) and interband transitions from the β-band (one boson pair, $\tilde{n} = 1$) to the yrast band are

$$B(E2;\ v+1,\ \tilde{n}=0,\ J=2v+2 \rightarrow v,\ \tilde{n}=0, J=2v) =$$

$$= \frac{v+1}{\omega_v} K_v^{2v+7} A_v(\varkappa) \tag{11a}$$

$$K_v = \frac{2\sqrt{\omega_v \omega_{v+1}}}{\omega_v + \omega_{v+1}}, \quad A_v(\varkappa) = 1 + \frac{2\varkappa}{\sqrt{5}} \frac{2v+7}{\omega_v + \omega_{v+1}} \tag{11b}$$

$$B(E2; v, \tilde{n}=1, J=2v \rightarrow v+1, \tilde{n}=0, J=2v) =$$

$$= \frac{4v+5}{4V+1} \frac{v+1}{v+5/2} \frac{1}{\omega_{v+1}} K_v^{2v+9} B_v(\varkappa) \tag{12a}$$

$$B_v(\varkappa) = 1 - (v+\frac{5}{2}) \frac{\omega_{v+1} - \omega_v}{2\,\omega_v} + \frac{4\varkappa}{\sqrt{5}} \frac{2v+7}{\omega_v + \omega_{v+1}} \Big[1 - (v+\frac{5}{2}) \frac{\omega_{v+1} - \omega_v}{4\,\omega_v}\Big] \tag{12b}$$

As an illustrative example we consider a typical soft spherical nuclei ^{104}Ru. Fig. 1 shows the comparison of calculations (11) for $\varkappa = -0.22$ (solid line) with the experimental data [16] . Dashed lines correspond to predictions of the complicated IBA versions (IBA-(2) takes into account s- and d-bosons of two kinds, "proton" and "neutron" ones, whereas IBA+g adds g-bosons with l=4).

In the following Table we have collected the reduced probabilities [16-18] of allowed E2-transitions in ^{104}Ru (columns 2 and 5; the experimental errors of the last digits are indicated in parentheses). Columns 3 and 6 show the results of present one-parameter model (10) for $\varkappa = -0.22$.

Energy levels (in units of the energy $E(2_1^+)$) of yrast states are compared in Fig. 2 with the calculations of

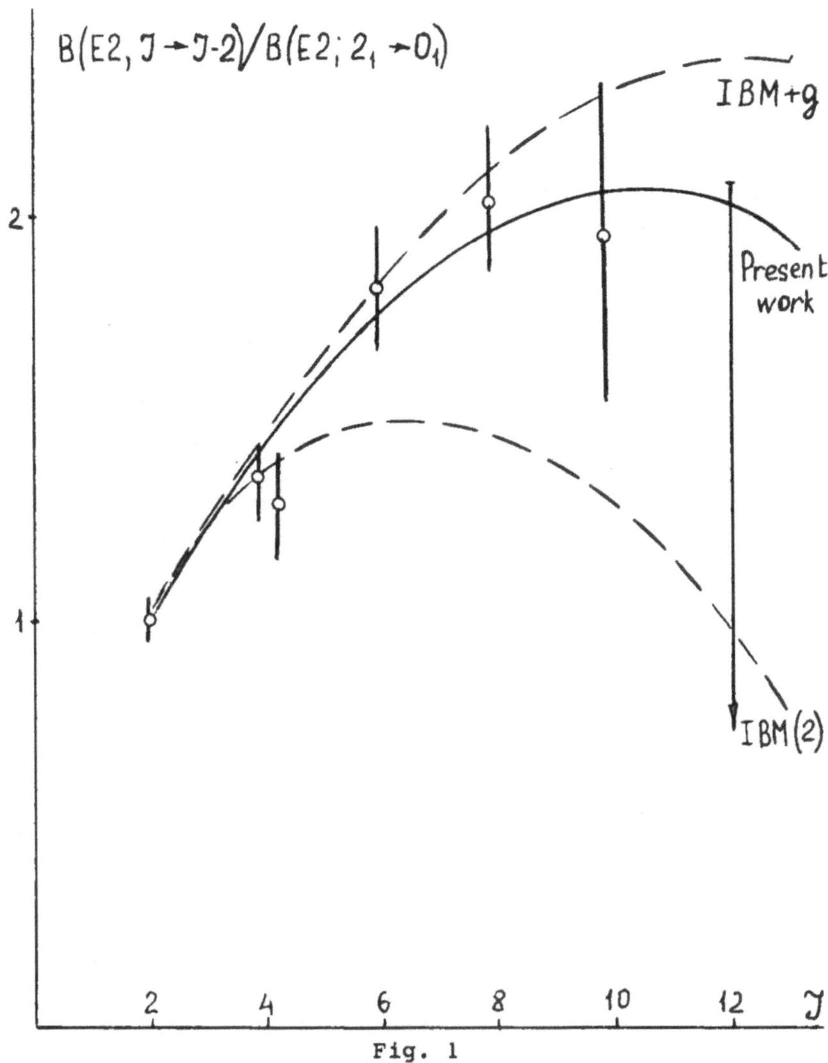

Fig. 1

QAAM model (the simplest adiabatic limit with one parameter γ = 0.026) and manyparameter models IBA-(2) and IBA+g [16,17]. Similarly, energies of side bands ($J<2v$) are given in Fig. 3. The number of fitted parameters for IBA calculations is indicated.

We see that the exposure of principal anharmonic ef-

Transition	$\frac{B(E2)}{B(E2;\ 2_1 \rightarrow 0_1)}$ (Exp.)	$\frac{B(E2)}{B(E2;\ 2_1 \rightarrow 0_1)}$ (Theor.)
$2_1 \rightarrow 0_1$	1.00	1.00
$4_1 \rightarrow 2_1$	1.44(23)	1.57
$6_1 \rightarrow 4_1$	2.03(28)	1.90
$8_1 \rightarrow 6_1$	2.20(27)	2.05
$10_1 \rightarrow 8_1$	2.00(31)	2.08
$12_1 \rightarrow 10_1$	2.09	2.02
$2_2 \rightarrow 2_1$	1.0(2)	2.57
$3_1 \rightarrow 2_2$	1.28(27)	1.36
$3_1 \rightarrow 4_1$	0.28(15)	0.54
$4_2 \rightarrow 2_2$	0.54(16)	0.99
$4_2 \rightarrow 4_1$	0.33(10)	0.91
$5_1 \rightarrow 3_1$	0.79(60)	1.07
$6_2 \rightarrow 4_2$	1.24(24)	1.40
$8_2 \rightarrow 6_2$	1.48(73)	1.58
$0_2 \rightarrow 2_1$	0.41(6)	0.47
$2_3 \rightarrow 0_2$	0.66(17)	1.22
$2_3 \rightarrow 4_1$	0.28(38)	0.11
$2_3 \rightarrow 2_2$	0.06-0.12	0.06
$2_3 \rightarrow 0_1$	0.012(11)	0.018
$5_1 \rightarrow 6_1$	0.27(17)	0.48
$5_1 \rightarrow 4_2$	0.44(18)	0.49
$3_1 \rightarrow 4_1$	0.31(12)	0.54
$7_1 \rightarrow 8_1$	0.26(21)	0.40
$7_1 \rightarrow 5_1$	1.30(67)	1.40
$7_1 \rightarrow 6_2$	0.26(15)	0.25
$4_3 \rightarrow 6_1$	0.08	0.04

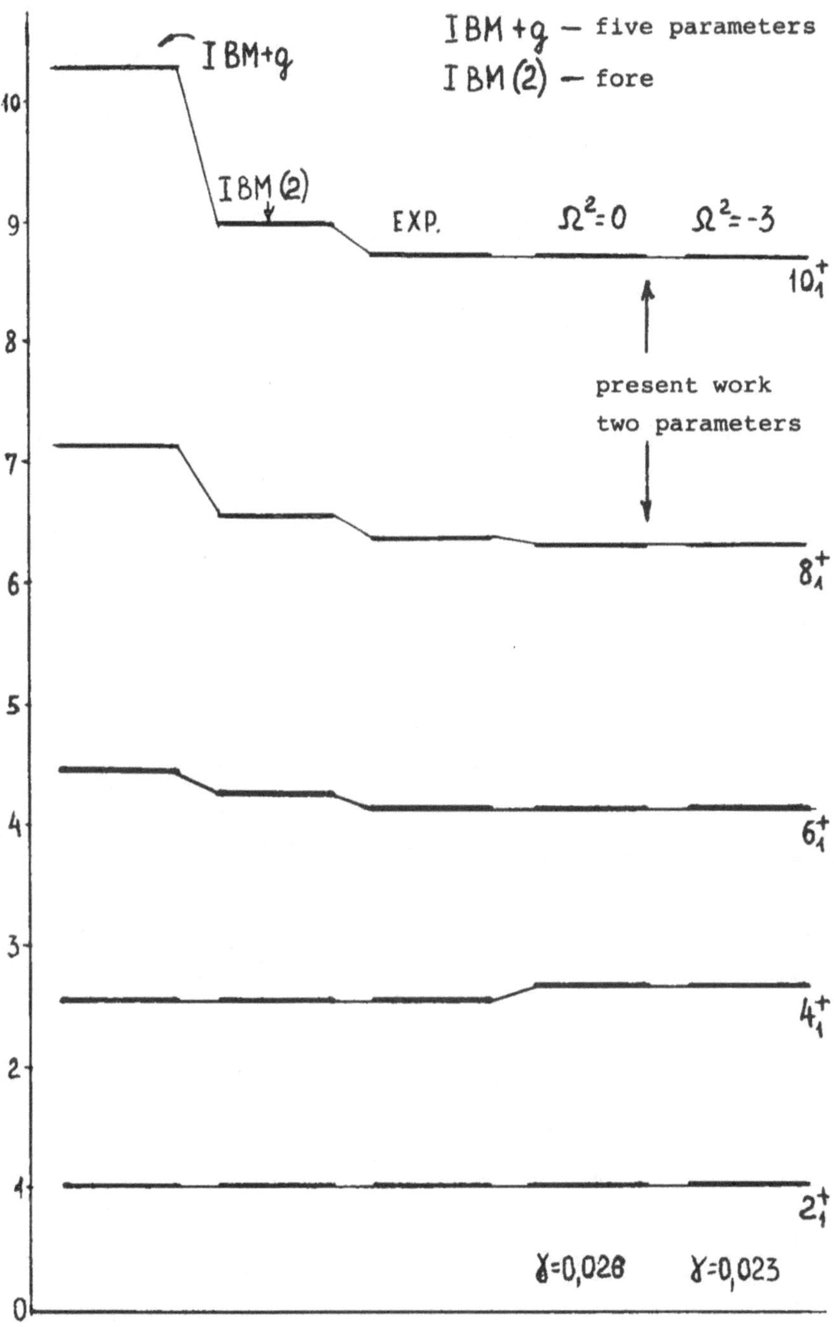
IBM+g — five parameters
IBM(2) — fore
IBM+g
IBM(2)
EXP.
$\Omega^2=0$
$\Omega^2=-3$
10_1^+
present work
two parameters
8_1^+
6_1^+
4_1^+
2_1^+
$\gamma=0{,}026$
$\gamma=0{,}023$
10
9
8
7
6
5
4
3
2
1
0

Fig. 2

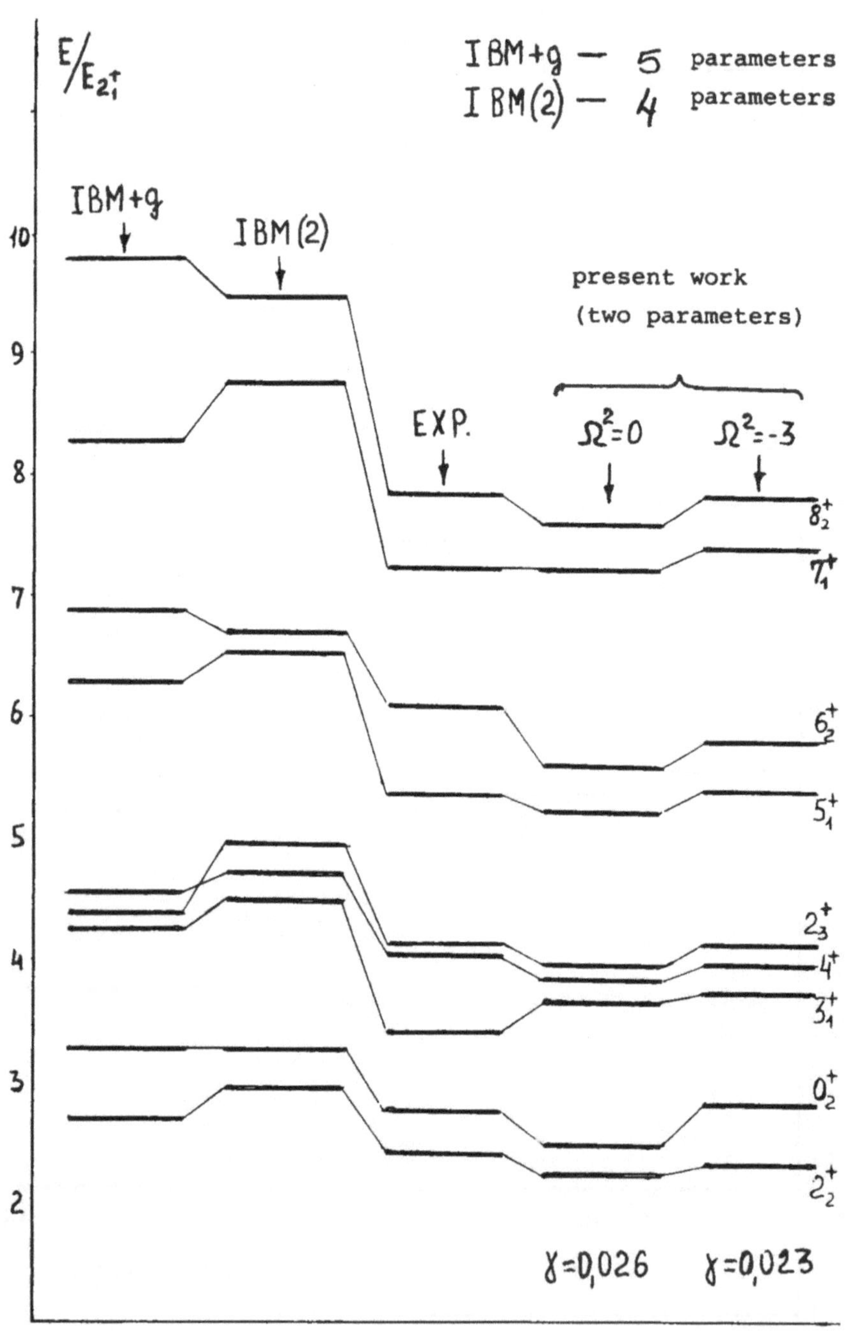
E/E$_{2_1^+}$
IBM+g — 5 parameters
IBM(2) — 4 parameters
IBM+g
IBM(2)
present work
(two parameters)
EXP.
Ω^2=0
Ω^2=-3
8_2^+
7_1^+
6_2^+
5_1^+
2_3^+
4^+
3_1^+
0_2^+
2_2^+
γ=0,026
γ=0,023
10
9
8
7
6
5
4
3
2

Fig. 3

fects makes possible to formulate a simple phenomenological approach reproducing main features of the real picture with the minimum number of free parameters. The scheme can be specified further introducing new parameters the total number of them being still essentially less than in traditional models. In such a way one obtains a good agreement for quantities forbidden in the harmonic approximation (expectation values of quadrupole moments and transition probabilities with $|\Delta v| \neq 1$. Moreover, the approach fits not very large negative values of Ω^2 where the standard methods are highly unstable. In some cases one can obtain better agreement introducing $\Omega^2 < 0$ as an additional parameter (see the last column of Fig.3) whereas results for the yrast band (Fig. 2) don't change.

The main conclusion of the analysis is that low-lying collective states of soft spherical nuclei manifest the new dynamic symmetry: the five-dimensional isotropic oscillator with the quartic anharmonicity.

References

1. Bohr.A., Mottelson,B. (1975). Nuclear Structure, 2, Benjamin, N.Y.
2. Zelevinsky,V.G. (1984). In: Group-Theoretical Methods in Physics, Harwood Academic Publishers.
3. Gneuss,G., Greiner,W (1971). Nucl. Phys. A171, 449.
4. Janssen,D., Jolos,R.V., Dönau,F. (1974). Nucl. Phys. A224, 93.
5. Arima,A., Iachello,F. (1976). Ann. Phys. 99, 253.
6. Beliaev,S.T., Zelevinsky, V.G. (1962). Nucl. Phys. 39, 582.
7. Tamura,T., Weeks,K.J., Kishimoto,T. (1979). Phys. Rev. C20, 307.
8. Marshalek,E.R. (1980). Nucl. Phys. A347, 253.
9. Belyaev,S.T. (1965). Nucl. Phys. 64, 17.
10. Zelevinsky,V.G. (1980). Nucl. Phys. A337, 40.
11. Zelevinsky,V.G. (1984). Izv. Akad. Nauk (ser. fiz.) 48, 2054 (Russian).

12. Zelevinsky,V.G. (1984). Izv. Akad. Nauk (ser. fiz.) 48, 79 (Russian).

13. Vorov,O.K., Zelevinsky,V.G. (1985). Nucl. Phys. A439, 207.

14. Zelevinsky,V.G. (1985). Izv. Akad. Nauk (ser. fiz.) 49, 65 (Russian).

15. Vorov,O.K., Zelevinsky,V.G. (1983). Yad. Fiz. 37, 1392 (Russian).

16. Stachel,J. et al. (1984). Nucl. Phys. A419, 589.

17. Stachel,J. et al. (1982). Nucl. Phys. A383, 429.

18. Heyde, K. et al. (1983). Nucl. Phys. A398, 235.

SOME GROUP-THEORETICAL ASPECTS OF SYMPLECTIC COLLECTIVE MODEL OF HEAVY NUCLEI

Yu.F. Smirnov, R.M. Asherova[x)]

Nuclear Physics Institute, Moscow State University, Moscow,
x) Institute of Physics and Power Engineering, Obninsk

1. Introduction. Recently various versions Sp(N,R), (N=2,4,6) of a simplectic collective model (SCM) have been more widely employed in the nuclear theory (see e.g. [1,2] and Refs. therein). The essence of this model is that a Hamiltonian $H = \sum_i T_i + \sum_{j>i} V_{ij}$ of the nucleus containing A nucleons is diagonalized in the basis

$$| n(\lambda_1\mu_1), (\lambda_o\mu_o) LMK\rangle = \text{const} \left[(A^+_{\alpha\beta})^n_{(\lambda_1\mu_1)} \Phi_{(\lambda_o\mu_o)} \right]^{(\lambda\mu)}_{LM\varepsilon} \quad (1)$$

of a single infinite-dimensional irreducible representation (IR) $D^{(\lambda_o\mu_o)}$ of Sp(G,R). Generators of this group are

$$A^+_{\alpha\beta} = \sum a^+_{\alpha i} a^+_{\beta i} \quad (\alpha,\beta = x,y,z),\quad A^-_{\alpha\beta} = (A^+_{\alpha\beta})^+,$$

$$A^o_{\alpha\beta} = \frac{1}{2} \Sigma (a^+_{\alpha i} a_{\beta i} + a_{\beta i} a^+_{\alpha i})$$

composed of conventional creation-annihilation operators of oscillator quanta $a^+_{\alpha i} = \frac{1}{\sqrt{2}} (v_{\alpha i} - ip_{\alpha i})$ and $a_{\alpha i}$ IR of interest belongs to a discrete series with a lowest vector and is characterized unambigously by an index $(\lambda_o\mu_o)$ of its lowest SU(3)-multiplet. The vectors of this multiplet are denoted as $\Phi_{(\lambda_o\mu_o)}$. The values of $(\lambda_o\mu_o)$ for concrete nucleus compatible with exclusion principle can be determined on the basis of Elliott's SU(3)-scheme [3]. The combination $(A^+_{\gamma\beta})^n_{(\lambda_1\mu_1)}$ denotes such an n-th symmetric degree of operators A^+ (each of them is transformed by IR (20) of the SU(3) group), that the resulting operator $(A^+)^n$ belongs to IR $(\lambda_1\mu_1)$ of SU(3). The values of angular momentum L and its projection M are mentioned, ε are the remaining quantum members required to unambi-

rotation-vibration model $W_6 \otimes T_5 \times SO(3)$ as $\mathcal{A} \to \infty$.

Thus, the development of SCM results in an urgent physical problem of finding the connection between the representations $D^{[f]}$ of a certain group G and representations $D^{(\phi)}$ of another group H into which the group G transforms under the contraction. The structure of H is simpler, as a rule, its $IRD^{(\phi)}$ are also somewhat simpler than $IRD^{[f]}$. If the relatonship between these representations is established, one can find asymptotic properties of matrices and other characteristics of representations $D^{[f]}$ of G in the limit case of large values of quantum numbers $[f] = [f_1, f_2, \dots, f_l]$, (l=rank G) labelling these representations.

In the present work we propose a new approach to the contraction problem. This approach suggests the initial realization of the representations $D^{[f]}$ of G on the universal enveloping algebra for H followed by a calculation of the limit at large values of the numbers f_i. The idea of this approach can be explained most clearly by consideration of examples, as it will be done below, where the physically relevant contractions $Sp(2,R) \to W_2$, $Sp(6,R) \to W_6 \otimes U(3)$, $V(3) \to T_5 \otimes SO(3)$, are described.

2. Bosonization as a method of representation contraction. An example: Sp(2,R). In [11] we formulated a general method for obtaining the realization of a representation for an arbitrary Lie algebra in terms of boson operators b_i^+, b_i (i=1,..., n) where n is a number of positive roots of G_1. In this case each generator E_α of G_1 is a polynomial of boson operators, i.e. the representation of G_1 is realized on the universal enveloping algebra of W_n. The method [11] is purely algebraic, it is based only on commutation relations of generators of G_1 and is not associated with coherent states technique unlike [12]. In [8] by an example of Sp(6,R) we show how mixed boson realizations may be obtained in the case when only a part of

guous labelling of the IR basic vectors in the reduction

$$Sp(6,R) \supset SU(3) \supset SO(3) \qquad (2)$$

The main problem when applying SCM for specific nuclei in the calculation of nuclear Hamiltonian matrix elements in the basis (1). Despite a certain progress here [2] the calculation of the realistic nuclear Hamiltonian matrix is still very laborious particularly for heavy nuclei when IR of very high dimension ($\lambda,\mu \sim 20$) is dealt with. Currently in this respect there is in fact a lack of calculations within the total SCM Sp(6,R) (even for light nuclei simplified SCM versions Sp(2,R), Sp(4,R) are used) and in the range of heavy nuclei the level spectra are analysed using a phenomenological collective Hamiltonian [4]

$$H_{eff} = H_o + a(Q\cdot Q) + b\rho^4 + c(Q\cdot Q\cdot Q) + d(Q\cdot Q)^2 \qquad (3)$$

Here $H_o = \sum_\alpha A^o_{\alpha\alpha}$ is a conventional oscillator shell model Hamiltonian, $\rho^2 = \sum_\alpha (A^+_{\alpha\alpha} + 2A^o_{\alpha\alpha} + A^-_{\alpha\alpha})$ is a "global" radius of nucleus, appearing in the K-harmonic method, $Q_{\alpha\beta} = A^+_{\alpha\beta} + 2\,A^o_{\alpha\beta} + A^-_{\alpha\beta}$ is a mass quadrupole momentum of the nucleus. Though the matrix elements of the Sp(6,R) group generator Q_{ij} in reduction (2) have been obtained not long ago [5-7], the presence of large values of λ_o, μ_o makes calculations bulky and the results noninterpretable. At the same time this fact creates however premises for an essential simplification of the problem because in the limit of large values λ_o, μ_o the group Sp(6,R) can be contracted to a semidirect product $W_6 \circledS SU(3)$ of the Heisenberg-Weyl group W_6 and SU(3) group. This possibility was investigated in [8,9] and resulted in the model of bosons with U(3)-symmetry representing a simplified version of SCM Sp(6,R) that is convenient for practical calculations. Also [10] points out that when considering nuclear states characterized by the values $\lambda_o,\mu_o \gg 1$, U(3) can be contracted to the dynamical group of a rigid top $T_5 \circledS SO(3)$. This enables us to pass from SCM Sp(6,R) to a

generators of G_1 are expressed in terms of boson operators, but the remaining generators are set in a standard operator/matrix form. This version of bosonization is especially convenient for $Sp(6,R) \rightarrow W_6 \otimes U(3)$ contraction and we describe it in detail. The essence of our approach to bosonization can be understood by an example of SCM $Sp(2,R)$. Generators of this algebra are $A^+ = \sum_\alpha A^+_{\alpha\alpha}$, $A^- = \sum_\alpha A^-_{\alpha\alpha}$, $A^0 = \sum_\alpha A^0_{\alpha\alpha}$, that satisfy

$$[A^0, A^\pm] = \pm 2A^\pm, \quad [A^-, A^+] = 4A^0 \qquad (4)$$

Our goal is to map IR space of D^{E_0} a positive discrete series of $Sp(2,R)$ with basis

$$|n\rangle = (A^+)^n |E_0\rangle \qquad (5)$$

(where $|E_0\rangle$ is the lowest vector) onto the boson states space

$$|n) = (b^+)^n |0) \quad (n = 0,1,2,\ldots) \qquad (6)$$

In this case the generators of $Sp(2,R)$ must be mapped onto boson operators b^+, b so that the boson images $A_B^{\pm,0}$ of $Sp(2,R)$ generators $A^{\pm,0}$ would obey (4). By [11] we map each vector (5) onto the appropriate boson states (6): $|n\rangle \rightarrow |n)$, and as a boson image of A^+ we chose the operator b^+

$$A^+ \rightarrow A_B^+ = b^+ \qquad (7)$$

Since $A^0|E_0\rangle = E_0|E_0\rangle$, $A^-|E_0\rangle = 0$, then $A^0|n\rangle = (E_0+2n)|n\rangle$

Let A_B^0 be chosen so that in the boson basis the same relation holds: $A_B^0 |n) = (E_0+2n) |n)$

Clearly $A_B^0 = E_0+2n$, where $n=b^+b$ is a boson number operator.

$$[A^-, (A^+)^n] = 4n(A^+)^{n-1}(A^0+n-1) \qquad (8)$$

implies

$$A^-|n\rangle = 4n(E_0+n-1) |n-1\rangle \qquad (9)$$

Comparing it with $b|n) = n|n-1)$ we see that to fulfil a relation similar to (9) in the boson space we are to assume that

$$A_B^- = 4b(E_0+b^+b-1) \qquad (10)$$

The way of constructing A_B^+, A_B^-, A_B^0 implies that they satisfy (4); this can be proved also by direct calculation. $|n\rangle \rightarrow |n)$ is a map between unnormalized bases. In this case non-Hermitian boson realization of the Dyson type [13] is available, where $(A_B^+)^+ \neq A^-$. Should the reflection $|n\rangle \to |n)$ be performed between two normalized bases

$$|n\rangle = \frac{1}{2^n}\sqrt{\frac{\Gamma(E_o)}{n!\,\Gamma(E_o+n)}}\quad (A^+)^n\, |E_o\rangle,$$

$$|n) = \frac{1}{\sqrt{n!}}\,(b^+)^n|0),$$

the result will be Holstein-Primakoff representation [14]:

$$A^o_{HP} = E_o+2b^+b,\ A^+_{HP} = 2b^+\sqrt{E_o+b^+b},\ A^-_{HP}=2b\sqrt{E_o+b^+b-1} \qquad (11)$$

with $(A^+_{HP})^+=A^-_{HP}$. If we pass to the limit $\hat{n}/E_o \ll 1$ in Eqs. (10), (11), then

$$A^o_B=A^o_{HP}\cong E_o,\ A^+_{HP}=A^+_B\cdot 2\sqrt{E_o}=2\sqrt{E_o}b^+,\ A^-_B\cong 2\sqrt{E_o}\ A^-_{HP}\cong 4E_ob \qquad (12)$$

Thus, in the considered approximation which corresponds to the contraction $Sp(2,R)\to W_2$ A^+, A^- coincide with usual boson operators b^+, b except for common constant factor $2\sqrt{E_o}$, and (4) transform into standard CR for boson operators $[b,b^+] = 1$. The transition from Dyson's realization $A^+_B=b^+$, $A^-_B=4E_ob$ to Holstein-Primakoff Hermitian realization within this limit consists in the fact that factor $(2\sqrt{E_o})^2$ followed by b is distributed between the b^+ and b so that the ratio $(A^+)^+=A^-$ is fulfilled.

This example demonstrates a principle for obtaining boson realization applicable to an arbitrary Lie group, and also shows that the boson realization of generators is actually a very convenient frame for performing group contraction. (This fact was also mentioned in [15]). A simple relation between Dyson and Holstein-Primakoff operators within the range $\hat{n}/E_o \ll 1$ will be used below to obtain Holstein-Primakoff operators for Sp(6,R).

3. Mixed boson realization of Sp(6,R). Each basis vector (1) of Sp(6,R)'s IR

$$|n_{\alpha\beta};\varkappa\rangle = \prod_{\alpha\beta} (A^+_{\alpha\beta})^{n_{\alpha\beta}}\ \Phi_{(\lambda_o\mu_o)\varkappa} \qquad (13)$$

where vectors of the lowest U(3)-multiplet satisfy

$$A^-_{\alpha\beta}\ \Phi_{(\lambda_o\mu_o)\varkappa} = 0 \qquad (14)$$

is mapped onto the boson vector

$$|n_{\alpha\beta};\varkappa\rangle = \prod_{\alpha\beta}(b^+)^{n_{\alpha\beta}}\ \Phi^B_{(\lambda_0\mu_0)\varkappa} \tag{15}$$

where the boson image $\Phi^B_{(\lambda_0\mu_0)\varkappa}$ of the vector $\Phi_{(\lambda_0\mu_0)\varkappa}$ is also transformed as the row of the IR $(\lambda_0\mu_0)$ of SU(3) and satisfies

$$b_{\alpha\beta}\ \Phi^B_{(\lambda_0\mu_0)\varkappa} = 0 \tag{16}$$

In the space of vectors Φ^B the generators $S^0_{\alpha\beta}$ of U(3) act and commute with b^+, b. By same methods as above we obtain the following boson images for generators of Sp(6,R)

$$\begin{aligned} &A^+_{\alpha\beta} \to b^+_{\alpha\beta} \\ &A^0_{\alpha\beta} \to S^0_{\alpha\beta} + \sum_\gamma b^+_{\alpha\gamma}b_{\gamma\beta} + b^+_{\alpha\beta}b_{\beta\beta}, \\ &A^-_{zz} \to 4b_{zz}(S^0_{zz} + 2\hat{n}_{zz} + \hat{n}_{zy} + \hat{n}_{zx} - 1) + b^+_{xx}b^2_{zz} + b^+_{yy}b^2_{zz} + \\ &\quad + 2b^+_{xy}b_{zx}b_{zy} + 2_{zx}\hat{n}_{zx}S_{xz} + 2b_{zy}\hat{n}_{zy}S^0_{yz} \end{aligned} \tag{17}$$

Boson images of the remaining generators $A^-_{\alpha\beta}$ can be obtained from (17) with the use of commutation relations, e.g. $A^-_{zx} = -\frac{1}{2}\left[A^0_{zx}, A^-_{zz}\right]$, etc. In Eq. (17) it means that $\hat{n}_{\alpha\beta} = b^+_{\alpha\beta}b_{\alpha\beta}$.

So we have obtained exact mixed boson realization of Sp(6,R). The term "mixed" means the fact that the generators (17) of Sp(6,R) are expressed not only in terms of boson operators $b^+_{\alpha\beta} = b^+_{\beta\alpha}$ and $b_{\alpha\beta}$, but through U(3) generators $S^0_{\alpha\beta}$ too. This realization differs from [8] since $\left[b_{\alpha\beta}, S^{0\ \alpha\beta}_{\gamma\delta}\right] = \left[b^+_{\alpha\beta}, S^0_{\alpha\beta}\right] = 0$ whereas in [8] the generators of U(3) non-commutating with boson operators were employed A realization similar to (17) was obtained by Quesne [16] employing a partially-coherent states technique.

In low-lying states of heavy nuclei the conditions $n_{\alpha\beta}, \lambda_0, \mu_0 \ll E^0_y$, are satisfied, where E^0_y is the minimal of signature components $\left[E^0_z, E^0_x, E^0_y\right]$ of the lowest U(3)-multiplet. In fact the oscillatory energy per one Cartesian degree of freedom in a nucleus (including a zero point energy) is rather high. For example, the low-lying levels os a $^{74}_{32}Gl_{42}$ nucleus is characterized by U(3)-signature $[f] = [111,85,75]$, i.e. $\lambda_0 = 26$, $\mu_0 = 10$, $E^0_y = 75$. In the wave function of the ground state the vectors (1) with $n \leq 20$, are dominating, therefore the above conditions ($\lambda_o \mu_o << 1, n_{\alpha\beta}, \lambda_o : \mu_o << E_y^o$) are fulfilled. In so doing the contraction $Sp(6,R) \to W_6 \circledS U(3)$ may be performed by retaining in (17) only low order terms with respect to small parameters $\hat{n}_{\alpha\beta}/E_y^o$. The result is the following asymptotic Holstein-Primakoff realization

$$A_{ii}^{+} \to 2\sqrt{E_y^o}\, b_{ii}^{+} \qquad A_{ii}^{-} \to 2\sqrt{E_y^o}\, b_{ii}$$

$$A_{ik}^{+} \to 2\sqrt{E_y^o}\, b_{ik}^{+} \qquad A_{ik}^{-} \to \sqrt{2E_y^o}\, b_{ik} \tag{18}$$

Taking this into account it is possible to change the calculations with effective Hamiltonian (3) in the basis (1) by simpler calculations in the boson basis

$$|n(\lambda_1\mu_1);\ (\lambda_o\mu_o)\ :\ (\lambda\mu)\ LMK\rangle$$

composed of orthonormalized vectors

$$|n_{\alpha\beta}, \varkappa\rangle = \prod_{\alpha\beta} \frac{1}{\sqrt{n_{\alpha\beta}!}} (b_{\alpha\beta}^{+})^{n_{\alpha\beta}} \Phi^{B}_{(\lambda_o\mu_o)\varkappa}$$

In this case the boson image of the effective Hamiltonian must be used that may be obtained from (3) substituting boson images

$$\rho^2 \cong \sum_i \sqrt{E_y^o}(b_{ii}^{+} + b_{ii}) + E_o$$

$$Q_{\alpha\beta} \cong \sqrt{2E_y^o}(b_{\alpha\beta}^{+} + b_{\alpha\beta}) + S_{\alpha\beta}^o$$

4. <u>$SU(3) \to T_5 \circledS SO(3)$ contraction</u>. The dynamic symmetry group $H = T_5 \circledS SO(3)$ of a rigid top is generated by quadrupole operators $Q_m (m=0,\pm1,\pm2)$ and angular momentum operators $L_{0,\pm1}$. The following CR hold:

$$[Q_m, Q_{m'}] = 0, \quad [L_\mu, Q_m] = \sum_{m'} \langle 2m' | L_\mu | 2m\rangle Q_{m'} \tag{19}$$

(CR for $L_{0,\pm1}$ are standard). The CR for q_m, l_μ of G=SU(3) differ from (19) only in one point. Namely, generators q_m do not commute

$$[q_m, q_{m'}] = -\sqrt{5}(2m2m' | 1_\mu) 1_\mu \tag{20}$$

In unitary IR for both G and H we have

$$(Q_m)^{+} = (-1)^m Q_{-m}, \quad (q_m)^{+} = -(-1)^m q_{-m} \tag{21}$$

IR's $T_5 \circledS SO(3)$ are divided into three classes [17,18] according to the eigenvalues ($Q_{\pm 1}=0$, $Q_0=\beta\cos\gamma$, $Q_{\pm 2} = \frac{1}{\sqrt{2}}\beta\sin\gamma$) of a quadrupole tensor in the system of main inertia axes.

That is why there are three alternative classes of IR's: 1) $\beta=0$, a spherical top; 2) $\beta>0$, $\gamma=0°$ or $60°$, a symmetric top; 3) $\beta>0$, $\gamma\neq 0$ or $60°$, an asymmetric top. Two Casimir operators $G_2 \cong (Q\cdot Q)$, $G_3 \cong (Q\,Q\,Q) = [[Q \times Q]^{(2)} \times Q]$ may be introduced. Their eigenvalues $g_2=\beta^2$; $g_3 = -\sqrt{\frac{2}{35}}\,\beta^3\cos^3\gamma$ can be used for the classification of IR's $D^{(\beta,\gamma)}$ of $T_5 \circledS SO(3)$. As IR's basis vectors the symmetric top eigenfunctions $\psi^{(\beta,\gamma)\pm}_{LMK} =$

$$= \sqrt{\frac{2L+1}{16\pi^2(1+\delta_{k0})}}\left[D^L_{MK}(\theta_i) \pm (-1)^L D^{L*}_{M-K}(\theta_i)\right]\phi^{(\beta,\gamma)} \tag{22}$$

can be adopted. Here $\phi^{(\beta,\gamma)}$ is an "intrinsic function" which is an eigenfunction of operators $Q_{0,\pm 2}$ in the system of main inertia axes with the corresponding eigenvalues. IR's $D^{(\beta,\gamma)}$ of $T_5 \circledS SO(3)$ are infinite-dimensional and contain in general the IRD^L of the subgroup SO(3) with multiplicity $\nu_L > 1$. That is why an additional quantum number should be introduced in order to distinguish between multiple representations D^L. The eigenvalues of the rigid top Hamiltonian

$$X = ([L \times L]^{(2)} \cdot Q =$$

$$= \frac{1}{\sqrt{6}}\{\beta\cos\gamma\,(3L_3^2 - L^2) + \sqrt{3}\beta\sin\gamma(L_1^2 - L_2^2)\} \tag{23}$$

may be taken as additional quantum numbers.

In Eg. (23) L_1, L_2, L_3 are angular momentum projections on the main axes of the tensor Q. IR's of the asymmetric top type can be subdivided into 4 classes according to the symmetry with respect to the stability subgroup D_2: 1) $D^{(\beta,\gamma)}_A$, the functions (22) ψ^+_{LMK} with even K constitute a basis; 2) $D^{(\beta,\gamma)}_{B_1}$, the basis is ψ^-_{LMK}, with even K ; 3) $D^{(\beta,\gamma)}_{B_2}$, the basis is ψ^+_{LMK}, with odd K ;

4) $D_{B_3}^{(\beta\ \gamma)}$, the basis is ψ_{LMK}^{-} with odd K ;

Generators Q_m trabsform the basis functions (22) as follows [17]:

$$Q_m\psi_{LMK}^{(\beta,\gamma)} =$$

$$= \sum_{L'} \sqrt{\frac{2L+1}{2L'+1}}(LM2m|L'M')\{(LK20|L'K)\beta\cos\gamma\psi_{L'M'K} + \quad (24)$$

$$+\frac{1}{\sqrt{2}}\beta\sin\gamma\left[(LK22|L'K+2)\psi_{L'M'K+2}+(LK2-2|L'K-2)\psi_{L'M'K-2}\right]\}$$

To obtain a realization of representations of SU(3) in the basis of symmetric top eigenfunctions, a correspondence between functions (22) and Elliott's basis functions [1,3] should be established

$$\psi_{LMK}^{(\beta,\gamma)} \to |\ 1 = LMK\ = P_{MK}^{L}\ \phi_{K}^{(\lambda\mu)}$$

The effect of generators q_m on $|LMK\rangle$ vectors is determined by the relation [1]

$$q_m|LMK\rangle =$$

$$= \sum_{L'\varkappa} (LM2m|L'M') \sqrt{\frac{2L+1}{2L'+1}} \left[(LK20|L'K') \frac{1}{\sqrt{6}}(2\lambda+\mu+3)|LMK\rangle + \right.$$

$$+ (LK22|L'K+2)\ \frac{1}{2}\ \sqrt{(\mu-K)(\mu+K+2)}\ \ |L'M'K+2\rangle +$$

$$\left. + (LK2-2|L'K-2)\ \frac{1}{2}\ \sqrt{(\mu+K)\ \ (\mu-K+2)}\ |L'M'K-2\rangle\right] \quad (25)$$

Comparing this expression with (24) we see that their structures are very similar but not identical. The operators

$$\hat{q}_{\pm 2} = \frac{1}{2}\sqrt{(\mu\pm 1_3)(\mu\pm 1_3+2)},\ \hat{q}_o = 2\lambda+\mu+3 \quad (26)$$

transform the functions $\psi_{LMK}^{(\beta,\gamma)}$ in agreement with (25). Therefore it is possible to consider the operators q_m as a realization of the representation of SU(3) on the enveloping algebra of $T_5 \circledS SO(3)$ ("rotor" realization). Contraction $SU(3)\to T_5 \circledS SO(3)$ corresponds to the transition to the limit $\hat{L}_3/\mu \ll 1$ (for $\lambda/\mu \sim$ const). In this case we have $q_{\pm 2} = \frac{1}{\sqrt{2}}\ \beta\sin\gamma \cong \frac{\sqrt{6}}{2}\mu$, $q_o = \beta\cos\gamma \cong 2\lambda+\mu$. It means that the IR's $D^{(\beta,\gamma)}$ and $D^{(\lambda,\mu)}$ of the $T_5 \circledS SO(3)$ group with the parameters $\beta^2 = 4\lambda^2+4\lambda\mu+4\mu^2$, $tg\gamma = \sqrt{3}\mu/(2\lambda+\mu)$. The type of D_2-symmetry of the representation $D^{(\lambda\mu)}$ is determined by the parity of λ and μ. From the symmetry of Elliott's functions $|LMK\rangle = (-1)^{\lambda+L+\mu}|LM-K\rangle$ it follows, that the case λ and μ even corresponds to the D_2-symmetry A . Similarly, a set λ-odd, μ-even corresponds to symmetry B_1; λ-even, μ-odd to symmetry B_3; λ-odd, μ-odd to symmetry B_2. Usually the eigenvalues of the Bargmann-Moshinsky operator $X = ([1 \times 1]^{(2)} \cdot q)$ are applied as additional quantum numbers to distinguish the multiple IR D^L of the subgroup SO(3) in a given IR $D^{(\lambda\mu)}$ of SU(3).

Compare X with (21) we see that in the limit $\lambda, \mu >> 1$ the eigenvalue spectrum of this operator is similar to the spectrum of asymmetric top levels with the corresponding quadrupole deformation β and γ and D_2-symmetry type. We have mentioned this fact in [19] where the asymptotic properties of Elliott's basis are considered.

Thus, the three above examples show how the contraction representation $D^{[f]}$ of G to representation $D^{(\phi)}$ of H can be carried out. For this map $\psi_\alpha^{[f]}$ to $\Phi_\alpha^{(\varphi)}$. Note that dimensions of IR $D^{[f]}$ and $D^{(\varphi)}$ are different, in particular, the last IR can be infinite dimensional; in principe the basis $\psi_\alpha^{[f]}$ can be non-orthonormalized (or even overcomplete as in the case of Elliott's basis for SU(3)). Then each generator j_i of G should be mapped onto its image J_i acting in the space of $\Phi_\alpha^{(\varphi)}$. This image is a polynomial or a power series in generators of H so that the action of J_i on $_\alpha^{(\phi)}$ is equivalent to the action of j_i on the corresponding vectors $\psi_\alpha^{[f]}$. An exact realization of the generators j_i on the enveloping algebra of H obtained in this way is interesting by itself. It is also a starting point for contraction G→H. For this purpose a limiting transition to large values of quantum numbers $[f] = f_1, f_2, \ldots, f_1$ (simultaneously for all f_i or for souce of them) must be performed. In such a manner the asymptotic properties of various group-theoretical quantities (matrices in IR, Clebsch-Gordan coefficients, etc.) for the initial group G within the range of large quantum numbers can be found.

References

1. Filippov,G.F., Ovcharenko,V.I., Smirnov,Yu,F. (1981). Microscopic Theory of Atomic Nuclei Collective Excitations, Kiev, Naukova Dumka.

2. Filippov,G.F., Vasilevsky,V.S., Chopovsky,L.L.(1984). El. Chastits At. Yadra, 15, 1338; (1985). El. Chastits At. Yadra, 16.

3. Elliott,J.P. (1958). Proc. Roy. Soc. a London, 245, 128, 562.

4. Peterson,D.R., Hecht,K.T. (1980). Nucl Phys. A344,361.

5. Rowe,D.J., Rosenstell,G., Carr,R. (1984). J.Phys. A, 17, L399.

6. Deenen,J., Quesne,C. (1984). J. Phys. A, 17, L405.

7. Castanos,O., Chacon,E., Moshinsky,M. (1984). J. Math. Phys., 25, 1211.

8. Tolstoy,V.N., Smirnov,Yu.F. (1983). Thesis and programm of 33-th all-union conference on nuclear spectroscopy and nuclear structure. Leningrad, Nauka, p. 160.

9. Rowe,D.J. (1984). Coherent states, contractions and classical limit of the non-compact symplectic group. Preprint University of Toronto.

10. Leblanc,R., Carvalho,I., Rowe,D. (1984). Phys. Lett. 140B, 155.

11. Smirnov,Yu.F., Tolstoy,V.N., Saharuck,A.A., Gavlicheck,M., Burdik,Ch. (1983). In: "Group theoretical methods in physics", M. Nauka, 2, p. 517.

12, Dobactewski,Y. (1981). Nucl. Phys. A369, 237: A380,1.

13. Dyson,F. (1956). Phys. Rev. 102, 1217.

14. Holstein,T., Primakoff,H. (1940). Phys. 58, 1098.

15. Celegini,E., Tarlini,V., Vitiello,G. (1984). Il. N. Cim. 84A, 19.

16. Deenne,J., Quesne,C. (1985). Partially Coherent states of the real symplectic group. Preprint PTM84-01. Université Libre de Bruxelles.

17. Ui,H. (1970). Progr. Theor. Phys. 44, 153.

18. Bargmann,V., Moshinsky,M. (1961). Nucl. Phys. 23,177.

19. Asherova,R.M., Smirnov,Yu.F. (1973). Rept. Math. Phys. 4, 83.

GROUP-THEORETICAL APPROACH TO THE THREE BODY PROBLEM

G.I. Kuznetsov
Kurchatov Institute of Atomic Energy, Moscow

1. Introduction. As it is known, a several particle system may be considered in a multidimensional space. Introducing in this space spherical coordinates one can describe the system with the help of hyperspherical functions. The application of hyperspherical functions in the several body problem leads to constructing particle states with prescribed permutational symmetry, i.e. here we have to calculate matrix elements for the particle permutation operators of hyperspherical functions, e.g. of functions of trees [1]. These matrix elements coincide with the matrix elements of rotations in many-dimensional space. In the case of several particles matrix elements of permutations can be reduced to matrix elements of rotation operators of three-particles [2].

We consider here only three particles for which permutation is a transformation described by an orthogonal matrix of two Jacobi vectors. This transformation, in its turn, induces a matrix acting in the space of trees. Papers [1-5] are devoted to calculating such a matrix for orthogonal Jacobi vectors. In [3] authors develop all the functions in series and then integrate the latter. In [4,5] the recurrence is used, i.e. first, one finds "simple" matrices, then a relation among them and then the "complete" matrix is established.

For constructing tree transformation matrices of particle permutations we use the addition theorem for

Jacobi polynomials proved by Koornwinder [6-8] and consider Jacobi vectors in general position because they, on one hand, correspond to a general position of three particles and, on the other hand, it can be straightforwardly used for finding the other representations for transformation matrix of $\mathcal{D}$-functions acting in the tree space [9,8]. Notice that $\mathcal{D}$-functions found in [11] are not fit for the realization of permutation groups.

2. Tree functions in three-body problem. Usually in the hyperspherical approach the Ψ-function may be found as

$$\Psi(\vec{\xi}_1,\vec{\xi}_2)=\sum_{\sigma\{l_i\}}\Psi_{\sigma\{l_i\}}(\varrho)\Gamma_{\sigma\{l_i\}}(\Omega_5,s,\tau)\ . \qquad (1)$$

The partical waves satisfy

$$\left[\frac{1}{\varrho^5}\frac{d}{d\varrho}\varrho^5\frac{d}{d\varrho}-\frac{\sigma(\sigma+4)}{\varrho^2}-\varkappa^2\right]\Psi_{\sigma\{l_i\}}(\varrho)=\frac{2m}{\hbar^2}\sum_{\sigma'\{l'_i\}}V^{\{l_i\}\{l'_i\}}_{\sigma\sigma'}(\varrho)\Psi_{\sigma'\{l'_i\}}(\varrho),$$

where

$$V^{\{l_i\}\{l'_i\}}_{\sigma\sigma'}(\varrho)=\sum_{s,\tau}\Gamma_{\sigma\{l_i\}}\hat{V}\,\Gamma_{\sigma'\{l'_i\}}, \qquad (3)$$

$\varrho^2=\vec{\xi}_1^{\,2}+\vec{\xi}_2^{\,2}$, $\quad\Omega_5$ is 5-dimensional sphere.

The antisymmetric functions $\Gamma_{\sigma\{l_i\}}(\Omega_5,s,\tau)$ can be completely constructed using the spin-isospin functions $\Phi(s,\tau)$ and the hyperspherical functions

$$\Psi^{[f]\mu f}_{\sigma LM:i}(\Omega_5)=\sum_{\{l_im_i\}}C_{\{l_im_i\}}\Psi^{l_1l_2m_1m_2}_{\sigma LM}(\Omega_5). \qquad (4)$$

Here μf enumerates the representation vectors $[f]$, i i indicates the moments constituting the given moment L. The orthonormal system of three-particle tree functions is

$$\Psi^{l_1 l_2 m_1 m_2}_{\sigma LM}(\vec{\xi}_1,\vec{\xi}_2)=\Psi^{\{l_i m_i\}}_{\sigma LM}(\Omega_5)=\Psi_{\sigma l_1 l_2}(\theta)Y^{l_1 l_2}_{LM}(\vec{n}_1,\vec{n}_2), \quad (5)$$

where

$$Y^{l_1 l_2}_{LM}(\vec{n}_1,\vec{n}_2)=\sum_{m_1+m_2=M}\langle l_1 m_1 l_2 m_2|LM\rangle Y_{l_1 m_1}(\vec{n}_1)Y_{l_2 m_2}(\vec{n}_2), \quad (6)$$

$$\Psi_{\sigma l_1 l_2}(\theta)=2^{3/2}\left\{N^{l_2+\frac{1}{2},\,l_1+\frac{1}{2}}_{\frac{\sigma-l_1-l_2}{2}}\right\}^{-\frac{1}{2}}(1-\cos 2\theta)^{l_2/2}\times$$

$$\times(1+\cos 2\theta)^{l_1/2}P^{l_2+1/2,\,l_1+1/2}_{(\sigma-l_1-l_2)/2}(\cos 2\theta), \quad (7)$$

$$N^{\alpha\beta}_{n}=\frac{2^{\alpha+\beta+1}\Gamma(n+\alpha+1)\Gamma(n+\beta+1)}{(2n+\alpha+\beta+1)\Gamma(n+1)\Gamma(n+\alpha+\beta+1)}$$

is squared norm, $2n=\sigma-l_1-l_2,\ \alpha=l_2+1/2,\ \beta=l_1+1/2,\ P^{\alpha\beta}_n(x)$ is the Jacobi polynomial.

The functions (5) are defined on the sphere $\vec{\xi}^{\,2}_1+\vec{\xi}^{\,2}_2=\rho^2$, where

$$\vec{\xi}_1=\rho\cos\theta\,\vec{n}_1,\ \vec{n}_1=(\cos\vartheta_1,\ \sin\vartheta_1\cos\varphi_1,\ \sin\vartheta_1\sin\varphi_1),$$

$$\vec{\xi}_2=\rho\sin\theta\,\vec{n}_2,\ \vec{n}_2=(\cos\vartheta_2,\ \sin\vartheta_2\cos\varphi_2,\ \sin\vartheta_2\sin\varphi_2). \quad (8)$$

Then $\vec{\xi}^{\,2}_1-\vec{\xi}^{\,2}_2=\rho^2\cos 2\theta$.

Since for the integration only an explicit form of functions is of importance, we choose not the basis (5) but the basis corresponding to the following tree

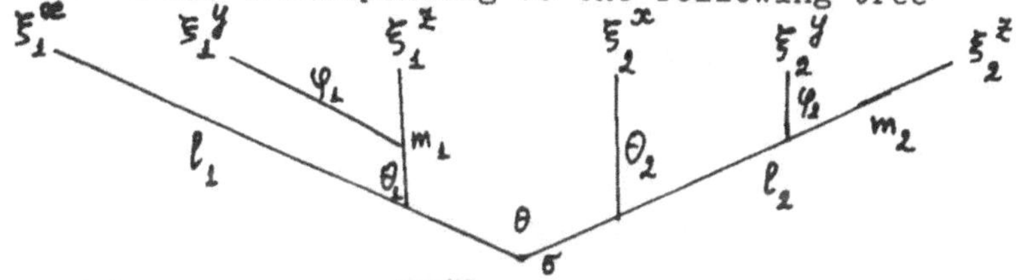

$$\Psi^{m_1 m_2}_{\sigma l_1 l_2}(\vec{\xi}_1,\vec{\xi}_2)=\Psi^{m_1 m_2}_{\sigma l_1 l_2}(\Omega_5)=\Psi_{\sigma l_1 l_2}(\theta)Y_{l_1 m_1}(\vec{n}_1)Y_{l_2 m_2}(\vec{n}_2). \quad (9)$$

We see from (5) and (9) that the transition from one basis to another one is simple and so is the transition between matrix elements.

3. Evaluating permutation matrix of general type for three-particle tree functions. Let

$$\vec{\xi}'_1 = \vec{\xi}_1 \cos\chi + \vec{\xi}_2 \sin\chi, \quad \vec{\xi}'_2 = -\vec{\xi}_1 \sin\chi + \vec{\xi}_2 \cos\chi \tag{10}$$

where

$$\vec{\xi}'_1 = \rho\cos\theta' \vec{n}'_1, \quad \vec{\xi}'_2 = \rho\sin\theta' \vec{n}'_2,$$
$$\vec{n}'_{1,2} = (\cos\vartheta'_{1,2},\ \sin\vartheta'_{1,2}\cos\varphi'_{1,2},\ \sin\vartheta'_{1,2}\sin\varphi'_{1,2}). \tag{11}$$

Then we have

$$\vec{\xi}'^2_1 - \vec{\xi}'^2_2 = \rho\cos 2\theta' = (\vec{\xi}^2_1 - \vec{\xi}^2_2)\cos 2\chi + \rho^2 \sin 2\theta \sin 2\chi (\vec{n}_1 \vec{n}_2) =$$
$$= \rho^2(\cos 2\theta \cos 2\chi + \sin 2\theta \sin 2\chi \cos\psi), \tag{12}$$

$$\cos\psi = (\vec{n}_1 \cdot \vec{n}_2) =$$
$$= \cos\vartheta_1 \cos\vartheta_2 + \sin\vartheta_1 \sin\vartheta_2 \cos(\varphi_1 - \varphi_2). \tag{13}$$

Now the transformed function can be developed in the series in initial functions

$$\Psi^{m'_1 m'_2}_{6 l'_1 l'_2}(\vec{\xi}'_1, \vec{\xi}'_2) = \sum_{\{l'_i\}\{m'_i\}} \langle l'_1 l'_2 m'_1 m'_2 | l_1 l_2 m_1 m_2 \rangle^{\chi}_{6} \Psi^{m_1 m_2}_{6 l_1 l_2}(\vec{\xi}_1, \vec{\xi}_2). \tag{14}$$

Our aim is calculating the transformation matrix

$$\langle l'_1 l'_2 m'_1 m'_2 / l_1 l_2 m_1 m_2 \rangle^{\chi}_{6} .$$

Using (10) we can show that the scalar product $(\xi, \xi') = \rho^2 \cos\chi$ where $\xi = (\vec{\xi}_1, \vec{\xi}_2)$ and $\xi' = (\vec{\xi}'_1, \vec{\xi}'_2)$. That means a rotation of ξ to ξ' by an angle χ corresponds to the transformation (10) and the transformation matrix (14) is a $\mathfrak{D}$-function in the tree space.

Basically all the constructions can be carried out in the plane $\vec{\xi}_1, \vec{\xi}_2$ which suffices to get three nucleon permutation matrix. However, we consider general position because we want to get also a D-function.

In primed variables the transformed function is

$$\Psi^{m_1' m_2'}_{\sigma l_1' l_2'}(\vec{\xi}_1', \vec{\xi}_2') = 2^{\frac{3+l_1'+l_2'}{2}} \left\{ N_{n'}^{l_2'+\frac{1}{2}, l_1'+\frac{1}{2}} \right\}^{-\frac{1}{2}} \sin^{l_2'}\theta' \cos^{l_1'}\theta' \times$$
$$\times P_{n'}^{l_2'+\frac{1}{2}, l_1'+\frac{1}{2}} (\cos 2\theta') Y_{l_1 m_1}(\vec{n}_1') Y_{l_2 m_2}(\vec{n}_2'), \quad (15)$$

where $2n' = \sigma - l_1' - l_2'$. Multiplying (14) by $\Psi^*_{\sigma l_1 l_2}(\theta)$ $Y^*_{l_1 m_1}(\vec{n}_1) Y^*_{l_2 m_2}(\vec{n}_2)$ and integrating over $d\rho(\theta, \vartheta_1, \vartheta_2, \varphi_1, \varphi_2) = 2^{-3}(1-\cos 2\theta)^{1/2}(1+\cos 2\theta)^{1/2} d\cos 2\theta\, d\vec{n}_1 d\vec{n}_2$ we get an integral representation of the matrix

$$\langle l_1' l_2' m_1' m_2' | l_1 l_2 m_1 m_2 \rangle^{\chi}_{\sigma} =$$
$$= \int_{\Omega_5} \Psi^{m_1' m_2'}_{\sigma l_1' l_2'}(\xi') \Psi^{m_1 m_2}_{\sigma l_1 l_2}(\xi) d\rho(\theta, \vec{n}_1, \vec{n}_2). \quad (16)$$

In order to integrate in (16) it is necessary to change the primed variables for unprimed ones in the Jacobi polynomial and in the spherical functions (15). The transition is quite easy if one takes into account the following relations for spherical functions [12]

$$\mathcal{Y}_{lm}(s\vec{z}_a + t\vec{z}_b) = \sum_{\lambda=0}^{l} \frac{\sqrt{4\pi}}{[\lambda]} \binom{2l+1}{2\lambda}^{1/2} \langle \lambda m_\lambda l-\lambda\, m-m_\lambda | lm \rangle$$
$$\times \mathcal{Y}_{\lambda m_\lambda}(s\vec{z}_a)\, \mathcal{Y}_{l-\lambda\, m-m_\lambda}(t\vec{z}_b), \quad (17)$$

where

$$\binom{x}{y} = \frac{x!}{y!(x-y)!}, \quad \mathcal{Y}_{lm}(s\vec{z}) = s^l |z|^l Y_{lm}(\vec{n}),$$
$$[\lambda] = (2\lambda+1)^{1/2}, \quad \mathcal{Y}^*_{lm}(\theta,\varphi) = s^l |z|^l (-)^m Y_{l,-m}(\theta,\varphi), \quad (18)$$
$$Y_{lm}(\theta\varphi) = (-)^m \left[\frac{2l+1}{4\pi} \cdot \frac{(l-m)!}{(l+m)!} \right]^{1/2} P_l^m(\cos\theta) e^{im\varphi},$$
$$Y_{lm}(\pi-\theta, \varphi+\pi) = (-)^l Y_{lm}(\theta,\varphi)$$

and the addition theorem for the Jacobi [6-8] and Gegenbauer [6-8,13] polynomials:

$$P_n^{\alpha,\beta}(2|\cos\theta_1\cos\theta_2+ze^{i\psi}\sin\theta_1\sin\theta_2|^2-1)=\sum_{k=0}^{n}\sum_{\ell}^{k}C_{nk\ell}^{\alpha,\beta}(\sin\theta_1\sin\theta_2)^{k+\ell}$$

$$\times(\cos\theta_1\cos\theta_2)^{k-\ell}P_{n-k}^{\alpha+k+\ell,\beta+k-\ell}(\cos 2\theta_1)P_{n-k}^{\alpha+k+\ell,\beta+k-\ell}(\cos 2\theta_2)\times \quad (19)$$

$$\times P_{\ell}^{\alpha-\beta-1,\beta+k-\ell}(2z^2-1)z^{k-\ell}\frac{\beta+k-\ell}{\beta}C_{k-\ell}^{\beta}(\cos\psi).$$

$$C_{k-\ell}^{\beta}(\cos\psi)=\sum_{x=0}^{k-\ell}a_{k-\ell,x}^{\beta}(\sin\vartheta_1)^{x}C_{k-\ell-x}^{\beta+x}(\cos\vartheta_1)(\sin\vartheta_2)^{x}C_{k-\ell-x}^{\beta+x}(\cos\vartheta_2)$$

$$\times C_{x}^{\beta-1/2}(\cos(\varphi_1-\varphi_2)). \quad (20)$$

$$C_{x}^{\beta-1/2}(\cos(\varphi_1-\varphi_2))=\sum_{x_1}^{x}a_{xx_1}^{\beta-1/2}\sin^{x_1}\varphi_1 C_{x-x_1}^{\beta-1/2+x_1}(\cos\varphi_1)\times$$

$$\times\sin^{x_1}\varphi_2\, C_{x-x_1}^{\beta-1/2+x_1}(\cos\varphi_2)C_{x_1}^{\beta-1}(1) \quad (20')$$

where

$$C_{x_1}^{\beta-1}(1)=\frac{\Gamma(2\beta-2+x_1)}{\Gamma(x_1+1)\Gamma(2\beta-2)}, \quad P_n^{\alpha\beta}(1)=\frac{(\alpha+1)n}{n!}, \quad (21)$$

$$a_{k-\ell,x}^{\beta}=\frac{\Gamma(2\beta-1)2^{2x}(k-\ell-x)!\,[\Gamma(\beta+x)]^2}{[\Gamma(\beta)]^2\Gamma(2\beta+k-\ell+x)}(2\beta+2x+1),$$

$$C_{nk\ell}^{\alpha\beta}=\frac{(\alpha+k+\ell)\Gamma(\alpha+\beta+n+k+1)\Gamma(\alpha+k)}{\Gamma(\alpha+\beta+n+1)\Gamma(\alpha+n+\ell+1)}\times \quad (22)$$

$$\times\frac{\Gamma(\beta+1)\Gamma(\beta+n+1)\Gamma(n-k+1)}{\Gamma(\beta+k+1)\Gamma(\beta+n-\ell+1)}.$$

First let us consider the spherical functions. From (17) and (18) we have

$$Y_{\ell_1' m_1'}(\vec{n}_1')Y_{\ell_2' m_2'}(\vec{n}_2')=\sum_{\lambda_1\lambda_2}^{\ell_1'\ell_2'}\frac{\sqrt{4\pi}}{[\lambda_1]}\binom{2\ell_1'+1}{2\lambda_1}^{1/2}\langle\lambda_1 m_{\lambda_1}\ell_1'-\lambda_1 m_1'-$$
$$-m_{\lambda_1}|\ell_1' m_1'\rangle(-)^{\lambda_2}\frac{\sqrt{4\pi}}{[\lambda_2]}\binom{2\ell_2'+1}{2\lambda_2}^{1/2}\langle\lambda_2 m_{\lambda_2}\ell_2'-\lambda_2 m_2'-m_{\lambda_2}|\ell_2' m_2'\rangle$$
$$\times(\cos\theta)^{\lambda_1+\lambda_2}(\sin\theta)^{\ell_1'+\ell_2'-\lambda_1-\lambda_2}\frac{(\cos\chi)^{\lambda_1+\ell_2'-\lambda}(\sin\chi)^{\lambda_2+\ell_1'-\lambda_1}}{(\sin\theta')^{\ell_2'}(\cos\theta')^{\ell_1'}}\times$$
$$\times Y_{\lambda_1 m_{\lambda_1}}(\vec{n}_1)Y_{\ell_1'-\lambda_1 m_1'-m_{\lambda_1}}(\vec{n}_2)Y_{\lambda_2 m_2}(\vec{n}_1)Y_{\ell_2'-\lambda_2 m_2'-m_{\lambda_2}}(\vec{n}_2)=$$
$$=\sum_{\lambda_1\lambda_2}^{\ell_1'\ell_2'}A^{\ell_1'\ell_2' m_1' m_2'}_{\lambda_1\lambda_2 m_{\lambda_1} m_{\lambda_2}}\frac{(\cos\theta)^{\lambda_1+\lambda_2}(\sin\theta)^{\ell_1'+\ell_2'-\lambda_1-\lambda_2}}{(\sin\theta')^{\ell_2'}(\cos\theta')^{\ell_1'}}(\cos\chi)^{\lambda_1+\ell_2'-\lambda_2}\times$$
$$\times(\sin\chi)^{\lambda_2+\ell_1'-\lambda_1}Y_{\lambda_1 m_{\lambda_1}}(\vec{n}_1)Y_{\lambda_2 m_{\lambda_2}}(\vec{n}_1)Y_{\ell_1'-\lambda_1 m_1'-m_{\lambda_1}}(\vec{n}_2)Y_{\ell_2'-\lambda_2 m_2'-m_{\lambda_2}}(\vec{n}_2).$$

The sines and cosines in the denominator of (23) can be cancelled out together with similar terms preceding the Jacobi polynomial in (15). The product of three spherical functions can be written as follows

$$Y^*_{\ell_1 m_1}(\vec{n}_1)Y_{\lambda_1 m_{\lambda_1}}(\vec{n}_1)Y_{\lambda_2 m_{\lambda_2}}(\vec{n}_1)=(-)^{m_1}\sum_{L_1\lambda}\frac{[\ell_1][\lambda_1][\lambda_2]}{[L_1]\,4\pi}\langle\lambda_1 0\lambda_2 0|\lambda 0\rangle\times$$
$$\times\langle\lambda_1 m_{\lambda_1}\lambda_2 m_{\lambda_2}|\lambda m_{\lambda_1}+m_{\lambda_2}\rangle\langle\ell_1 0\lambda 0|L_1 0\rangle\langle\ell_1 -m_1\lambda m_{\lambda_1}+m_{\lambda_2}|L_1 M_1\rangle$$
$$\times Y_{L_1 M_1}(\vec{n}_1)=\sum_{L_1\lambda}B^{L_1 M_1 m_{\lambda_1} m_{\lambda_2} m_1}_{\lambda_1\lambda_2\ell_1\lambda}Y_{L_1 M_1}(\vec{n}_1). \quad (24)$$

$$Y^*_{\ell_2 m_2}(n_2)Y_{\ell_1'-\lambda m_1'-m_{\lambda_1}}(\vec{n}_2)Y_{\ell_2'-\lambda_2 m_2'-m_{\lambda_2}}(\vec{n}_2)=$$
$$=\sum_{L_2,\ell_{12}'}B^{L_2 M_2 m_1'-m_\lambda,\, m_2'-m_{\lambda_2} m_2}_{\ell_1'-\lambda\ \ell_2'-\lambda_2\ \ell_2\ell_{12}'}Y_{L_2 M_2}(\vec{n}_2), \quad (24')$$

$$M_1=m_{\lambda_1}+m_{\lambda_2}-m_1,\quad M_2=m_1'-m_{\lambda_1}+m_2'-m_{\lambda_2}-m_2. \quad (25)$$

Further we shall also use the following formula

$$Y_{LM}(\vec{n}) = \frac{(2M)!}{M!\,2^M}\left[\frac{2L+1}{4\pi}\,\frac{(L-M)!}{(L+M)!}\right](\sin\vartheta)^M C^{M+\frac{1}{2}}_{L-M}(\cos\vartheta)\,e^{iM\varphi},$$
$$Y_{L0}(1) = \left(\frac{2L+1}{4\pi}\right)^{1/2}. \qquad (26)$$

Let us write the Jacobi polynomial with argument $\cos 2\theta'$ as a function of θ, χ, ψ. The addition theorem (19) corresponds then to $\theta_1 = \theta$, $\theta_2 = \chi$, $z = 1$, i.e.

$$P^{\ell_2'+\frac{1}{2},\,\ell_1'+\frac{1}{2}}_{n'}(\cos 2\theta') = \sum_{k=0}^{n'}\sum_{\ell}^{k} C^{\alpha\beta}_{n'k\ell}(\sin\theta\sin\chi)^{k+\ell}(\cos\theta\cos\chi)^{k-\ell}\times$$
$$\times P^{\ell_2'+\frac{1}{2}+k+\ell,\,\ell_1'+\frac{1}{2}+k-\ell}_{n'-k}(\cos 2\theta)\,P^{\ell_2'+\frac{1}{2}+k+\ell,\,\ell_1'+\frac{1}{2}+k-\ell}_{n'-k}(\cos 2\chi)\times$$
$$\times P^{\ell_2'-\ell_1'-\ell,\,\ell_1'+k-\ell+\frac{1}{2}}_{\ell}(1)\,\frac{\beta+k-\ell}{\beta}\,C^{\beta}_{k-\ell}(\cos\psi), \qquad (27)$$

where

$$\alpha = \ell_2'+1/2,\quad \beta = \ell_1'+1/2,\quad 2n' = 6-\ell_1'-\ell_2'. \qquad (28)$$

Note that if $\vec{\xi}_1 \perp \vec{\xi}_2$ then $\cos\psi = 0$, and θ' is defined only from θ and χ. Under these conditions the integration of any two triples of spherical functions over $d\vec{n}_1$, and $d\vec{n}_2$ gives a 9_j -symbol. This is not a surprise because 9_j -symbol is a Clebsch-Gordan coefficient of O(4) in its realization as O(3)xO(3) and our basis is expressed in terms of $Y_{\ell_1 m_1}(\vec{n}_1)Y_{\ell_2 m_2}(\vec{n}_2)$only. Integrating over θ with the development in series of Jacobi polynomials gives the final expression for the matrix of permutation group in the form of a triple sum.

Taking into account (16), (20)-(28) it is possible to write the sought matrix as follows [14]:

$$\langle \ell_1'\ell_2'm_1'm_2' / \ell_1\ell_2 m_1 m_2\rangle_6^{\chi} = \left\{N^{\ell_2'+\frac{1}{2},\,\ell_1'+\frac{1}{2}}_{n'}\right\}^{-\frac{1}{2}}\left\{N^{\ell_2+\frac{1}{2},\,\ell_1+\frac{1}{2}}_{n}\right\}^{-\frac{1}{2}}\times$$

$$\times \sum_{k=0}^{n'} \sum_{l=0}^{k} C_{n'kl}^{\alpha\beta} \sum_{\lambda_1\lambda_2}^{l_1'l_2'} A_{\lambda_1\lambda_2 m_{\lambda_1} m_{\lambda_2}}^{l_1'l_2'm_1'm_2'} \frac{1}{2^k} \frac{l_1'+\frac{1}{2}+k-l}{l_1'+\frac{1}{2}} P_l^{l_2'-l_1'-1,\, l_1'+k-l+\frac{1}{2}}(1)$$

$$\times (\sin\chi)^{\lambda_2-\lambda_1+l_1'-l_2'} (\cos\chi)^{\lambda_1-\lambda_2+l_2'-l_1'} (\sin\chi)^{l_2+k-l} (\cos\chi)^{l_1'+k-1} \times$$

$$\times P_{n'-k}^{l_2'+\frac{1}{2}+k+l,\, l_1'+\frac{1}{2}+k-l} (\cos 2\chi) I_{nn'k\lambda_1\lambda_2}^{l_1l_2ll_1'l_2'} (\text{Vertex } \sigma) \times \quad (29)$$

$$\times \sum_{x=0}^{k-l} a_{k-l,x}^{l_1'+\frac{1}{2}} \sum_{L_1\lambda} B_{\lambda_1\lambda_2 l_1\lambda}^{L_1M_1m_{\lambda_1}m_{\lambda_2}m_1} I_{klL_1M_1}^{xl_1'+\frac{1}{2}} (\text{Vertex } l_1) \times$$

$$\times \sum_{L_2 l_{12}} B_{l_1'-\lambda_1\, l_2'-\lambda_2\, l_2\; l_{12}}^{L_2M_2m_1'-m_{\lambda_1}m_2-m_{\lambda_2}m_2} I_{klL_2M_2}^{xl_1'+\frac{1}{2}} (\text{Vertex } l_2) \times$$

$$\times \sum_{x_1}^{x} a_{xx_1}^{l_1'} C_{x_1}^{l_1'-\frac{1}{2}}(1) \prod_{i=1}^{2} I_{xM_i}^{x_1l_1'+\frac{1}{2}} (\text{Vertex } m_i),$$

where

$$I_{n'nk\lambda_1\lambda_2}^{l_1l_2l_1'l_2'}(\text{Vertex } \sigma) = \int_{-1}^{1} (1-x)^{\frac{l_2+1+l_1'+l_2'-\lambda_1-\lambda_2+k+l}{2}} \times \quad (30)$$

$$\times (1+x)^{\frac{l_1+1+\lambda_1+\lambda_2+k-l}{2}} P_{n'k}^{l_2'+\frac{1}{2}+k+l,\, l_1'+\frac{1}{2}+k-l}(x) P_n^{l_2+\frac{1}{2},\, l_1+\frac{1}{2}}(x)\, dx,$$

$$I_{klL_{1,2}M_{1,2}}^{x\beta}(\text{Vertex } l_{1,2}) = \int_{-1}^{1} \sin^x \vartheta_{1,2}\, C_{k-l-x}^{\beta+x}(\cos\vartheta_{1,2}) \times$$

$$\times Y_{L_{1,2}}(\vartheta_{1,2}, \varphi_{1,2}=0)\, d\cos\vartheta_{1,2}. \quad (31)$$

$$I_{xM_{1,2}}^{x\beta}(\text{Vertex } m_{1,2}) = \int_0^{2\pi} \sin^{x_1}\varphi_{1,2}\, C_{x-x_1}^{\beta-1/2+x_1}(\cos\varphi_{1,2}) e^{iM_{1,2}\varphi_{1,2}} d\varphi_{1,2}. \quad (32)$$

Developing Jacobi polynomials in series

$$P_n^{l_2+\frac{1}{2},\, l_1+\frac{1}{2}}(x) = \frac{1}{2^n} \sum_{m=0}^{n} \binom{n+l_2+\frac{1}{2}}{m} \binom{n+l_1+\frac{1}{2}}{n-m} (-)^{n-m} (1-x)^{n-m} (1+x)^m \quad (33)$$

and evaluating the integral [15 , p. 201], we bring (30) to the form

$$I^{l_1 l_2 l l_1' l_2'}_{nn'k\lambda_1\lambda_2}(\text{Vertex } 6) = \sum_{m=0}^{n} \binom{n+l_2+\frac{1}{2}}{m} \binom{n+l_1+\frac{1}{2}}{n-m} (-)^{n-m} 2^{\mu} \times$$

$$\times \frac{\Gamma(\alpha_3)\Gamma(n'+l_2'+l+\frac{3}{2})}{\Gamma(\beta_2)\Gamma(1-\alpha_1)} \cdot \Gamma\left(\frac{l_1+\lambda_1+\lambda_2+k-l+2m+3}{2}\right) {}_3F_2\left(\begin{matrix}\alpha_1\alpha_2\alpha_3\\ \beta_1\beta_2\end{matrix}; 1\right), \tag{34}$$

where

$$\begin{aligned}
2\mu &= l_1+l_2+l_1'+l_2'+2k+4,\\
2\alpha_1 &= 2(k-n') = 2k+l_1'+l_2'-6,\\
2\alpha_2 &= 6+l_1'+l_2'+2k+4,\\
2\alpha_3 &= l_1'+l_2'-\lambda_1-\lambda_2+k+l+6-l_1-2m+3,\\
\beta_1 &= l_2'+l+k+3/2, \quad 2\beta_2 = 6_1+l_1'+l_2'+2k+6,\\
6_1 &= l_1+l_2.
\end{aligned} \tag{35}$$

Substituting in (31) the spherical function for its expression in terms of Gegenbauer polynomial (26), we get the integral evaluated earlier 14, 16 . Therefore, we have

$$I^{\varkappa\beta}_{klL_1M_1}(\text{Vertex } l_1) = \left[\frac{2L_1+1}{4\pi} \cdot \frac{(L_1-M_1)!}{(L_1+M_1)!}\right]^{1/2} \frac{(2M_1)!}{2^{M_1} M_1!} \times$$

$$\times \begin{cases} \mathfrak{M}_{\text{even}}, & \text{if } k-l-\varkappa = 2p,\ L_1-M_1 = 2q,\\ \mathfrak{M}_{\text{odd}}, & \text{if } k-l-\varkappa = 2p+1,\ L_1-M_1 = 2q+1, \end{cases} \tag{36}$$

where

$$\mathfrak{M}_{\text{even}} = \frac{(-)^{\frac{L_1-M_1}{2}} \pi^{\frac{1}{2}} (2\beta+2\varkappa)_{2p} (M_1+\frac{1}{2})_q \Gamma\left(\frac{\varkappa+M_1}{2}+1\right)\Gamma\left(\frac{\varkappa-M_1}{2}+1\right)}{(k-l-\varkappa)!\, q!\, \Gamma\left(\frac{\varkappa+L_1+3}{2}\right)\Gamma\left(\frac{\varkappa-L_1+2}{2}\right)}$$

$$\times {}_4F_3\left(\begin{matrix} -p,\ p+\beta+\varkappa,\ \frac{\varkappa+M}{2},\ \frac{\varkappa-M}{2}+1;\\ \beta+\varkappa+\frac{1}{2},\ \frac{\varkappa+L_1+3}{2},\ \frac{\varkappa-L_1+2}{2}; \end{matrix} 1\right), \tag{37}$$

$$\mathfrak{M}_{\text{odd}} = \frac{(-)^{\frac{L_1-M_1-1}{2}} \pi^{1/2} (2\beta+2\varkappa)_{2p+1} (M_1+\frac{1}{2})_{q+1} \Gamma\left(\frac{\varkappa+M_1}{2}+1\right)\Gamma\left(\frac{\varkappa-M_1}{2}+1\right)}{(k-l-\varkappa)!\, q!\, \Gamma\left(\frac{\varkappa+L_1+4}{2}\right)\Gamma\left(\frac{\varkappa-L_1+3}{2}\right)}$$

$$\times {}_4F_3\left(\begin{matrix} -p,\ p+\beta+\varkappa+1,\ \frac{\varkappa+M_1}{2}+1,\ \frac{\varkappa-M_1+2}{2};\\ \beta+\varkappa+\frac{1}{2},\ \frac{\varkappa+L_1+4}{2},\ \frac{\varkappa-L_1+3}{2}; \end{matrix} 1\right), \tag{38}$$

$$I^{x\beta}_{\ell\ell L_2 M_2}(\text{Vertex } \ell_2) = I^{x\beta}_{\ell\ell L_1\to L_2;\, M_1\to M_2}(\text{Vertex } \ell_1). \quad (39)$$

As for (32) we have [14, 16]

$$I^{x_1\beta}_{xM_1}(\text{Vertex } m_1) = \frac{(-)^{\frac{x-x_1+M_1}{2}}(2\pi)x_1!\left(\beta-\frac{1}{2}+x_1\right)_{x-x_1}}{2^{x_1}(x-x_1)!\left(\frac{x+M_1}{2}\right)!\left(\frac{2x_1-x-M_1}{2}\right)!}\times$$

$$\times {}_3F_2\left(\begin{matrix}-x+x_1,\ -\frac{x+M_1}{2},\ \beta-1/2+x_1\,; \\ -\beta+\frac{1}{2}-x+1,\ \frac{2x_1-x-M}{2}+1\,;\end{matrix}\ 1\right), \quad (40)$$

$$I^{x_1\beta}_{xM_2}(\text{Vertex } m_2) = I^{x_1\beta}_{xM_1\to M_2}(\text{Vertex } m_1). \quad (41)$$

$${}_4F_3\left(\begin{matrix}\alpha_1\alpha_2\alpha_3\alpha_4\,; \\ \beta_1\beta_2\beta_3\,;\end{matrix}\ 1\right) \quad \text{and} \quad {}_3F_2\left(\begin{matrix}\alpha_1\alpha_2\alpha_3\,; \\ \beta_1\ \beta_2\,;\end{matrix}\ 1\right)$$

are the hypergeometrical functions of unit argument in (34), (37)-(40).

4. Conclusion. The relations (29), (34)-(41) give complete solution of the stated problem: for orthogonal Jacobi vectors there remains only the triple sum defined by (27) and (33).

References

1. Vilenkin N.Ya., Kuznetsov G.I., Smorodinsky Ya.A. (1965). Yad. Fiz. (Nucl. Phys.), 2, 906 (Russian).
2. Kildyushov M.S., (1972). Yad. Fiz. (Nucl. Phys.), 16, 217 (Russian).
3. Raynal J. Revai J., (1970). Nuovo Cimento, A68, 612.
4. Efros V.D.(1972). Yad. Fiz. (Nucl. Phys.), 16, 226 (Russian).
5. Efros V.D., Smorodinsky Ya.A. (1973), Yad. Fiz. (Nucl. Phys.), 17, 210 (Russian).
6. Koornwinder T.H. (1972), Indag. Math., 34, 188.
7. Koornwinder T.H. (1973), SIAM. Jour. Appl. Math. 25, 236.
8. Koornwinder T.H. (1972), The addition formula for Jacobi polynomials. 2,3. Math. Centrum Afd. Toegepaste

Wisk. Rep. 133, 135.

9. Kuznetsov G.I. (1979), Yad. Fiz. (Nucl. Phys.), 30, 1158 (Russian).

10. Kuznetsov G.I. (1980), Yad. Fiz. (Nucl. Phys.), 32, 554 (Russian).

11. Kuznetsov G.I. (1977) Inter. Jour. Theor. Phys. 16, 345.

12. Baz E.E., Castel B. (1972). Graphical methods of spin algebras, Marcel. Dekker, Inc. N.Y.

13. Gradshteyn I.S., Ryzhik I.M. (1965). Tables of integrals, series and products, Acad. Press, N.Y.

14. Kuznetsov G.I., (1978). Preprint, IAE-2977, Moscow.

15. Bateman H., Erdelyi A. (1954) Tables of integral transforms, McGraw-Hill Book Company, Inc. N.Y. et al.

16. Kuldyushov M.S., Kuznetsov G.I. (1973) Preprint IAE-2263, Moscow.

SU(4)-SYMMETRY IN NUCLEAR PHYSICS

D.M. Vladimirov, Yu.V. Gaponov

I.V. Kurchatov Institute of Atomic Energy, Moscow

1. Background. In 1931 W. Heisenberg conjectured a realization of an isotopic $SU_T(2)$-symmetry of nucleons in nuclei. In 1937 his hypothesis was generalized by E. Wigner to embrace the spin-isospin SU(4)-symmetry [1]. In his first papers Wigner proposed the SU(4)-symmetry mass formula universally describing nuclear isobar series with a number of nucleons A

$$M(A)=a(A)+b(A)\left(p(p+4)+p'(p'+2)+(p'')^2\right) \quad (1)$$

and conjectured quantum numbers of nuclei's ground states A(N,Z) having assumed them to belong to supermultiplets

$$(pp'p'')=\begin{cases}(T_z\ 0\ 0) & \text{for even } N,Z \\ (T_z\ 1\ 0) & \text{for odd } N,Z \\ (T_z\ \tfrac{1}{2}\ \pm\tfrac{1}{2}) & \text{for A odd in neutrons (protons)}\end{cases} \quad (2)$$

The values a(A) and b(A) are universal functions of A independent only of N and Z. The development of the nucleus theory and the appearance of the shell model made the scientists abandon the Wigner symmetry as fundamental one. This was the result of the discovery of strong spin-orbital forces in nuclei (the j-j coupling) manifesting themselves in splitting components $j=\ell\pm 1/2$ of spin-orbit doublets (the SU(4)-symmetry breaking) at high values of the Coulomb energy which in principle, was enough to break the isotopic $SU_T(2)$-symmetry and the discovery of pairing effects. At the same time, some mathematical difficulties in describing representations of SU(4) in the Wigner reduction $SU(4) \supset SU_T(2)\times SU_S(2)$ were revealed. In this nonstandard reduction an arbitrary supermultiplet breaks into spin-isospin multiplets requiring degenerate states some additional quantum numbers for their distinction. At present the problem of describing degenerate states is not solved in general [2] and even the Clebsch-Gordan coefficients of SU(4) in the Wigner reduction are known only for some special multiplets containing no degenerate states [3].

An interest in the Wigner symmetry was revived in the early sixties due to the discovery of an isobaric analogue resonance (AR) of nuclei. The experimental discovery of the AR reaffirmed the reality of isomultiplets and, thus, approximately resorted the $SU_T(2)$-symmetry in nuclei regardless of the great value of the Coulomb field. From the sixties the study of the SU(4)-symmetry in nuclei went on in two main directions: the investigation of mass relations and the search for collective states of the spin-isospin type similar to and connected with the SU(4)-symmetry. The discovery of such a resonance, i.e. the Gamow-Teller resonance (GTR) in 1979 in experiments on the (p,n) reaction, raised the whole problem to a new level and made it possible to make a direct quantitative analysis of a degree of the SU(4)-symmetry restoration (at the present time in concrete nuclei) on the basis of polarization experiments with the excitation of the AR and GTR. Here we review the results based on our recent investigations [4] and the new data.

2. <u>Mass relations in SU(4)-symmetry scheme</u>. The use of the Wigner expression of the symmetry term (1) in analyzing empiric mass relations may serve only as an indirect test of the SU(4)-symmetry realization in nuclei. Such an analysis was performed by many authors (see e.g. [5]). Danos, Gillet et al. carried out the most detailed ana-

lysis [6]. They showed that for $A \lesssim 164$ the Wigner form of the symmetry term is preferable to the Weizsacker one, and for $A \gtrsim 164$ the descriptions of masses have comparable accuracy. However, the Wigner form is not the general possible form in the SU(4)-symmetry scheme, since, together with the Casimir operator C_2, there are independent operators C_3 and C_4. Burdet et al. [7] proposed to take into account their contribution phenomenologically along with that of C_2. Thus, the problem was set to generalize the Wigner mass formula to the highest Casimir operator of SU(4).

The investigation of the problem of the SU(4)-symmetry realization in the framework of Wigner's mass formula generalization for a heavy nuclei ($A \gtrsim 216$) was made in /8/. The mass formula containing the terms proportional to C_2, C_3, C_4 and $(C_2)^2$ and two terms breaking the SU(4) symmetry, i.e. the Coulomb E_c and the pairing E_p, has the form:

$$M(A,T)=a(A)+b(A)\,1/2\,(t^2+s^2+y^2)+c(A)\,2/3\,tsy+d(A)\big(t^2s^2+t^2y^2+s^2y^2-5/2\,(t^2+s^2+y^2)\big)+e(A)\,1/4\,(t^4+s^4+y^4)+E_c+\Delta E_p \qquad (3)$$

$$(t=p+2,\ s=p'+1,\ y=p'')$$

From the comparison of experimental data on mass differences with (3) for $A \geqslant 216$ the phenomenological universal parameters b(A), c(A), d(A) and e(A) were determined. The Coulomb difference ΔE_c was found from independent data on AR positions, the term ΔE_p has a standard phenomenological form. It turned out that the mass difference always is of the form

$$d_{ef}(AT)=d(A)+2\Delta(A)/T(T+3) \qquad (4)$$

whose mean value was found from the analysis (Δ(A) is a pairing constant). Practically, for each concrete A and T the mass differences depend only on three parameters.

The results of the investigation suggest that the mass differences of nuclei-isobars with $A \geqslant 216$ are reproduced by eq. (3) with an accuracy $\sim$170 keV. The main contribution to the symmetry term is determined by the term containing b(A) and proportional to C_2. The unified parameter $b(A, T_0)$ has an empirical form

$$b(A,T_0)=b(A)+e(A)T_0(T_0+1)=735-3(A-230)(keV)\ (T_0=26) \quad (5)$$

with a root-mean-square error ~ 10 keV. The constant C(A) sets the contribution of the term proportional to C_3 and connected with three-particle correlations of nucleons. Its value is

$$C(A)=-6.0\pm 2.0\ (keV) \quad (6)$$

The parameter $d_{ef}(A,T)=d_{ef}(A,T_0)$ characterizes even-odd oscillations of masses in isobars even in A and is of the form:

$$d_{ef}(A,T_0)=0.81-0.05(A-230),\ \delta d_{ef}=0.15\ (keV) \quad (7)$$

The mass analysis does not allow us to separate the contribution of four-particle correlations (d(A)) from that of pairing (Δ(A)), therefore, an alternative solution is possible. The parameter e(A) is

$$e(A)=0.15\pm 0,08\ (keV) \quad (8)$$

However, its estimate depends on the assumptions on the behaviour of the Coulomb differences ΔE_c in heavy nuclei. On the whole, the analysis of heavy nuclei masses proved self-consistency of the SU(4) hypothesis, the soundness of studying fine effects the Casimir operators and the possibility of describing mass differences with accuracy no worse than that of the best mass formulas.

3. Gamow-Teller resonance and SU(4)-symmetry problem. Hypothesis of SU(4)-symmetry restoration in charged $p\bar{n}$-branch of heavy nuclei. The GTR existence is a necessary condition for the SU(4)-symmetry, moreover, a consequence of the SU(4)-symmetry is a proximity of all the characteristics of the GTR and AR: the energy parameters, i.e. the AR and GTR positions ($E_{AR} \sim E_{GTR}$), and matrix elements of their β -decay $g^2M^2_{AR} \sim g^2M^2_{GTR}$. The GTR

theory was developed in the early papers of Japanese authors [9] and in [10, 11]. The final experimental proof of its existence was given only in 1979 [12], further investigation of its characteristics allowed one to check the degree of the SU(4)-symmetry realization at the level of microscopic parameters of the theory verified by modern experiments.

Within the microscopic theory of finite Fermi systems GTR is a collective isobaric 1^+ state of the $p\bar{n}$-type of the nucleus A(N,Z) positioned in the nucleus A(N-1, Z+1) near the AR. Its parameters are determined by an equation of the theory [10] including the parameters of the spin-isospin interaction g_0^- and effective spin-isospin charge e_q. The AR equation includes the isospin interaction f_0^-. In a simple two-level model the GTR parameters are determined through mean spin-orbit energy of excess neutrons $\bar{E}_{es}$ and the difference between the Fermi energy of neutrons and protons ΔE

$$E_{GTR} \sim E_{AR} \approx (f_0^- - g_0^-)\Delta E + C_1(g_0^-)\bar{E}_{es}(\bar{E}_{es}/\Delta E),$$

$$g_A^2 M^2_{GTR} \simeq g_A^2 e_q^2 (N-Z)(1-C_2(g_0^-))\bar{E}^2_{es}/(\Delta E^2+\bar{E}^2_{es}) \qquad (9)$$

From here the necessary conditions for the SU(4)-symmetry realization follow: the proximity of spin-isospin and isospin interaction constants ($f_0^- \approx g_0^-$), the necessity of the spin-orbit contribution suppression (the second term in $E_{GTR} - E_{AR}$), the proximity of the A and V interaction effective constants ($g_A e_q \approx g_V$) in suppression of the second term in the second formula. The spin-orbit correction suppression is naturally reached with the increase of ΔE proportional to N - Z in the region of heavy nuclei on the stability line. The treatment of the experimental data on the GTR within the frame of the TFFS demonstrated the proximity of interaction constants /10/:

$$g_0^- / f_0^- \approx 0.93 \qquad (10)$$

and that of effective constants of β-decay in nuclei [13]

$$(g_A/g_V)_{nucl} = g_A^{eq}/g_V \approx 1.0$$

Note that for vacuum nucleons the SU(4)-symmetry is broken heavily since $g_A/g_V = 1.26 \pm 0.01$. Finally, there exists an experimental region of the GTR and AR degeneration in the region of ^{208}Pb ($E_{GTR} - E_{AR} \approx 0.4$ MeV) - ^{238}U($E_{GTR} - E_{AR} \approx -0.2$ MeV), where the effects of the SU(4)-symmetry breaking of the spin-orbit type are compensated for by the effects of breaking since $f_0^- \neq g_0^-$. Thus, this region is that of the SU(4) symmetry restoration. All these facts allow one to formulate a conjecture on the SU(4)-symmetry restoration in the charged $p\bar{n}$-branch of heavy nuclei excitation [10]. Note that it is precisely in that region where the confirmation of the mass formula of the SU(4)-type was obtained (Sect.2).

4. SU(4)-symmetry study in charge-exchange reactions on beams of ^{6}Li and polarized protons. Direct charge-exchange reactions with the GTR and AR excitations of target nuclei provide us with the unique possibility of the experimental measurement of a degree of the SU(4)-symmetry break in concrete nuclei. Clearly, that from the SU(4)-symmetry point of view the processes depending on spin properties of incident particles are particularly important since they characterize spin-isospin properties in a charged channel. The amplitude of such processes in the Born approximation can be obtained from a phenomenological SU(4)-symmetry interaction potential of an incident particle with a target nucleus [4]. This potential mean value at the $t_z s_z y_z$ characteristics of the incident particle and in the target nucleus state with the quantum numbers $T_z S_z Y_z$ is

$$\bar{V}_{t_z s_z y_z}(T_z S_z Y_z) = V_0 + V_1/A(t_z T_z + s_z S_z' + y_z Y_z) \quad (12)$$

and gets the Lane form for even-even nuclei (T00). Clearly, it is different for different polarizations of incident particles and for a nuclei with $S_z \neq 0$ (odd nuclei). Proceeding from this form of potentials and known expressions for the Clebsch-Gordan coefficients (which requires special mathematical calculations even in the case of odd nuclei made e.g. in [14]), one can calculate amplitudes of recharge processes of polarized nucleons on polarized nuclei with the AR and GTR excitation of the target nucleus [4]. As for even-even nuclei the AR excitation is determined only by processes without any spin flip:

$$\sigma(AR)=1/2\sum_{\sigma\sigma'}|F_{p\sigma-n\sigma'}|^2 = T|F^1|^2, \quad |F_{p\uparrow-n\uparrow}|^2 : |F_{p\uparrow-n\downarrow}|^2 = 1:1 \qquad (13)$$

The GTR excitation takes place both in the processes without and with the nucleon spin flip:

$$\sigma(GTR)=3T|F^1|^2, \quad |F_{p\uparrow-n\uparrow}|^2 : |F_{p\downarrow-n\uparrow}|^2 = 1:2 \qquad (14)$$

and their contributions' ratio is 2:1. The amplitude F^1 is a phenomenological part of the amplitude connected with the potential V_1 (12). The AR excitation is contributed in neutron-odd nuclei by the spin-flip processes

$$\sigma(AR)=(T+3/4\,T^{-1})|F^1|^2, \quad |F_{p\uparrow-n\uparrow}|^2 : |F_{p\uparrow-n\downarrow}|^2 = (T+(4T)^{-1}) : (2T)^{-1} \qquad (15)$$

The contribution of the GTR or AR excitation channels with and without spin flip can be experimentally investigated in the processes with a polarized proton beam, and the relations (14) or (15), immediately checked. To study the AR excitation effect in the spin-flip processes in odd nuclei, the (^6Li, ^6He) reaction seems most convenient. There a transition of a proton into a neutron with a spin flip practically takes place, and the channel without spin flip is strictly prohibited [15]. The reaction $\sigma(AR)=3/4\,T^{-1}|F^1|^2$ can be observed in odd nuclei with a small T. At present, polarization experiments on recharge with the GTR and AR excitation have not been realized and are only in project. Comparison of data on total cross sections of the AR and

GTR excitation show that, apparently, the SU(4)-symmetry can be realized for nuclear nucleons with the kinetic energy of 40-50 MeV but will be considerably broken with its increase up to the range 100 to 200 MeV.

5. M1 resonances and SU(4)-symmetry problem for nuclei excitation neutral branch. The study of the GTR and its characteristics yields information on the part played by the SU(4)-symmetry in the charged channel of nuclei excitation. According to the modern concepts the neutral channel of nuclei excitation contains a strong spin-orbit interaction which is the reason for the dropping of the SU(4)-symmetry hypothesis in developing the shell model of a nucleus. Now no mechanism effectively suppressing the spin-orbit forces is available, therefore, one should not expect the SU(4)-symmetry restoration in the neutral branch. At the same time, a recent experimental discovery of collective M1 resonances of isovector and scalar types raises a number of new questions pertaining to the development of the SU(4)-symmetry scheme in the excitation neutral branch and to further study of mechanisms of its breaking in this branch.

Indeed, let us state the problem of inclusion of the isovector M1 resonance in the scheme of the Wigner SU(4) multiplet constructed using the ground state of the nucleus A(N,Z). It turns out that one can do that only at the price of dropping the Wigner hypothesis (2) and transition to the general supermultiplet (pp'p '') where $p' \geq 1$ for even-even nucleus A(N,Z) and this could mean the necessity of introducing a new approximate quantum number p' of the spin type which should have been determined from experiment. Moreover, the labelling of the states of the SU(4)-multiplets indicates that in the general case in this supermultiplet one should expect three collective 1^+ resonance of the GTR type of the characteristics $(T, S) = (T_0 - 1, p' - 1)$ (three-fold degeneracy), two 1^+ resonances of the characteristics $T_0 - 1, p')$ (two-fold degeneracy) and one 1^+ resonance of the characteristics $(T_0 - 1, p' + 1)$ (non degenerate). At first glance their number is arbitrary. However, thorough analysis based on [14] shows that of the three possible states one must have the structure 1p-1h (particle-hole); another, 2p-2h (two particles-two holes) and the third, 3p-3h, so that only the first one is observable and the other two states form the background of compound states of close structure, which is in charge of the GTR width. Analogously, of the two states $(T_0 - 1, p')$ one should have the structure 1p-1h and the other 2p-2h. An experimental candidate for the first one is the known observable j-satellite of the GTR which usually accompanies the latter but lies below the AR and which arises simultaneously with the GTR in theoretical models [14]. The existence in heavy nuclei of the state $(T_0 - 1, p' + 1)$ with 1p-1h, a candidate for which is a state of the type of a reciprocal spin flip, is predicted [4] but so far has not been observed experimentally. The characteristics of the isovector resonance in the neutral branch are $(T_0, p'-1)$. Thus, the experimental situation does not contradict a generalization (2) of the Wigner hypothesis and even supports it. So, in the generalized multiplet the description of the states is united as of those of the same nature: the ground state of A(N,z), its AR, GTR and its j-satellite, the state of the reciprocal spin flip and M1 resonance of A(N,Z) with $\Delta T = 1$.

If we assume such a hypothesis of the SU(4)-multiplet, it is possible in the Born approximation to calculate the amplitudes of direct nuclear reactions (p,n) and (p,p') with the formation of all these states on the basis of the SU(4)-symmetry mean potential (12) as well as its broken form

$$\bar{V}_{t_z s_z y_z}(T_z S_z Y_z) = V_0 + (V_\tau t_z T_z + V_\sigma \sigma_z S_z + V_{\tau\sigma} y_z Y_z)/A \quad (16)$$

Mathematically, we deal here with the description of degenerate ≤ 3 states yet not solved in general. However, using the fact that the states under investigation lie in the vicinity of the highest weight, one can construct a recurrent procedure and determine all the necessary Clebsch-Gordan coefficients [14]. As a result, the following expressions for the cross-sections of excitation processes of the studied collective states in direct nucleon reactions on polarized beams of protons are obtained:

$$\sigma(AR) = T|F_\tau|^2 + T^{-1}(p'+1)(p')^{-1}(p'')^2|F_{\tau\sigma}|^2, \quad |F_{p\uparrow-n\uparrow}|^2 : |F_{p\uparrow-n\downarrow}|^2 = (T|F_\tau|^2 + \tfrac{1}{3}T^{-1}(p'+1)(p')^{-1}(p'')^2|F_{\tau\sigma}|^2) : \tfrac{2}{3}T^{-1}(p'+1)(p')^{-1}(p'')^2|F_{\tau\sigma}|^2$$

$$\sigma(GTR) = ((2p'-1)/(2p'+1)(T+p'+1) - (Tp')^{-1}((p')^2+(p'')^2))|F_{\tau\sigma}|^2,$$
$$|F_{p\uparrow-n\uparrow}|^2 : |F_{p\uparrow-n\downarrow}|^2 = 1:2,$$
$$\sigma(j\text{-stat}) = (T+1-(p'+1)/(Tp'))|F_{\tau\sigma}|^2 \quad (17)$$
$$|F_{p\uparrow-n\uparrow}|^2 : |F_{p\uparrow-n\downarrow}|^2 = 1:2,$$
$$\sigma(\text{spin-flip}) = (T-p')|F_{\tau\sigma}|^2, \quad |F_{p\uparrow-n\uparrow}|^2 : |F_{p\uparrow-n\downarrow}|^2 = 1:2.$$
$$\sigma(M1, \Delta T=1) = 1/3((p')^2-(p'')^2)/p'|F_{\tau\sigma}|^2, \quad |F_{p\uparrow-p'\uparrow}|^2 : |F_{p\uparrow-p'\downarrow}|^2 = 1:2$$

As one can see from these expressions, most interesting are the experiments on relative contribution of the channel with and without spin flip, the AR excitation in the processes with the spin flip and excitation of the isovector ML resonance of the neutral branch. The latter ones yield important information on the p' and p'' quantum numbers value. The excitation neutral branch analysis is not complete without a brief discussion of the data on the isoscalar M1 resonance which was also discovered in inelastic scattering reactions. Treatment of experimental data in ^{48}Ca, ^{90}Zr by the TFFS methods shows that, apparently, the spin-isospin constant of quasiparticle interaction is small and, consequently, differs greatly from f_o and g_o. This conclusion considerably depends on the estimates of a single-particle spin-orbit energy and, therefore, is a tentative one. But being accepted, it means a heavy break of the SU(4)-symmetry in the isoscalar neutral channel, the mechanism of which is still to be understood.

References

1. Wigner E.P., (1937). Phys. Rev. 51, 106, 947.
2. Jeugt J. Van der, et al. (1983). Ann. Phys. 147, 85-139.
3. Hecht K.T., Pang S.C. (1969). J. Math. Phys. 10, 9, 1571.
4. Gaponov, Yu.V. (1984). Yad. Fiz. (Nucl. Phys.) 40, 1(7), 85 (Russian).
5. Gaponov Yu.V. (1981). Yad. Fiz. 34, 65 (Nucl. Phys.) (Russian).
6. Danos M. et al. (1981). Nucl. Phys. A361, 192.
7. Burdet J. et al. (1968). Nuovo Cim. 54B, 1.
8. Gaponov Yu.V. et al. (1982). Nucl. Phys. A391, 93.
9. Ikeda K. et al. (1962), Phys. Lett. 2, 169; (1963) 3, 271.
10. Gaponov, Yu.V., Lyutostansky Yu.S. (1974). Yad. Fiz. (Nucl. Phys.), 19, 62 (Russian). (1974) Sov. J. Nucl. Phys. 19, 33.
11. Gabracov S.I., et al. (1971) Phys. Lett. B36, 275.
12. IUCF-1979. (1979). Indiana (USA), p. 27.
13. Gaponov Yu.V. (1983). Pisma v ZhETF (ZhETF Lettr.) 38, 4, 204-208 (Russian).
14. Vladimirov D.M. (1984) Preprint IAE-3949/1, Moscow.
15. Bang J., Gaponov Yu.V., (1984). Physica Scripta 30, 104-111.